Computational River Dynamics

Computational River Dynamics

Weiming Wu

National Center for Computational Hydroscience and Engineering,
University of Mississippi, MS, USA

Taylor & Francis
Taylor & Francis Group
LONDON / LEIDEN / NEW YORK / PHILADELPHIA / SINGAPORE

Cover Illustration Credit:

Sediment laden drainage, Betsiboka River, Madagascar (2002)
Courtesy of NASA, National Aeronautics and Space Administration, Houston, TX 77058, USA

Taylor & Francis is an imprint of the Taylor & Francis Group, an informa business

© 2008 Taylor & Francis Group, London, UK

Typeset by Vikatan Publishing Solutions (P) Ltd, Chennai, India.
Printed and bound in Great Britain by Anthony Rowe Ltd (CPI-group),
Chippenham, Wiltshire

Published by: Taylor & Francis/Balkema
    P.O. Box 447, 2300 AK Leiden, The Netherlands
    e-mail: Pub.NL@tandf.co.uk
    www.balkema.nl, www.taylorandfrancis.co.uk, www.crcpress.com

Library of Congress Cataloging-in-Publication Data
Wu, Weiming.
    Computational river dynamics / Weiming Wu.
        p. cm.
    Includes bibliographical references.
    ISBN 978-0-415-44961-8 (hardcover : alk. paper) – ISBN 978-0-415-44960-1
    (pbk. : alk. paper) 1. Sediment transport. I. Title.

TC175.2 .W82 2007
627'.042–dc22   2007040342

ISBN: 978-0-415-44961-8 (hardback)
ISBN: 978-0-415-44960-1 (paperback)
ISBN: 978-0-203-93848-5 (e-book)

Contents

Preface

Rivers, as part of the nature, have been a focus of human activities since the beginning of civilization. Through engineering practices, such as flood control, water supply, irrigation, drainage, channel design, river regulation, navigation improvement, power generation, environment enhancement, and aquatic habitat protection, humans have come to understand more about rivers and established basic principles and analytical methodologies for river engineering. With the help of computation and information techniques, numerical modeling of flow and sediment transport in rivers has improved greatly in recent decades and been applied widely as a major research tool in solving river engineering problems. These advances motivated me to write this book on the physical principles, numerical methods, and engineering applications of computational river dynamics.

Most of the topics included in this book have been the central theme of my research work. I developed a simple 1-D quasi-steady sediment transport model for my bachelor's degree in 1986, a width-averaged 2-D unsteady open-channel flow model in my master's thesis in 1988, and an integrated 1-D and depth-averaged 2-D sediment transport model under quasi-steady flow conditions in my Ph.D. dissertation in 1991 at the Department of River Engineering, Wuhan University of Hydraulic and Electric Engineering, China. In 1995–1997, I established a 3-D sediment transport model at the Institute for Hydromechanics, University of Karlsruhe, supported by the Alexander von Humboldt Foundation, Germany. Since 1997, I have revisited 1-D and 2-D models and developed a 1-D channel network model and a depth-averaged 2-D model for unsteady flow and non-uniform sediment transport at the National Center for Computational Hydroscience and Engineering, University of Mississippi, USA, through a Specific Research Agreement between the USDA Agricultural Research Service and the University of Mississippi. I have also reviewed sediment transport theories, established several sediment transport formulas, and developed models for dam-break fluvial processes, vegetation effects, cohesive sediment transport, and contaminant transport. All these model developments and studies contributed to this book.

This book is intended primarily as a reference book for river scientists and engineers. It is also useful for professionals in hydraulic, environmental, agricultural, and geological engineering. It can be used as a textbook for civil engineering students at the graduate level.

My fascination with river engineering and computational river dynamics began with my first supervisor, Prof. Jianheng Xie. Later I learned a great deal about turbulence models and computational techniques in CFD from Prof. Wolfgang Rodi.

I also would like to acknowledge Prof. Sam S.Y. Wang for his long-term support and encouragement. I am greatly indebted to these three scientists.

I sincerely thank Drs. Mustafa S. Altinakar, Xiaobo Chao, George S. Constantinescu, Blair Greimann, Eddy J. Langendoen, Wolfgang Rodi, Steve H. Scott, F. Douglas Shields, Jr., Pravi Shrestha, Dalmo A. Vieira, Thomas Wenka, Keh-Chia Yeh, Xinya Ying, and Tingting Zhu for reviewing this book. I also thank my colleagues in Wuhan, Karlsruhe, and Ole Miss and my friends all over the world for their care and encouragement.

I would like to thank Taylor & Francis for publishing this book. In particular, many thanks are due to Dr. Germaine Seijger and Mr. Lukas Goosen for their professional handling of this project, Ms. Maartje Kuipers for designing the cover, Mrs. Shyamala Ravishankar and her team for carefully typesetting the manuscript, and the Anthony Rowe Ltd for printing it. I also thank my assistants Dr. Zhiguo He and Miss Podjanee Inthasaro for their help in proofreading of this book.

Special thanks go to my wife Ling and daughter Siyuan who gave me tremendous support during this endeavor.

Weiming Wu
Ole Miss, October 2007

Notations

Symbol	Meaning
A	Cross-sectional area of flow in 1-D model
A_b	Bed area at the cross-section
B	Channel width at the water surface
b	Flow width at height z in width-averaged 2-D model
C	Depth-, width- or section-averaged suspended-load concentration
	Contaminant concentration
C_d	Drag coefficient of sediment particle or vegetation
C_d, C_s, C_t	Concentrations of dissolved, sorbed, and total contaminants
C_h	Chezy coefficient
C_t	Depth- or cross-section-averaged concentration of total load
C_*	Depth- or cross-section-averaged equilibrium suspended-load concentration
c	Local sediment concentration
c_b, c_{b*}	Actual and equilibrium near-bed suspended-load concentrations
c_{v0}, c_v	local and depth-averaged concentrations of vegetation
D	Diameter of vegetation stem
D_b, E_b	Near-bed deposition and entrainment fluxes of sediment
D_{sx}, D_{sy}	Dispersion fluxes of suspended load
$D_{xx}, D_{xy}, D_{yx}, D_{yy}$	Dispersion transports of momentum
d	Sediment diameter
d_{50}	Median diameter of sediment mixture
d_k	Sediment diameter of size class k
d_m	Arithmetic mean diameter of sediment
$\vec{F}_d, \vec{f}_d$	Drag forces on vegetation
$F_w, F_s, F_b, \ldots$	Fluxes across cell faces
F_i	External force per unit volume
f_c	Coriolis coefficient
g	Gravitational acceleration
h	Flow depth
h_v	Vegetation height
J	Jacobian determinant
K	Conveyance of channel
k	Turbulent kinetic energy

Symbol	Meaning
k_s	Equivalent (effective) roughness height
Subscript k	Sediment size class index
	Associated with turbulent kinetic energy
L, L_t	Adaptation length of sediment
l_m	Mixing length
N_a	Vegetation density
n	Manning roughness coefficient
P_k	Production of turbulence by shear
p	Pressure
p'	Pressure correction
p_{bk}	Bed-material gradation in mixing layer
p'_m	Porosity of sediment deposit
Q	Flow discharge
Q_b, Q_{b*}	Actual and equilibrium bed-load transport rates
Q_t, Q_{t*}	Actual and equilibrium total-load transport rates
q	Unit flow discharge
q_b, q_{b*}	Actual and equilibrium unit bed-load transport rates
$q_l, q_{blk}, q_{slk}, q_{tlk}$	Side discharges of flow and sediment
R	Hydraulic radius of channel
R_b	Hydraulic radius of channel bed
R_s	Hydraulic radius of vegetated bed
S	Source term
S_f	Energy slope, friction slope
T	Temperature or transport stage number
$T_{xx}, T_{xy}, T_{yx}, T_{yy}$	Depth-averaged stresses
t	Time
$U(\hat{U})$	Depth- or section-averaged flow velocity
U_x, U_y	Depth-averaged flow velocities in x- and y-directions
U_x, U_z $(\tilde{U}_x, \tilde{U}_z)$	Width-averaged flow velocities in x- and z-directions
U_c	Critical average velocity for incipient motion
U_*	bed shear velocity
u_b, U_b	Bed-load velocities
$u_i(u_x, u_y, u_z)$	Flow velocities in $x_i(x, y, z)$ directions
$\hat{u}_m(\hat{u}_\xi, \hat{u}_\eta, \hat{u}_\zeta)$	Flow velocities in $\xi_m(\xi, \eta, \zeta)$ directions
x, y	Horizontal Cartesian coordinates
x_i	i-coordinate in the Cartesian coordinate system
z	Vertical coordinate above a datum (or bed)
z_b	Bed surface elevation
z_s	Water surface elevation
α	Adaptation coefficient of sediment
α_{bx}, α_{by}	Direction cosines of bed-load movement
β	Correction factor for momentum
β_s, β_t	Correction factors for suspended and total loads
χ	Wetted perimeter at the cross-section
Δ	Sand dune height
ΔA_P	Area of the control volume centered at P
Δt	Time step
$\Delta x, \Delta y$	Grid spacings
$\Delta z_b, \Delta A_b$	Changes in bed elevation and area
δ	Thickness of bed-load layer
δ_{ij}	Kronecker delta

Symbol	Meaning
δ_m	Mixing layer thickness
δh, δQ	Increments of water stage and flow discharge
ε	Dissipation rate of turbulent energy
ε_s	Turbulent diffusivity of sediment
ϕ_r	Repose angle of sediment
γ	Specific weight of water and sediment mixture
γ_f, γ_s	Specific weights of water and sediment
κ	Von Karman constant
λ	Darcy-Weisbach friction factor
μ, ν	Dynamic and kinematic viscosities of water
ν_t	Turbulent or eddy viscosity
π	Circumference-diameter ratio ≈ 3.14159
Θ	Shields number
Θ_c	Critical Shields number
θ, ψ	Temporal, spatial weighting factors
ρ	Density of water and sediment mixture
ρ_0	Density of flow density at water surface
ρ_b	Density of water and sediment mixture at bed surface layer
ρ_d	Dry density of sediment deposit
ρ_f, ρ_s	Densities of water and sediment
σ_s	Schmidt number
τ, τ_{ij}	Shear stresses
τ_b, τ_b'	Bed shear stress, grain shear stress
τ_c	Critical shear tress for incipient motion
τ_{ce}	Critical shear stress for erosion
τ_s	Wind driving force at the water surface
ω_s	Sediment settling velocity
ω_{sf}	Floc settling velocity
ξ, η, ζ	Logical or curvilinear coordinates

Chapter I

Introduction

1.1 OVERVIEW OF RIVER ENGINEERING

The origin of river engineering dates back to ancient times. According to historical records, the Chinese began building levees along the Yellow River about six thousand years ago. Approximately around the same period, irrigation systems and flood control structures were constructed in Mesopotamia, and also some ten centuries later in Egypt. During the Renaissance period, the observation of water flow and sediment transport was carried out by the Italian artist and engineer Leonardo da Vinci (1452–1519). Since then, scientists and engineers have performed a great number of studies on rivers, and constructed dams, levees, dikes, bridges, river training works, navigation facilities, and water supply facilities along rivers. This section briefly highlights the key issues in river engineering.

River dynamics

The study on the flow, sediment transport, and channel evolution processes in rivers began centuries ago, but river dynamics emerged as a distinct discipline of science and technology only after M. P. DuBoys established a bed-load formula in 1879 and H. Rouse proposed a function for the vertical distribution of suspended sediment in 1937. River dynamics deals with river flow and sediment problems, such as turbulent flow in alluvial channels, movable bed roughness, sediment settling, incipient motion, transport, deposition, and erosion. River dynamics also incorporates the study of fluvial processes, including river pattern classification, channel evolution laws, and regime theory. It provides physical principles and analysis methods for river engineering.

Flood control and mitigation

Flood is one of the biggest disasters rivers can cause. A river system is usually in balance – to a certain degree – with the hydrological and geological conditions of its basin. When the amount of runoff generated from uplands due to overwhelming rainfall exceeds the transport and storage capacity of the river system, the flow will overtop or break banks and flood neighboring areas. Owing to thousands of years' struggling against flood threats, humans have developed many flood control technologies, such as levees, river training works, storage areas, and diversion structures. Levees are one of the major measures used to control flood in many rivers. River training works, such as spur-dikes, weirs, and bank revetments, are often constructed to control the flow and protect banks and levees. Flood storage areas, such as reservoirs, lakes, detention ponds, and floodplains, help detain the flood propagation speed and reduce the peak of the flood. Diversion areas or channels are usually designated for emergency purposes when flood threatens the safety of backbone structures and key areas. As a new technology, flood forecasting and warning systems have been established in many regions to mitigate the flood disaster.

Reservoir sedimentation

Sediment deposition reduces the storage capacity and life span of reservoirs. With time, the deposition will extend upstream and submerge more land, while sediments, especially coarse particles, will be detained by reservoirs, causing erosion in downstream channels. The deposition and erosion processes and the ultimate equilibrium profiles in reservoirs and downstream channels are topics of concern. After reservoirs reach equilibrium states, their efficiency in terms of flood control, power generation, and sediment detention may be significantly reduced, and then problems with dam decommission and rehabilitation and their impacts on the environment become important.

Sediment control in low-head hydro-projects

Low-head hydro-projects include low dams, sluice gates, spillways, power generation facilities, water diversion structures, water intake structures, and navigation facilities. Because the reservoirs formed by low dams are small, sediment transport and morphological evolution in the reservoirs and downstream channels reach new equilibrium states relatively quickly. The appropriate design of sluice gates, spillways, and power generators can prevent coarse sediments from entering into turbines. In principle, navigation and water intake structures should be placed at locations such as the outer bank of the channel bend where less sediment deposition occurs. Flows around hydro-projects should be controlled with certain river training works. Sometimes it may be necessary to dredge and flush the deposits.

River restoration

Because of the impact of human activities or the variation of natural environment conditions, river systems change their forms through bed aggradation, degradation, and bank migration. These changes may be undesirable. For example, channel meandering and main flow displacement may cause land loss, bridge failure, levee breach, and difficulty in water intake. Serious erosion and deposition may impair aquatic habitats. Once adverse impacts occur, training, mitigation, and restoration are needed to change river systems to more favorable stable states.

Protection of structure foundations

In-stream structures, such as bridge piers, abutments, spur-dikes, and weirs, change the flow significantly and may induce considerable erosion. Erosion also occurs due to jet impingement at the downstream of sluice gates, spillways, and overfalls. Local erosion is the major reason for the failures of many structures. Because of the complexity of the processes involved, the prediction and prevention of local erosion around structures are very challenging.

Sediment problems in estuaries

Morphodynamic processes under the actions of river runoff, tidal flow, and wave currents in estuaries are extremely complex. A large amount of fine-grained cohesive sediments coming from rivers are deposited in estuarine regions, forming mouth bars and reducing the flow depth in navigation channels. Salinity intrusion intensifies the deposition of cohesive sediments and affects the water quality. Fine-grained sediments also enter harbors and cause significant deposition there. Training works and dredging are necessary to maintain the navigation channels.

Watershed management

Water bodies, such as rivers, lakes, and reservoirs, receive water runoff and sediment load from uplands. Serious erosion in the uplands increases the downstream sediment load, causing sedimentation and reducing the storage and transport capacities of downstream river systems. Conversely, a reduction in the erosion of uplands decreases the downstream sediment supply, causing channel degradation, headcutting, and bank instability. Rational watershed management is essential for both uplands and river systems.

Sediment-related environmental problems

Environmental quality is an important global issue. Wastes from industry and agriculture impair not only the water quality, but also the sediment quality in the receiving river systems. Sediments, especially clay and silt, are associated with the transport of many pollutants. The impaired sediments also accumulate on the channel bed with time, and later become a major source of pollution through resuspension.

1.2 ROLE OF COMPUTATIONAL SIMULATION IN RIVER ENGINEERING ANALYSIS

River flow and sediment transport are among the most complex and least understood processes or phenomena in nature. It is very difficult to find analytical solutions for most problems in river engineering, and it is very tedious to obtain numerical solutions without the help of high-speed computers. Therefore, before the 1970s, many river engineering problems had to be solved through field investigations and laboratory physical models (also called scale models). With the recent advancements in computer technology, computational models have been greatly improved and widely applied to solve real-life problems. One-dimensional (1-D) models have been used in short- and long-term simulations of flow and sediment transport processes in rivers, reservoirs, and estuaries. Two-dimensional (2-D) and three-dimensional (3-D) models have been used to predict in more detail the morphodynamic processes under complex flow conditions in curved and braided channels and around river training works, bridge piers, spur-dikes, and water intake structures.

Physical modeling and computational simulation are the two major tools used in river engineering analysis. Both have their advantages and disadvantages. Physical modeling can provide directly visible results, but it is expensive and time-consuming. Because the flow, sediment transport, and bed change processes in rivers are very complicated, it is difficult to ensure similarity between a physical model and its prototype. Errors may arise due to distortions of model scale and variations in experimental environments such as temperature. Computational simulation gives direct, real-scale predictions without any scale distortion and is cost-effective. However, the reliability of computational simulation relies on how well the physical processes are mathematically described through governing equations, boundary conditions, and empirical formulas; how accurately the differential governing equations are discretized using numerical schemes; how effectively the discretized algebraic equations are solved using direct or iterative solution methods; and whether the numerical solution procedures are correctly coded using computer languages. If the mathematical description is unreasonable, the numerical discretization incorrect, the solution method ineffective, or if the computed code has bugs, the results from a numerical model cannot be trusted. Because many empirical formulas are used to close the mathematical problems, the applicability of computational simulation is still somehow limited. Before a numerical model can be applied to a real-life project, it needs to be verified and validated using analytical solutions and data measured in laboratories and fields.

To solve a real-life engineering problem correctly and effectively, the integration of field investigation, physical modeling, and computational simulation is needed. Field investigation is the first thing to do for a comprehensive understanding of the problem. It provides the necessary hydrologic and sediment information on the study domain and boundary conditions, which are required in both physical modeling and computational simulation. It also provides data to calibrate physical and computational models. If the study reach is not long, either physical modeling or computational simulation can be chosen to analyze the problem. The most cost-effective method is to use physical models to study a few scenarios and collect enough data to calibrate computational models, and then use the calibrated computational models to analyze more scenarios. If the study reach is too long, 1-D numerical models are often used in the entire reach; they provide boundary conditions for 2-D and 3-D numerical models as well as physical models for detailed analyses in important subreaches.

1.3 SCOPE, PROBLEMS, AND STRATEGIES OF COMPUTATIONAL RIVER DYNAMICS

Computational river dynamics is a branch of computational fluid dynamics (CFD). It solves river engineering problems using numerical methods. River flow is an incompressible flow; therefore, many successful numerical methods developed in CFD can be applied here. However, river flow has a free surface and movable bed, which make computational river dynamics relatively complicated and difficult. Many assumptions and empirical formulas must be used to close the mathematical systems, and the approximate solutions sought may not be unique. Thus, computational river dynamics is an engineering science rather than applied mathematics. Not only must a successful numerical modeler possess knowledge about numerical techniques, but he or she must also have enough experience in river engineering.

In computational river dynamics, the flow, sediment transport, and morphological change processes in rivers are described by a set of coupled, non-linear algebraic and differential equations that usually cannot be solved in closed form. The analysis of river morphodynamic phenomena thus requires an approximation process, the end result of which is a field of discrete property values at a finite number of locations ("points" or "nodes") distributed over the study domain. The general procedure for developing a computational model consists, essentially, of the following steps:

(1) Conceptualize the complicated physical phenomena of study, with the necessary simplifications and assumptions that express our understanding of the nature of the system and its behavior (e.g., dimensionality; steady, quasi-steady, or unsteady; laminar or turbulent; subcritical, supercritical, or mixed; gradually or rapidly varied flow; fixed or movable bed; bed load, suspended load, or total load; low or high sediment concentration; uniform or multiple sediment sizes; equilibrium or non-equilibrium transport; cohesive or non-cohesive; bank erosion; channel meandering; contaminants; solution domain; initial and boundary conditions);
(2) Describe the physical phenomena of study using a set of algebraic and differential equations that are subject to the conservation laws of mass, momentum, and energy;
(3) Divide the study domain into a mesh of points, finite volumes, or finite elements corresponding to the used numerical methods;
(4) Discretize the differential equations to equivalent algebraic equations by introducing 'trial functions', held to approximate the exact solution locally;
(5) Solve the coupled algebraic equations, which are subject to case-specific boundary conditions, using an iteration or elimination algorithm to find the property values at the grid points, and
(6) Code the established solution procedures using computer languages, such as FORTRAN, C, or C++, and package the model with a graphical user interface for pre- and post-processing, if possible.

The major problems in computational river dynamics include:

(1) Adequacy of the (simplified) conceptual models representing the complicated real system and its behavior;
(2) Realism of the mathematical models describing the complex hydrodynamic and morphodynamic processes that cannot be represented exactly (e.g., turbulence, bed roughness, and the interaction between flow and sediment), and reliability of the empirical formulas used to close the mathematical systems;
(3) Ability to generate adequate meshes over complex domains;
(4) Accuracy and consistency of numerical approximations;
(5) Numerical stability and computational efficiency of solution methods;
(6) Correctness of computer coding, and

(7) Reliability of numerical solutions and applicability of computational models in different situations.

To insure the quality of the simulation results, a computational model of flow and sediment transport should be verified and validated before application in solving real-life problems. Model verification and validation usually follow three steps (Wang and Wu, 2005):

(1) *Verification by Analytic Solutions.* The agreement between analytic and numerical solutions certifies the correctness of the mathematical formulation, numerical methods, and computer programming. It can also determine errors of numerical solution quantitatively.
(2) *Validation by Laboratory Experiments.* Because laboratory experiments conducted in controlled environments can eliminate many unnecessary complications, the numerical model should be able to reproduce the same physical phenomena measured in laboratories.
(3) *Validation by Field Measurements.* One portion of the field data should be used to calibrate the physical parameters in the model, and the remaining data can be used to determine whether the computational model can simulate the real-life problem. Researchers must realize that the numerical results may only approximately agree with the measured data, because the computational model only represents a simplified version of the physical processes in natural rivers. However, the realistic trend of spatial and temporal variations should be predicted correctly.

The application of a computational model to the solution of a real-life problem involves the following five major tasks:

(1) *Data Preparation.* Data should be collected and analyzed to understand the physical processes of study, determine initial and boundary conditions, estimate model parameters, and calibrate the model. The required data should include, but are not limited to, geomorphic, hydrological, hydraulic, and sediment information, largely depending on the model used and the study case. They can be obtained via in-situ field survey and from historical records.
(2) *Estimation of Model Parameters.* Model parameters can be classified as numerical and physical. Numerical parameters, such as time step, grid spacing, number of size classes, and relaxation coefficient, result from numerical discretization and solution methods. They should be determined by considering the accuracy the study problem requires and the stability of the numerical schemes used. Physical parameters can be subdivided into two groups. One group represents the physical properties of water and sediment, such as water density, viscosity, sediment density, particle size, particle shape factor, and bed-material porosity. These physical properties can be measured. The other group results from the conceptualization of physical processes and represents the characteristics of flow and sediment transport, including channel roughness coefficient, sediment transport capacity, sediment adaptation length, and mixing layer thickness. These physical

process parameters are often calibrated using measured data or determined using empirical formulas.

(3) *Model Calibration.* The computational model should be calibrated using the available data measured in the study reach to insure that the aforementioned parameters are estimated correctly, that the empirical formulas are chosen appropriately, and that the observed physical processes are generally well reproduced by the model. The calibrated model can then be applied to predict the physical processes in various scenarios.

(4) *Interpretation of Simulation Results.* Because sediment transport models are highly empirical and the model development and application processes are not infallible, engineering judgment should be used in the interpretation of simulation results. Consulting with model developers, senior scientists, and local engineers who are familiar with the study channel can enhance confidence in the end results. In addition, efficiently grouping important results using attractive graphs and tables permits an easy understanding and communication among model developers, users, and report readers.

(5) *Analysis of Uncertainties.* Sources of uncertainties include model conceptualization, boundary conditions, model parameters, and data. Uncertainties may be reduced by using a more adequate model, selecting appropriate boundary conditions, calibrating model parameters carefully, and collecting more reliable data. Sensitivity analysis and stochastic modeling may also be conducted to resolve uncertainties.

As described above, the development and application of a computational model is a long process consisting of many steps. The accuracy and reliability of the end results rely on manipulations at every step. The developer should approximate the physical processes reasonably via the mathematical model, derive and code the numerical discretization and solution methods correctly, and verify and validate the model thoroughly. The user should prepare the data carefully, estimate model parameters correctly, necessarily calibrate the model, reasonably interpret results, and consider possible uncertainties.

1.4 CLASSIFICATION OF FLOW AND SEDIMENT TRANSPORT MODELS

Flow and sediment transport models can be classified in various ways, as described below.

According to their *dimensionality*, flow and sediment transport models are classified as 1-D, vertical 2-D, horizontal 2-D, and 3-D. Flow and sediment transport in natural rivers are usually 3-D phenomena, which should be more realistically simulated using 3-D models. However, 3-D models are more time-consuming. Therefore, 1-D and 2-D models have been established via simplifications, such as section-, depth-, and width-averaging, to achieve feasible solutions in engineering practices. 1-D models study the longitudinal profiles of the cross-section-averaged properties of flow and sediment transport in rivers. The vertical 2-D models, which may be idealized or width-averaged, study the (width-averaged) properties of flow and sediment transport in the longitudinal section. Because of the complexity of channel geometries, the width-averaged 2-D models are preferable to the idealized vertical 2-D models in natural situations. The horizontal 2-D models, which also are often called depth-averaged 2-D models, study the horizontal distributions of the depth-averaged quantities of flow and sediment. 1-D models are widely applied in the simulation of long-term sedimentation processes in long channels, 3-D models are often used in local fields with strong 3-D features, and 2-D models are in between them.

Based on *flow states*, flow and sediment transport models are often categorized as steady, quasi-steady, or unsteady. Steady models do not include the time-derivative terms in flow and sediment transport equations, but consider temporal changes in bed elevation and bed-material gradation. Quasi-steady models divide an unsteady hydrograph into many time intervals, each of which is represented by a steady flow discharge. Quasi-steady models are often used in the simulation of long-term fluvial processes in rivers, but they cannot be used in cases with strong unsteadiness, such as tidal flow in estuaries and flash floods in small watersheds. Unsteady models are more general and can be used to simulate unsteady fluvial processes as well as steady and quasi-steady processes.

As for the *number of sediment size classes* simulated, sediment transport models can be uniform (single-sized) or non-uniform (multiple-sized). Uniform sediment models represent the entire sediment mixture using a single-sized class, whereas non-uniform sediment models divide the sediment mixture into a number of size classes and study the behavior of each size class. Because sediments in natural rivers are usually non-uniform in size and experience interaction among different size classes, non-uniform sediment models are more realistic.

In accordance with *sediment transport modes*, sediment transport models are often grouped as bed-load, suspended-load, and total-load models. Many early developed models considered only bed-load or suspended-load transport. Because sediment may change from bed load to suspended load or vice versa depending on flow conditions, total-load models are more preferable.

Based upon *sediment transport states*, sediment transport models are classified as equilibrium (saturated) and non-equilibrium (unsaturated). In many of the early models, it is assumed that the actual sediment transport rate is equal to the capacity of flow carrying sediment at equilibrium conditions at each computational point (cross-section or vertical line). The models based on this local equilibrium assumption are called equilibrium transport models. However, alluvial river systems always change in time and space due to many reasons; therefore, absolute equilibrium states rarely exist in natural conditions. The local equilibrium assumption is not realistic, particularly in cases of strong erosion and deposition. Non-equilibrium sediment transport models renounce this assumption and adopt transport equations to determine the actual bed-load and suspended-load transport rates. Non-equilibrium transport models are being more widely applied in river engineering these days.

In terms of *numerical methods*, flow and sediment transport models are categorized as finite difference, finite volume, finite element, finite analytic, or efficient element models. Since each of these numerical methods has its advantages and disadvantages, numerical models based on all them exist in the literature. The choice of a specific model depends on the nature of the problem, the experience of the modeler, and the capacity of the computer being used.

Depending on the *calculation procedure*, flow and sediment transport models can be classified as fully decoupled, semi-coupled, or fully coupled. Fully decoupled models ignore the influence of sediment transport and bed change on the flow field by assuming a low sediment concentration and a small bed change, and calculate the flow and sediment transport separately at each time step. Fully coupled models compute all the flow and sediment quantities simultaneously. Semi-coupled models calculate some quantities in coupled form and the others separately. For example, flow and sediment modules may be decoupled, whereas sediment transport, bed change, and bed material sorting in the sediment module may be coupled. Because flow, sediment, and bed material always interact with each other in an alluvial river system, fully coupled models are more general and physically reasonable, whereas the applicability of fully decoupled and semi-coupled models is limited. However, coupled models are more sophisticated and may require more computational effort than decoupled models. In addition, the results from decoupled models may be justified due to the difference in time scales of flow and sediment transport and the use of empirical formulas for bed roughness and sediment transport capacity. Fully decoupled and semi-coupled models are still used by many investigators.

Depending on how to *conceptualize* sediment, sediment transport models can be discerned as particulate and continuous-medium models. Particulate models treat sediment as a group of particulate entities and describe the movement of single particles, whereas continuous-medium models assume sediment as a kind of pseudo-continuous medium. The assumption of continuous-medium models is only valid when the characteristic size of the sediment particles is much shorter than the characteristic length of the processes of study and the volume under consideration has enough sediment particles. Apparently, particulate models are not limited in this way. From a strictly theoretical point of view, particulate models should be preferred. However, because of the limitations of computer capacity, considerable difficulties are encountered in the simulation of the behavior of millions or even billions of irregularly shaped particles that may collide randomly. In reality, particulate models are only feasible when the sediment concentration is extremely low. Therefore, continuous-medium models are more widely applied in the study of sediment transport in rivers. A typical continuous-medium model is the diffusion model that is most often used for suspended-load transport.

1.5 COVERAGE AND FEATURES OF THIS BOOK

The subjects of this book include physical principles, numerical methods, model closures, and application examples in computational river dynamics. It is organized into twelve chapters.

Chapter 1 provides a general overview of computational river dynamics and the arrangement of this book. Chapter 2 introduces the mathematical descriptions of flow, sediment transport, and morphological change processes in rivers. Chapter 3 presents the fundamentals of sediment transport. Chapter 4 introduces the numerical techniques widely used to solve open-channel flows with sediment transport, such as the finite difference method and the finite volume method. These methods are applied and extended in the remaining chapters of this book. Chapter 5 describes the 1-D modeling approaches that are widely used in computational river dynamics. Chapter 6 explains the depth-averaged and width-averaged 2-D models for flow and sediment transport. It also includes a discussion of the enhancement of the depth-averaged 2-D model in order to take the effects of secondary flows in curved channels into account. Chapter 7 illustrates the 3-D modeling approaches for turbulent flow, general sediment transport in rivers, and local scour around in-stream hydraulic structures. Chapter 8 covers the general techniques used to integrate and couple various models, such as domain decomposition; the coupling of 1-D, 2-D, and 3-D channel models; and the integration of channel and watershed models. Chapters 9–12 introduce several special topics related to river engineering, such as dam-break fluvial processes, vegetation effects on fluvial processes, cohesive sediment transport, and contaminant transport.

This book is one of the first to present a complete picture of the physical principles and numerical methods used in computational river dynamics. It covers the fundamentals of flow and sediment transport in rivers, including many newly developed non-uniform sediment transport formulas. It is unique in presenting multidimensional numerical models, including 1-D, depth-averaged 2-D, width-averaged 2-D, and 3-D models, as well as integration and coupling of these models. It introduces many recently developed numerical methods for solving open-channel flows, such as the SIMPLE(C) algorithms with Rhie and Chow's momentum interpolation method on non-staggered grids, the projection method, and the extended stream function and vorticity method. It presents state-of-the-art sediment transport modeling approaches, such as non-equilibrium transport models, non-uniform total-load transport models, and semi-coupled and coupled procedures for flow and sediment calculations. It includes many engineering applications, such as reservoir sedimentation, channel erosion (due to dam construction), channel widening and meandering, local scour around in-stream hydraulic structures, vegetation effects on channel morphodynamic processes, and cohesive sediment transport.

Chapter 2

Mathematical description of flow and sediment transport

This chapter presents a mathematical basis for computational river dynamics, including definition of water and sediment properties, sediment diffusion theory, Reynolds-averaged flow and sediment transport equations and their turbulence closures, derivation of 1-D and 2-D model equations from 3-D model equations, formulation of equilibrium and non-equilibrium sediment transport models, and equations of non-uniform sediment transport and bed material sorting.

2.1 PROPERTIES OF WATER AND SEDIMENT

2.1.1 Properties of water

Density and specific weight of water

Water density, ρ_f, is the mass of water per unit volume, often in $\text{kg} \cdot \text{m}^{-3}$ (kilograms per cubic meter) in the international unit (SI) system. It is $1{,}000 \text{ kg} \cdot \text{m}^{-3}$ at $4°\text{C}$ and varies slightly with temperature, as shown in Table 2.1.

The specific weight of water, γ_f, is the weight of water per unit volume, often in $\text{N} \cdot \text{m}^{-3}$ (Newtons per cubic meter). It is related to the water density by

$$\gamma_f = \rho_f g \tag{2.1}$$

where g is the gravitational acceleration and equals about $9.80665 \text{ m} \cdot \text{s}^{-2}$ (meters per square second).

Viscosity of water

Water deforms under the action of shear. The dynamic viscosity of water, μ, is the constant of proportionality relating the shear stress, τ, to the deformation rate, du/dz, as follows:

$$\tau = \mu \frac{du}{dz} \tag{2.2}$$

where u is the flow velocity, and z is the coordinate normal to the flow direction.

Table 2.1 Density and viscosity of water

Temperature (°C)	Density (kg · m^{-3})	Dynamic viscosity (N · s · m^{-2})	Kinematic viscosity (m^{2}s^{-1})
0	1000	1.79×10^{-3}	1.79×10^{-6}
5	1000	1.51×10^{-3}	1.51×10^{-6}
10	1000	1.31×10^{-3}	1.31×10^{-6}
15	999	1.14×10^{-3}	1.14×10^{-6}
20	998	1.00×10^{-3}	1.00×10^{-6}
25	997	8.91×10^{-4}	8.94×10^{-7}
30	996	7.97×10^{-4}	8.00×10^{-7}
35	994	7.20×10^{-4}	7.25×10^{-7}
40	992	6.53×10^{-4}	6.58×10^{-7}

The kinematic viscosity of water, ν, is the ratio of the dynamic viscosity to the water density:

$$\nu = \frac{\mu}{\rho_f} \tag{2.3}$$

The units often used for viscosities μ and ν are N · s · m^{-2} and m^{2}s^{-1}, respectively.

Water viscosity is directly related to molecular interactions. It decreases as water temperature increases, as shown in Table 2.1. For common temperatures in rivers, the kinematic viscosity can be approximated by

$$\nu = (1.785 - 0.0584T + 0.00116T^2 - 0.0000102T^3) \times 10^{-6} \text{m}^2\text{s}^{-1} \tag{2.4}$$

where T is the temperature in °C (degrees Celsius).

2.1.2 Properties of sediment

2.1.2.1 Physical properties of single particles

Density and specific weight of sediment

Sediment density, ρ_s, is the mass of sediment per unit volume, often in kg · m^{-3}. It depends on the material of sediment. The density of quartz particles is about 2,650 kg · m^{-3} and does not vary significantly with temperature. In most natural rivers, the density of sediment can be assumed to be constant.

The specific weight of sediment, γ_s, is the weight of sediment per unit volume, often in N · m^{-3}. It is related to the sediment density by

$$\gamma_s = \rho_s g \tag{2.5}$$

Due to the buoyancy effect, the specific weight of sediment particles submerged in water is lighter than the actual specific weight exposed to air. According to Archimedes' principle, the specific weight of submerged sediment is the difference between the specific weights of sediment and water, $\gamma_s - \gamma_f$.

The specific gravity of sediment, G, is the ratio of the specific weights of sediment and water at a standard reference temperature that is commonly set at 4°C. The specific gravity of quartz particles is

$$G = \frac{\gamma_s}{\gamma_f} = \frac{\rho_s}{\rho_f} = 2.65 \tag{2.6}$$

Particle size and grade scale

Sediment particle size may be represented by nominal diameter, sieve diameter, and fall diameter. The nominal diameter, d, is defined as the diameter of a sphere that has the same volume as the given particle, i.e.,

$$d = \sqrt[3]{\frac{6V}{\pi}} \tag{2.7}$$

where V is the volume of the sediment particle, and π is the circumference-diameter ratio (≈ 3.14159). The SI units often used for sediment size are mm (millimeters) and m (meters).

A sediment particle may be considered as an ellipsoid. Denote a, b, and c as its diameters in the longest, the intermediate, and the shortest mutually perpendicular axes, respectively. Thus, as an approximation, the particle volume may be estimated as $V \approx \pi abc/6$, and then substituting this formula into Eq. (2.7) yields the following relation for the nominal diameter:

$$d \approx \sqrt[3]{abc} \tag{2.8}$$

The sieve diameter is the length of the side of a square sieve opening through which the given particle will just pass. It is approximately equal to the intermediate diameter b. The sieve diameter is slightly smaller than the nominal diameter. For naturally worn sediment particles over the range of about 0.2 to 20 mm, the sieve diameter is approximately 0.9 times the nominal diameter on average (U. S. Interagency Committee, 1957; Raudkivi, 1990).

The standard fall diameter is the diameter of a sphere that has a specific gravity of 2.65 and has the same terminal settling velocity as the given particle in quiescent, distilled water at a temperature of 24°C.

Sediment size may be measured by calipers, by optical methods, by photographic methods, by sieving, or by sedimentation methods (Vanoni, 1975; Simons and Senturk, 1992). For coarse particles, such as boulders, cobbles, and coarse gravel, size may be determined by direct measurement of the volume or the diameters a, b, and c in the longest, the intermediate, and the shortest axes, which are usually converted to the nominal diameter by Eq. (2.7) or (2.8). For fine gravel and sand, size may be determined by sieving or visual accumulation tube. For silt and clay, size is measured by hydraulic settling methods, such as the pipet method, bottom withdrawal method, and hydrometer method. The fall diameter is often obtained by these methods for silt and clay.

Table 2.2 Sediment grade scale

Class	Size range (mm)	Class	Size range (mm)
Very large boulders	4,000–2,000	Very coarse sand	2–1
Large boulders	2,000–1,000	Coarse sand	1–0.5
Medium boulders	1,000–500	Medium sand	0.5–0.25
Small boulders	500–250	Fine sand	0.25–0.125
		Very fine sand	0.125–0.062
Large cobbles	250–130	Coarse silt	0.062–0.031
Small cobbles	130–64	Medium silt	0.031–0.016
		Fine silt	0.016–0.008
Very coarse gravel	64–32	Very fine silt	0.008–0.004
Coarse gravel	32–16	Coarse clay	0.004–0.002
Medium gravel	16–8	Medium clay	0.002–0.001
Fine gravel	8–4	Fine clay	0.001–0.0005
Very fine gravel	4–2	Very fine clay	0.0005–0.00024

The aforementioned boulders, cobbles, gravel, sand, silt, and clay are classi-fied based on the grade scale listed in Table 2.2, which is commonly used in river engineering. Each class may be further divided into several subclasses.

Shape factor

The shape of sediment particles in natural rivers is very irregular. It is often described by the Corey shape factor:

$$S_p = \frac{c}{\sqrt{ab}} \tag{2.9}$$

The Corey shape factor of naturally worn particles is usually about 0.7.

2.1.2.2 Bulk properties of sediment mixtures

Size distribution

A mixture that consists of sediment particles with non-uniform sizes can be represented by a suitable number of size classes. Each size class, numbered as k, is defined by the lower and upper bound diameters and represented by a characteristic diameter, d_k. If the lower and upper bound diameters of size class k are denoted as d_{lk} and d_{uk}, the characteristic diameter may be determined using $d_k = \sqrt{d_{lk}d_{uk}}$, $d_k = (d_{lk} + d_{uk})/2$, or $d_k = (d_{lk} + d_{uk} + \sqrt{d_{lk}d_{uk}})/3$.

The fraction, p_k, of each size class is the ratio of its weight (volume or number) to the total weight (volume or number) of the mixture, ranging from 0 to 1. It should be noted that p_k is also often defined by percent, ranging from 0 to 100.

The size distribution (composition, gradation) of a sediment mixture can be mea-sured by sieving analysis. It is often represented by the frequency histogram (pyramid) and cumulative size frequency curve. The histogram is constructed by plotting the sizes representing size class intervals on the abscissa and the actual percent (by weight, vol-ume or number) of the total sample contained in each size class on the ordinate, as

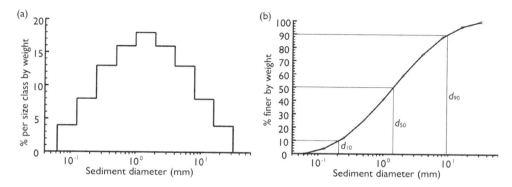

Figure 2.1 Size distribution: (a) histogram and (b) cumulative frequency curve.

shown in Fig. 2.1(a). The cumulative size frequency curve shows the percent of material finer than a given sediment size in the total sample, as shown in Fig. 2.1(b). For a sediment mixture with a normal size distribution, the cumulative size frequency curve is a straight line on the normal probability paper.

Characteristic diameters

The median diameter, d_{50}, is the particle size at which 50% by weight of the sample is finer. Likewise, d_{10} and d_{90} are the particle sizes at which 10% and 90% by weight of the sample are finer, respectively. The diameters d_{10}, d_{50}, and d_{90} can be read from the cumulative size frequency curve, as shown in Fig. 2.1(b).

The arithmetic mean diameter is determined by

$$d_m = \sum_{k=1}^{N} p_k d_k / 100 \tag{2.10}$$

where p_k is by percent, and N is the total number of size classes.

The geometric mean diameter is given by

$$d_g = d_1^{p_1/100} \cdot d_2^{p_2/100} \cdot \ldots \cdot d_N^{p_N/100} \tag{2.11}$$

Uniformity

The uniformity of a sediment mixture can be described by the standard deviation:

$$\sigma_g = \left(\frac{d_{84.1}}{d_{15.9}} \right)^{1/2} \tag{2.12}$$

or the gradation coefficient:

$$Gr = \frac{1}{2} \left(\frac{d_{84.1}}{d_{50}} + \frac{d_{50}}{d_{15.9}} \right) \tag{2.13}$$

where $d_{15.9}$ and $d_{84.1}$ are the particle sizes at which 15.9% and 84.1% by weight of the sample are finer, respectively. For a normal size distribution, $\sigma_g = d_{84.1}/d_{50} = d_{50}/d_{15.9}$.

Kramer (1935) defined a uniformity parameter as the ratio of the mean sizes of the two portions in the cumulative size frequency curve separated by d_{50}:

$$M = \sum_{A_k=0}^{50} p_k d_k \bigg/ \sum_{A_k=50}^{100} p_k d_k \qquad (2.14)$$

where A_k is the cumulative percentage of sediment finer than size d_k.

For uniform sediment, $M = 1$. A smaller value of M corresponds to a more non-uniform sediment mixture.

Porosity and dry density

A sediment deposit is a porous material and has voids among solid particles. Its porosity, p'_m, is a measure of the volume of voids per unit volume of the deposit:

$$p'_m = \frac{V_v}{V_v + V_s} \qquad (2.15)$$

where V_v and V_s are the volumes of voids and solids, respectively.

The dry density, ρ_d, and dry specific weight, γ_d, of a sediment deposit are the mass and weight of the solids per unit total volume. They are related to the porosity by

$$\rho_d = \rho_s(1 - p'_m), \quad \gamma_d = \gamma_s(1 - p'_m) \qquad (2.16)$$

Han *et al.* (1981) proposed the following semi-empirical formula to calculate the initial porosity of a uniform sediment deposit:

$$p'_m = \begin{cases} 1 - 0.525 \left(\frac{d}{d+4\delta_1}\right)^3 & d < 1 \text{ mm} \\ 0.3 + 0.175 e^{-0.095(d-d_0)/d_0} & d \geq 1 \text{ mm} \end{cases} \qquad (2.17)$$

where d is the sediment size in mm; d_0 is a reference size, set to be 1 mm; and δ_1 is the thickness of the water film attaching to sediment particles, given a value of about 0.0004 mm.

In a non-uniform sediment deposit, fine particles probably fill the voids among coarse particles. Han *et al.* (1981) investigated this filling phenomenon and proposed a method for the overall porosity of the deposit. However, their method is relatively complicated and inconvenient to use. If a sediment deposit is composed of only fine particles or if its size range is narrow, the filling phenomenon is negligible and the overall porosity can be calculated using the Colby (1963) method:

$$\frac{1}{1 - p'_m} = \sum_{k=1}^{N} \frac{p_k}{1 - p'_{mk}} \qquad (2.18)$$

where p_k is the fraction of the kth size class in the sediment deposit; and p'_{mk} is the porosity of size class k, which can be calculated using Eq. (2.17) or another method.

Komura (1963) proposed an empirical formula for the initial porosity of a sediment deposit:

$$p'_m = 0.245 + \frac{0.0864}{(0.1d_{50})^{0.21}} \qquad (2.19)$$

where d_{50} is the median diameter of the deposit (mm).

Wu and Wang (2006) revalidated the Komura formula (2.19) using extensive laboratory and field data, as shown in Fig. 2.2. It can be seen that the Komura formula is quite close to the trend of the data set, slightly underestimating the dry density for sand and gravel and overestimating it for silt and coarse clay. The Han *et al.* formula has more errors. A more accurate curve was obtained in Fig. 2.2 and expressed as

$$p'_m = 0.13 + \frac{0.21}{(d_{50} + 0.002)^{0.21}} \qquad (2.20)$$

where d_{50} is in mm.

In addition, the porosity and dry density of a fine-grained sediment deposit may vary with deposit depth and residence time due to consolidation. This is discussed in Section 11.1.6.

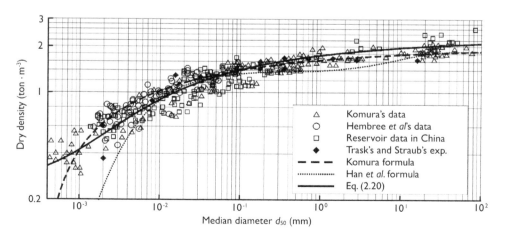

Figure 2.2 Initial dry density of deposit as function of median diameter (Wu and Wang, 2006).

Repose angle

The repose angle is the angle, with respect to the horizontal, of the slope formed by sediment particles submerged in water under incipient sliding conditions. According to laboratory experiments by Tianjing University (see Zhang *et al.*, 1989), the repose angle, ϕ_r, is related with sediment size as follows:

$$\phi_r = 32.5 + 1.27d \qquad (2.21)$$

where ϕ_r is in degrees, and d is in mm. Eq. (2.21) was calibrated using the data in the sediment size range between 0.2 and 4.4 mm.

The repose angle is also related to other properties of sediment particles, such as density, shape, gradation, compaction, and material. It may range from 30° to 42° for non-cohesive sediment particles. More discussion on the repose angles of various sediments can be found in Simons and Senturk (1992).

2.1.2.3 Definition of sediment loads

All sediment particles moving with flowing water are called total load. The total load can be divided into bed load and suspended load as per sediment transport mode or bed-material load and wash load as per sediment source, as depicted in Fig. 2.3.

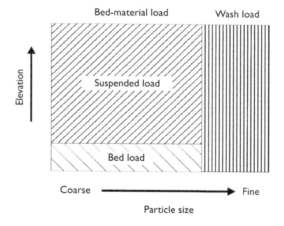

Figure 2.3 Definition of sediment loads.

The bed load consists of sediment particles that slide, roll, or saltate in the layer several particle sizes above the bed surface. It usually accounts for 5–25% of the total load for fine particles and more for coarse particles in natural rivers.

The suspended load is composed of sediment particles that move in suspension in the water column above the bed-load layer. Its weight is continuously supported by the turbulence of flow.

The bed-material load is made up of moving sediment particles that are found in appreciable quantities in the channel bed. It constantly exchanges with the bed material and has significant contribution to the channel morphology.

The wash load is comprised of moving sediment particles that are derived from upstream sources other than the channel bed. It is not found in appreciable quantities in the bed. It is finer than the bed-material load and rarely exchanges with the bed material. Einstein (1950) defined wash load as the grain size of which 10% of the bed material mixture is finer.

It should be noted that the definition of wash load and bed-material load depends on flow and sediment conditions. Some wash load in upstream channels may become bed-material load in downstream channels due to the weakening of flow strength. Some sediment particles are wash load in the main channel but may be bed-material load in flood plains.

By definition, the bed-material load is the sum of bed load and suspended load. So is the wash load. However, the wash load consists of fine particles that move mainly in suspension, and thus dividing it into bed load and suspended load does not make much sense in practice.

2.1.3   Properties of the water and sediment mixture

Fig. 2.4 shows a sketch of a mixture consisting of a volume of water, V_w, and a volume of sediment, V_s. It is termed as "mixture" for short. Sediment concentration is defined as

$$c = \frac{V_s}{V_w + V_s} \quad \text{or} \quad \hat{c} = \frac{\gamma_s V_s}{\gamma_f V_w + \gamma_s V_s} \tag{2.22}$$

where c is the concentration by volume, and $\hat{c}$ is the concentration by weight (mass). They are related by $\hat{c} = Gc/[1 + (G - 1)c]$. Both them are unitless. In addition, sediment concentration is sometimes given by weight or mass per unit volume of the mixture ($\mathrm{N \cdot m^{-3}}$ or $\mathrm{kg \cdot m^{-3}}$), which is obtained by $\gamma_s c$ or $\rho_s c$. It is also given in parts per million by weight (ppm), which is equivalent to $10^6 \hat{c}$.

Note that the volumetric sediment concentration c is used in this book, except where stated otherwise.

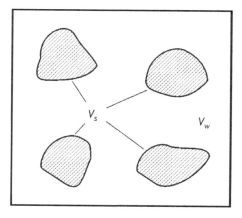

Figure 2.4 Sketch of the water and sediment mixture.

The density of the mixture, ρ, is determined by

$$\rho = \rho_f(1 - c) + \rho_s c \tag{2.23}$$

and the specific weight of the mixture is correspondingly given by $\gamma = \rho g$.

The velocity of the mixture, u_i, is defined as

$$u_i = \frac{1}{\rho}[\rho_f(1 - c)u_{fi} + \rho_s c u_{si}] \tag{2.24}$$

where u_{fi} is the i-component of water velocity, u_{si} is the i-component of sediment velocity, and i denotes three spatial directions ($= 1, 2, 3$).

2.2 GOVERNING EQUATIONS OF WATER AND SEDIMENT TWO-PHASE FLOW

Because the stochastically averaged properties of a group of sediment particles are mainly concerned in river engineering, sediment is often assumed to be a kind of continuous medium. Two mathematical models can be used to describe the water and sediment two-phase flow based on this assumption. One is the two-fluid model that considers water and sediment as two fluids and establishes the continuity and momentum equations for each phase. The other is the diffusion model that considers the movement of sediment particles as a phenomenon of diffusion in the water flow and hence establishes the continuity and momentum equations for the water-sediment mixture and the transport (diffusion) equation for sediment particles. The two-fluid model is more general, from which the diffusion model can be derived, as described by Wu and Wang (2000). Detailed discussions on the two-fluid model can also be found in Soo (1967), Ni et al. (1991), Liu (1993), and Greimann and Holly (2001). However, the two-fluid model is not introduced here because it is quite complex. The flow and sediment transport equations used in this book are based on the diffusion model.

2.2.1 Hydrodynamic equations

Applying the mass and momentum conservation laws leads to the continuity and momentum equations for the instantaneous movement of the water-sediment mixture. These equations are written in Cartesian tensor notations as follows:

$$\frac{\partial \rho}{\partial t} + \frac{\partial(\rho u_i)}{\partial x_i} = 0 \tag{2.25}$$

$$\frac{\partial(\rho u_i)}{\partial t} + \frac{\partial(\rho u_i u_j)}{\partial x_j} = F_i - \frac{\partial p}{\partial x_i} + \frac{\partial \tau_{ij}}{\partial x_j} \tag{2.26}$$

where t is the time; x_i is the i-coordinate in the Cartesian coordinate system; p is the pressure of the mixture; τ_{ij} ($i,j = 1$, 2, 3) are the stresses of the mixture; and F_i is the i-component of the external force on the mixture, such as gravity.

Note that the spatial direction indices i and j are subject to Einstein's summation convention: Subscript (or superscript) repeated twice in any product or quotient of terms is summed over the entire range of values of that subscript (or superscript). For example, $a_{ij}b_j = \sum_{j=1}^{3} a_{ij}b_j = a_{i1}b_1 + a_{i2}b_2 + a_{i3}b_3$ in the 3-D system.

According to numerous experimental studies, the water-sediment mixture with low sediment concentration (less than about 200 $kg \cdot m^{-3}$) is a kind of Newtonian fluid, the constitutive relation of which is given by the Navier-Poisson law:

$$\tau_{ij} = 2\mu_m D_{ij} - \frac{2}{3}\mu_m D_{kk}\delta_{ij} \tag{2.27}$$

where μ_m is the dynamic viscosity of the mixture; D_{ij} is the tensor of deformation rate, defined as $D_{ij} = (\partial u_i/\partial x_j + \partial u_j/\partial x_i)/2$; and δ_{ij} is the Kronecker delta, with $\delta_{ij} = 1$ when $i = j$ and $\delta_{ij} = 0$ when $i \neq j$.

When the sediment concentration is high, the mixture becomes a non-Newtonian fluid, such as Bingham fluid. The relation between shear stress and deformation rate for the one-directional shear flow of Bingham fluid is written as

$$\tau_{13} = \tau_B + \eta\frac{du}{dz} \tag{2.28}$$

where τ_{13} is the shear stress, τ_B is the yield stress, and η is the plastic viscosity.

Extending Eq. (2.28) to the multi-directional shear flow yields the general constitutive relation of Bingham fluid (Prager, 1961; Wu and Wang, 2000):

$$\tau_{ij} = \left(2\mu_m + \frac{\tau_B}{I_2^{1/2}}\right)\left(D_{ij} - \frac{1}{3}D_{kk}\delta_{ij}\right) \tag{2.29}$$

where $I_2 = \frac{1}{2}(D_{ij}D_{ij} - \frac{1}{3}D_{kk}^2)$, and μ_m is η.

The single-directional shear flow field can be divided into two zones by $\tau_{13} \geq \tau_B$ and $\tau_{13} < \tau_B$. Eq. (2.28) is only applicable to the zone of $\tau_{12} \geq \tau_B$. Similarly, Eq. (2.29) is valid in the zone of $\tau_{ij}\tau_{ij} \geq 2\tau_B^2$ for the multi-directional shear flow.

2.2.2 Sediment transport equation

Sediment transport is governed by the following mass balance equation:

$$\frac{\partial c}{\partial t} + \frac{\partial(u_{si}c)}{\partial x_i} = 0 \tag{2.30}$$

Because the sediment velocity u_{si} is not a dependent variable in the diffusion model, Eq. (2.30) is rewritten as

$$\frac{\partial c}{\partial t} + \frac{\partial(u_i c)}{\partial x_i} = -\frac{\partial}{\partial x_i}[(u_{si} - u_i)\,c] \tag{2.31}$$

where $u_{si} - u_i$ is the "diffusion" velocity of sediment in the mixture. It can be related to the interphase velocity difference $u_{fi} - u_{si}$ by $u_{si} - u_i = -\rho_f(1 - c)(u_{fi} - u_{si})/\rho$, based on Eq. (2.24). Wu and Wang (2000) derived a differential equation for the interphase velocity difference $u_{fi} - u_{si}$ from the momentum equations of the two-fluid model, but the derived equation is complex and inconvenient to use. For fine sediments, the particle inertia or the lag between (local) flow and sediment movement is very small, and nearly no relative motion exists except for the settling due to gravity; thus, the following relation is assumed:

$$u_{si} - u_i = -\omega_{sm}\delta_{3i} \tag{2.32}$$

where ω_{sm} the settling velocity of sediment particles in the water-sediment mixture, and the subscript "3" in δ_{3i} denotes the vertical direction defined by gravity.

Substituting Eq. (2.32) into the sediment transport equation (2.31) yields the closed sediment transport equation:

$$\frac{\partial c}{\partial t} + \frac{\partial(u_i c)}{\partial x_i} = \frac{\partial}{\partial x_i}(\omega_{sm}c\delta_{3i}) \tag{2.33}$$

2.2.3 Simplification in the case of low sediment concentration

If the sediment concentration is low, $\rho \approx \rho_f \approx$ constant, $1 - c \approx 1$, $\mu_m \approx \mu$, and ω_{sm} is close to ω_s, the settling velocity of a single particle in clear water. Then the continuity equation (2.25) and momentum equation (2.26) of the mixture can be simplified to

$$\frac{\partial u_i}{\partial x_i} = 0 \tag{2.34}$$

$$\frac{\partial u_i}{\partial t} + \frac{\partial(u_i u_j)}{\partial x_j} = \frac{1}{\rho}F_i - \frac{1}{\rho}\frac{\partial p}{\partial x_i} + \frac{1}{\rho}\frac{\partial \tau_{ij}}{\partial x_j} \tag{2.35}$$

and the constitutive relation (2.27) of Newtonian fluid can be written as

$$\tau_{ij} = \mu\left(\frac{\partial u_i}{\partial x_j} + \frac{\partial u_j}{\partial x_i}\right) \tag{2.36}$$

Substituting Eq. (2.36) into Eq. (2.35) leads to the following Navier-Stokes equation widely used in the single-phase fluid mechanics for laminar flows or instantaneous motions of turbulent flows:

$$\frac{\partial u_i}{\partial t} + \frac{\partial(u_i u_j)}{\partial x_j} = \frac{1}{\rho}F_i - \frac{1}{\rho}\frac{\partial p}{\partial x_i} + \frac{\mu}{\rho}\frac{\partial^2 u_i}{\partial x_j \partial x_j} \tag{2.37}$$

The sediment transport equation (2.33) can be further simplified to

$$\frac{\partial c}{\partial t} + \frac{\partial (u_i c)}{\partial x_i} = \frac{\partial}{\partial x_i}(\omega_s c \delta_{3i})$$

(2.38)

In principle, Eq. (2.38) is applicable only for fine sediments and low concentrations (see Greimann and Holly, 2001). It is commonly accepted that if the sediment size is finer than 1 mm and the sediment concentration is lower than 0.1 by volume, Eq. (2.38) can be approximately used.

In summary, the water and sediment two-phase flow model in the case of low sediment concentration can be simplified to the model of clear water flow with sediment transport. Because the sediment concentration usually is not high in most natural rivers, the simplified diffusion model has been widely adopted in river dynamics.

2.3   TIME-AVERAGED MODELS OF TURBULENT FLOW AND SEDIMENT TRANSPORT

2.3.1   Mean movement equations

Eqs. (2.34), (2.37), and (2.38), which are the exact equations for instantaneous motions of flow and sediment, cannot be solved directly in most cases, because of limited computer capacity. Since engineers usually are not interested in the details of the turbulent fluctuating motions, how to describe and solve the mean motions of turbulent flow is important in practice. As suggested by Osborne Reynolds, the instantaneous quantity of a variable ϕ can be divided into mean and fluctuating quantities as

$$\phi = \bar{\phi} + \phi'$$

(2.39)

where "−" denotes the mean quantity, and "′" denotes the fluctuating quantity. The mean quantity is defined as

$$\bar{\phi} = \frac{1}{T} \int_t^{t+T} \phi \, d\tau$$

(2.40)

where T is the time period considered, which should be much longer than the fluctuation period of turbulence, as shown in Fig. 2.5.

The fluctuating quantity satisfies

$$\overline{(\phi')} = \frac{1}{T} \int_t^{t+T} \phi' \, d\tau = 0$$

(2.41)

Reynolds-averaging Eqs. (2.34), (2.37), and (2.38) yields the mean continuity and momentum equations of flow and the mean transport equation of sediment:

$$\frac{\partial \bar{u}_i}{\partial x_i} = 0$$

(2.42)

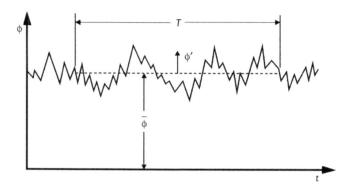

Figure 2.5 Reynolds' time-averaging procedure.

$$\frac{\partial \bar{u}_i}{\partial t} + \frac{\partial (\bar{u}_i \bar{u}_j)}{\partial x_j} = \frac{1}{\rho}\bar{F}_i - \frac{1}{\rho}\frac{\partial \bar{p}}{\partial x_i} + \frac{\mu}{\rho}\frac{\partial^2 \bar{u}_i}{\partial x_j \partial x_j} - \frac{\partial \overline{u_i' u_j'}}{\partial x_j} \tag{2.43}$$

$$\frac{\partial \bar{c}}{\partial t} + \frac{\partial (\bar{u}_i \bar{c})}{\partial x_i} = \frac{\partial}{\partial x_i}(\omega_s \bar{c}\delta_{3i}) - \frac{\partial \overline{u_i' c'}}{\partial x_i} \tag{2.44}$$

where $\overline{u_i' u_j'}$ is the correlation between the fluctuating velocities in the x_i- and x_j-directions, and $\overline{u_i' c'}$ is the correlation between the fluctuating sediment concentration and velocity in the x_i-direction. Physically, $-\rho\overline{u_i' u_j'}$ represents the momentum transport due to turbulent motions; it is called the turbulent or Reynolds stress. $-\overline{u_i' c'}$ is the turbulent sediment flux, representing the sediment transport due to turbulence.

The set of equations (2.42)–(2.44) is not closed, due to the appearance of high-order correlation terms. In the next subsections, methods are introduced briefly to close this equation set on the levels of zero-, one-, and two-equation turbulence models. A detailed review can be found in Rodi (1993).

2.3.2 Zero-equation turbulence models

Boussinesq's eddy viscosity concept is widely used to model the turbulent or Reynolds stresses in Eq. (2.43). This concept assumes that, in analogy to the viscous stresses in Eq. (2.36), the turbulent stresses are proportional to the mean velocity gradients:

$$-\overline{u_i' u_j'} = v_t \left(\frac{\partial \bar{u}_i}{\partial x_j} + \frac{\partial \bar{u}_j}{\partial x_i} \right) - \frac{2}{3}k\delta_{ij} \tag{2.45}$$

where v_t is the turbulent or eddy viscosity; and k is the turbulent kinetic energy, defined as $k = \overline{u_i' u_i'}/2$. In contrast to the molecular viscosity v, the eddy viscosity v_t is not a fluid property but strongly depends on the state of turbulence, perhaps varying largely in time and space. In most flow regions, the eddy viscosity is much larger than the molecular viscosity; thus, the latter usually is negligible.

In direct analogy to the turbulent momentum transport, the turbulent sediment flux is assumed to be proportional to the gradient of sediment concentration:

$$-\overline{u_i'c'} = \varepsilon_s \frac{\partial \bar{c}}{\partial x_i} \tag{2.46}$$

where ε_s is the turbulent diffusivity of sediment. ε_s is often related to the eddy viscosity by $\varepsilon_s = \nu_t / \sigma_c$, with σ_c being the Schmidt number (between 0.5 and 1.0), which is discussed in detail in Section 3.5.1.

Note that the last term of Eq. (2.45) can be combined with the pressure-gradient term when Eq. (2.45) is inserted into the momentum equation (2.43). Therefore, the appearance of k in Eq. (2.45) does not necessitate the determination of it, and the set of equations (2.42)–(2.44) is closed using relations (2.45) and (2.46). Then the problem becomes how to determine the eddy viscosity. The eddy viscosity is usually assumed to be proportional to the velocity scale $\hat{u}$ and the length scale L_m of (large-scale) turbulent motions:

$$\nu_t \propto \hat{u} L_m \tag{2.47}$$

One of the zero-equation turbulence models widely used for the eddy viscosity is the Prandtl mixing length model, which postulates that the velocity scale $\hat{u}$ for two-dimensional shear flows is equal to the mean-velocity gradient times the mixing length, thus yielding

$$\nu_t = l_m^2 \left| \frac{\partial \bar{u}}{\partial z} \right| \tag{2.48}$$

where l_m is the mixing length. Commonly used relations are: $l_m = \kappa z$ for boundary layer flows, and $l_m = \kappa z \sqrt{(1 - z/h)}$ for open-channel flows, in which κ is the von Karman constant, z is the distance to the wall boundary or the bed, and h is the flow depth.

The mixing length model is suitable for flows where turbulence is in local equilibrium, rather than where the convective and/or diffusive transport of turbulence is important. Generally, the mixing length model is often used for simple shear-layer flows where l_m can be specified empirically. It is rarely used for rapidly varied flows, such as recirculating flows, in channels with complex geometry, due to difficulties in specifying l_m.

Another often used zero-equation turbulence model is the parabolic eddy viscosity model:

$$\nu_t = \kappa U_* z \left(1 - \frac{z}{h} \right) \tag{2.49}$$

where U_* is the bed shear velocity.

Eq. (2.49) may be derived by substituting the log distribution of flow velocity along the flow depth and the relation $l_m = \kappa z\sqrt{(1 - z/h)}$ into Eq. (2.48). Therefore, model (2.49) can be seen as a special case of model (2.48).

2.3.3 One-equation turbulence models

In the often used one-equation turbulence model, the eddy viscosity is determined using the Kolmogorov-Prandtl expression, which adopts $\sqrt{k}$ as the velocity scale and reads

$$v_t = c'_\mu \sqrt{k} L_m \tag{2.50}$$

where c'_μ is a coefficient of about 0.084.

Unlike the mixing length model, the one-equation turbulence model uses a transport equation to determine the turbulent energy k and, in turn, the fluctuating velocity scale. The transport equation of k can be derived in exact form from the continuity and Navier-Stokes equations. For high Reynolds numbers, this equation reads

$$\frac{\partial k}{\partial t} + \frac{\partial}{\partial x_i}(\bar{u}_i k) = -\frac{\partial}{\partial x_i}\left[\overline{u'_i\left(\frac{u'_j u'_j}{2} + \frac{p'}{\rho}\right)}\right] - \overline{u'_i u'_j}\frac{\partial \bar{u}_i}{\partial x_j} - v\overline{\frac{\partial u'_i}{\partial x_j}\frac{\partial u'_i}{\partial x_j}} \tag{2.51}$$

The three terms on the right-hand side of Eq. (2.51) represent the diffusion, production, and dissipation of k, respectively. To close this equation, the diffusion term is treated in analogy to Eq. (2.46), and the dissipation term is determined as $c_D k^{3/2}/L_m$, thus yielding the modeled k equation:

$$\frac{\partial k}{\partial t} + \frac{\partial}{\partial x_i}(\bar{u}_i k) = \frac{\partial}{\partial x_i}\left(\frac{v_t}{\sigma_k}\frac{\partial k}{\partial x_i}\right) + P_k - c_D\frac{k^{3/2}}{L_m} \tag{2.52}$$

where P_k is the production of turbulence by shear, defined as $P_k = -\overline{u'_i u'_j}\partial\bar{u}_i/\partial x_j$; σ_k is a coefficient of about 1.0; and c_D is a coefficient, usually set as $c_D \approx 0.08/c'_\mu$.

For the turbulence in a state of local equilibrium, its production is equal to dissipation, and then Eq. (2.52) can be simplified as $v_t(\partial\bar{u}/\partial z)^2 - c_D k^{3/2}/L_m = 0$ in the case of shear flows. By using Eqs. (2.48) and (2.50), the following relation can be derived (Rodi, 1993):

$$\frac{L_m}{l_m} = \frac{c_D^{1/4}}{c_\mu'^{3/4}} \tag{2.53}$$

Therefore, L_m can be determined using simple empirical formulas similar to those for the mixing length l_m.

2.3.4   Two-equation turbulence models

Linear k-ε turbulence models

Because of difficulties in specifying the length scale of turbulence in complex flows, the one-equation turbulence model described above has limitations. Thus, a differential equation for the length scale is often added, which in conjunction with the velocity scale equation constitute a two-equation turbulence model. The widely used two-equation models include the k-ε turbulence models, which replace the length scale with the dissipation rate $\varepsilon = v\overline{\partial u_i'/\partial x_j \cdot \partial u_i'/\partial x_j}$ and assume

$$v_t = c_\mu \frac{k^2}{\varepsilon} \tag{2.54}$$

where c_μ is a coefficient.

The k equation (2.52) is rewritten as

$$\frac{\partial k}{\partial t} + \frac{\partial}{\partial x_i}(\bar{u}_i k) = \frac{\partial}{\partial x_i}\left(\frac{v_t}{\sigma_k}\frac{\partial k}{\partial x_i}\right) + P_k - \varepsilon \tag{2.55}$$

An exact ε equation can be derived from the continuity and Navier-Stokes equations, but it includes several terms that have little known physical meanings and have to be modeled drastically (Rodi, 1971). The final modeled ε equation is expressed as

$$\frac{\partial \varepsilon}{\partial t} + \frac{\partial}{\partial x_i}(\bar{u}_i \varepsilon) = \frac{\partial}{\partial x_i}\left(\frac{v_t}{\sigma_\varepsilon}\frac{\partial \varepsilon}{\partial x_i}\right) + \frac{\varepsilon}{k}(c_{\varepsilon 1}P_k - c_{\varepsilon 2}\varepsilon) \tag{2.56}$$

where σ_ε, $c_{\varepsilon 1}$, and $c_{\varepsilon 2}$ are coefficients.

Launder and Spalding (1974) suggested a set of values for the coefficients: $c_\mu = 0.09$, $c_{\varepsilon 1} = 1.44$, $c_{\varepsilon 2} = 1.92$, $\sigma_k = 1.0$, and $\sigma_\varepsilon = 1.3$, as listed in Table 2.3. The k-ε model using this set of coefficients is often called the standard k-ε turbulence model.

However, the standard k-ε model overpredicts the spread rate of axisymmetric jet by about 30% (Rodi, 1993) and underpredicts the flow reattachment length downstream of a backward-facing step by 15–20% (Abe *et al.*, 1994). Many modifications of it have been suggested in the literature. Several examples are given below, and more can be found in Rodi (1993) and other references.

In the standard k-ε model, the ε equation is modeled drastically and may have limitations. Chen and Kim (1987) added a second time scale of the production range of turbulence kinetic energy spectrum and modified the ε equation to consider

Table 2.3 Coefficients in linear k-ε turbulence models

k-ε Model	c_μ	$c_{\varepsilon 1}$	$c_{\varepsilon 2}$	σ_k	σ_ε
Standard	0.09	1.44	1.92	1.0	1.3
Non-equilibrium	0.09	$1.15 + 0.25P_k/\varepsilon$	1.90	0.8927	1.15
RNG	0.085	$1.42 - \eta(1 - \eta/\eta_0)/(1 + \beta\eta^3)$	1.68	0.7179	0.7179

non-equilibrium between turbulence generation and dissipation. The modified k and ε equations are still formulated as Eqs. (2.55) and (2.56), with a functional form of the coefficient $c_{\varepsilon 1}$ as $c_{\varepsilon 1} = 1.15 + 0.25 P_k/\varepsilon$. The other coefficients are recalibrated as $c_\mu = 0.09$, $c_{\varepsilon 2} = 1.90$, $\sigma_k = 0.8927$, and $\sigma_\varepsilon = 1.15$. The modified model is called the non-equilibrium k-ε turbulence model, which has been tested in the case of recirculating flows with improved performance over the standard version (Shyy et al., 1997).

Yakhot et al. (1992) rederived the ε equation using the renormalized group (RNG) theory. One new term was introduced to take into account the highly anisotropic features of turbulence, usually associated with regions of large shear, and to modify the viscosity accordingly. This term was claimed to improve the simulation accuracy of the RNG k-ε turbulence model for highly strained flows. It can be included in the coefficient $c_{\varepsilon 1}$ by $c_{\varepsilon 1} = 1.42 - \eta(1 - \eta/\eta_0)/(1 + \beta\eta^3)$. Here, $\beta = 0.015$, $\eta = (2S_{ij}S_{ij})^{1/2}k/\varepsilon$, $S_{ij} = (\partial \bar{u}_i/\partial x_j + \partial \bar{u}_j/\partial x_i)/2$, and $\eta_0 = 4.38$. The other coefficients are $c_\mu = 0.085$, $c_{\varepsilon 2} = 1.68$, $\sigma_k = 0.7179$, and $\sigma_\varepsilon = 0.7179$, as listed in Table 2.3.

The standard k-ε turbulence model is restricted to high-Reynolds-number flows and is not applicable in the viscous sublayer near a wall. Jones and Launder (1972) proposed a low-Reynolds-number k-ε turbulence model, and later many investigators, e.g., Chien (1982) and Abe et al. (1994), suggested revisions. Usually, a damping function is introduced in the eddy viscosity equation (2.54) to mimic the direct effect of molecular viscosity on the shear stress, while two damping functions are multiplied to the production and destruction terms in the equation (2.56) to increase the magnitude of ε (for additional dissipation) near the wall and to incorporate the low Reynolds number effect on the decay of isotropic turbulence, respectively. Expressions of these damping functions and their performance can be found in Srikanth and Majumdar (1992) and Abe et al. (1994).

All the above k-ε turbulence models based on the Boussinesq assumption are often called linear k-ε turbulence models. In addition to them, another frequently used two-equation model is the k-ω turbulence model established by Wilcox (1993) by replacing the length scale with the specific dissipation rate ω. Because ω is related to ε by $\omega = \varepsilon/(\beta^* k)$ with $\beta^* = 0.09$, the k-ω model is similar to the standard k-ε model, with different coefficients. Its details are left to interested readers.

Nonlinear k-ε turbulence models

The Boussinesq assumption, which adopts an isotropic eddy viscosity concept for all Reynolds stresses, fails for flows with sudden changes in mean-strain rate or with "extra" strain rates, e.g., curved flows, because the Reynolds stresses adjust to such changes at a rate unrelated to the mean flow processes. Lumley (1970), Rodi (1976), Saffman (1976), Wilcox and Rubesin (1980), and Speziale (1987) derived more general relations for the Reynolds stresses. For example, the relation of Speziale (1987) reads

$$\frac{\tau_{ij}}{\rho} = 2v_t S_{ij} - \frac{2}{3}k\delta_{ij} + 4C_D c_\mu^2 \frac{k^3}{\varepsilon^2}\left(S_{ik}S_{kj} - \frac{1}{3}S_{mk}S_{km}\delta_{ij}\right)$$

$$+ 4C_E c_\mu^2 \frac{k^3}{\varepsilon^2}\left(\hat{S}_{ij} - \frac{1}{3}\hat{S}_{kk}\delta_{ij}\right) \tag{2.57}$$

where τ_{ij} are the Reynolds stresses, $\hat{S}_{ij} = \partial S_{ij}/\partial t + \bar{u}_k \partial S_{ij}/\partial x_k - S_{kj}\partial \bar{u}_i/\partial x_k - S_{ki}\partial \bar{u}_j/\partial x_k$, and $C_D = C_E = 1.68$.

Rodi (1976) assumed that the transport of turbulent stresses is proportional to the transport of turbulent energy, thus simplifying the full Reynolds stress equation to an algebraic expression:

$$\tau_{ij} = -\rho k \left[\frac{2}{3}\delta_{ij} + \frac{(1-\gamma)(P_{ij} - \frac{2}{3}P_k\delta_{ij})}{P_k - (1-c_1)\varepsilon} \right] \tag{2.58}$$

where γ and c_1 are coefficients; and P_{ij} is the stress production, defined as $P_{ij} = -\overline{u_i'u_k'}\partial u_j/\partial x_k - \overline{u_j'u_k'}\partial u_i/\partial x_k$.

Eqs. (2.57) and (2.58) include the Boussinesq approximation as leading terms and need to be coupled with a k-ε model. Thus, they are often called nonlinear k-ε turbulence models. Speziale showed model (2.57) yields more accurate predictions for the Reynolds normal stress in turbulent channel flows and homogenous shear flows than the standard k-ε turbulence model. For example, for a unidirectional uniform duct flow, the standard k-ε model predicts $\tau_{zz} - \tau_{yy} = 0$, whereas the nonlinear k-ε turbulence model yields

$$\tau_{zz} - \tau_{yy} = \rho C_D c_\mu^2 \frac{k^3}{\varepsilon^2}\left[\left(\frac{\partial \bar{u}}{\partial z}\right)^2 - \left(\frac{\partial \bar{u}}{\partial y}\right)^2 \right] \tag{2.59}$$

As a result, the nonlinear k-ε turbulence model (2.57) is able to simulate the turbulence-driven secondary flows (Speziale, 1987; Pezzinga, 1994). So is Rodi's algebraic stress model (2.58).

2.3.5 Other turbulence models and simulations

A more advanced turbulence model is the Reynolds stress model, in which the transport equations of $\overline{u_i'u_j'}$ are derived in exact form from the continuity and Navier-Stokes equations and modeled to obtain a closed system (see Rodi, 1993). The large eddy simulation (LES) and direct numerical simulation (DNS) of turbulent flows have also advanced recently in CFD. However, the Reynolds stress model, LES, and DNS have been little tested so far and are not yet in use for practical applications in river engineering.

2.4 DERIVATION OF 1-D AND 2-D FLOW AND SEDIMENT TRANSPORT EQUATIONS

Because of the limitation of computer capacity, solving a full 3-D model is time-consuming; this was particularly true several decades ago. Thus, the development of 1-D and 2-D models has been an important task in computational river dynamics (e.g., de Saint Venant, 1871; Kuipers and Vreugdenhil, 1973). The derivation of

1-D and 2-D model equations from 3-D model equations via section-, depth-, and width-integrating (averaging) approaches is introduced in this section.

Before deriving the spatially-integrated models, let us introduce the 3-D flow and sediment transport equations and the associated boundary conditions at the water surface, channel bottom, and banks. In the Cartesian coordinate system shown in Fig. 2.6, the 3-D continuity and momentum equations (2.42) and (2.43) of flow with low sediment concentration are rewritten as

$$\frac{\partial u_x}{\partial x} + \frac{\partial u_y}{\partial y} + \frac{\partial u_z}{\partial z} = 0 \tag{2.60}$$

$$\frac{\partial u_x}{\partial t} + \frac{\partial (u_x^2)}{\partial x} + \frac{\partial (u_y u_x)}{\partial y} + \frac{\partial (u_z u_x)}{\partial z} = \frac{1}{\rho} F_x - \frac{1}{\rho} \frac{\partial p}{\partial x} + \frac{1}{\rho} \frac{\partial \tau_{xx}}{\partial x}$$
$$+ \frac{1}{\rho} \frac{\partial \tau_{xy}}{\partial y} + \frac{1}{\rho} \frac{\partial \tau_{xz}}{\partial z} \tag{2.61}$$

$$\frac{\partial u_y}{\partial t} + \frac{\partial (u_x u_y)}{\partial x} + \frac{\partial (u_y^2)}{\partial y} + \frac{\partial (u_z u_y)}{\partial z} = \frac{1}{\rho} F_y - \frac{1}{\rho} \frac{\partial p}{\partial y} + \frac{1}{\rho} \frac{\partial \tau_{yx}}{\partial x}$$
$$+ \frac{1}{\rho} \frac{\partial \tau_{yy}}{\partial y} + \frac{1}{\rho} \frac{\partial \tau_{yz}}{\partial z} \tag{2.62}$$

$$\frac{\partial u_z}{\partial t} + \frac{\partial (u_x u_z)}{\partial x} + \frac{\partial (u_y u_z)}{\partial y} + \frac{\partial (u_z^2)}{\partial z} = \frac{1}{\rho} F_z - \frac{1}{\rho} \frac{\partial p}{\partial z} + \frac{1}{\rho} \frac{\partial \tau_{zx}}{\partial x}$$
$$+ \frac{1}{\rho} \frac{\partial \tau_{zy}}{\partial y} + \frac{1}{\rho} \frac{\partial \tau_{zz}}{\partial z} \tag{2.63}$$

where $x \ (= x_1)$ and $y \ (= x_2)$ are the horizontal coordinates; $z \ (= x_3)$ is the vertical coordinate above a datum; u_x, u_y, and u_z are the components of mean velocity in the x-, y- and z-directions; τ_{xx}, τ_{xy}, $\ldots$, and τ_{zz} are the stresses (including both molecular and turbulent effects); and F_x, F_y, and F_z are the components of the resultant external force in the x-, y- and z-directions. As gravity is assumed to be the only external force, $F_x = F_y = 0$ and $F_z = -\rho g$.

Note that the bar "-", denoting time-averaged quantities, is omitted in Eqs. (2.60)–(2.63) for simplicity.

For gradually varied (shallow water) flows, the inertia and diffusion effects in the vertical momentum equation (2.63) are usually neglected, yielding the hydrostatic pressure equation:

$$\frac{\partial p}{\partial z} = -\rho g \tag{2.64}$$

Under the assumption of constant ρ along the depth, Eq. (2.64) has an analytic solution:

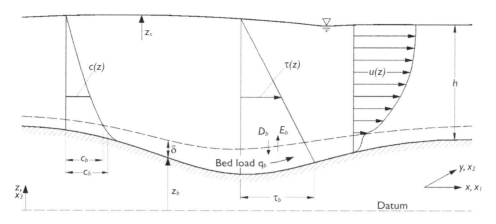

Figure 2.6 Configuration of flow and sediment transport.

$$p = p_a + \rho g(z_s - z) \tag{2.65}$$

where z_s is the water surface elevation, and p_a is the atmospheric pressure at the water surface. A constant p_a is assumed here for a short river reach.

Substituting Eq. (2.65) into Eqs. (2.61) and (2.62) yields the x- and y-momentum equations for gradually varied flows:

$$\frac{\partial u_x}{\partial t} + \frac{\partial (u_x^2)}{\partial x} + \frac{\partial (u_y u_x)}{\partial y} + \frac{\partial (u_z u_x)}{\partial z} = -g\frac{\partial z_s}{\partial x} + \frac{1}{\rho}\frac{\partial \tau_{xx}}{\partial x} + \frac{1}{\rho}\frac{\partial \tau_{xy}}{\partial y} + \frac{1}{\rho}\frac{\partial \tau_{xz}}{\partial z} \tag{2.66}$$

$$\frac{\partial u_y}{\partial t} + \frac{\partial (u_x u_y)}{\partial x} + \frac{\partial (u_y^2)}{\partial y} + \frac{\partial (u_z u_y)}{\partial z} = -g\frac{\partial z_s}{\partial y} + \frac{1}{\rho}\frac{\partial \tau_{yx}}{\partial x} + \frac{1}{\rho}\frac{\partial \tau_{yy}}{\partial y} + \frac{1}{\rho}\frac{\partial \tau_{yz}}{\partial z} \tag{2.67}$$

Because the channel bed and banks generally vary in much lower speed than the flow, the following non-slip condition is applied at these solid boundaries:

$$u_{bx} = 0, u_{by} = 0, u_{bz} = 0 \tag{2.68}$$

The water surface is a free moving boundary, the location of which is part of the solution. For a particle on the free surface, its location (x, y, z) can be described by

$$z = z_s(x, y, t) \tag{2.69}$$

and it moves with the free surface, i.e.,

$$\frac{dx}{dt} = u_{bx}, \frac{dy}{dt} = u_{by}, \frac{dz}{dt} = u_{bz} \tag{2.70}$$

where u_{bx}, u_{by}, and u_{bz} are the components of flow velocity at the water surface in the x-, y- and z-directions. Thus, differentiating Eq. (2.69) with respect to t leads to the following kinematic condition of free surface:

$$\frac{\partial z_s}{\partial t} + u_{bx}\frac{\partial z_s}{\partial x} + u_{by}\frac{\partial z_s}{\partial y} = u_{bz} \tag{2.71}$$

The three-dimensional sediment transport equation (2.44) closed using Eq. (2.46) is rewritten as

$$\frac{\partial c}{\partial t} + \frac{\partial(u_x c)}{\partial x} + \frac{\partial(u_y c)}{\partial y} + \frac{\partial(u_z c)}{\partial z} - \frac{\partial(\omega_s c)}{\partial z} = \frac{\partial}{\partial x}\left(\varepsilon_s \frac{\partial c}{\partial x}\right) + \frac{\partial}{\partial y}\left(\varepsilon_s \frac{\partial c}{\partial y}\right)$$
$$+ \frac{\partial}{\partial z}\left(\varepsilon_s \frac{\partial c}{\partial z}\right) \tag{2.72}$$

In general, Eq. (2.72) is approximately applicable to all sediment loads (if fine enough) in the entire water column. However, because bed load and suspended load behave differently, the water column is often divided into two zones: a bed-load zone from the bed elevation z_b to $z_b + \delta$ and a suspended-load zone from $z_b + \delta$ to z_s, as shown in Fig. 2.6. Here, δ is the thickness of the bed-load zone, which is usually assumed to be about twice the sediment diameter (Einstein, 1950) or half the bed-form height.

The net vertical sediment flux across the water surface should be zero and, thus, the suspended-load boundary condition at the water surface is

$$\left(\varepsilon_s \frac{\partial c}{\partial z} + \omega_s c\right)_{z=z_s} = 0 \tag{2.73}$$

There are usually two approaches to specify the suspended-load boundary condition at the interface between the suspended-load and bed-load zones. One approach is to assume the near-bed suspended-load concentration to be at equilibrium:

$$c|_{z=z_b+\delta} = c_{b*} \tag{2.74}$$

where c_{b*} is the equilibrium (capacity) sediment concentration at the interface.

The other approach is to assume that the near-bed sediment entrainment flux is at the capacity of flow picking up sediment under the considered flow conditions and bed sediment configurations:

$$E_b = -\varepsilon_s \left.\frac{\partial c}{\partial z}\right|_{z=z_b+\delta} = \omega_s c_{b*} \tag{2.75}$$

where E_b is the entrainment flux of sediment at the interface. Correspondingly, the deposition flux at the interface is defined as $D_b = \omega_s c_b$, in which c_b is the suspended-load concentration at the interface between the suspended-load and bed-load zones. Note that E_b and D_b are defined per unit area of horizontal plane rather than bed surface; the bed surface may be curved, whereas E_b and D_b are along the vertical direction.

Eqs. (2.74) and (2.75) are often called "concentration" and "gradient" boundary conditions, respectively. Eq. (2.74) is applicable for equilibrium sediment transport at the interface, while Eq. (2.75) is applicable for both equilibrium and non-equilibrium sediment transports. In particular, for equilibrium transport, $D_b = E_b$ and Eq. (2.75) becomes Eq. (2.74). Therefore, Eq. (2.75) is more general than Eq. (2.74). More discussions about the near-bed suspended-load boundary condition are given in Sections 2.5.2 and 7.3.1.

2.4.1 Depth-averaged 2-D model equations

Depth-averaged hydrodynamic equations

The depth-averaged quantity Φ of a three-dimensional variable ϕ is defined by

$$\Phi = \frac{1}{h} \int_{z_b}^{z_s} \phi \, dz \tag{2.76}$$

Integrating the continuity equation (2.60) over the flow depth yields

$$\int_{z_b}^{z_s} \frac{\partial u_x}{\partial x} dz + \int_{z_b}^{z_s} \frac{\partial u_y}{\partial y} dz + \int_{z_b}^{z_s} \frac{\partial u_z}{\partial z} dz = 0 \tag{2.77}$$

which is reformulated using the Leibniz integral rule as

$$\frac{\partial}{\partial x} \int_{z_b}^{z_s} u_x dz - u_{bx}\frac{\partial z_s}{\partial x} + u_{bx}\frac{\partial z_b}{\partial x} + \frac{\partial}{\partial y} \int_{z_b}^{z_s} u_y dz - u_{by}\frac{\partial z_s}{\partial y} + u_{by}\frac{\partial z_b}{\partial y}$$
$$+ u_{hz} - u_{bz} = 0 \tag{2.78}$$

Substituting boundary conditions (2.68) and (2.71) into Eq. (2.78) leads to the depth-integrated 2-D continuity equation:

$$\frac{\partial h}{\partial t} + \frac{\partial (hU_x)}{\partial x} + \frac{\partial (hU_y)}{\partial y} = 0 \tag{2.79}$$

where U_x and U_y are the depth-averaged quantities of local velocities u_x and u_y, defined by Eq. (2.76). Note that $\partial h/\partial t$ in Eq. (2.79) may be replaced by $\partial z_s/\partial t$, because the bed change is omitted.

Integrating the x-momentum equation (2.66) over the flow depth yields

$$
\int_{z_b}^{z_s} \frac{\partial u_x}{\partial t} dz + \int_{z_b}^{z_s} \frac{\partial (u_x^2)}{\partial x} dz + \int_{z_b}^{z_s} \frac{\partial (u_y u_x)}{\partial y} dz + \int_{z_b}^{z_s} \frac{\partial (u_z u_x)}{\partial z} dz
$$

$$
= -g \int_{z_b}^{z_s} \frac{\partial z_s}{\partial x} dz + \frac{1}{\rho} \int_{z_b}^{z_s} \frac{\partial \tau_{xx}}{\partial x} dz + \frac{1}{\rho} \int_{z_b}^{z_s} \frac{\partial \tau_{xy}}{\partial y} dz + \frac{1}{\rho} \int_{z_b}^{z_s} \frac{\partial \tau_{xz}}{\partial z} dz \quad (2.80)
$$

and then applying the Leibniz rule to this equation yields

$$
\frac{\partial}{\partial t} \left(\int_{z_b}^{z_s} u_x dz \right) - u_{bx} \frac{\partial z_s}{\partial t} + u_{bx} \frac{\partial z_b}{\partial t} + \frac{\partial}{\partial x} \left(\int_{z_b}^{z_s} u_x^2 dz \right) - u_{bx}^2 \frac{\partial z_s}{\partial x} + u_{bx}^2 \frac{\partial z_b}{\partial x}
$$

$$
+ \frac{\partial}{\partial y} \left(\int_{z_b}^{z_s} u_y u_x dz \right) - u_{by} u_{bx} \frac{\partial z_s}{\partial y} + u_{by} u_{bx} \frac{\partial z_b}{\partial y} + u_{bz} u_{bx} - u_{bz} u_{bx}
$$

$$
= -gh \frac{\partial z_s}{\partial x} + \frac{1}{\rho} \frac{\partial}{\partial x} \left(\int_{z_b}^{z_s} \tau_{xx} dz \right) - \frac{1}{\rho} \tau_{xx,s} \frac{\partial z_s}{\partial x} + \frac{1}{\rho} \tau_{xx,b} \frac{\partial z_b}{\partial x}
$$

$$
+ \frac{1}{\rho} \frac{\partial}{\partial y} \left(\int_{z_b}^{z_s} \tau_{xy} dz \right) - \frac{1}{\rho} \tau_{xy,s} \frac{\partial z_s}{\partial y} + \frac{1}{\rho} \tau_{xy,b} \frac{\partial z_b}{\partial y} + \frac{1}{\rho} (\tau_{xz,s} - \tau_{xz,b}) \quad (2.81)
$$

Substituting boundary conditions (2.68) and (2.71) into Eq. (2.81) results in the depth-integrated x-momentum equation:

$$
\frac{\partial (h U_x)}{\partial t} + \frac{\partial (h U_x^2)}{\partial x} + \frac{\partial (h U_y U_x)}{\partial y} = -gh \frac{\partial z_s}{\partial x} + \frac{1}{\rho} \frac{\partial [h(T_{xx} + D_{xx})]}{\partial x}
$$

$$
+ \frac{1}{\rho} \frac{\partial [h(T_{xy} + D_{xy})]}{\partial y} + \frac{1}{\rho} (\tau_{sx} - \tau_{bx})
$$

$$
(2.82)
$$

where T_{xx} and T_{xy} are the depth-averaged normal and shear stresses; D_{xx} and D_{xy} account for the dispersion momentum transports due to the vertical non-uniformity of velocity, defined as $D_{xx} = -\frac{\rho}{h} \int_{z_b}^{z_s} (u_x - U_x)^2 dz$ and $D_{xy} = -\frac{\rho}{h} \int_{z_b}^{z_s} (u_x - U_x)(u_y - U_y) dz$; τ_{sx} is the x-component of shear force per unit horizontal area, usually due to wind driving at the water surface, defined as $\tau_{sx} = \tau_{xz,s} - \tau_{xx,s} \partial z_s/\partial x - \tau_{xy,s} \partial z_s/\partial y$; and τ_{bx} is the x-component of bed shear force per unit horizontal area, defined as $\tau_{bx} = \tau_{xz,b} - \tau_{xx,b} \partial z_b/\partial x - \tau_{xy,b} \partial z_b/\partial y$. Note that τ_{bx} may be written as $\tau_{bx} = m_b \hat{\tau}_{bx}$, in which $\hat{\tau}_{bx}$ is the x-component of bed shear force per unit bed surface area, and m_b is the bed slope coefficient defined as $m_b = [1 + (\partial z_b/\partial x)^2 + (\partial z_b/\partial y)^2]^{1/2}$.

Similarly, integrating Eq. (2.67) over the flow depth leads to the depth-integrated y-momentum equation:

$$\frac{\partial(hU_y)}{\partial t} + \frac{\partial(hU_xU_y)}{\partial x} + \frac{\partial(hU_y^2)}{\partial y} = -gh\frac{\partial z_s}{\partial y} + \frac{1}{\rho}\frac{\partial[h(T_{yx} + D_{yx})]}{\partial x}$$

$$+ \frac{1}{\rho}\frac{\partial[h(T_{yy} + D_{yy})]}{\partial y} + \frac{1}{\rho}(\tau_{sy} - \tau_{by})$$

(2.83)

where T_{yx} and T_{yy} are the depth-averaged shear and normal stresses; D_{yx} and D_{yy} account for the dispersion momentum transports due to the vertical non-uniformity of velocity, defined as $D_{yx} = D_{xy}$ and $D_{yy} = -\frac{\rho}{h}\int_{z_b}^{z_s}(u_y - U_y)^2 dz$; τ_{sy} is the y-component of wind shear force per unit horizontal area at the water surface; and τ_{by} is the y-component of bed shear force per unit horizontal area.

The depth-averaged stresses $T_{ij}(i, j = x, y)$ can be related to the gradients of the depth-averaged velocities by the Bossinesq assumption similar to Eq. (2.45) in a turbulence model, such as the depth-averaged k-ε turbulence model proposed by Rastogi and Rodi (1978). However, there is not a general method to handle the dispersion terms D_{ij}. D_{ij} are not related to turbulence, but both D_{ij} and T_{ij} represent momentum transports as effective stresses. In nearly straight channels, the dispersion transports are usually combined with the turbulent stresses. In curved channels, secondary flows, especially the helical flow, play an important role in fluvial processes, and thus the dispersion transports become important and should be taken into account through additional model closures. This is discussed in Section 6.3.

Depth-averaged sediment transport equations

Unlike the depth-averaged quantities defined by Eq. (2.76), the depth-averaged suspended-load concentration, C, is defined by

$$C = \frac{1}{(h - \delta)U_s}\int_{z_b+\delta}^{z_s} u_s c\, dz$$

(2.84)

where U_s is the streamwise depth-averaged velocity, and u_s is the local flow velocity projected to the streamwise direction. By definition, $U_s = \int_{z_b+\delta}^{z_s} u_s dz/(h - \delta)$, but U_s is approximately set as the resultant depth-averaged velocity $U = \sqrt{U_x^2 + U_y^2}$ at each horizontal point.

Integrating the three-dimensional sediment transport equation (2.72) over the suspended-load zone leads to

$$\int_{z_b+\delta}^{z_s}\frac{\partial c}{\partial t}dz + \int_{z_b+\delta}^{z_s}\frac{\partial(u_xc)}{\partial x}dz + \int_{z_b+\delta}^{z_s}\frac{\partial(u_yc)}{\partial y}dz + \int_{z_b+\delta}^{z_s}\frac{\partial(u_zc)}{\partial z}dz - \int_{z_b+\delta}^{z_s}\frac{\partial(\omega_sc)}{\partial z}dz$$

$$= \int_{z_b+\delta}^{z_s}\left[\frac{\partial}{\partial x}\left(\varepsilon_s\frac{\partial c}{\partial x}\right)\right]dz + \int_{z_b+\delta}^{z_s}\left[\frac{\partial}{\partial y}\left(\varepsilon_s\frac{\partial c}{\partial y}\right)\right]dz + \int_{z_b+\delta}^{z_s}\left[\frac{\partial}{\partial z}\left(\varepsilon_s\frac{\partial c}{\partial z}\right)\right]dz$$

(2.85)

By applying the Leibniz integral rule and boundary conditions (2.68), (2.71), (2.73), and (2.75) to Eq. (2.85) and assuming that the bed-load zone is very thin, i.e., $\delta \ll h$, the depth-integrated suspended-load transport equation is obtained:

$$
\frac{\partial}{\partial t}\left(\frac{hC}{\beta_s}\right) + \frac{\partial(hU_xC)}{\partial x} + \frac{\partial(hU_yC)}{\partial y}
$$
$$
= \frac{\partial}{\partial x}\left[h\left(\varepsilon_s\frac{\partial C}{\partial x} + D_{sx}\right)\right] + \frac{\partial}{\partial y}\left[h\left(\varepsilon_s\frac{\partial C}{\partial y} + D_{sy}\right)\right] + E_b - D_b \qquad (2.86)
$$

where β_s is a correction factor for suspended load:

$$
\beta_s = \int_{z_b+\delta}^{z_s} u_s c\,dz \Big/ \left(U_s \int_{z_b+\delta}^{z_s} c\,dz\right) \qquad (2.87)
$$

Note that the coefficient β_s in Eq. (2.86) should not be zero; otherwise, no suspended load moves. β_s should also appear in the diffusion terms in Eq. (2.86), but it is lumped into the diffusivity ε_s for simplicity. However, if the depth-averaged sediment concentration C is defined using Eq. (2.76) rather than (2.84), β_s should appear in the convection terms rather than the storage term, i.e.,

$$
\frac{\partial(hC)}{\partial t} + \frac{\partial(\beta_s hU_xC)}{\partial x} + \frac{\partial(\beta_s hU_yC)}{\partial y}
$$
$$
= \frac{\partial}{\partial x}\left[h\left(\varepsilon_s\frac{\partial C}{\partial x} + D_{sx}\right)\right] + \frac{\partial}{\partial y}\left[h\left(\varepsilon_s\frac{\partial C}{\partial y} + D_{sy}\right)\right] + E_b - D_b \qquad (2.88)
$$

It should be clarified that defining the depth-averaged suspended-load concentration C by Eq. (2.76) results in the unit suspended-load discharge $q_s = \beta_s UhC$, while the definition (2.84) yields $q_s = UhC$. If mass balance is respected, either definition can be used. The definition (2.84) is adopted in this book, except where stated otherwise.

The coefficient β_s is actually the ratio of the depth-averaged sediment and flow velocities and accounts for the temporal lag between flow and suspended-load transport in the depth-averaged 2-D model. As demonstrated later, β_s also appears in the 1-D model. However, this lag due to difference between the depth-averaged flow and sediment velocities can be automatically taken into account in the 3-D (or vertical 2-D) model, which directly uses the local flow velocity and sediment concentration as dependent variables. The evaluation of β_s is discussed in Section 3.8.

D_{sx} and D_{sy} in Eq. (2.86) are called dispersion sediment fluxes, which account for the dispersion effect due to the non-uniform distributions of flow velocity and sediment concentration over the flow depth, defined as $D_{sx} = -\frac{1}{h}\int_{z_b}^{z_s}(u_x - U_x)(c - C)dz$ and $D_{sy} = -\frac{1}{h}\int_{z_b}^{z_s}(u_y - U_y)(c - C)dz$. In nearly straight channels, the dispersion fluxes may be combined with the (turbulent) diffusion fluxes, with ε_s replaced by a mixing coefficient to represent the diffusion and dispersion effects together. In curved channels, the dispersion fluxes become important and need to be modeled, as discussed in Section 6.3.

Integrating the three-dimensional sediment transport equation (2.72) over the bed-load zone leads to the bed-load mass balance equation:

$$(1 - p'_m)\frac{\partial z_b}{\partial t} + \frac{\partial(\delta c_\delta)}{\partial t} + \frac{\partial(\alpha_{bx}q_b)}{\partial x} + \frac{\partial(\alpha_{by}q_b)}{\partial y} = D_b - E_b \qquad (2.89)$$

where p'_m is the porosity of bed material at the bed surface, c_δ is the average volumetric concentration of sediment at the bed-load zone, q_b is the bed-load transport rate by volume per unit time and width (m^2s^{-1}), and α_{bx} and α_{by} are the direction cosines of bed-load movement. The bed load is usually assumed to move along the direction of bed shear stress but may be affected by secondary flows in curved channels and gravity in channels with steep bed and bank slopes.

The first term on the left-hand side of Eq. (2.89) represents the bed change, which results from the exchange between moving sediment and bed material. The second term accounts for the storage effect. In general, the average bed-load concentration c_δ is related to the bed-load transport rate q_b and velocity u_b by $c_\delta = q_b/(\delta u_b)$, thus yielding

$$(1 - p'_m)\frac{\partial z_b}{\partial t} + \frac{\partial}{\partial t}\left(\frac{q_b}{u_b}\right) + \frac{\partial(\alpha_{bx}q_b)}{\partial x} + \frac{\partial(\alpha_{by}q_b)}{\partial y} = D_b - E_b \qquad (2.90)$$

Because the bed-load velocity u_b is usually slower than the flow velocity, Eq. (2.90) accounts for the temporal lag between flow and bed-load transport.

Summing Eqs. (2.86) and (2.90) leads to the overall sediment balance equation:

$$(1 - p'_m)\frac{\partial z_b}{\partial t} + \frac{\partial}{\partial t}\left(\frac{hC_t}{\beta_t}\right) + \frac{\partial q_{tx}}{\partial x} + \frac{\partial q_{ty}}{\partial y} = 0 \qquad (2.91)$$

where C_t is the depth-averaged concentration of total load; q_{tx} and q_{ty} are the total-load fluxes: $q_{tx} = \alpha_{bx}q_b + hU_xC - \varepsilon_s h\partial C/\partial x - hD_{sx}$ and $q_{ty} = \alpha_{by}q_b + hU_yC - \varepsilon_s h\partial C/\partial y - hD_{sy}$; and β_t is a correction factor for total load, related to β_s and u_b by

$$\beta_t = \frac{hC_t}{hC/\beta_s + q_b/u_b} = \frac{1}{r_s/\beta_s + (1 - r_s)U/u_b} \qquad (2.92)$$

where r_s is the ratio of suspended load to total load.

2.4.2 Width-averaged 2-D model equations

Fig. 2.7 shows the configuration of a cross-section. The width-averaged quantity $\tilde{\Phi}$ of a three-dimensional variable ϕ is defined by

$$\tilde{\Phi} = \frac{1}{b}\int_{b_1}^{b_2} \phi \, dy \qquad (2.93)$$

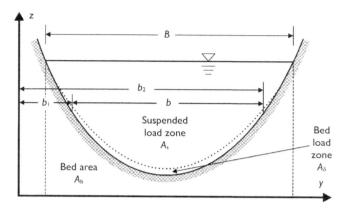

Figure 2.7 Configuration of cross-section.

where b_1 and b_2 are the y-coordinate values of two banks; and $b = b_2 - b_1$, i.e., the flow width at height z. Note that x and y are herein set along the longitudinal and transverse directions of the channel, respectively, for deriving the width-averaged 2-D and 1-D model equations. However, this arrangement is not necessary in the depth-averaged 2-D and 3-D models.

Integrating the three-dimensional continuity equation (2.60) along the y-coordinate axis over the flow width yields

$$\int_{b_1}^{b_2} \frac{\partial u_x}{\partial x} dy + \int_{b_1}^{b_2} \frac{\partial u_y}{\partial y} dy + \int_{b_1}^{b_2} \frac{\partial u_z}{\partial z} dy = 0 \qquad (2.94)$$

and applying the Leibniz integral rule and the non-slip boundary condition at banks to this equation yields the width-integrated continuity equation of flow:

$$\frac{\partial(b\widetilde{U}_x)}{\partial x} + \frac{\partial(b\widetilde{U}_z)}{\partial z} = 0 \qquad (2.95)$$

where $\widetilde{U}_x$ and $\widetilde{U}_z$ are the width-averaged velocities in the x- and z-directions, defined by Eq. (2.93).

In a similar manner, integrating the x- and z-momentum equations (2.61) and (2.63) over the flow width leads to the width-integrated momentum equations:

$$\frac{\partial(b\widetilde{U}_x)}{\partial t} + \frac{\partial(b\widetilde{U}_x^2)}{\partial x} + \frac{\partial(b\widetilde{U}_z\widetilde{U}_x)}{\partial z} = -\frac{1}{\rho}b\frac{\partial\widetilde{p}}{\partial x} + \frac{1}{\rho}\frac{\partial[b(\widetilde{T}_{xx} + \widetilde{D}_{xx})]}{\partial x}$$
$$+ \frac{1}{\rho}\frac{\partial[b(\widetilde{T}_{xz} + \widetilde{D}_{xz})]}{\partial z} - \frac{1}{\rho}(m_1\tau_{x1} + m_2\tau_{x2})$$

$$(2.96)$$

$$\frac{\partial(b\widetilde{U}_z)}{\partial t} + \frac{\partial(b\widetilde{U}_x\widetilde{U}_z)}{\partial x} + \frac{\partial(b\widetilde{U}_z^2)}{\partial z} = -bg - \frac{1}{\rho}b\frac{\partial\tilde{p}}{\partial z} + \frac{1}{\rho}\frac{\partial[b(\widetilde{T}_{zx} + \widetilde{D}_{zx})]}{\partial x}$$
$$+ \frac{1}{\rho}\frac{\partial[b(\widetilde{T}_{zz} + \widetilde{D}_{zz})]}{\partial z} - \frac{1}{\rho}(m_1\tau_{z1} + m_2\tau_{z2})$$

(2.97)

where $\tilde{p}$ and $\widetilde{T}_{ij}(i,j = x,z)$ are the width-averaged pressure and stresses, respectively; $\widetilde{D}_{ij}$ are the dispersion momentum transports due to the lateral non-uniformity of flow velocity, defined as $\widetilde{D}_{xx} = -\frac{\rho}{b}\int_{b_1}^{b_2}(u_x - \widetilde{U}_x)^2 dy$, $\widetilde{D}_{xz} = \widetilde{D}_{zx} = -\frac{\rho}{b}\int_{b_1}^{b_2}(u_x - \widetilde{U}_x)(u_z - \widetilde{U}_z)dy$, and $\widetilde{D}_{zz} = -\frac{\rho}{b}\int_{b_1}^{b_2}(u_z - \widetilde{U}_z)^2 dy$; τ_{xl} and $\tau_{zl}(l = 1,2)$ are the shear stresses in the x- and z-directions on the two bank surfaces; and m_l are the bank slope coefficients, defined as $m_l = [1 + (\partial b_l/\partial x)^2 + (\partial b_l/\partial z)^2]^{1/2}$.

For gradually varied flows, the effects of inertia, diffusion, and dispersion in the vertical momentum equation (2.97) can be neglected, yielding the hydrostatic pressure equation (2.65). The x-momentum equation (2.96) is then turned to

$$\frac{\partial(b\widetilde{U}_x)}{\partial t} + \frac{\partial(b\widetilde{U}_x^2)}{\partial x} + \frac{\partial(b\widetilde{U}_z\widetilde{U}_x)}{\partial z} = -gb\frac{\partial\tilde{z}_s}{\partial x} + \frac{1}{\rho}\frac{\partial[b(\widetilde{T}_{xx} + \widetilde{D}_{xx})]}{\partial x}$$
$$+ \frac{1}{\rho}\frac{\partial[b(\widetilde{T}_{xz} + \widetilde{D}_{xz})]}{\partial z} - \frac{1}{\rho}(m_1\tau_{x1} + m_2\tau_{x2})$$

(2.98)

where $\tilde{z}_s$ is the laterally-averaged water surface elevation.

Integrating Eq. (2.72) over the flow width leads to the width-integrated suspended-load transport equation:

$$\frac{\partial(b\widetilde{C})}{\partial t} + \frac{\partial(b\widetilde{U}_x\widetilde{C})}{\partial x} + \frac{\partial(b\widetilde{U}_z\widetilde{C})}{\partial z} - \frac{\partial(b\omega_s\widetilde{C})}{\partial z}$$
$$= \frac{\partial}{\partial x}\left[b\left(\varepsilon_s\frac{\partial\widetilde{C}}{\partial x} + \widetilde{D}_{sx}\right)\right] + \frac{\partial}{\partial z}\left[b\left(\varepsilon_s\frac{\partial\widetilde{C}}{\partial z} + \widetilde{D}_{sz}\right)\right] + S_c$$

(2.99)

where $\widetilde{C}$ is the width-averaged concentration of suspended load; $\widetilde{D}_{sx}$ and $\widetilde{D}_{sz}$ are the dispersion fluxes, defined as $\widetilde{D}_{sx} = -\frac{1}{b}\int_{b_1}^{b_2}(u_x - \widetilde{U}_x)(c - \widetilde{C})dy$ and $\widetilde{D}_{sz} = -\frac{1}{b}\int_{b_1}^{b_2}(u_z - \widetilde{U}_z)(c - \widetilde{C})dy$; and S_c includes the sediment exchange at banks and the side discharge from tributaries.

The bed-load zone is so thin that it is not necessary to consider the vertical variation of sediment concentration in this zone. The width-integrated bed-load transport is determined using the 1-D transport equation introduced in the next subsection.

2.4.3 Section-averaged 1-D model equations

The cross-section-averaged quantity $\widehat{\Phi}$ of a three-dimensional variable ϕ is defined by

$$\widehat{\Phi} = \frac{1}{A} \iint_A \phi \, dA = \frac{1}{A} \int_{z_b}^{z_s} \int_{b_1}^{b_2} \phi \, dy \, dz \tag{2.100}$$

where A is the flow area in the cross-section, as shown in Fig. 2.7.

Integrating the 3-D continuity equation (2.60) over the cross-section leads to

$$\int_{z_b}^{z_s} \int_{b_1}^{b_2} \frac{\partial u_x}{\partial x} dy \, dz + \int_{z_b}^{z_s} \int_{b_1}^{b_2} \frac{\partial u_y}{\partial y} dy \, dz + \int_{z_b}^{z_s} \int_{b_1}^{b_2} \frac{\partial u_z}{\partial z} dy \, dz = 0 \tag{2.101}$$

which is reformulated to the following 1-D continuity equation by applying the Leibniz rule, the non-slip condition (2.68) at the channel bed and banks, and the kinematic condition (2.71) at the water surface:

$$\frac{\partial A}{\partial t} + \frac{\partial (A\widehat{U})}{\partial x} = 0 \tag{2.102}$$

where $\widehat{U}$ is the flow velocity averaged over the cross-section, defined by Eq. (2.100).

Integrating the 3-D momentum equation (2.66) over the cross-section yields the 1-D momentum equation:

$$\frac{\partial (A\widehat{U})}{\partial t} + \frac{\partial (A\widehat{U^2})}{\partial x} = -gA\frac{\partial \tilde{z}_s}{\partial x} + \frac{1}{\rho}\frac{\partial [A(\widehat{T}_{xx} + \widehat{D}_{xx})]}{\partial x} + \frac{1}{\rho}(B\hat{\tau}_{sx} - \chi \hat{\tau}_{bx}) \tag{2.103}$$

where $\widehat{T}_{xx}$ is the normal stress averaged over the cross-section, $\widehat{D}_{xx}$ is the dispersion momentum transport, B is the channel width at the water surface, χ is the wetted perimeter, $\hat{\tau}_{sx}$ is the wind driving force per unit horizontal area at the water surface, and $\hat{\tau}_{bx}$ is the shear force per unit area of bed and bank surfaces.

The turbulent stress term in Eq. (2.103) is usually ignored, because it is much weaker than the convection term. The dispersion term is often combined with the convection term by introducing a correction factor. In inland rivers, the wind driving force usually is negligible. Therefore, the resulting 1-D momentum equation is

$$\frac{\partial (A\widehat{U})}{\partial t} + \frac{\partial (\hat{\beta}A\widehat{U^2})}{\partial x} = -gA\frac{\partial \tilde{z}_s}{\partial x} - \frac{1}{\rho}\chi \hat{\tau}_{bx} \tag{2.104}$$

where $\hat{\beta}$ is the correction factor for momentum, defined as $\hat{\beta} = \iint_A u^2 dA/(A\widehat{U^2})$, with u being the streamwise flow velocity in the 3-D model.

The average suspended-load concentration, $\widehat{C}$, at the cross-section is defined as

$$\widehat{C} = \frac{1}{A_s \widehat{U}} \iint_{A_s} ucdA \tag{2.105}$$

where c is the local suspended-load concentration, and A_s is the flow area in the suspended-load zone, as shown in Fig. 2.7. It is often assumed that $A_s \approx A$. Integrating Eq. (2.72) over the suspended-load zone leads to the 1-D suspended-load transport equation:

$$\frac{\partial}{\partial t}\left(\frac{A\widehat{C}}{\hat{\beta}_s}\right) + \frac{\partial}{\partial x}(A\widehat{U}\widehat{C}) = \frac{\partial}{\partial x}\left(A\varepsilon_s \frac{\partial \widehat{C}}{\partial x}\right) + B(\widehat{E}_b - \widehat{D}_b) \tag{2.106}$$

where $\widehat{E}_b$ and $\widehat{D}_b$ are the width-averaged sediment entrainment and deposition fluxes at the interface between the suspended-load and bed-load zones, and $\hat{\beta}_s$ is the correction factor for suspended load:

$$\hat{\beta}_s = \iint_{A_s} ucdA \Big/ \left(\widehat{U}\iint_{A_s} cdA\right) \tag{2.107}$$

Note that no dispersion term appears in Eq. (2.106), due to the definition of $\widehat{C}$ in Eq. (2.105). However, if $\widehat{C}$ is defined by Eq. (2.100), a dispersion term should appear in Eq. (2.106). Normally, the diffusion term in Eq. (2.106) is ignored, yielding

$$\frac{\partial}{\partial t}\left(\frac{A\widehat{C}}{\hat{\beta}_s}\right) + \frac{\partial(A\widehat{U}\widehat{C})}{\partial x} = B(\widehat{E}_b - \widehat{D}_b) \tag{2.108}$$

Integrating Eq. (2.72) over the bed-load zone yields the 1-D bed-load mass balance equation:

$$(1 - p'_m)\frac{\partial A_b}{\partial t} + \frac{\partial(A_\delta C_\delta)}{\partial t} + \frac{\partial Q_b}{\partial x} = B(\widehat{D}_b - \widehat{E}_b) \tag{2.109}$$

where $\partial A_b/\partial t$ is the rate of change in bed area; A_b is the cross-sectional area of the bed above a reference datum, as shown in Fig. 2.7; A_δ is the cross-sectional area of the bed-load zone; Q_b is the bed-load transport rate at the cross-section; and C_δ is the laterally-averaged bed-load concentration.

In analogy to Eq. (2.90), by using $C_\delta = Q_b/(A_\delta U_b)$, Eq. (2.109) can be rewritten as

$$(1 - p'_m)\frac{\partial A_b}{\partial t} + \frac{\partial}{\partial t}\left(\frac{Q_b}{U_b}\right) + \frac{\partial Q_b}{\partial x} = B(\widehat{D}_b - \widehat{E}_b) \tag{2.110}$$

where U_b is the laterally-averaged velocity of bed load.

Summing Eqs. (2.108) and (2.110) leads to the 1-D mass balance equation of total load:

$$(1 - p'_m)\frac{\partial A_b}{\partial t} + \frac{\partial}{\partial t}\left(\frac{A\widehat{C_t}}{\hat{\beta}_t}\right) + \frac{\partial(A\widehat{U}\widehat{C_t})}{\partial x} = 0 \qquad (2.111)$$

where $\widehat{C}_t$ is the total-load concentration averaged over the cross-section, defined as $\widehat{C}_t = (Q_b + A_s\widehat{U}\widehat{C})/(A\widehat{U})$; and $\hat{\beta}_t$ is a correction factor for total load, related to $\hat{\beta}_s$ and U_b by $\hat{\beta}_t = A\widehat{C}_t/(A\widehat{C}/\hat{\beta}_s + Q_b/U_b) = 1/[r_s/\hat{\beta}_s + (1 - r_s)\widehat{U}/U_b]$, which is similar to Eq. (2.92).

2.4.4 Effects of sediment transport and bed change on flow

Recall that the aforementioned 1-D, 2-D, and 3-D hydrodynamic equations ignore the effects of flow density and bed change by assuming that the sediment concentration is low and that the bed varies much more slowly than the flow. This assumption is not valid for high shear flows with strong sediment transport. In addition, the flow density varies with salinity, temperature, and other factors. In general, the 3-D hydrodynamic equations with a variable flow density ρ are Eqs. (2.25) and (2.26), which are rewritten in the Cartesian coordinate system shown in Fig. 2.6 as

$$\frac{\partial \rho}{\partial t} + \frac{\partial(\rho u_x)}{\partial x} + \frac{\partial(\rho u_y)}{\partial y} + \frac{\partial(\rho u_z)}{\partial z} = 0 \qquad (2.112)$$

$$\frac{\partial(\rho u_x)}{\partial t} + \frac{\partial(\rho u_x^2)}{\partial x} + \frac{\partial(\rho u_y u_x)}{\partial y} + \frac{\partial(\rho u_z u_x)}{\partial z} = F_x - \frac{\partial p}{\partial x} + \frac{\partial \tau_{xx}}{\partial x} + \frac{\partial \tau_{xy}}{\partial y} + \frac{\partial \tau_{xz}}{\partial z}$$

$$(2.113)$$

$$\frac{\partial(\rho u_y)}{\partial t} + \frac{\partial(\rho u_x u_y)}{\partial x} + \frac{\partial(\rho u_y^2)}{\partial y} + \frac{\partial(\rho u_z u_y)}{\partial z} = F_y - \frac{\partial p}{\partial y} + \frac{\partial \tau_{yx}}{\partial x} + \frac{\partial \tau_{yy}}{\partial y} + \frac{\partial \tau_{yz}}{\partial z}$$

$$(2.114)$$

$$\frac{\partial(\rho u_z)}{\partial t} + \frac{\partial(\rho u_x u_z)}{\partial x} + \frac{\partial(\rho u_y u_z)}{\partial y} + \frac{\partial(\rho u_z^2)}{\partial z} = F_z - \frac{\partial p}{\partial z} + \frac{\partial \tau_{zx}}{\partial x} + \frac{\partial \tau_{zy}}{\partial y} + \frac{\partial \tau_{zz}}{\partial z}$$

$$(2.115)$$

Like Eq. (2.63), the z-momentum equation (2.115) can be simplified to the hydrostatic pressure equation (2.64) for gradually varied (shallow water) flows. When the flow density is variable in the vertical direction, Eq. (2.64) has the following solution:

$$p = p_a + \int_z^{z_s} \rho g \, dz \qquad (2.116)$$

Substituting Eq. (2.116) into Eqs. (2.113) and (2.114) and assuming a constant p_a yields

$$\frac{\partial(\rho u_x)}{\partial t} + \frac{\partial(\rho u_x^2)}{\partial x} + \frac{\partial(\rho u_y u_x)}{\partial y} + \frac{\partial(\rho u_z u_x)}{\partial z} = -\rho_0 g \frac{\partial z_s}{\partial x} - g \int_z^{z_s} \frac{\partial \rho}{\partial x} dz$$

$$+ \frac{\partial \tau_{xx}}{\partial x} + \frac{\partial \tau_{xy}}{\partial y} + \frac{\partial \tau_{xz}}{\partial z} \quad (2.117)$$

$$\frac{\partial(\rho u_y)}{\partial t} + \frac{\partial(\rho u_x u_y)}{\partial x} + \frac{\partial(\rho u_y^2)}{\partial y} + \frac{\partial(\rho u_z u_y)}{\partial z} = -\rho_0 g \frac{\partial z_s}{\partial y} - g \int_z^{z_s} \frac{\partial \rho}{\partial y} dz$$

$$+ \frac{\partial \tau_{yx}}{\partial x} + \frac{\partial \tau_{yy}}{\partial y} + \frac{\partial \tau_{yz}}{\partial z} \quad (2.118)$$

where ρ_0 is the flow density at the water surface.

Integrating Eqs. (2.112), (2.117), and (2.118) over the flow depth leads to the depth-integrated 2-D flow equations:

$$\frac{\partial(\rho h)}{\partial t} + \frac{\partial(\rho h U_x)}{\partial x} + \frac{\partial(\rho h U_y)}{\partial y} + \rho_b \frac{\partial z_b}{\partial t} = 0 \quad (2.119)$$

$$\frac{\partial(\rho h U_x)}{\partial t} + \frac{\partial(\rho h U_x^2)}{\partial x} + \frac{\partial(\rho h U_y U_x)}{\partial y} = -\rho g h \frac{\partial z_s}{\partial x} - \frac{1}{2} g h^2 \frac{\partial \rho}{\partial x} + \frac{\partial[h(T_{xx} + D_{xx})]}{\partial x}$$

$$+ \frac{\partial[h(T_{xy} + D_{xy})]}{\partial y} + \tau_{sx} - \tau_{bx} \quad (2.120)$$

$$\frac{\partial(\rho h U_y)}{\partial t} + \frac{\partial(\rho h U_x U_y)}{\partial x} + \frac{\partial(\rho h U_y^2)}{\partial y} = -\rho g h \frac{\partial z_s}{\partial y} - \frac{1}{2} g h^2 \frac{\partial \rho}{\partial y} + \frac{\partial[h(T_{yx} + D_{yx})]}{\partial x}$$

$$+ \frac{\partial[h(T_{yy} + D_{yy})]}{\partial y} + \tau_{sy} - \tau_{by} \quad (2.121)$$

where ρ_b is the density of the water-sediment mixture in the bed surface layer, determined by $\rho_b = \rho_f p'_m + \rho_s (1 - p'_m)$, with p'_m being the porosity of the surface-layer bed material. Note that in the derivation of Eqs. (2.119)–(2.121), it is assumed that the flow density is constant along the flow depth but varies horizontally.

Integrating Eqs. (2.112), (2.113), and (2.115) over the flow width yields the width-integrated 2-D equations of flow with a density varying in the longitudinal section:

$$\frac{\partial(\rho b)}{\partial t} + \frac{\partial(\rho b \tilde{U}_x)}{\partial x} + \frac{\partial(\rho b \tilde{U}_z)}{\partial z} = 0 \quad (2.122)$$

$$\frac{\partial(\rho b \tilde{U}_x)}{\partial t} + \frac{\partial(\rho b \tilde{U}_x^2)}{\partial x} + \frac{\partial(\rho b \tilde{U}_z \tilde{U}_x)}{\partial z} = -b \frac{\partial \tilde{p}}{\partial x} + \frac{\partial[b(\tilde{T}_{xx} + \tilde{D}_{xx})]}{\partial x}$$

$$+ \frac{\partial[b(\tilde{T}_{xz} + \tilde{D}_{xz})]}{\partial z} - (m_1 \tau_{x1} + m_2 \tau_{x2})$$

$$(2.123)$$

$$\frac{\partial(\rho b \widetilde{U}_z)}{\partial t} + \frac{\partial(\rho b \widetilde{U}_x \widetilde{U}_z)}{\partial x} + \frac{\partial(\rho b \widetilde{U}_z^2)}{\partial z} = -\rho b g - b\frac{\partial \widetilde{p}}{\partial z} + \frac{\partial[b(\widetilde{T}_{zx} + \widetilde{D}_{zx})]}{\partial x}$$

$$+ \frac{\partial[b(\widetilde{T}_{zz} + \widetilde{D}_{zz})]}{\partial z} - (m_1 \tau_{z1} + m_2 \tau_{z2})$$

(2.124)

By applying the hydrostatic pressure assumption, the vertical momentum equation (2.124) can be simplified to Eq. (2.116), and then the streamwise momentum equation (2.123) is turned to

$$\frac{\partial(\rho b \widetilde{U}_x)}{\partial t} + \frac{\partial(\rho b \widetilde{U}_x^2)}{\partial x} + \frac{\partial(\rho b \widetilde{U}_z \widetilde{U}_x)}{\partial z} = -\left(\rho_0 b g \frac{\partial \widetilde{z}_s}{\partial x} + g b \int_z^{z_s} \frac{\partial \rho}{\partial x} dz\right)$$

$$+ \frac{\partial[b(\widetilde{T}_{xx} + \widetilde{D}_{xx})]}{\partial x} + \frac{\partial[b(\widetilde{T}_{xz} + \widetilde{D}_{xz})]}{\partial z}$$

$$- (m_1 \tau_{x1} + m_2 \tau_{x2})$$

(2.125)

Integrating Eqs. (2.112) and (2.117) over the cross-section yields the general 1-D equations of flow with a longitudinally variable density:

$$\frac{\partial(\rho A)}{\partial t} + \frac{\partial(\rho A \widehat{U})}{\partial x} + \rho_b \frac{\partial A_b}{\partial t} = 0$$

(2.126)

$$\frac{\partial}{\partial t}(\rho A \widehat{U}) + \frac{\partial}{\partial x}\left(\rho \beta A \widehat{U}^2\right) = -\rho g A \frac{\partial \widetilde{z}_s}{\partial x} - \frac{1}{2} g A h_p \frac{\partial \rho}{\partial x} - \chi \hat{\tau}_{bx}$$

(2.127)

where $h_p = \int_0^B h_{2d}^2 dy/A$, with h_{2d} being the local flow depth.

Note that the effect of sediment concentration on the flow field is taken into account in Eqs. (2.112)–(2.127) through the density of the water-sediment mixture defined in Eq. (2.23). The effect of bed change is considered in the 1-D and depth-averaged 2-D models by including the bed change terms in Eqs. (2.119) and (2.126), whereas this is done in the width-averaged 2-D and 3-D models by specifying the near-bed fluxes and changing the computational domains at the bed boundary.

2.5 NET EXCHANGE FLUX OF SUSPENDED LOAD NEAR BED

2.5.1 Exchange model using near-bed capacity formula

In the depth-averaged 2-D (or 1-D) model, the near-bed sediment exchange flux $D_b - E_b$ in the suspended-load transport equation must be modeled, because the near-bed concentration c_b is not a dependent variable to be solved. The deposition flux

D_b $(= \omega_s c_b)$ is usually determined by relating c_b to the depth-averaged suspended-load concentration C through $c_b = \alpha_c C$, in which α_c is the adaptation or recovery coefficient.

The entrainment flux E_b $(= \omega_s c_{b*})$ can be determined by directly using an empirical formula for the near-bed suspended-load transport capacity c_{b*}. The net exchange flux thus reads

$$D_b - E_b = \alpha_c \omega_s C - \omega_s c_{b*} \qquad (2.128)$$

Examples of this model can be found in Spasojevic and Holly (1990) and Minh Duc (1998).

The coefficient α_c in non-equilibrium sediment transport states is little known and very difficult to determine theoretically. It is often approximately evaluated using the Rouse, Lane-Kalinske, or another distribution of suspended-load concentration introduced in Section 3.5.1, established under equilibrium conditions. For example, the use of the Rouse distribution yields (Minh Duc, 1998)

$$\alpha_c = (h - \delta) \Bigg/ \int_\delta^h \left(\frac{h - z}{z} \frac{\delta}{h - \delta} \right)^{\omega_s / \kappa U_*} dz \qquad (2.129)$$

Lin (1984) proposed the following relation for α_c:

$$\alpha_c = 3.25 + 0.55 \ln \left(\frac{\omega_s}{\kappa U_*} \right) \qquad (2.130)$$

which was used by Spasojevic and Holly (1990).

2.5.2 Exchange model using average capacity formula

The entrainment flux E_b can also be determined by relating c_{b*} to the equilibrium (capacity) depth-averaged suspended-load concentration C_* through $c_{b*} = \alpha_{c*} C_*$, in which α_{c*} is the adaptation coefficient under the equilibrium condition and C_* is determined using an empirical formula. Therefore, the net exchange flux is determined by

$$D_b - E_b = \alpha_c \omega_s C - \alpha_{c*} \omega_s C_* \qquad (2.131)$$

In the equilibrium sediment transport state, $\alpha_c = \alpha_{c*}$, but in a non-equilibrium state, $\alpha_c \neq \alpha_{c*}$. Because equilibrium is acquired through exchange between bed material and moving sediment near the bed, the sediment in the lower layer near the bed usually reaches equilibrium more promptly than the sediment in the upper layer near the water surface. In other words, the relative difference between the actual and equilibrium sediment concentrations in the lower layer is usually smaller than that in the upper layer. Therefore, one may expect that for erosion, $C/c_b \leq C_*/c_{b*}$ and $\alpha_c \geq \alpha_{c*}$; for deposition, $C/c_b \geq C_*/c_{b*}$ and $\alpha_c \leq \alpha_{c*}$.

However, the difference between α_c and α_{c*} is often assumed to be negligible, for simplicity. Thus, the net exchange flux can be determined by (Han, 1980; Wu, 1991)

$$D_b - E_b = \alpha \omega_s (C - C_*) \tag{2.132}$$

where α is a new adaptation coefficient.

Equating Eqs. (2.131) and (2.132) leads to $\alpha \omega_s (C - C_*) = \alpha_c \omega_s C - \alpha_{c*} \omega_s C_*$ and then

$$\alpha = \alpha_c + (\alpha_c - \alpha_{c*}) \frac{C_*}{C - C_*} \tag{2.133}$$

$$\alpha = \alpha_{c*} + (\alpha_c - \alpha_{c*}) \frac{C}{C - C_*} \tag{2.134}$$

When erosion occurs, $\alpha_c \geq \alpha_{c*}$ and $C < C_*$; when deposition occurs, $\alpha_c \leq \alpha_{c*}$ and $C > C_*$. Substituting these relations into Eqs. (2.133) and (2.134) results in $\alpha \leq \alpha_c$ and $\alpha \leq \alpha_{c*}$. Therefore, the coefficient α in Eq. (2.132) is usually less than the two coefficients α_c and α_{c*} in Eq. (2.131) (Wu, 1991).

Galappatti and Vreugdenhil (1985) derived a function for α through an approximate analytical integration of the pure vertical 2-D convection-diffusion equation of suspended load. They used the "concentration" boundary condition (2.74), which assumes equilibrium sediment transport near the bed. Armanini and di Silvio (1986) argued that the "concentration" boundary condition may result in large errors for fine sediments. They derived a different function for α through the integration of Galappatti and Vreugdenhil by specifying the "gradient" boundary condition (2.75). In addition, Armanini and de Silvio performed a sensitivity analysis of the approximate solutions by applying the procedure of Galappatti and Vreugdenhil directly to the transport (cu) instead of to the concentration (c). Armanini and de Silvio's function can be approximated as

$$\frac{1}{\alpha} = \frac{a}{h} + \left(1 - \frac{a}{h}\right) \exp\left[-1.5 \left(\frac{a}{h}\right)^{-1/6} \frac{\omega_s}{U_*}\right] \tag{2.135}$$

where a is the thickness of the bottom layer, defined as $a = 33z_0 = 33h/\exp(1 + \kappa C_h/\sqrt{g})$, in which z_0 is the zero-velocity distance in the logarithmic velocity distribution, and C_h is the Chezy resistance coefficient of the channel. The thickness of the bottom layer has the order of magnitude of the grain diameter when the bed is flat, and the order of magnitude of the bed form height in the presence of bed forms.

Zhou and Lin (1998) also established a formula for α using the analytical solutions of the pure vertical 2-D convection-diffusion equation of suspended load with constant diffusivity in steady, uniform flow. They adopted the analytical solution with the "concentration" boundary condition for erosion case, and that with the "gradient" boundary condition for deposition case. The coefficient α is determined by

$$\alpha = \frac{R}{4} + \frac{\sigma_1^2}{R} \tag{2.136}$$

where $R = 6\omega_s/(\kappa U_*)$, and σ_1 is the first positive root of the following equations:

$$tg(\sigma) = -\frac{\sigma}{R} \text{ (for erosion)}, \quad 2ctg(\sigma) = \frac{2\sigma}{R} - \frac{R}{2\sigma} \text{ (for deposition)} \quad (2.137)$$

Eq. (2.136) represents two curves for α in cases of erosion and deposition, respectively, as plotted in Fig. 2.8. The difference between these two curves is significant for small Rouse numbers $\omega_s/(\kappa U_*)$, but gradually decreases as the Rouse number increases. It should be noted that because the "concentration" boundary condition is used, the curve for erosion case may have large errors for fine sediments (small Rouse numbers), as discussed by Armanini and di Silvio (1986).

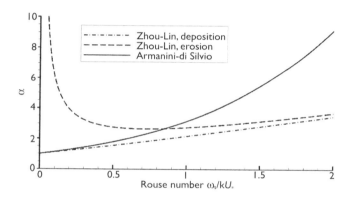

Figure 2.8 Relation between adaptation coefficient and Rouse number.

Armanini and di Silvio's function, Eq. (2.135) with $a/h = 0.017$ is also plotted in Fig. 2.8. It is shown that for small Rouse numbers Eq. (2.135) is close to Zhou and Lin's curve for deposition case, and as the Rouse number increases, the difference between Eqs. (2.135) and (2.136) increases. It is also shown that the values of α given by these two methods are always larger than 1.

It should be noted that Eqs. (2.135) and (2.136) were derived for a pure vertical 2-D case under many assumptions and simplifications. Their application in natural rivers should be done with caution, because the adaptation coefficient α is affected by many other factors, as discussed in Section 2.5.3.

2.5.3 Complexity of adaptation coefficient of sediment

Effect of cross-sectional shape

The value of α in the 1-D model is related to the cross-sectional shape. This is demonstrated by the following analysis suggested by Zhou and Lin (1998).

Width-integrating the steady depth-averaged 2-D suspended-load transport equation leads to

$$\int_0^B Uh\frac{\partial C}{\partial x}dy = -\int_0^B \alpha\omega_s(C - C_*)dy \qquad (2.138)$$

and the 1-D formulation of Eq. (2.138) is

$$\widehat{U}H\frac{\partial\widehat{C}}{\partial x} = -\alpha_{1d}\omega_s(\widehat{C} - \widehat{C}_*) \qquad (2.139)$$

where $H, \widehat{U}, \widehat{C}$, and $\widehat{C}_*$ are the flow depth, velocity, actual and equilibrium suspended-load concentrations averaged over the cross-section, respectively; and α_{1d} is the adaptation coefficient in the 1-D model.

The equilibrium depth-averaged suspended-load concentration at each vertical line may be determined using the Zhang (1961) formula introduced in Section 3.5.3:

$$C_* = K_*\left(\frac{U^3}{gh\omega_s}\right)^m \qquad (2.140)$$

where K_* is a coefficient, and m is an exponent.

In analogy to Eq. (2.140), the actual depth-averaged suspended-load concentration at each vertical line is assumed to have the relation:

$$C = K\left(\frac{U^3}{gh\omega_s}\right)^m \qquad (2.141)$$

where K is a coefficient similar to K_*.

The depth-averaged flow velocity at each vertical line is assumed to be proportional to the local flow depth:

$$U \propto h^r \qquad (2.142)$$

where r is an exponent and has a value of 2/3 if the Manning equation is used.

Substituting relations (2.140)–(2.142) into Eq. (2.138) and comparing the resulting equation with Eq. (2.139) leads to (Zhou and Lin, 1998)

$$\alpha_{1d} = \frac{\int_0^B h^{r+1}dy \int_0^B \alpha h^{(3r-1)m}dy}{B\int_0^B h^{(3r-1)m+r+1}dy} \qquad (2.143)$$

Eq. (2.143) shows that α_{1d} is related to the cross-sectional shape and varies with exponents m and r. After α has been determined using Eq. (2.136), α_{1d} can be calculated using Eq. (2.143). As an approximation, α may be assumed to be constant along the channel width, thus yielding

$$\frac{\alpha_{1d}}{\alpha} = \lambda_\alpha = \frac{\int_0^B h^{r+1} dy \int_0^B h^{(3r-1)m} dy}{B \int_0^B h^{(3r-1)m+r+1} dy} \tag{2.144}$$

where λ_α is considered as a correction factor to account for the influence of cross-sectional shape. Normally, λ_α is in the range of 0.25–1.0.

Effects of other factors

The settling velocity ω_s in Eqs. (2.72) and (2.132) is often set as that of a single particle in quiescent, distilled water. This is valid if the sediment concentration is very low, but in general the effect of sediment concentration on ω_s should be considered. Moreover, ω_s considers only the actions of drag force and submerged weight in still water. In reality, sediment particles also experience other forces exerted by moving water (Li, 1993; Wu and Wang, 2000). In particular, the Saffman (1965) lift force, which might be important near the bed where the velocity gradient is high, may reduce the settling velocity. These effects should be lumped in the adaptation coefficient α, if no corrections are made to the settling velocity. This usually leads to reduction in α values.

In addition, the above analyses of α consider only the flat bed without bed forms. Bed forms often exist in natural rivers and affect the sediment exchange near the bed and, in turn, the values of α. However, this effect is little understood. The bed-load layer may become thicker because of bed forms, so that reduction in α values may be expected based on Eq. (2.129) or (2.135).

Therefore, the adaptation coefficient α lumps the effects of many factors on sediment transport. Tests in many rivers and reservoirs conducted by Han (1980) and Wu (1991) suggest that α is about 1 for strong erosion, 0.5 for mild erosion and deposition, and 0.25 for strong deposition in 1-D models. These values differ from those (larger than 1) predicted by Eqs. (2.129), (2.130), (2.135), and (2.136), but they are qualitatively reasonable if these corrections due to the effects of cross-sectional shape, sediment concentration, Saffman lift force, and bed forms are considered. However, these values are given for reference only, and calibrating α using measurement data is preferable for a specific case study.

2.6 EQUILIBRIUM AND NON-EQUILIBRIUM SEDIMENT TRANSPORT MODELS

2.6.1 Formulation of equilibrium transport model

Each of the 1-D, 2-D, and 3-D sediment transport models described in Section 2.4 has only two governing equations, namely the suspended-load transport equation and the bed-load mass balance equation, but there are three unknowns: suspended-load concentration, bed-load transport rate, and bed change rate. Thus, one more equation is required to close each model. Most of the first sediment transport models adopt the assumption of local (instantaneous) equilibrium for bed-load transport, which assumes that the actual bed-load transport rate is equal to the transport capacity under the equilibrium condition at every computational point (cross-section or vertical line), i.e.,

$$q_b = q_{b*}(U, h, \tau, d, \gamma_s, \ldots) \tag{2.145}$$

where q_{b*} is the equilibrium (capacity) bed-load transport rate, which can be determined using an empirical formula introduced in Section 3.4.

Eq. (2.145) can be used to close the 1-D, 2-D, and 3-D sediment transport models. For example, the depth-averaged 2-D model is closed using Eq. (2.145) for bed-load transport rate, Eq. (2.86) for suspended-load concentration, and Eq. (2.90) or (2.91) for bed change. This approach is often called the equilibrium (or saturated) sediment transport model.

2.6.2 Formulation of non-equilibrium transport model

Because of variations in flow conditions and channel properties, the sediment transport in natural rivers usually is not in states of equilibrium. Sediment cannot reach new equilibrium states instantaneously, due to the temporal and spatial lags between flow and sediment transport. Therefore, the assumption of local equilibrium transport is usually unrealistic and may have significant errors in cases of strong erosion and deposition. A more realistic and general approach is the non-equilibrium (or unsaturated) sediment transport model, which is described below.

For only suspended-load transport, the bed change is attributed to the net sediment flux at the lower boundary of the suspended-load zone and thus determined by

$$(1 - p_m')\frac{\partial z_b}{\partial t} = D_b - E_b$$
$$= \alpha \omega_s (C - C_*) \tag{2.146}$$

For only bed-load transport, Bell and Sutherland (1983) proposed a loading law based on their analysis of laboratory tests:

$$\frac{\partial q_b}{\partial x} = K_l(q_{b*} - q_b) + \frac{q_b}{q_{b*}}\frac{\partial q_{b*}}{\partial x} \tag{2.147}$$

where K_l is the loading-law coefficient. However, because Eq. (2.147) is an observation of steady bed-load transport, its application to unsteady total-load sediment transport is not straightforward. In addition, the last term on the right-hand side of Eq. (2.147) lacks a physical basis. Daubert and Lebreton (1967), Wellington (1978), Nakagawa and Tsujimoto (1980), Phillips and Sutherland (1989), and Thuc (1991) used the following more general bed-load exchange model near the bed:

$$(1 - p_m')\frac{\partial z_b}{\partial t} = \frac{1}{L_b}(q_b - q_{b*}) \tag{2.148}$$

where L_b is the adaptation length of bed load. Eq. (2.148) is based on theoretical reasoning similar to that of Einstein (1950) but for bed load at a non-equilibrium state.

On the right-hand side of Eq. (2.148), q_b/L_b and q_{b*}/L_b represent the deposition and entrainment rates of bed load, respectively.

Wu *et al.* (2000a) extended relation (2.148) to the total-load transport when bed load is dominant. However, for general cases in which bed load and suspended load are equivalently important, Wu (2004) suggested the following relation for bed change:

$$(1 - p'_m)\frac{\partial z_b}{\partial t} = \frac{1}{L_t}(q_t - q_{t*}) \tag{2.149}$$

where L_t is the adaptation length of total load; and q_t and q_{t*} are the actual and equilibrium (capacity) total-load transport rates: $q_t = q_b + UhC$ and $q_{t*} = q_{b*} + UhC_*$.

The term on the right-hand side of Eq. (2.149) was also adopted by Armanini and di Silvo (1988) for the exchange flux between the bed-load layer and the bed. Eq. (2.149) can be conveniently used in models that compute total (bed-material) load transport without discerning bed load and suspended load. When suspended load and bed load are calculated separately, Eq. (2.149) may be rewritten as

$$(1 - p'_m)\frac{\partial z_b}{\partial t} = \alpha_t \omega_s(C - C_*) + \frac{1}{L_t}(q_b - q_{b*}) \tag{2.150}$$

where α_t is the adaptation coefficient of total load: $\alpha_t = (Uh)/(L_t\omega_s)$.

Because the bed-load layer usually is very thin, it can be assumed that $\alpha \approx \alpha_t$. Thus, substituting Eq. (2.150) into Eq. (2.90) yields the bed-load transport equation (Wu, 2004):

$$\frac{\partial}{\partial t}\left(\frac{q_b}{u_b}\right) + \frac{\partial(\alpha_{bx}q_b)}{\partial x} + \frac{\partial(\alpha_{by}q_b)}{\partial y} = \frac{1}{L_t}(q_{b*} - q_b) \tag{2.151}$$

An alternative is to use Eqs. (2.146) and (2.148) to compute the bed changes due to suspended load and bed load, respectively, and then sum them to obtain the bed change due to total load, i.e.,

$$(1 - p'_m)\frac{\partial z_b}{\partial t} = D_b - E_b + \frac{1}{L}(q_b - q_{b*}) \tag{2.152}$$

Note that L_b in Eq. (2.148) is replaced by the adaptation length L in Eq. (2.152). L is approximately equal to L_t in general cases and reduces to L_b in the case of bed load. However, L_t and L are noted differently in this book for use of L_t in bed-material load models and L in models computing bed load and suspended load separately. Further discussion on the relation between L_t and L is given in Section 5.1.2.1.

Eq. (2.152) is written in a general form, so that it can be used in the 1-D, 2-D, and 3-D models that compute bed load and suspended load separately. Substituting it into

Eq. (2.90) yields the following bed-load transport equation:

$$\frac{\partial}{\partial t}\left(\frac{q_b}{u_b}\right) + \frac{\partial(\alpha_{bx}q_b)}{\partial x} + \frac{\partial(\alpha_{by}q_b)}{\partial y} = \frac{1}{L}(q_{b*} - q_b) \tag{2.153}$$

which is similar to Eq. (2.151) except that L_t is replaced by L.

By using Eq. (2.153) for bed-load transport, Eq. (2.86) for suspended-load transport, and Eq. (2.152) or the overall sediment continuity equation (2.91) for bed change, the depth-averaged 2-D sediment transport model is closed. Similar closures can be derived for the 1-D, width-averaged 2-D, and 3-D models, which are explained in detail in Chapters 5–7.

2.6.3 Adaptation length of sediment

The adaptation length is a characteristic distance for sediment to adjust from non-equilibrium to equilibrium transport. It is a very important parameter in the non-equilibrium sediment transport model. For suspended load, the adaptation length L_s is calculated by

$$L_s = \frac{Uh}{\alpha\omega_s} \tag{2.154}$$

where α is the adaptation coefficient described in Section 2.5.

For bed load, the adaptation length L_b has been given significantly different values in the literature. Bell and Sutherland (1983) found that L_b was a function of time t in an experimental case of bed degradation downstream of a dam due to clear water inflowing. In numerical modeling studies, Nakagawa and Tsujimoto (1980), Phillips and Sutherland (1989), Thuc (1991), and Wu et al. (2000a) set L_b as the average saltation step length of sand on the bed for laboratory cases, whereas Rahuel et al. (1989) and Fang (2003) gave much larger values, such as one or two times the grid spacing for field cases.

One reason for the aforementioned differences in values of L_b is that the bed-load movement is closely associated with bed forms, which are usually on a small scale in laboratory experiments and on a larger scale in natural rivers. Naturally, L_b may take the value related to the length scale of the dominant bed form (Wu et al., 2004a; Wu, 2004). For example, in Bell and Sutherland's (1983) experiments of channel degradation due to clear water, the transport of sediment (mainly bed load) was significantly influenced by the scour hole near the flume inlet, and thus the adaptation length was related to the dimension of the scour hole as a function of time t. In the case where the bed is mainly covered by sand ripples, which usually occurs in laboratory experiments, L_b may take the average saltation step length of sand or the length of sand ripples, as adopted by Nakagawa and Tsujimoto (1980), Phillips and Sutherland (1989), Thuc (1991), and Wu et al. (2000a). If sand dunes are the dominant bed form, L_b may take the length of sand dunes, which is about 5–10 times the flow depth. If alternate bars are the dominant bed form, L_b may take the length of alternate bars, which is about 6.3 times the channel width (Yalin, 1972).

On the other hand, considering numerical accuracy and sometimes stability in the solution of bed-load transport equation, e.g., Eq. (2.153), the grid spacing should be (several times) smaller than the adaptation length. However, because of limited computer capacity, the grid spacing has to be given large values (sometimes much larger than the length of the dominant bed form) in field cases, and to obtain feasible solutions, L_b is hence set to one or two times the grid spacing (Rahuel *et al.*, 1989; Fang, 2003). This treatment perhaps is the choice under certain circumstances, but it may give grid-dependent solutions.

Because bed-material load is a combination of bed load and suspended load, its adaptation length can be given the larger of L_b and L_s (Wu *et al.*, 2004a):

$$L_t = \max\{L_b, L_s\} \tag{2.155}$$

or a weighted average of L_b and L_s:

$$L_t = (1 - r_s)L_b + r_s L_s \tag{2.156}$$

where r_s is the ratio of suspended load to bed-material (total) load.

In cases where bed load and suspended load coexist, L_s is usually larger than L_b, and thus Eq. (2.155) gives $\alpha = \alpha_t$, which is required in the derivation of Eq. (2.151). Therefore, Eq. (2.155) was used by Wu (2004) in many cases. However, because L_b and α are usually treated as calibrated parameters, the difference between Eqs. (2.155) and (2.156) is not important.

Because wash load does not have significant exchange with the bed, its adaptation coefficient α and length L can be set to be zero and infinitely large, respectively.

It should be pointed out that because the values of L_b and α vary by case, the methods discussed above and in Section 2.5 are only empirical guidance for evaluating these two parameters. Their calibration using available measurement data is recommended to obtain more reliable results for real-life problems. Sensitivities of sediment transport models to these parameters are demonstrated in Sections 5.6 and 9.2.

2.7 TRANSPORT AND SORTING OF NON-UNIFORM SEDIMENT MIXTURES

2.7.1 Non-uniform sediment transport

In the case of non-uniform sediment transport, moving sediment particles collide and interact; bed sediment particles experience the hiding and exposure effects, because fine particles are more likely to be hidden and coarse particles have more chance to be exposed to flow. However, if the sediment concentration is low, interactions among the moving sediment particles are usually negligible, so that each size class of the moving sediment mixture can be assumed to have the same transport behavior as uniform sediment. This assumption is adopted in this book, except where stated otherwise. As an example, a depth-averaged 2-D non-uniform sediment transport model based on it is presented below.

As described in Section 2.1.2.2, the non-uniform sediment mixture is divided into N size classes. Under the assumption of low sediment concentration, Eq. (2.86) is applied to determine the transport of each size class of suspended load:

$$\frac{\partial}{\partial t}\left(\frac{hC_k}{\beta_{sk}}\right) + \frac{\partial(hU_xC_k)}{\partial x} + \frac{\partial(hU_yC_k)}{\partial y}$$

$$= \frac{\partial}{\partial x}\left[h\left(\varepsilon_s\frac{\partial C_k}{\partial x} + D_{sxk}\right)\right] + \frac{\partial}{\partial y}\left[h\left(\varepsilon_s\frac{\partial C_k}{\partial y} + D_{syk}\right)\right] + \alpha\omega_{sk}(C_{*k} - C_k)$$

$$(k = 1, 2, \ldots, N) \tag{2.157}$$

where subscript k is the sediment size class index; C_k and C_{*k} are the actual and equilibrium (capacity) depth-averaged concentrations of the kth size class of suspended load, respectively; D_{sxk} and D_{syk} are the dispersion fluxes; β_{sk} is the correction factor defined by Eq. (2.87) for size class k; and ω_{sk} is the settling velocity of the kth size class of sediment.

Note that the size class index k in this book is not subject to Einstein's summation convention.

Extending Eq. (2.153) for the transport of each size class of bed load yields

$$\frac{\partial}{\partial t}\left(\frac{q_{bk}}{u_{bk}}\right) + \frac{\partial(\alpha_{bx}q_{bk})}{\partial x} + \frac{\partial(\alpha_{by}q_{bk})}{\partial y} = \frac{1}{L}(q_{b*k} - q_{bk}) \tag{2.158}$$

where q_{bk} and q_{b*k} are the actual and equilibrium (capacity) transport rates of the kth size class of bed load, respectively, and u_{bk} is the bed-load velocity.

Note that the values of α and L may vary with size classes. However, for simplicity, they are not explicitly noted with the subscript k, because they are often treated as calibrated parameters and each is given the same value for all size classes in most cases.

Extending Eq. (2.152) for the fractional bed change yields

$$(1 - p'_m)\left(\frac{\partial z_b}{\partial t}\right)_k = \alpha\omega_{sk}(C_k - C_{*k}) + \frac{1}{L}(q_{bk} - q_{b*k}) \tag{2.159}$$

where $(\partial z_b/\partial t)_k$ is the rate of change in bed elevation due to size class k.

The total rate of change in bed elevation, $\partial z_b/\partial t$, is determined by

$$\frac{\partial z_b}{\partial t} = \sum_{k=1}^{N}\left(\frac{\partial z_b}{\partial t}\right)_k \tag{2.160}$$

Note that even though the sediment concentration is assumed to be low, the hiding and exposure phenomena in non-uniform bed materials always exist. However, these phenomena affect only the entrainment of sediment from the bed. Such effects are accounted for through the fractional sediment transport capacities C_{*k} and q_{b*k}. This is discussed in Sections 3.4–3.6.

2.7.2 Bed material sorting

The size gradation (composition) of bed material may vary along the vertical direction due to historical sedimentation. To consider this variation, the bed material above the nonerodible layer is often divided into multiple layers, as shown in Fig. 2.9. The top layer is the mixing layer. All sediment particles in the mixing layer are subject to exchange with those moving with flow, i.e., entraining from the mixing layer to the water column or depositing from the water column to the mixing layer. The second layer is a subsurface layer. More underlying subsurface layers can be added, if needed. However, the sediment particles in the subsurface layers do not directly exchange with the moving particles.

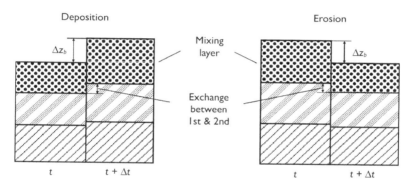

Figure 2.9 Multiple-layer model of bed material sorting.

The mixing layer concept was adopted by Hirano (1971), Bayazit (1975), Karim and Kennedy (1982), Rahuel *et al.* (1989), Armanini and di Silvio (1988), Wu (1991), and van Niekerk *et al.* (1992). The temporal variation of the bed-material gradation in the mixing layer can be determined by (Wu, 1991; also see Wu, 2004)

$$\frac{\partial(\delta_m p_{bk})}{\partial t} = \left(\frac{\partial z_b}{\partial t}\right)_k + p_{bk}^* \left(\frac{\partial \delta_m}{\partial t} - \frac{\partial z_b}{\partial t}\right) \tag{2.161}$$

where δ_m is the mixing layer thickness; p_{bk} is the fraction of the kth size class of bed material contained in the mixing layer; and p_{bk}^* is p_{bk} when $\partial z_b/\partial t - \partial \delta_m/\partial t \geq 0$ and the fraction of the kth size class of bed material contained in the second layer when $\partial z_b/\partial t - \partial \delta_m/\partial t < 0$.

The first term on the right-hand side of Eq. (2.161) represents the exchange between moving sediment and bed material, while the last term accounts for the exchange between the mixing and second layers, due to rise or descent of the lower bound of the mixing layer.

The bed-material gradation in the second layer is calculated by

$$\frac{\partial(\delta_s p_{sbk})}{\partial t} = -p_{bk}^* \left(\frac{\partial \delta_m}{\partial t} - \frac{\partial z_b}{\partial t}\right) \tag{2.162}$$

where δ_s is the second layer thickness, and p_{sbk} is the fraction of the kth size class of bed material contained in the second layer.

Eq. (2.162) assumes no exchange between the second and third layers. This is physically right. In addition, changing the layer thickness or moving the layer divisions up or down during the simulation may induce numerical mixing of sediment between layers and thus should be avoided, except that the division between the mixing and second layers may change due to variations in flow, sediment, and bed conditions.

Rahuel et $al.$ (1989) treated the bed-load layer and the mixing layer together as an active layer in a 1-D bed-load model, and Spasojevic and Holly (1990; 1993) extended this concept to 2-D and 3-D total-load models. The sediment balance in the active layer is described as

$$(1 - p'_m)\frac{\partial(\delta_m p_{bk})}{\partial t} + \nabla \cdot \vec{q}_{bk} + E_{bk} - D_{bk} - (1 - p'_m)p_{bk}^*\left(\frac{\partial \delta_m}{\partial t} - \frac{\partial z_b}{\partial t}\right) = 0$$

(2.163)

where D_{bk} and E_{bk} are the deposition and entrainment fluxes of the kth size class of sediment at the lower bound of the suspended-load zone.

It can be seen that Eq. (2.163) is the sum of Eqs. (2.158), (2.159), and (2.161), with only the storage term in Eq. (2.158) omitted.

2.7.3 Mixing layer thickness

The mixing layer thickness is related to the time scale under consideration (Bennett and Nording, 1977; Rahuel et $al.$, 1989; Wu, 1991). If a very short, nearly instantaneous time scale is considered, the mixing layer should be a thin bed surface layer containing particles susceptible to entrainment due to a momentary increase in the local bed shear stress. This is called the instantaneous mixing layer. If the time scale is longer, e.g., in the order of magnitude of the time it takes for a bed form (ripple or dune) to traverse its own wavelength, the mixing layer can be the order of magnitude of the bed form height. If the time scale is much longer, e.g., in the order of magnitude of the computational time step, the mixing layer includes the layer of material eroded or deposited and the instantaneous mixing layer.

Since the sand dune height is generally relative to the flow depth, Karim and Kennedy (1982) evaluated the mixing layer thickness as 0.1–0.2 times the flow depth. Borah et $al.$ (1982) determined the mixing layer thickness under armoring conditions by

$$\delta_m = \frac{d_L}{(1 - p'_m)p_{bm}}$$

(2.164)

where d_L is the smallest size of the sediment particles that are immobile, and p_{bm} is the fraction of all the immobile particles in the mixing layer.

Van Niekerk *et al.* (1992) related the mixing layer thickness to the dimensionless bed shear stress as follows:

$$\delta_m = 2d_{50} \frac{\tau_b'}{\tau_{c50}} \qquad (2.165)$$

where d_{50} is the median size of bed material in the mixing layer, τ_b' is the skin friction component of bed shear stress, and τ_{c50} is the critical shear stress for incipient motion corresponding to d_{50}.

Wu and Vieira (2002) set the mixing layer thickness as the larger of half the sand dune height and twice the sediment size:

$$\delta_m = \max\,[0.5\Delta, 2d_{50}] \qquad (2.166)$$

where Δ is the sand dune height, which can be calculated using the van Rijn (1984c) formula.

The mixing layer thickness is an important parameter in non-uniform sediment transport models. The sensitivity of model results to it is demonstrated in Section 5.6.

Chapter 3

Fundamentals of sediment transport

Introduced in this chapter are basic theories and empirical formulas of sediment transport, which are essentially used to close the mathematical models of flow, sediment transport, and morphological change in alluvial rivers. Some of them can be found in Graf (1971), Vanoni (1975), Chien and Wan (1983), Chang (1988), Zhang *et al.* (1989), Raudkivi (1990), Simons and Senturk (1992), Julien (1995), and Yang (1995). However, many recently developed non-uniform sediment transport formulas are particularly included here.

3.1 SETTLING OF SEDIMENT PARTICLES

3.1.1 General considerations

Settling or fall velocity is the average terminal velocity that a sediment particle attains in the settling process in quiescent, distilled water. It is related to particle size, shape, submerged specific weight, water viscosity, sediment concentration, etc.

A sediment particle experiences gravity, buoyant force, and drag force during its settling. Its submerged weight, which is the difference between the gravity and buoyant force, is expressed as

$$W_s = (\rho_s - \rho)ga_1d^3 \tag{3.1}$$

where d is the sediment size, a_1d^3 is the volume of the sediment particle, and a_1 has a value of $\pi/6$ for a spherical particle. Note that ρ is actually given as the pure water density ρ_f because a single particle (or low concentration) is considered.

The drag force is the result of the tangential shear stress exerted by the fluid (skin drag) and the pressure difference (form drag) on the particle. It is written in the general form:

$$F_d = C_d\rho a_2d^2\frac{\omega_s^2}{2} \tag{3.2}$$

where C_d is the drag coefficient, ω_s is the settling velocity, a_2d^2 is the projected area of the particle on the plane normal to the direction of settling, and a_2 has a value of $\pi/4$ for a spherical particle.

The drag force should be equal to the submerged weight in the terminal stage of settling, yielding

$$\omega_s = \left(\frac{a_1}{a_2} \frac{2}{C_d} \frac{\rho_s - \rho}{\rho} gd\right)^{1/2} \tag{3.3}$$

3.1.2 Settling velocity of spherical particles

In the laminar (streamline) settling region (i.e., the particle Reynolds number $R_e = \omega_s d/\nu < 1.0$), Stokes derived the drag force on a spherical particle by solving the Navier-Stokes equations without inertia terms. The derived drag coefficient is

$$C_d = \frac{24}{R_e} \tag{3.4}$$

Inserting Eq. (3.4) into Eq. (3.3) leads to the Stokes law for the settling velocity of spherical particles:

$$\omega_s = \frac{1}{18} \frac{\rho_s - \rho}{\rho} g \frac{d^2}{\nu} \tag{3.5}$$

where ω_s and d are in $m \cdot s^{-1}$ (meters per second) and m (meters), respectively.

Oseen (1927) solved the Navier-Stokes equations, including some inertia terms, and obtained the following relation:

$$C_d = \frac{24}{R_e} \left(1 + \frac{3}{16} R_e\right) \tag{3.6}$$

Goldstein (1929) found a relatively complete solution of Oseen's approximation as follows:

$$C_d = \frac{24}{R_e} \left(1 + \frac{3}{16} R_e - \frac{19}{1280} R_e^2 + \frac{71}{20480} R_e^3 + \cdots\right) \tag{3.7}$$

Eq. (3.7) is valid for R_e up to 2.0. Beyond this range, the drag coefficient usually has to be determined by experiments rather than theoretical solutions. Rouse (1938) summarized the available experimental data and obtained the relation between C_d and R_e shown in Fig. 3.1, which can be used to determine C_d and, in turn, the settling velocity of spherical particles.

Fig. 3.1 shows that when $R_e > 1,000$ — i.e., in the turbulent settling region — the drag coefficient is no longer related to the particle Reynolds number and has a value of about 0.45, thus yielding

$$\omega_s = 1.72 \sqrt{\frac{\rho_s - \rho}{\rho} gd} \tag{3.8}$$

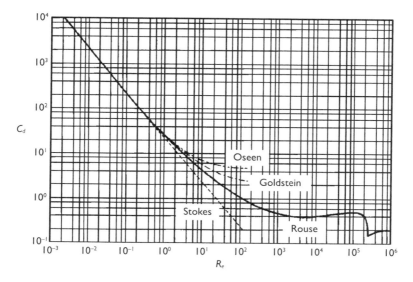

Figure 3.1 Relation between C_d and R_e for spheres.

3.1.3 Settling velocity of sediment particles

Sediment particles in natural rivers, which usually have irregular shapes and rough surfaces, exhibit differences in settling velocity in comparison with spherical particles. Rubey (1933) derived the following formula for the settling velocity of natural sediment particles:

$$\omega_s = F \sqrt{\left(\frac{\rho_s}{\rho} - 1\right) gd} \tag{3.9}$$

where $F = 0.79$ for particles larger than 1 mm settling in water with temperatures between 10 and 25°C. For smaller grain sizes, F is determined by

$$F = \left[\frac{2}{3} + \frac{36\nu^2}{gd^3 (\rho_s/\rho - 1)}\right]^{1/2} - \left[\frac{36\nu^2}{gd^3 (\rho_s/\rho - 1)}\right]^{1/2} \tag{3.10}$$

Zhang (1961; also see Zhang and Xie, 1993) assumed the drag force on a sediment particle in the transition between laminar and turbulent settling regions as

$$F_d = C_1 \rho \nu d \omega_s + C_2 \rho d^2 \omega_s^2 \tag{3.11}$$

where C_1 and C_2 are coefficients. Based on many laboratory data, Zhang obtained the formula for the settling velocity of naturally worn sediment particles:

$$\omega_s = \sqrt{\left(13.95\frac{v}{d}\right)^2 + 1.09\left(\frac{\rho_s}{\rho} - 1\right)gd} - 13.95\frac{v}{d} \qquad (3.12)$$

The Zhang formula can be used in a wide range of sediment sizes from laminar to turbulent settling regions. It can be simplified to Eq. (3.5) with a coefficient of 1/25.6 in the laminar settling region, and to Eq. (3.8) with a coefficient of 1.044 in the turbulent settling region.

Van Rijn (1984b) suggested the use of the Stokes law, Eq. (3.5), in computing the settling velocity for sediment particles smaller than 0.1 mm, the Zanke (1977) formula for particles from 0.1 to 1 mm:

$$\omega_s = 10\frac{v}{d}\left\{\left[1 + 0.01\left(\frac{\rho_s}{\rho} - 1\right)\frac{gd^3}{v^2}\right]^{1/2} - 1\right\} \qquad (3.13)$$

and the following formula for particles larger than 1 mm:

$$\omega_s = 1.1\left[\left(\frac{\rho_s}{\rho} - 1\right)gd\right]^{1/2} \qquad (3.14)$$

In fact, the formulas of Rubey, Zhang, and Zanke have the same formulation with different coefficients. Similar formulas were also proposed by Cancharov (1954; see Cheng, 1997), Sha (1965), Graf (1971), Hallermeier (1981), Raudkivi (1990), Julien (1995), and Ahrens (2000). In general, the drag coefficient can be approximated by (Cheng, 1997)

$$C_d = \left[\left(\frac{M}{R_e}\right)^{1/n} + N^{1/n}\right]^n \qquad (3.15)$$

where M, N, and n are coefficients. Table 3.1 lists the values of these three coefficients given by different investigators for naturally worn sediment particles. The coefficient M was given a value of 24 by Rubey (1933), Zanke (1977), and Julien (1995), and values between 32–34 by Zhang (1961), Raudkivi (1990), and Cheng (1997). The tests against measurement data performed by Cheng have shown that for natural sediment particles the values of 32–34 for M give better predictions than the value of 24. Note that the latter corresponds to the Stokes law, Eq. (3.5), for spherical particles. Rubey gave the coefficient N a value of 2.1, which significantly underestimates the settling velocity for coarse sediment particles.

Cheng (1997) used $M = 32$, $N = 1$, and $n = 1.5$, and derived the following formula for the settling velocity of naturally worn sediment particles:

$$\omega_s = \frac{v}{d}\left(\sqrt{25 + 1.2D_*^2} - 5\right)^{1.5} \qquad (3.16)$$

where $D_* = d[(\rho_s/\rho - 1)g/v^2]^{1/3}$.

Table 3.1 Values of M, N, and n

Author	M	N	n
Rubey (1933)	24	2.1	1
Zhang (1961)	34	1.2	1
Zanke (1977)	24	1.1	1
Raudkivi (1990)	32	1.2	1
Julien (1995)	24	1.5	1
Cheng (1997)	32	1	1.5

Eqs. (3.9), (3.12), (3.13), and (3.16) are valid for naturally worn sediment particles, the Corey shape factors of which usually are about 0.7. Krumbein (1942), Corey (1949), McNown *et al.* (1951), Wilde (1952), and Schulz *et al.* (1954) experimentally investigated the effect of particle shape on settling velocity. Based on these experiments, the Subcommittee on Sedimentation of the U.S. Interagency Committee on Water Resources (1957) recommended a series of curves shown in Fig. 3.2 to determine the settling velocity of sediment particles for given particle size, Corey shape factor, and water temperature. However, this graphical relation is inconvenient to use, because several interpolations must be conducted to obtain the sought solution. In addition, all the data used in the calibration were in the range of $R_e > 3$, and the relation was extended to the range of $R_e < 3$ based on the assumption that it approaches the Stokes law, Eq. (3.5), for spheres.

Romanovskii (1972) also performed experiments to investigate the effect of particle shape on settling velocity and obtained a relation of settling velocity with particle size

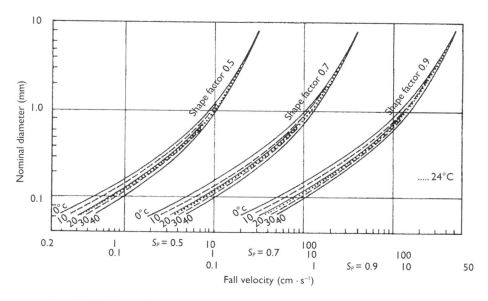

Figure 3.2 Relation of fall velocity with particle size, shape factor, and temperature (U.S. Interagency Committee, 1957).

and shape factor in the turbulent settling region. Dietrich (1982) proposed an empirical formula to determine the settling velocity of sediment from laminar to turbulent settling regions, considering the effects of sediment size, density, shape factor, and roundness factor. However, the roundness factor used in the Dietrich formula is rarely measured in practice, and his formula is very complicated and relatively difficult to use. Jimenez and Madsen (2003) simplified the Dietrich formula, but still graphically related two coefficients to the shape factor.

For more generality and convenience to use, Wu and Wang (2006) calibrated the coefficients M, N, and n in Eq. (3.15) as follows by using the natural sediment settling data of Krumbein (1942), Corey (1949), Wilde (1952), Schulz *et al.* (1954), and Romanovskii (1972):

$$M = 53.5e^{-0.65S_P}, \quad N = 5.65e^{-2.5S_P}, \quad n = 0.7 + 0.9S_P \qquad (3.17)$$

where S_P is the Corey shape factor defined in Eq. (2.9). Fig. 3.3 compares the measured drag coefficients and those calculated using Eq. (3.15) with coefficients determined by Eq. (3.17). Because the data in Fig. 3.3 were in the range of $R_e > 3$, the trend of the $C_d - R_e$ relation in the range of $R_e < 3$ was determined using the data sets of Zegzhda, Arkhangel'skii, and Sarkisyan compiled by Cheng (1997). Because naturally worn sediment particles were used in these three sets of experiments, their Corey shape factors were assumed to be 0.7. The relationship between C_d and R_e in the range of these data is shown in Fig. 3.4.

It should be noted that when $S_P = 1.0$, the proposed Eq. (3.15) with coefficients determined by Eq. (3.17) deviates from the relation of spheres obtained by Rouse (1938). The reason is that the naturally worn sediment particles with a Corey shape

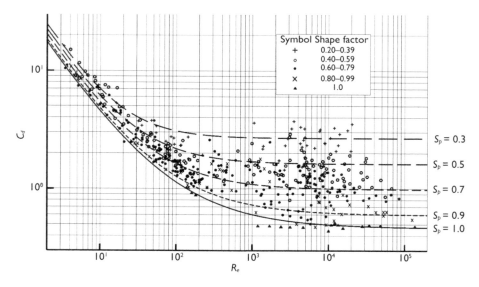

Figure 3.3 Drag coefficient as function of Reynolds number and particle shape (Wu and Wang, 2006).

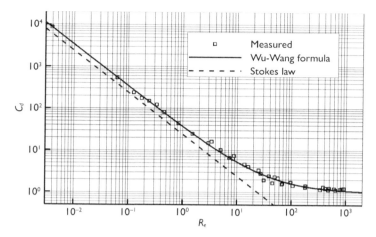

Figure 3.4 Drag coefficient as function of Reynolds number for naturally worn sediment particles ($S_P = 0.7$) (Wu and Wang, 2006).

factor of 1.0 may not be exactly spherical, and other factors, such as particle surface roughness, also affect the settling process.

Inserting Eq. (3.15) into Eq. (3.3) yields the general relation of settling velocity (Wu and Wang, 2006):

$$\omega_s = \frac{Mv}{Nd}\left[\sqrt{\frac{1}{4} + \left(\frac{4N}{3M^2}D_*^3\right)^{1/n}} - \frac{1}{2}\right]^n \tag{3.18}$$

Note that the sediment size d in Eq. (3.18) should be the nominal diameter (in meters), on which the drag coefficient C_d in Fig. 3.3 was based.

Eq. (3.18) is applied with coefficients M, N, and n determined using Eq. (3.17). It is an explicit relation of settling velocity with sediment size and shape factor; thus, it can be easily used. The predictions using Eq. (3.18) and the curves recommended by the U.S. Interagency Committee (1957) are compared in Fig. 3.5. Here, the temperature is 24°C, the Corey shape factors are in the range of 0.3–0.9, and the sediment sizes are between 0.2 and 64 mm. It can be seen that these two methods give very close predictions. The average deviation between them is about 2.75%. However, larger deviations are expected for fine sediments (less than 0.2 mm in diameter). The reason, which has been mentioned above, is that the U.S. Interagency Committee's curves approach the Stokes law, Eq. (3.5), that might result in 30% error for the settling velocity of natural sediment particles as shown in Fig. 3.4. Eq. (3.18) has been validated using measurement data and should have better accuracy than the U.S. Interagency Committee's curves for fine sediment particles.

In addition, Wu and Wang (2006) compared more than ten sediment settling velocity formulas, and found that the formulas of Zhang (1961), Hallermeier (1981), Dietrich (1982), Cheng (1997), Ahrens (2000), Jimenez and Madsen (2003), and Wu and Wang (2006) have comparable and reasonable reliabilities for predicting the

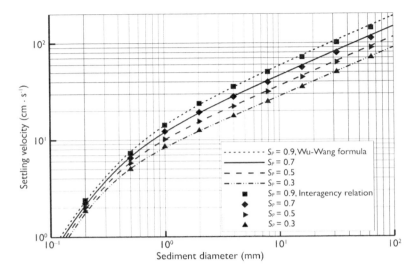

Figure 3.5 Comparison of Eq. (3.18) and the method of U.S. Interagency Committee (Wu and Wang, 2006).

settling velocity of naturally worn sediment particles (with a Corey shape factor of about 0.7). The average errors normally are less than 9%. If the shape factor is concerned, the formula of Wu and Wang is more convenient and has better accuracy on average.

3.1.4   Influence of sediment concentration on settling velocity

The settling velocity of a sediment particle in turbid water is influenced by the presence of other particles. Experiments have shown that when the sediment concentration is high, the settling velocity in turbid water is strongly reduced in comparison with that in clear water. This effect, known as hindered settling, is largely caused by the return flow of water induced by the settling of sediment. According to Richardson and Zaki (1954), the sediment settling velocity in turbid water, ω_{sm}, can be determined by

$$\omega_{sm} = (1 - c)^n \omega_s \tag{3.19}$$

where ω_s is the settling velocity in clear water, c is the volumetric sediment concentration, and n is an empirical exponent that varies from 4.9 to 2.3 for $R_e = \omega_s d / v$ increasing from 0.1 to 1000. For particles in the range of 0.05 to 0.5 mm under normal flow conditions, the coefficient n is about 4.

Based on his and McNown and Lin's (1952) experiments, Oliver (1962) proposed a formula for ω_{sm}:

$$\omega_{sm} = (1 - 2.15c)(1 - 0.75c^{0.33})\omega_s \tag{3.20}$$

Sha (1965) proposed a similar formula:

$$\omega_{sm} = \left(1 - \frac{c}{2\sqrt{d_{50}}}\right)^n \omega_s \tag{3.21}$$

where $n = 3$ according to the experimental data for sediment with a diameter of 0.01 mm.

For cohesive sediments, the settling process is more complex. This is discussed in Section 11.1.3.

3.2 INCIPIENT MOTION OF SEDIMENT

3.2.1 Equilibrium of a single sediment particle at incipient motion

Consider sediment particles on the channel bed, as shown in Fig. 3.6. The forces acting on them include the drag force F_D, lift force F_L, and submerged weight W_s. If the sediment particles are cohesive, a cohesion force also exists. However, quantification of the cohesion force is quite difficult because it is related to the physical and chemical properties of water and sediment. For simplicity, only non-cohesive sediment particles are considered here, so that the cohesion force is excluded.

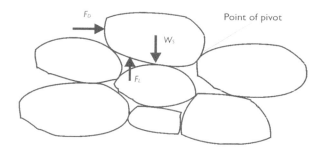

Figure 3.6 Forces on a sediment particle on the bed.

As the flow strength increases, the sediment particles on the bed will start moving. This is termed as "incipient motion." The modes of incipient motion can be sliding, rolling, and saltating. In the case of rolling, the force balance for a sediment particle at incipient motion can be expressed as

$$-k_1 dW_s + k_2 dF_D + k_3 dF_L = 0 \tag{3.22}$$

where $k_1 d$, $k_2 d$, and $k_3 d$ are the distances from the lines of action of forces W_s, F_D, and F_L to the point of pivot.

The drag and lift forces acting on the particle are usually determined by

$$F_D = C_D a_2 d^2 \rho \frac{u_b^2}{2} \tag{3.23}$$

$$F_L = C_L a_3 d^2 \rho \frac{u_b^2}{2} \tag{3.24}$$

where u_b is the bottom flow velocity acting on the particle; $a_2 d^2$ and $a_3 d^2$ are the projected areas of the particle on the planes normal to the flow direction and the vertical direction, respectively; and C_D and C_L are the drag and lift coefficients, related to particle shape, position on the bed, etc.

Inserting Eqs. (3.1), (3.23), and (3.24) into Eq. (3.22) yields the critical bottom velocity for sediment incipient motion:

$$u_{bc} = \left(\frac{2k_1 a_1}{k_2 a_2 C_D + k_3 a_3 C_L} \frac{\rho_s - \rho}{\rho} gd \right)^{1/2} \tag{3.25}$$

3.2.2 Incipient motion criteria for a group of sediment particles

Eq. (3.25) is a criterion for the incipient motion of an individual particle on the bed. For a group of sediment particles, there are two approaches to determine the threshold criterion of incipient motion: stochastic and deterministic. The stochastic approach considers the sediment incipient motion as a random phenomenon due to the stochastic properties of turbulent flow and sediment transport. This approach usually does not adopt a threshold value of sediment transport rate as the criterion at which the sediment particles start moving. The pioneer using the stochastic approach for sediment transport is Einstein (1942, 1950).

The deterministic approach usually adopts a certain amount of sediment particles in motion as the incipient motion criterion. Theoretically, a zero bed-load transport rate should be used, but this is not meaningful in practice. Numerous experiments have shown that even when the flow strength is much weaker than the critical condition proposed by Shields (1936), there are still some sediment particles moving on the bed. Kramer (1935) defined three types of motion of bed material: weak movement (only a few particles are in motion on the bed), medium movement (the grains of mean diameter begin to move), and general movement (all the mixture is in motion). However, his criterion is only qualitative and difficult to use. Therefore, several low levels of bed-load transport rate were suggested as the quantitative critical condition for incipient motion — for instance, $q_{b*} = 14 \text{ cm}^3 \text{m}^{-1} \text{min}^{-1}$ by Waterways Experiment Station, U.S. Army Corps of Engineers, and $q_{b*}/(\rho_s d\omega_s) = 0.000317$ by Han and He (1984). Yalin (1972) also proposed a quantitative criterion related to the number of particles moving on the bed. For a non-uniform sediment mixture, the threshold criterion for incipient motion is more complex because of interactions among different size classes. Parker et al. (1982) suggested the following threshold condition for the incipient motion of non-uniform sediment particles on gravel beds:

$$W_k^* = \frac{q_{b*k}(\rho_s/\rho - 1)}{p_{bk}(ghS_f)^{1/2}hS_f} = 0.002 \qquad (3.26)$$

where W_k^* is a dimensionless bed-load transport rate, q_{b*k} is the volumetric transport rate per unit width for the kth size class of bed load, p_{bk} is the fraction by weight of the kth size class in bed material, h is the flow depth, and S_f is the energy slope.

3.2.3 Incipient motion of uniform sediment particles

Critical average velocity

Using Eq. (3.25) and the power-law distribution of velocity

$$u = \frac{m+1}{m}\left(\frac{z}{h}\right)^{1/m} U \qquad (3.27)$$

yields the critical average velocity for sediment incipient motion:

$$U_c = K\left(\frac{\rho_s - \rho}{\rho}gd\right)^{1/2}\left(\frac{h}{d}\right)^{1/m} \qquad (3.28)$$

where U_c is the critical velocity averaged over the cross-section or flow depth ($\mathrm{m \cdot s^{-1}}$), and K is the coefficient determined by experiments. For example, Shamov (1959; see Zhang and Xie, 1993) used $m = 6$ and $K = 1.14$, while Zhang (1961) used $m = 7$ and $K = 1.34$.

The similarity between Eqs. (3.3) and (3.28) yields the following formula for the critical average velocity (Yang, 1973):

$$\frac{U_c}{\omega_s} = \begin{cases} 0.66 + 2.5/[\log(U_* d/v) - 0.06] & 1.2 < U_* d/v < 70 \\ 2.05 & U_* d/v \geq 70 \end{cases} \qquad (3.29)$$

where U_* is the bed shear velocity.

Critical shear stress

Using Eq. (3.25) and the logarithmic distribution of velocity

$$u = 5.75 U_* \log\left(30.2\frac{z\chi_s}{k_s}\right) \qquad (3.30)$$

yields

$$\frac{\tau_c}{(\gamma_s - \gamma)d} = \frac{2k_1 a_1}{k_2 a_2 C_D + k_3 a_3 C_L} \frac{1}{\left[5.75\log(30.2z_d\chi_s/k_s)\right]^2} \qquad (3.31)$$

where τ_c is the critical shear stress for sediment incipient motion, z_d is the height at which the bottom velocity acts on the particle, k_s is the bed roughness height, and χ_s is a correction factor related to the roughness Reynolds number $k_s U_* / \nu$ in general situations and has a value of 1 for a hydraulic rough bed.

Because C_D, C_L, and χ_s are related to flow conditions, Eq. (3.31) can be rewritten as

$$\frac{\tau_c}{(\gamma_s - \gamma)d} = f(U_* d / \nu) \tag{3.32}$$

Eq. (3.32) was first proposed by Shields (1936). The dimensionless parameter $\tau_c / [(\gamma_s - \gamma)d]$, denoted as Θ_c, is often called the critical Shields number. Shields drew a curve of Θ_c and $R_{e*} = U_* d / \nu$ using his experimental data. However, the original Shields curve did not have any measurement data in the range of small R_{e*}. Therefore, many investigators, such as Yalin and Karahan (1979) and Chien and Wan (1983), modified the original Shields curve using wider ranges of data. Fig. 3.7 shows the Shields curve modified by Chien and Wan.

Because the relation between Θ_c and R_{e*} in Fig. 3.7 is not explicit, iteration is needed to obtain the critical shear stress for a given sediment size. However, an explicit relation between Θ_c and the non-dimensional particle size $D_* = d[(\rho_s / \rho - 1)g / \nu^2]^{1/3}$ can be obtained from Fig. 3.7. It is approximated by (Wu and Wang, 1999)

$$\frac{\tau_c}{(\gamma_s - \gamma)d} = \begin{cases} 0.126 D_*^{-0.44}, & D_* < 1.5 \\ 0.131 D_*^{-0.55}, & 1.5 \le D_* < 10 \\ 0.0685 D_*^{-0.27}, & 10 \le D_* < 20 \\ 0.0173 D_*^{0.19}, & 20 \le D_* < 40 \\ 0.0115 D_*^{0.30}, & 40 \le D_* < 150 \\ 0.052, & D_* \ge 150 \end{cases} \tag{3.33}$$

where τ_c and d are in $N \cdot m^{-2}$ and m, respectively.

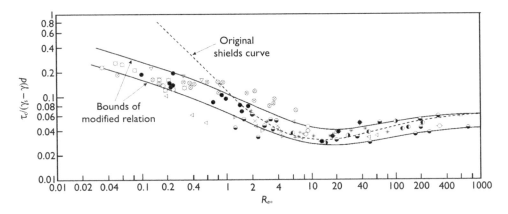

Figure 3.7 Shields curve modified by Chien and Wan (1983).

3.2.4   Incipient motion of non-uniform sediment particles

Interactions exist among different size classes of a non-uniform sediment mixture on the bed. Coarse particles have higher chances of exposure to flow, while fine particles are more likely sheltered by coarse particles. Therefore, it is necessary to consider the effect of this hiding and exposure mechanism on non-uniform sediment transport. The widely used approach is to introduce correction factors into the existing formulas of uniform sediment incipient motion and transport, as discussed below.

Qin formula

Qin (1980) proposed the following formula for the incipient motion of non-uniform sediment particles:

$$U_{ck} = 0.786 \left(\frac{h}{d_{90}} \right)^{1/6} \sqrt{ \frac{\gamma_s - \gamma}{\gamma} g d_k \left(1 + 2.5m \frac{d_m}{d_k} \right) } \qquad (3.34)$$

where U_{ck} is the critical average velocity for the incipient motion of size class k of bed material $(\mathrm{m \cdot s^{-1}})$, d_k is the diameter of size class $k \, (\mathrm{m})$, d_m is the arithmetic mean diameter of bed material (m), and m represents the compactness of non-uniform bed material:

$$m = \begin{cases} 0.6, & \eta_d < 2 \\ 0.76059 - 0.68014/(\eta_d + 2.2353), & \eta_d \geq 2 \end{cases}$$

where $\eta_d = d_{60}/d_{10}$. A formula similar to Eq. (3.34) was also proposed by Xie and Chen (1982; see Zhang and Xie, 1993).

Methods of Egiazaroff and others

Egiazaroff (1965), Ashida and Michiue (1971), Hayashi *et al.* (1980), and Parker *et al.* (1982) developed formulas to determine the incipient motion of non-uniform sediment particles by introducing correction factors as functions of the non-dimensional sediment size d_k/d_m or d_k/d_{50}. The Egiazaroff formula can be written as

$$\frac{\Theta_{ck}}{\Theta_c} = \left[\frac{\log 19}{\log(19 d_k/d_m)} \right]^2 \qquad (3.35)$$

where $\Theta_{ck} = \tau_{ck}/[(\gamma_s - \gamma)d_k]$, with τ_{ck} being the critical shear stress for the incipient motion of particle d_k in bed material; and Θ_c can be interpreted as the critical Shields number corresponding to d_m. Θ_c was given 0.06 by Egiazaroff. This value is too large in general. Misri *et al.* (1984) found that Θ_c should be 0.023–0.0303.

Ashida and Michiue (1971) modified the Egiazaroff formula as

$$\frac{\Theta_{ck}}{\Theta_c} = \begin{cases} [\log 19/\log(19d_k/d_m)]^2 & d_k/d_m \geq 0.4 \\ d_m/d_k & d_k/d_m < 0.4 \end{cases} \qquad (3.36)$$

and Hayashi et al. (1980) proposed a similar modification:

$$\frac{\Theta_{ck}}{\Theta_c} = \begin{cases} [\log 8/\log(8d_k/d_m)]^2 & d_k/d_m \geq 1 \\ d_m/d_k & d_k/d_m < 1 \end{cases} \qquad (3.37)$$

The formulas proposed by Parker et al. (1982) and others can be written as

$$\Theta_{ck} = \Theta_{c50} \left(\frac{d_k}{d_{50}}\right)^{-m} \qquad (3.38)$$

where Θ_{c50} is the critical Shields number corresponding to the medium size d_{50} of bed material, and m is an empirical coefficient between 0.5–1.0.

Method of Wu et al.

Consider a mixture of sediment particles with various diameters on the bed, as shown in Fig. 3.8. For simplicity, the sediment particles are assumed to be spheres. The drag and lift forces acting on a particle depend on how it is resting on the bed, i.e., whether it is hidden by other particles or exposed to flow. Its position on the bed can be represented by its exposure height Δ_e, which is defined as the difference between the apex elevations of it and the upstream particle. If $\Delta_e > 0$, the particle is considered to be at an exposed state; if $\Delta_e < 0$, it is at a hidden state. For a particle with diameter d_k in the bed surface layer, the value of Δ_e is in the range between $-d_j$ and d_k. Here, d_j is the diameter of the upstream particle. Because the sediment particles randomly distribute on the bed, Δ_e is a random variable. Δ_e is herein assumed to have a uniform probability distribution function:

$$f = \begin{cases} 1/(d_k + d_j), & -d_j \leq \Delta_e \leq d_k \\ 0, & \text{otherwise} \end{cases} \qquad (3.39)$$

The probability of particles d_j staying in front of particles d_k is assumed to be the fraction, p_{bj}, of particles d_j in bed material. Therefore, the probabilities of particles d_k hidden and exposed due to particles d_j are obtained from Eq. (3.39) as follows:

$$p_{hk,j} = p_{bj}\frac{d_j}{d_k + d_j} \qquad (3.40)$$

$$p_{ek,j} = p_{bj}\frac{d_k}{d_k + d_j} \qquad (3.41)$$

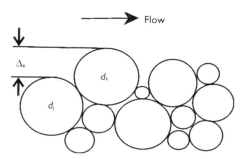

Figure 3.8 Definition of exposure height of bed material.

The total hidden and exposed probabilities, p_{hk} and p_{ek}, of particles d_k are then obtained by summing Eqs. (3.40) and (3.41) over all size classes, respectively:

$$p_{hk} = \sum_{j=1}^{N} p_{bj} \frac{d_j}{d_k + d_j} \tag{3.42}$$

$$p_{ek} = \sum_{j=1}^{N} p_{bj} \frac{d_k}{d_k + d_j} \tag{3.43}$$

where N is the total number of particle size classes in the non-uniform sediment mixture.

A relation $p_{hk} + p_{ek} = 1$ exists. For uniform sediment particles, $p_{hk} = p_{ek} = 0.5$, which means the hidden and exposed probabilities are equal. In a non-uniform sediment mixture, $p_{ek} \geq p_{hk}$ for coarse particles, and $p_{ek} \leq p_{hk}$ for fine particles. This can be demonstrated with a simple example. For a sediment mixture with two size classes $d_1 = 1$ mm, $p_{b1} = 0.4$ and $d_2 = 5$ mm, $p_{b2} = 0.6$, one can obtain $p_{h1} = 0.7 > p_{e1} = 0.3$, $p_{h2} = 0.3667 < p_{e2} = 0.6333$. It is shown that more coarse particles are exposed and more fine particles are hidden.

By using the hidden and exposed probabilities, a hiding and exposure correction factor is defined as (Wu *et al.*, 2000b)

$$\eta_k = \left(\frac{p_{ek}}{p_{hk}} \right)^{-m} \tag{3.44}$$

where m is an empirical parameter. The criterion for sediment incipient motion proposed by Shields (1936) is then modified as

$$\frac{\tau_{ck}}{(\gamma_s - \gamma)d_k} = \Theta_c \left(\frac{p_{ek}}{p_{hk}} \right)^{-m} \tag{3.45}$$

where $\Theta_c = 0.03$ and $m = 0.6$, which are calibrated using laboratory and field data, as

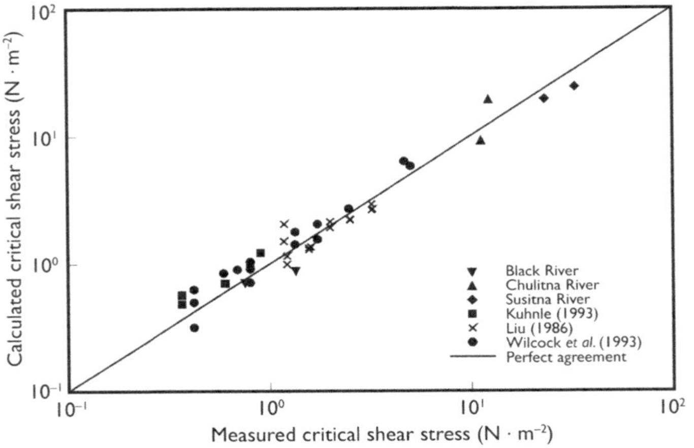

Figure 3.9 Comparison of measured and calculated critical shear stresses (Wu *et al.*, 2000b).

shown in Fig. 3.9. The measured critical shear stresses in Fig. 3.9 were determined using Eq. (3.26) as the reference transport threshold. The agreement between measurements and predictions is generally good.

3.2.5 Incipient motion of sediment particles on slopes

For a sediment particle on a sloped bed or bank, its incipient motion is affected not only by the drag and lift forces, but also by the component of gravity along the slope. Brooks (1963) suggested the following method to determine the critical shear stress $\tau_{c\varphi}$ for the incipient motion of sediment on a sloped bed:

$$\frac{\tau_{c\varphi}}{\tau_c} = -\frac{\sin\varphi \sin\theta_s}{\tan\phi_r} + \sqrt{\cos^2\varphi - \frac{\sin^2\varphi \cos^2\theta_s}{\tan^2\phi_r}} \qquad (3.46)$$

where φ is the slope angle with positive values for downslope beds, θ_s is the angle between the flow direction and the horizontal line of the slope, and ϕ_r is the repose angle.

Van Rijn (1989) also suggested a method to determine $\tau_{c\varphi}$:

$$\tau_{c\varphi} = k_1 k_2 \tau_c \qquad (3.47)$$

where k_1 is the correction factor for the streamwise-sloped bed (in the flow direction), determined by $k_1 = \sin(\phi_r - \varphi_L)/\sin\phi_r$; and k_2 is the correction factor for the sideward-sloped bed (normal to the flow direction), determined by $k_2 = \cos\varphi_T\sqrt{1 - \tan^2\varphi_T/\tan^2\phi_r}$. Here, φ_L and φ_T are the slope angles in the flow and sideward directions, respectively.

3.3 MOVABLE BED ROUGHNESS IN ALLUVIAL RIVERS

3.3.1 Bed forms in alluvial rivers

Bed forms in alluvial rivers are closely related to flow conditions. As the flow strength increases, a stationary flat bed may evolve to sand ripples, sand dunes, moving plane bed, anti-dunes, and chutes/pools (Richardson and Simons, 1967; Zhang *et al.*, 1989), as shown in Fig. 3.10. This process is explained below in more detail:

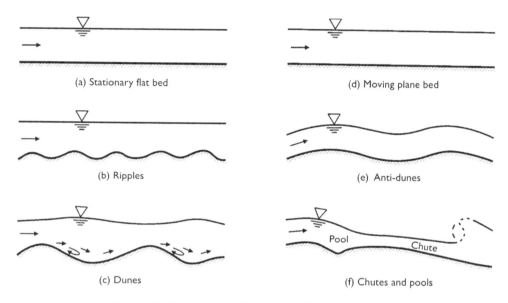

Figure 3.10 Bed forms in alluvial rivers (Zhang *et al.*, 1989).

(a) In the stage of stationary flat bed, the flow is weak and only a small amount of sediment particles move on the bed.

(b) As the flow strength increases, more and more sediment particles participate in motion, and sand ripples occur. The generation of sand ripples mainly depends on the stability of the movable bed under the action of turbulent shear flow. Their dimension is highly related to the bed-material size d, and they are about $100\ d$ in length and 50–$100\ d$ in height.

(c) Due to the effect of large-scale flow eddies, bed shear stress decreases and increases, and sediment deposits and erodes at alternate patterns, thus resulting in generation of sand dunes on the bed. In the upstream slope of a sand dune, flow acceleration usually causes sediment erosion; in the downstream slope, flow deceleration and separation cause sediment deposition. Therefore, the sand dunes migrate downstream in certain shapes. Their dimension is highly related to the flow depth h. They are usually about 5–$10\ h$ long and 0.1–$0.5\ h$ high.

(d) When the flow strength continually increases, sediment particles may be suspended and transported far downstream; thus, sand dunes are washed out, and the bed may become plane again. Although the bed is plane, sediment particles are still moving on the bed.

(e) Further increase in flow strength will induce anti-dunes. In the anti-dune stage, the flow Froude number usually is larger than 1, and the sediment movement is strongly influenced by the free surface flow. While water and sediment move downstream, the bed and water surface waves actually propagate upstream in phase. They may break like sea surfs or subside as standing waves.

(f) Chutes and pools occur at relatively large slopes, with high flow velocities and sediment concentrations. Sediment particles move intensively in this stage.

The stationary flat bed, ripples, and dunes are usually called the lower flow regime, while the moving plane bed, anti-dunes, and chutes/pools are called the upper flow regime. Anti-dunes and chutes/pools are mostly observed in laboratory flumes but rarely found in natural rivers.

In addition, other large-scale bed forms, such as point bars, alternate bars, and islands, often exist in natural rivers. They are usually generated by channel meandering, expansion, and contraction as well as tributary confluence. Their dimensions are thus related to channel width, depth, curvature, etc.

3.3.2 Division of grain and form resistances

For a channel bed with sand grains and bed forms (such as sand ripples and dunes), the bed shear stress, τ_b, may be divided into the grain (skin or frictional) shear stress, τ_b', and the form shear stress, τ_b'':

$$\tau_b = \tau_b' + \tau_b'' \tag{3.48}$$

The bed shear stress is usually calculated by

$$\tau_b = \gamma R_b S_f \tag{3.49}$$

where R_b is the hydraulic radius of the channel bed.

Einstein (1942) suggested the division of the hydraulic radius R_b into two parts R_b' and R_b'', corresponding to the grain and form roughnesses, and determined the grain and form shear stresses as

$$\tau_b' = \gamma R_b' S_f, \quad \tau_b'' = \gamma R_b'' S_f \tag{3.50}$$

The assumption of equal velocity: $U = R_b^{2/3} S_f^{1/2}/n$, $U = R_b'^{2/3} S_f^{1/2}/n'$, and $U = R_b''^{2/3} S_f^{1/2}/n''$ yields $R_b' = R_b(n'/n)^{3/2}$ and $R_b'' = R_b(n''/n)^{3/2}$. Here, U is the average flow velocity, n is the Manning roughness coefficient of channel bed, and n' and n'' are the Manning coefficients corresponding to the grain and form roughnesses, respectively. Therefore, from these two relations and Eqs. (3.49) and (3.50), the following relations for the grain and form shear stresses are obtained:

$$\tau_b' = \left(\frac{n'}{n}\right)^{3/2} \tau_b, \quad \tau_b'' = \left(\frac{n''}{n}\right)^{3/2} \tau_b \tag{3.51}$$

Eq. (3.51) is similar to the method adopted by Meyer-Peter and Mueller (1948). It should be noted that the grain roughness coefficient n' can be calculated using several methods, such as $n' = d^{1/6}/21.5$ (Strickler, 1923), $n' = d_{90}^{1/6}/26$ (Meyer-Peter and Mueller, 1948), $n' = d_{65}^{1/6}/24$ (Patel and Ranga Raju, 1996), and $n' = d_{50}^{1/6}/20$ (Li and Liu, 1963; Wu and Wang, 1999). Here, the units of sediment sizes and n' are m and $s \cdot m^{-1/3}$, respectively.

Inserting Eq. (3.51) into Eq. (3.48) leads to

$$n^{3/2} = (n')^{3/2} + (n'')^{3/2} \tag{3.52}$$

Unlike the above Einstein's method, Engelund (1966) suggested the division of the bed shear stress according to the energy slope and determined the grain and form shear stresses as

$$\tau_b' = \gamma R_b S_f', \quad \tau_b'' = \gamma R_b S_f'' \tag{3.53}$$

where S_f' and S_f'' are the parts of the energy slope corresponding to the grain and form roughnesses, respectively.

Applying the equal velocity assumption and the Manning equations $U = R_b^{2/3} S_f^{1/2}/n$, $U = R_b^{2/3} S_f'^{1/2}/n'$, and $U = R_b^{2/3} S_f''^{1/2}/n''$ yields $S_f' = S_f(n'/n)^2$ and $S_f'' = S_f(n''/n)^2$. Then substituting these two relations into Eq. (3.53) and using Eq. (3.49) results in

$$\tau_b' = \left(\frac{n'}{n}\right)^2 \tau_b, \quad \tau_b'' = \left(\frac{n''}{n}\right)^2 \tau_b \tag{3.54}$$

Inserting Eq. (3.54) into Eq. (3.48) leads to

$$n^2 = (n')^2 + (n'')^2 \tag{3.55}$$

Note that the exponents are 3/2 in Eq. (3.52), but 2 in Eq. (3.55). However, both Einstein's and Engelund's methods give the following relation for the Chezy coefficient:

$$\frac{1}{C_b^2} = \frac{1}{C_b'^2} + \frac{1}{C_b''^2} \tag{3.56}$$

where C_b is the total Chezy coefficient; and C_b' and C_b'' are the fractional Chezy coefficients corresponding to the grain and form roughnesses, respectively.

3.3.3 Movable bed roughness formulas

Einstein and Barbarossa (1952), Engelund and Hansen (1967), and Alam and Kennedy (1969) proposed empirical methods for separately calculating the grain and form resistances to flow. Li and Liu (1963), Richardson and Simons (1967), and Wu and Wang

(1999) suggested direct calculation of the total roughness coefficient of a movable bed. Van Rijn (1984c) and Karim (1995) established empirical relations to predict the height of bed forms and then the roughness coefficient on a movable bed. Brownlie (1983) proposed a formula to determine the flow depth rather than the roughness coefficient in an alluvial river. The van Rijn, Karim, and Wu-Wang formulas are introduced below as examples.

Van Rijn formula

Van Rijn (1984c) established a relation for the sand-dune height, Δ, as shown in Fig. 3.11 and expressed as

$$\frac{\Delta}{h} = 0.11 \left(\frac{d_{50}}{h}\right)^{0.3} (1 - e^{-0.5T})(25 - T) \tag{3.57}$$

where T is the non-dimensional excess bed shear stress or the transport stage number, defined as $T = (U'_*/U_{*cr})^2 - 1$; U'_* is the effective bed shear velocity related to grain roughness, determined by $U'_* = Ug^{0.5}/C'_h$, with $C'_h = 18\log(4h/d_{90})$; U_{*cr} is the critical bed shear velocity for sediment incipient motion, given by the Shields diagram; and d_{50} and d_{90} are the characteristic diameters of bed material.

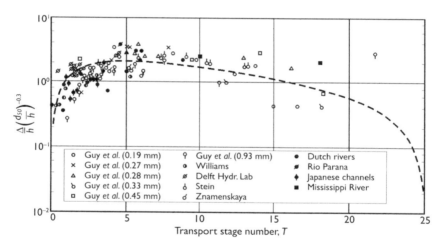

Figure 3.11 Relation of sand-dune height (van Rijn, 1984c).

In van Rijn's method, the length of sand dunes is set as $\lambda_d = 7.3h$, the grain roughness is $3d_{90}$, and the form roughness is $1.1\Delta(1 - e^{-25\Delta/\lambda_d})$. Therefore, the effective bed roughness is calculated by means of

$$k_s = 3d_{90} + 1.1\Delta(1 - e^{-25\Delta/\lambda_d}) \tag{3.58}$$

and the Chezy coefficient is then computed by

$$C_h = 18 \log \left(\frac{12 R_b}{k_s} \right) \tag{3.59}$$

where R_b is determined using Vanoni and Brooks' (1957) method.

Karim formula

Karim (1995) proposed the following formula to determine the Manning roughness coefficient on a movable bed:

$$n = 0.037 d_{50}^{0.126} \left(1.20 + 8.92 \frac{\Delta}{h} \right)^{0.465} \tag{3.60}$$

where n is in $\text{s} \cdot \text{m}^{-1/3}$; d_{50} is in m; and h is the hydraulic depth, which is the flow area divided by water surface width. The graphical relation between Δ and U_*/ω_s is shown in Fig. 3.12. In the range of $0.15 < U_*/\omega_s < 3.64$, Δ is determined by

$$\frac{\Delta}{h} = -0.04 + 0.294 \left(\frac{U_*}{\omega_s} \right) + 0.00316 \left(\frac{U_*}{\omega_s} \right)^2 - 0.0319 \left(\frac{U_*}{\omega_s} \right)^3 + 0.00272 \left(\frac{U_*}{\omega_s} \right)^4$$

$$\tag{3.61}$$

where ω_s is the settling velocity of sediment particles with size d_{50}.

Eq. (3.61) was calibrated using experimental data reported by Guy *et al.* (1966) and field data measured in the Missouri River.

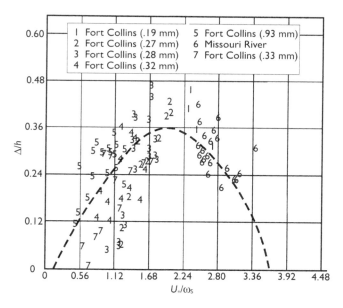

Figure 3.12 Relative roughness height as function of U_*/ω_s (Karim, 1995).

Wu-Wang formula

The Manning roughness coefficient n for a movable bed is often related to the bed sediment size d by

$$n = \frac{d^{1/6}}{A_n} \qquad (3.62)$$

where A_n is a roughness parameter related to bed-material size composition, particle shape, bed forms, flow conditions, etc.

For a stationary flat bed covered with uniform sediment particles, Strickler (1923) suggested $A_n = 21.1$. Here, the units of n and d are $s \cdot m^{-1/3}$ and m, respectively. For a stationary flat bed with non-uniform sediment particles, d is usually set as the median size d_{50}, and A_n is about 20 (Li and Liu, 1963; Zhang and Xie, 1993; Wu and Wang, 1999). If the sediment particles with slightly irregular shapes are tightly placed on the bed, A_n may have a larger value up to 24 (i.e., lower resistance to flow). If the sediment particles with rather irregular shapes are loosely placed on the bed, A_n has a smaller value between 17 and 20. In addition, if d is set as d_{65} or d_{90} rather than d_{50}, A_n has a value of 24 (Patel and Ranga Raju, 1996) or 26 (Meyer-Peter and Mueller, 1948), respectively.

For a movable bed with sand waves, the effect of bed forms on A_n should be included. Li and Liu (1963) proposed a relation of $A_n \sim U/U_c$ for natural rivers:

$$A_n = \begin{cases} 20(U/U_c)^{-3/2} & 1 < U/U_c \leq 2.13 \\ 3.9(U/U_c)^{2/3} & U/U_c > 2.13 \end{cases} \qquad (3.63)$$

However, Eq. (3.63) does not agree with most of the flume and field data used in the test performed by Wu and Wang (1999). To improve this shortcoming, Wu and Wang established a relation between $A_n/(g^{1/2}Fr^{1/3})$ and τ_b'/τ_{c50}, as shown in Fig. 3.13. Here, Fr is the Froude number $U/\sqrt{gh}$. The values of $A_n/(g^{1/2}Fr^{1/3})$ decrease, and then, increase as τ_b'/τ_{c50} increases. Physically, this trend represents the fact that sand ripples and dunes are formed first, and then, washed away gradually. For the convenience of users, the relation between $A_n/(g^{1/2}Fr^{1/3})$ and τ_b'/τ_{c50} in the range of $1 \leq \tau_b'/\tau_{c50} \leq 55$ is approximated by

$$\frac{A_n}{g^{1/2}Fr^{1/3}} = \frac{8[1 + 0.0235(\tau_b'/\tau_{c50})^{1.25}]}{(\tau_b'/\tau_{c50})^{1/3}} \qquad (3.64)$$

The critical shear stress τ_{c50} in Eq. (3.64) is calculated using the Shields curve modified by Chien and Wan (1983), and the grain shear stress τ_b' is calculated using Eq. (3.51), with n' calculated by $n' = d_{50}^{1/6}/20$ and τ_b by Eq. (3.49). The bed hydraulic radius R_b is determined using Williams' (1970) method: $R_b = h/(1 + 0.055h/B^2)$, in which B is the channel width.

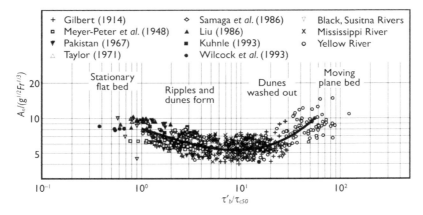

Figure 3.13 Relation between $A_n/(g^{1/2}Fr^{1/3})$ and τ'_b/τ_{c50} (Wu and Wang, 1999).

3.3.4 Comparison of movable bed roughness formulas

The movable bed roughness formulas of Li and Liu (1963), van Rijn (1984c), Karim (1995), and Wu and Wang (1999) were tested against 4,376 sets of flume and field data collected by Brownlie (1981). These data sets were measured by many investigators in several decades, covering flow discharges of 0.00263–28825.7 m^3s^{-1}, flow depths of 0.04–17.3 m, flow velocities of 0.2–3.32 $m \cdot s^{-1}$, bed slopes of 0.00002–0.067, sediment median diameters of 0.011–76.1 mm, and sediment size standard deviations up to 9.8. Table 3.2 compares the measured and predicted flow depths. It can be seen that the van Rijn, Karim, and Wu-Wang formulas almost have the same level of reliability for predicting the flow depth. As compared with the Li-Liu formula, the Wu-Wang formula has much improvement.

Table 3.2 Comparison of measured and predicted flow depths

Error range	% of calculated flow depths in error range			
	Li-Liu	van Rijn	Karim	Wu-Wang
±10%	21.8	44.0	41.0	41.5
±20%	41.8	77.9	74.9	75.9
±30%	58.8	91.4	91.0	94.4

3.4 BED-LOAD TRANSPORT

Laboratory experiments and field measurements have revealed that the sediment transport rate (or concentration) at an equilibrium state in a steady, uniform flow, which is often termed as the sediment transport capacity or the capacity of flow-carrying sediment, is a function of flow conditions and sediment properties. A variety of such functions for bed load, suspended load, and bed-material load have been established in the literature. Some of them are introduced in Sections 3.4–3.6.

3.4.1 Total transport rate of bed load

Many investigators — e.g., Duboys (1879), Schoklitsch (1930), Meyer-Peter and Mueller (1948), Bagnold (1966, 1973), Dou (1964), Graf (1971), Yalin (1972), Engelund and Fredsøe (1976), and van Rijn (1984a) — established formulas to calculate the total transport rate of bed load. The following formulas are presented as examples.

Meyer-Peter-Mueller formula

Meyer-Peter and Mueller (1948) related the bed-load transport rate to the excess grain shear stress:

$$\frac{q_{b*}}{\gamma_s\sqrt{(\gamma_s/\gamma - 1)gd_m^3}} = 8\left[\frac{(k/k')^{3/2}\gamma RS_f}{(\gamma_s - \gamma)d_m} - 0.047\right]^{3/2} \tag{3.65}$$

where q_{b*} is the bed-load transport rate by weight per unit time and width ($\mathrm{N\cdot m^{-1}s^{-1}}$); d_m is the arithmetic mean diameter of the bed sediment mixture (m); k is the reciprocal of the Manning roughness coefficient n of channel bed; k' is the reciprocal of the Manning coefficient n' due to grain roughness, calculated by $k' = 26/d_{90}^{1/6}$; and R is the hydraulic radius of the channel (m).

Bagnold formula

Bagnold (1966, 1973) related the sediment transport rate to the stream power $\tau_b U$ and derived a bed-load transport formula:

$$q_{b*} = \frac{\rho_s}{\rho_s - \rho}\frac{\tau_b U}{\tan\alpha}\frac{U_* - U_{*c}}{U_*}\left(1 - \frac{5.75U_*\log\left(\frac{0.37b}{nd}\right) + \omega_s}{U}\right) \tag{3.66}$$

where q_{b*} is by weight per unit time and width ($\mathrm{N\cdot m^{-1}s^{-1}}$), τ_b is in $\mathrm{N\cdot m^{-2}}$, $\tan\alpha$ is the friction coefficient of about 0.63, nd is the average height of acting force during a saltation, d is the sediment size (m), and $n = 1.4(U_*/U_{*c})^{0.6}$.

Dou formula

Dou (1964) also established an empirical formula for bed-load transport rate based on the stream power concept:

$$q_{b*} = K_0\frac{\rho_s}{\rho_s - \rho}\tau_b(U - U_c')\frac{U}{g\omega_s} \tag{3.67}$$

where q_{b*} is by mass per unit time and width $(\text{kg} \cdot \text{m}^{-1}\text{s}^{-1})$, U_c' is the critical average velocity for sediment particles to cease motion, and K_0 is an empirical coefficient with a value of 0.01 for sand.

Eq. (3.67) can also be used to determine bed-material load, for which $K_0 = 0.1$ as calibrated using Gilbert's data.

Yalin formula

Yalin (1972) analyzed the bed-load velocity and weight and then established the following bed-load formula:

$$\frac{q_{b*}}{\gamma_s d U_*} = 0.635 s \left[1 - \frac{1}{as} \ln(1 + as) \right] \tag{3.68}$$

where q_{b*} is by weight per unit time and width $(\text{N} \cdot \text{m}^{-1}\text{s}^{-1})$, $s = (\Theta - \Theta_c)/\Theta_c$, $a = 2.45\sqrt{\Theta_c}(\gamma/\gamma_s)^{0.4}$, and Θ is the Shields number $\tau_b/[(\gamma_s - \gamma)d]$.

Engelund-Fredsøe formula

Engelund and Fredsøe (1976) related the bed-load transport rate to the bed-load velocity and the probability for bed material to start moving, and obtained

$$\Phi_b = 11.6(\Theta - \Theta_c)\left(\sqrt{\Theta} - 0.7\sqrt{\Theta_c}\right) \tag{3.69}$$

where $\Phi_b = q_{b*}/[\gamma_s\sqrt{(\gamma_s/\gamma - 1)gd^3}]$, and q_{b*} is by weight per unit time and width $(\text{N} \cdot \text{m}^{-1}\text{s}^{-1})$.

Van Rijn formula

Van Rijn (1984a) determined bed load as

$$q_{b*} = 0.053 \left(\frac{\rho_s - \rho}{\rho} g \right)^{0.5} \frac{d_{50}^{1.5} T^{2.1}}{D_*^{0.3}} \tag{3.70}$$

where q_{b*} is by volume per unit time and width $(\text{m}^2\text{s}^{-1})$, D_* is the particle parameter defined in Eq. (3.16), and T is the transport stage number defined in Eq. (3.57). Eq. (3.70) was calibrated using data with a size range of 0.2–2 mm.

In addition, several bed-material load formulas, such as those of Ackers and White (1973) and Engelund and Hansen (1967), can be used to calculate the bed-load transport rate for coarse sediments. Yang (1984) modified his 1973 bed-material load formula for gravel transport, which is primarily in bed load.

Note that the bed-load formulas introduced above calculate the transport rate of uniform bed load or the total transport rate of non-uniform bed load as a single size class. Thus, they may be used for narrowly graded sediment mixtures.

3.4.2 Fractional transport rate of bed load

The pioneering research on the fractional transport rate of non-uniform sediment is attributed to Einstein (1950). After that, Ashida and Michiue (1972), Parker *et al.* (1982), Misri *et al.* (1984), Samaga *et al.* (1986a), Bridge and Bennett (1992), Patel and Ranga Raju (1996), and Wu *et al.* (2000b) proposed several methods to calculate the fractional transport rate of non-uniform bed load. Hsu and Holly (1992) proposed a method to compute the size composition of non-uniform bed load by considering the probability and availability of moving sediment. Some of these methods are introduced below.

Einstein formula

Einstein (1942, 1950) considered the probability of sediment transport due to the fluctuation of turbulent flow and established sediment transport functions based on fluid mechanics and probability theory. His bed-load function is graphically shown in Fig. 3.14 and expressed as

$$1 - \frac{1}{\sqrt{\pi}} \int_{-(1/7)\Psi_{*k}-2}^{(1/7)\Psi_{*k}-2} e^{-t^2} dt = \frac{43.5\Phi_{*k}}{1 + 43.5\Phi_{*k}} \qquad (3.71)$$

where $\Phi_{*k} = q_{b*k}/[p_{bk}\gamma_s\sqrt{(\gamma_s/\gamma - 1)gd_k^3}]$, and $\Psi_{*k} = \xi_b Y(\beta^2/\beta_x^2)\Psi$, in which q_{b*k} is the bed-load transport rate of size class k by weight per unit time and width, $\Psi = (\gamma_s - \gamma)d_k/(\gamma R'S_f)$, ξ_b and Y are the hiding and pressure correction factors for non-uniform sediment, $\beta = \log 10.6$, and $\beta_x = \log(10.6X/\Delta_s)$. R' is the hydraulic radius due to grain roughness, determined using Einstein's movable bed roughness method. Δ_s is the apparent roughness of bed surface, and $\Delta_s = k_s/\chi_s$, with $k_s = d_{65}$ and χ_s

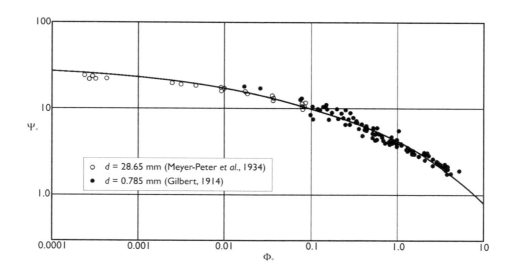

Figure 3.14 Einstein's (1950) bed-load function compared with uniform sediment data.

being the correction coefficient defined in Eq. (3.31). X is the characteristic grain size of bed material, defined as $X = 0.77\Delta_s$ if $\Delta_s/\delta > 1.8$ and $X = 1.39\delta$ if $\Delta_s/\delta \leq 1.8$, with δ being the laminar sublayer thickness ($= 11.6\nu/U_*'$).

Parker et al. formula

Based on the equal mobility concept, Parker *et al.* (1982) developed a gravel transport function, as shown in Fig. 3.15, in which the dimensionless bed-load transport rate W_k^* is defined in Eq. (3.26) and the dimensionless shear stress θ_k is

$$\theta_k = \frac{hS_f}{(\rho_s/\rho - 1)d_k\tau_{rk}^*} \tag{3.72}$$

where $\tau_{rk}^* = 0.0875d_{50}/d_k$, with d_{50} being the subpavement size.

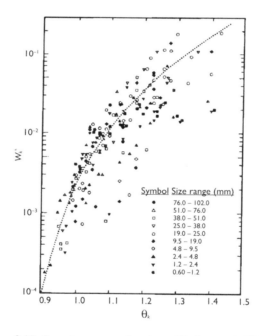

Figure 3.15 Gravel transport function of Parker et al. (1982).

Since all grain sizes are assumed to have approximately equal mobility, only one grain size, the subpavement size d_{50}, is used to characterize the bed-load transport rate as

$$W_k^* = \begin{cases} 0.0025 \exp[14.2(\theta_{50} - 1) - 9.28(\theta_{50} - 1)^2] & 0.95 < \theta_{50} < 1.65 \\ 11.2(1 - 0.822/\theta_{50})^{4.5} & \theta_{50} \geq 1.65 \end{cases}$$

$$\tag{3.73}$$

where θ_{50} is the dimensionless shear stress defined in Eq. (3.72) corresponding to the subpavement size d_{50}. This formula was verified using field data with sediment sizes ranging from 18 to 28 mm.

Considering the fact that the bed-load transport in gravel-bed rivers is accomplished by means of mobilization of grains exposed on the bed surface rather than substrate particles, Parker (1990) transformed Eq. (3.73) into a surface-based relation. The details can be found in his paper.

Hsu and Holly's method

The method proposed by Hsu and Holly (1992) first determines the size distribution of the transported sediment and then the total transport rate. The fraction of each size class in the transported material is postulated to be proportional to the joint probability of two factors: (1) its mobility under the prevailing hydraulic conditions, and (2) its availability on the bed surface (active layer).

If the fluctuation of flow velocity is assumed to have the Gaussian probability distribution, the mobility of size class k is derived as

$$P_{mo,k} = \frac{1}{\sigma\sqrt{2\pi}} \int_{U_{ck}/\overline{U}-1}^{\infty} \exp\left(-\frac{x^2}{2\sigma^2}\right) dx \qquad (3.74)$$

where $\overline{U}$ is the mean velocity of flow; U_{ck} is the incipient velocity of size class k, determined using the Qin (1980) formula (3.34) modified by recalibrating the coefficient 0.786 as 1.5; and σ is the standard deviation of the normalized fluctuating velocity $U'/\overline{U}$ and has a value of about 0.2.

The availability of size class k is equivalent to its fractional representation on the bed surface (active layer), p_{bk}. Thus, the fraction of size class k in the transported material is

$$p_k = \frac{P_{mo,k}\, p_{bk}}{\sum_{d_{min}}^{d_{max}} P_{mo,k}\, p_{bk}} \qquad (3.75)$$

After the size distribution of the transported material is obtained, the mean size d_{mt} and mean incipient velocity U_{ct} are calculated. The total bed-load transport rate can then be evaluated using any appropriate predictor. The Shamov formula was suggested and modified as

$$q_{b*} = 12.5\sqrt{d_{mt}} \left(\overline{U} - U_{c\,min}\right)\left(\frac{\overline{U}}{U_{ct}}\right)^3 \left(\frac{d_{mt}}{h}\right)^{1/4} \qquad (3.76)$$

where q_{b*} is the total transport rate of bed load per unit channel width ($\mathrm{kg \cdot m^{-1} s^{-1}}$), and $U_{c\,min}$ is the incipient velocity of the smallest size class ($\mathrm{m \cdot s^{-1}}$).

Methods of Ranga Raju and his co-workers

Ranga Raju and his co-workers (Misri *et al.*, 1984; Samaga *et al.*, 1986a; Patel and Ranga Raju, 1996) extended the Paintal (1971) uniform bed-load formula to computing the fractional transport rate of non-uniform bed load. Based on the assumption that the motion of fine particles is dominated by the lift force while the motion of coarse particles is by the drag force, Misri *et al.* (1984) proposed a semi-theoretical hiding-exposure correction factor. This correction factor was revised subsequently by Samaga *et al.* (1986a) and Patel and Ranga Raju (1996). In the latest version published by Patel and Ranga Raju, the bed-load function is shown in Fig. 3.16, in which the dimensionless bed-load transport rate and the effective shear stress are

$$\Phi_{bk} = \frac{q_{b*k}}{p_{bk}\gamma_s\sqrt{(\gamma_s/\gamma - 1)gd_k^3}} \tag{3.77}$$

$$\tau_{eff} = \xi_b\tau_b' \tag{3.78}$$

where $\tau_b' = \gamma R_b' S_f$, $R_b' = (Un'/S_f^{1/2})^{3/2}$, $n' = d_{65}^{1/6}/24$, and ξ_b is the hiding-exposure correction factor for the effective shear stress determined by

$$C_m\xi_b = 0.0713(C_s\tau_{*k}')^{-0.75144} \tag{3.79}$$

with $\tau_{*k}' = \tau_b'/[(\gamma_s - \gamma)d_k]$,

$$\log C_s = -0.1957 - 0.9571\log(\tau_b'/\tau_c) - 0.1949[\log(\tau_b'/\tau_c)]^2$$
$$+ 0.0644[\log(\tau_b'/\tau_c)]^3,$$

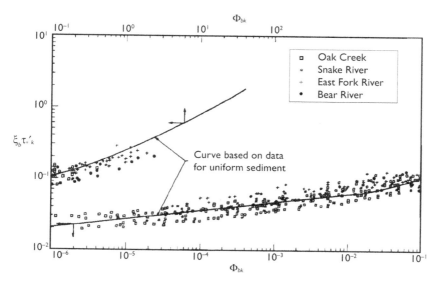

Figure 3.16 Fractional bed-load function (Patel and Ranga Raju, 1996).

$$C_m = \begin{cases} 1.0, & M > 0.38 \\ 0.7092 \log M + 1.293, & 0.05 < M \le 0.38 \end{cases}$$

where M is the Kramer uniformity coefficient, and τ_c is the critical shear stress for the arithmetic mean size d_m.

Wu et al. formula

Wu et al. (2000b) related the bed-load transport rate to the non-dimensional excess grain shear stress $T_k = \tau'_b / \tau_{ck} - 1$, with τ_{ck} and τ'_b determined using Eqs. (3.45) and (3.51), respectively. The established relation for the fractional transport rate of non-uniform bed load is graphically shown in Fig. 3.17 and expressed as

$$\Phi_{bk} = 0.0053 \left[\left(\frac{n'}{n} \right)^{3/2} \frac{\tau_b}{\tau_{ck}} - 1 \right]^{2.2} \tag{3.80}$$

where $\Phi_{bk} = q_{b*k} / [p_{bk} \sqrt{(\gamma_s/\gamma - 1) g d_k^3}]$, q_{b*k} is by volume per unit time and width $(\mathrm{m^2 s^{-1}})$, $n' = d_{50}^{1/6}/20$, and n is the Manning roughness coefficient of channel bed. Note that the hiding and exposure effect in non-unifrom bed material is accounted for through τ_{ck} determined using Eq. (3.45).

Eq. (3.80) was verified by using laboratory data for non-uniform bed load measured by Samaga et al. (1986a), Liu (1986), Kuhnle (1993), and Wilcock and McArdell (1993), as well as field data from five natural rivers: the Susitna, Chulitna,

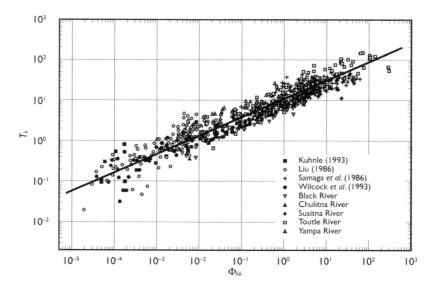

Figure 3.17 Relation of fractional bed-load transport rate (Wu et al., 2000b).

Black, Toutle, and Yampa Rivers compiled by Williams and Rosgen (1989). In each set of the selected field data, flow and sediment parameters were measured at the same time, and the bed-load rate and bed-material size composition were averaged from multiple samples across the same cross-section. These data sets cover a wide range of flow and sediment conditions, with flow discharges up to 2,800 m³s⁻¹ and sediment sizes from 0.062 to 128 mm.

3.4.3 Comparison of bed-load formulas

Because of the complexity of sediment transport processes, all existing sediment transport formulas are empirical or semi-empirical. Large discrepancies may exist among these formulas when they are applied in real-life engineering. Therefore, evaluation of their performances in various situations is very important.

Comparison of bed-load formulas using single-fraction data

Chien (1980; also see Chien and Wan, 1983) compared the formulas of Einstein (1942), Meyer-Peter and Mueller (1948), Bagnold (1966), and Yalin (1972) with measured data, as shown in Fig. 3.18. For weak sediment transport ($\Psi_* > 2$), the Yalin formula underpredicts the bed-load transport rate, and other formulas provide reasonably good predictions. The Meyer-Peter-Mueller formula seems to predict better than the Einstein formula in the weak transport stage, but the situation is reversed in the middle transport stage. However, for strong sediment transport ($\Psi_* < 2$), the predictions of these formulas are significantly different. Because in this range bed load and suspended load are very difficult to discern, and the measured data may have large errors, it is hard to judge which formula is better (Chien, 1980).

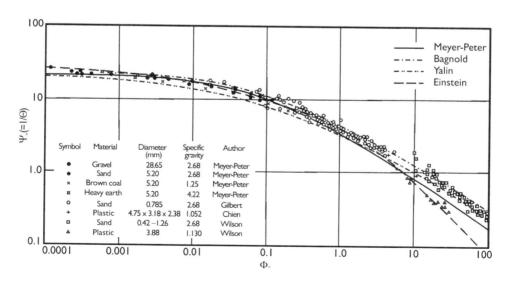

Figure 3.18 Comparison of bed-load formulas (Chien, 1980).

Wu *et al.* (2000b) compared their bed-load transport formula (3.80) and the formulas of Meyer-Peter and Mueller (1948), Bagnold (1966), and Engelund and Fredøse (1976) against 1,345 sets of uniform bed-load data. These data were selected from Brownlie's (1981) compilation by limiting the standard deviation of bed material $\sigma < 1.2$, the Shields number $\Theta > 0.055$, and the Rouse number $\omega_s/\kappa u_* > 2.5$. They were observed in several decades by many investigators, covering flow discharges of 0.00094–297 m^3s^{-1}, flow depths of 0.01–2.56 m, flow velocities of 0.086–2.88 $m \cdot s^{-1}$, surface slopes of 0.0000735–0.0367, and sediment sizes of 0.088–28.7 mm. None of them was used to calibrate the Wu *et al.* formula. As shown in Table 3.3, the Wu *et al.* formula provides the best results.

Many other investigators, such as Yang (1984) and van Rijn (1984a), have also compared bed-load transport formulas. The conclusions are usually different because different data have been used. However, it has been shown that the existing formulas have better predictions for flume data than for field data. The reasons are that the bed-load transport is more complex and the measurement instruments are less efficient in natural rivers. As recognized by van Rijn (1984a), it is hardly possible to predict the bed-load transport rate with accuracy less than a factor of 2. Perhaps his remark is useful for sediment engineers to judge the prediction capability of the existing sediment transport formulas.

Table 3.3 Calculated versus measured transport rates of uniform bed load

Error range	% of calculated transport rates in error range			
	Engelund-Fredøse	Bagnold	Meyer-Peter-Mueller	Wu et al.
$0.8 \leq r \leq 1.25$	21.4	21.4	21.3	38.7
$0.667 \leq r \leq 1.5$	37.4	38.9	39.4	59.3
$0.5 \leq r \leq 2$	54.1	57.2	66.2	80.1

Note: r is the ratio of calculated and measured transport rates.

Comparison of bed-load formulas using multi-fraction data

Ribberink *et al.* (2002) tested the performances of several multi-fraction bed-load transport formulas, including the Parker (1990) formula, the Wu *et al.* (2000b) formula, the Ackers-White (A&W, 1973) formula with the hiding-exposure correction factors of Day (1980) and Proffitt and Sutherland (P&S, 1983) (to be introduced in Section 3.6.2), and the Meyer-Peter-Mueller (MP&M, 1948) formula with the hiding-correction factors of Egiazaroff (1965) and Ashida and Michiue (A&M, 1972). The "single-size" Engelund-Hansen (E&H, 1967) and van Rijn (1984a) formulas without any hiding and exposure correction were added as reference. The data used cover the bed-load transport of widely graded sediment mixtures in the lower Shields regime. The results are summarized in Table 3.4 and expressed in mean under- or overestimation scores (factor n over/underestimation gives a score of $1/n$). Separate scores are made for the predicted total transport rate and mean transported diameter, and an average score for both.

Of all the compared multi-fraction formulas, the Wu *et al.* (2000b) formula gives the highest scores, followed by the Ackers-White formula with the hiding-exposure

Table 3.4 Verification scores of multi-fraction bed-load formulas (Ribberink et al., 2002)

Formula	Score for transport rate	Score for mean diameter	Average score
Wu et al.	0.43	0.86	0.64
E&H	0.34	0.63	0.49
A&W + Day	0.37	0.59	0.48
Parker (surface)	0.23	0.73	0.48
A&W + P&S	0.34	0.49	0.42
Van Rijn	0.18	0.54	0.36
MP&M + Egiaz.	0.26	0.34	0.30
MP&M + A&M	0.29	0.29	0.29

correction factor of Day (1980). Surprisingly, also the Engelund-Hansen formula, which was not developed for multi-fraction use for widely graded sediment mixtures, is the second-best formula in the list. All the Meyer-Peter-Mueller formulas give the worst scores, mainly due to many cases with zero predicted transport rate.

3.5 SUSPENDED-LOAD TRANSPORT

3.5.1 Vertical distribution of suspended-load concentration

For equilibrium sediment transport under steady, uniform flow conditions, the suspended-load transport equation (2.72) is simplified to

$$-\frac{\partial(\omega_s c)}{\partial z} = \frac{\partial}{\partial z}\left(\varepsilon_s \frac{\partial c}{\partial z}\right) \tag{3.81}$$

By using the sediment condition (2.73) at the water surface, Eq. (3.81) is further simplified to

$$\omega_s c + \varepsilon_s \frac{\partial c}{\partial z} = 0 \tag{3.82}$$

The diffusion coefficient ε_s is often assumed to be proportional to the eddy viscosity of turbulent flow. By using Eq. (2.49), a parabolic distribution of ε_s can be obtained:

$$\varepsilon_s = \frac{1}{\sigma_s}\kappa U_* z\left(1 - \frac{z}{h}\right) \tag{3.83}$$

where σ_s is the Schmidt number, related to sediment size, concentration, etc. Note that z is defined here as the vertical coordinate above the bed, for simplicity.

With ε_s determined using Eq. (3.83) with constant ω_s and σ_s along the flow depth, Eq. (3.82) can be solved to derive the following vertical distribution of suspended-load concentration:

$$\frac{c}{c_{b*}} = \left(\frac{h/z - 1}{h/\delta - 1}\right)^{\frac{\sigma_s \omega_s}{\kappa U_*}} \tag{3.84}$$

where δ is the reference level near the bed, and c_{b*} is the sediment concentration at δ. Eq. (3.84), which was first derived by Rouse (1937), is called the Rouse distribution.

Fig. 3.19 shows the profile of suspended-load concentration calculated using Eq. (3.84) with $\sigma_s = 1$. One can see that the calculated concentration is zero at the water surface and tends to be infinitely large as z is close to the bed. These are not physically reasonable. Therefore, the reference level δ is usually set at a certain height — e.g., $2d$, $0.05h$, and half the dune height — above the bed rather than directly at the bed.

Zhang (1961) derived a distribution function of suspended-load concentration by using the eddy viscosity determined from the mixing length measured by Nikuradse in uniform pipe flow:

$$\frac{l_m}{h} = 0.14 - 0.08\eta^2 - 0.06\eta^4 \tag{3.85}$$

where l_m is the mixing length, h is the radius of pipe or the flow depth, and $\eta = 1 - z/h$.

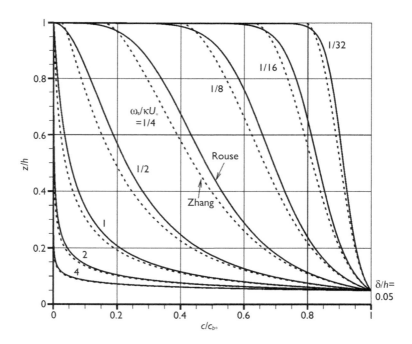

Figure 3.19 Distribution of suspended-load concentration.

The derived distribution function is

$$\frac{c}{c_{b*}} = \exp\left\{\frac{\omega_s}{\kappa U_*}[f(\eta) - f(\eta_b)]\right\}$$ (3.86)

where $\eta_b = 1 - \delta/h$, and

$$f(\eta) = 2\arctan\sqrt{\eta} + \ln\frac{1+\sqrt{\eta}}{1-\sqrt{\eta}} + \frac{\sqrt{2}}{a^{3/2}}\left[\ln\frac{\eta + \sqrt{2a\eta} + a}{\sqrt{a^2 + \eta^2}}\right.$$
$$\left. + \arctan\left(1 + \sqrt{\frac{2\eta}{a}}\right) - \arctan\left(1 - \sqrt{\frac{2\eta}{a}}\right)\right]$$

with $a = 1.526$.

The Zhang distribution is shown in Fig. 3.19 as dashed lines. It improves the sediment concentration near the water surface, but the formulation is more complicated and inconvenient to use.

Lane and Kalinske (1941) assumed $\sigma_s = 1$ and averaged the sediment diffusivity in Eq. (3.83) over the flow depth as

$$\bar{\varepsilon}_s = \frac{\kappa}{6}U_*h$$ (3.87)

and then introduced this value into Eq. (3.82) and derived

$$\frac{c}{c_{b*}} = \exp\left[-\frac{6\omega_s}{\kappa U_*}\left(\frac{z-\delta}{h}\right)\right]$$ (3.88)

Van Rijn (1984b) also derived a vertical distribution of suspended-load concentration using the following two-layer relation of sediment diffusivity:

$$\varepsilon_s = \begin{cases} \kappa U_*(1 - z/h)z/\sigma_s & z/h < 0.5 \\ 0.25\kappa U_*h/\sigma_s & z/h \geq 0.5 \end{cases}$$ (3.89)

In the case of small concentration ($c < c_{b*} < 0.001$), the van Rijn distribution is

$$\frac{c}{c_{b*}} = \begin{cases} [(h/z - 1)/(h/\delta - 1)]^r & z/h < 0.5 \\ (h/\delta - 1)^{-r}\exp[-4r(z/h - 0.5)] & z/h \geq 0.5 \end{cases}$$ (3.90)

where $r = \omega_s/(\kappa U_*)$.

The parameter $\omega_s/(\kappa U_*)$ is called the suspension or Rouse number. Physically, the Rouse number represents the effect of gravity (ω_s) against the effect of turbulent diffusion (κU_*). When the Rouse number is larger, the effect of gravity is stronger and the distribution of sediment concentration along the flow depth is less uniform. When the

Rouse number is smaller, the effect of turbulent diffusion is stronger and the distribution of sediment concentration is more uniform. It can be seen from Fig. 3.19 that when the Rouse number is larger than about 5.0, the relative concentration of suspended load is very small, and thus $\omega_s/(\kappa U_*) \approx 5.0$ can be used as the critical condition for suspension. When the Rouse number is less than about 0.06, the suspended-load concentration almost uniformly distributes along the flow depth, and thus $\omega_s/(\kappa U_*) \approx 0.06$ may be used as a condition to divide wash load and bed-material load.

Brush et al. (1962), Matyukhin and Prokofyev (1966), and Majumdar and Carstens (1967) experimentally showed that for fine particles $\sigma_s \cong 1$, and for coarse particles $\sigma_s > 1$. However, Einstein and Chien (1954) obtained the relation between $\omega_s/(\kappa U_*)$ and $\sigma_s\omega_s/(\kappa U_*)$ shown in Fig. 3.20 by comparing the measured suspended-load distribution with Eq. (3.84), and suggested that σ_s should be smaller than 1. This contradiction might be due to differences in flow and sediment conditions in which the data were measured.

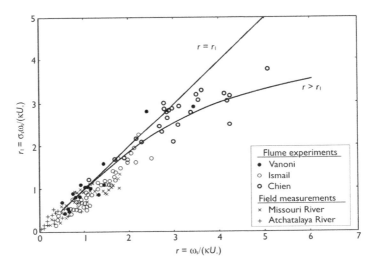

Figure 3.20 Relation between $\omega_s/(\kappa U_*)$ and $\sigma_s\omega_s/(\kappa U_*)$ (Einstein and Chien, 1954).

Van Rijn (1984b) proposed a formula to determine the Schmidt number σ_s:

$$\frac{1}{\sigma_s} = 1 + 2\left(\frac{\omega_s}{U_*}\right)^2 \quad \text{for} \ 0.1 < \frac{\omega_s}{U_*} < 1 \tag{3.91}$$

The von Karman constant has a value of about 0.4 for clear water flow and is a function of the depth-averaged concentration, settling velocity, and bed shear velocity for sediment-laden flow (Einstein and Chien, 1955). Yalin and Finlayson (1972) introduced a damping factor for the von Karman constant:

$$\kappa_m = \phi_\kappa \kappa \tag{3.92}$$

where κ is the von Karman constant for clear water flow, and κ_m is the one for sediment-laden flow. Van Rijn (1984b) determined the damping factor as

$$\phi_\kappa = 1 + \left(\frac{c}{c_0}\right)^{0.8} - 2\left(\frac{c}{c_0}\right)^{0.4} \tag{3.93}$$

where c is the local sediment concentration (by volume), and c_0 is the maximum sediment concentration ($= 0.65$).

3.5.2 Near-bed concentration of suspended load

Empirical formulas were established by Engelund and Fredsøe (1976), Smith and McLean (1977), van Rijn (1984b), Celik and Rodi (1988), Zyserman and Fredsøe (1994), and Cao (1999) for the near-bed concentration of single-sized suspended load, and by Einstein (1950), Garcia and Parker (1991), and Hu and Wang (1999) for the near-bed fractional concentration of multi-sized (non-uniform) suspended load. The Einstein, van Rijn, and Zyserman-Fredsøe formulas are introduced below as examples.

Einstein formula

Einstein (1950) set the reference level of suspended-load concentration at two grain diameters above the channel bed and related the near-bed concentration of suspended load to the bed-load transport rate q_{b*k} as follows:

$$c_{b*k} = \frac{1}{11.6} \frac{q_{b*k}}{\delta U'_*} \tag{3.94}$$

where c_{b*k} is the concentration of the kth size class of suspended load at the reference level δ (by weight per unit volume), and U'_* is the skin friction velocity.

Van Rijn formula

Van Rijn (1984b) set the reference level δ at the equivalent roughness height k_s or half the bed-form height and established

$$c_{b*} = 0.015\frac{d_{50}T^{1.5}}{\delta D_*^{0.3}} \tag{3.95}$$

where c_{b*} is the volumetric concentration of suspended load at the reference level, and T and D_* are defined in Eqs. (3.57) and (3.70).

Zyserman-Fredsøe formula

Zyserman and Fredsøe (1994) set the reference level at two grain diameters above the bed and determined the near-bed volumetric concentration of suspended load as

$$c_{b*} = \frac{0.331(\Theta' - 0.045)^{1.75}}{1 + 0.72(\Theta' - 0.045)^{1.75}} \tag{3.96}$$

where $\Theta' = U_*'^2/[(\rho_s/\rho - 1)gd]$.

Two issues regarding the aforementioned formulas of near-bed suspended-load concentration should be pointed out. One is that the suspended-load concentration near the channel bed is very difficult to measure at the present time and has to be extrapolated from those measured in the upper flow layer with the aid of an assumed vertical distribution of sediment concentration. The accuracy and reliability of this analysis highly depend on the used distribution function of sediment concentration near the bed. The often used Rouse distribution is not reliable near the bed, and some later modifications introduced in Section 3.5.1 do not improve much indeed. Therefore, the calibration of these formulas using direct measurement data near the bed should be carried out in the future.

The other issue is that the near-bed concentration is defined at different reference levels in different formulas. Each formula should be applied only at the height where the near-bed concentration is defined. This makes comparison of these formulas very difficult. For sediment transport modeling, it is more convenient to set the reference level at the interface between the bed-load and suspended-load layers.

3.5.3 Suspended-load transport rate

Einstein's method

Einstein's (1950) method determines the suspended-load transport rate by integrating the product of local sediment concentration c_k and flow velocity u over the suspended-load zone from $\delta(= 2d)$ to h:

$$q_{s*k} = \int_{\delta}^{h} c_k u\, dz \tag{3.97}$$

where q_{s*k} is the transport rate of the kth size class of suspended load.

Using the Rouse distribution of sediment concentration ($\sigma_s = 1$) and the logarithmic distribution of flow velocity in Eq. (3.30) (replacing U_* by U_*') yields

$$
\begin{aligned}
q_{s*k} &= \int_{\delta}^{h} c_{b*k} \left(\frac{h/z - 1}{h/\delta - 1}\right)^{\frac{\omega_{sk}}{\kappa U_*}} * 5.75 U_*' \log\left(30.2\frac{z}{\Delta_s}\right) dz \\
&= 11.6 U_*' c_{b*k} \delta \left[2.303 \log\left(\frac{30.2h}{\Delta_s}\right) * I_{1k} + I_{2k}\right]
\end{aligned} \tag{3.98}
$$

where $I_{1k} = 0.216\frac{\zeta_b^{r_k-1}}{(1-\zeta_b)^{r_k}} \int_{\zeta_b}^{1} (\frac{1-\zeta}{\zeta})^{r_k} d\zeta$, and $I_{2k} = 0.216\frac{\zeta_b^{r_k-1}}{(1-\zeta_b)^{r_k}} \int_{\zeta_b}^{1} (\frac{1-\zeta}{\zeta})^{r_k} \ln \zeta\, d\zeta$, with $\zeta = z/h$, $\zeta_b = \delta/h$, $r_k = \omega_{sk}/(\kappa U_*)$, and Δ_s is defined in Eq. (3.71).

Inserting Eq. (3.94) into Eq. (3.98) leads to

$$q_{s*k} = q_{b*k} \left[2.303 \log \left(\frac{30.2h}{\Delta_s} \right) * I_{1k} + I_{2k} \right] \qquad (3.99)$$

The fractional transport rate of bed-material load is then obtained by $q_{t*k} = q_{b*k} + q_{s*k}$.

Einstein's method is an important contribution to sediment research. However, it is laborious because numerical integrations are involved and U'_* is determined using his movable bed roughness formula that needs to be solved iteratively. Many tests have shown that Einstein's method can provide reasonable results for narrowly graded sediment mixtures, but not for those widely graded (Misri *et al.*, 1984; Samaga *et al.*, 1986a&b). Modifications were proposed by several investigators, such as Colby and Hembree (1955), Toffaleti (1968), and Shen and Hung (1983). Van Rijn (1984b) also established a similar method to calculate the suspended-load transport rate using Eq. (3.97) with his distribution function (3.90) and near-bed concentration formula (3.95).

Bagnold formula

Based on his stream power concept, Bagnold (1966) established the following formula to calculate the suspended-load transport rate:

$$q_{s*} = 0.01 \frac{\rho_s}{\rho_s - \rho} \frac{\tau_b U^2}{\omega_s} \qquad (3.100)$$

where q_{s*} is the suspended-load transport rate by weight per unit time and width $(N \cdot m^{-1}s^{-1})$.

Zhang formula

Based on the energy balance of sediment-laden flow, Zhang (1961; also see Zhang and Xie, 1993) derived the relation between suspended-load transport capacity C_* and parameter $U^3/(gR\omega_s)$, as shown in Fig. 3.21, using measured data from the Yangtze River, the Yellow River, etc. Here, C_* is the average suspended-load concentration $(kg \cdot m^{-3})$. One may write the Zhang formula as Eq. (2.140) with variable coefficients K_* and m. For convenience, Guo (2002) approximated the $C_* \sim U^3/(gR\omega_s)$ curve in Fig. 3.21 by the following equation:

$$C_* = \frac{1}{20} \left(\frac{U^3}{gR\omega_s} \right)^{1.5} \bigg/ \left[1 + \left(\frac{1}{45} \frac{U^3}{gR\omega_s} \right)^{1.15} \right] \qquad (3.101)$$

Wu and Li (1992) extended the Zhang formula to determine the fractional concentration of non-uniform suspended load as $C_{*k} = p_{bk}C_k^*$. Here, p_{bk} is the bed-material gradation, and C_k^* is the potential equilibrium concentration of size class k determined

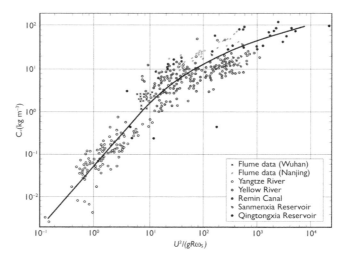

Figure 3.21 Relation of C_* and $U^3/(gR\omega_s)$ (Zhang, 1961).

by the $C_k^* \sim U^3/(gR\omega_{sk})$ curve calibrated using multiple-sized sediment data. However, Wu and Li's (1992) method does not explicitly consider the hiding and exposure effect among non-uniform sediment particles.

Wu et al. formula

Based on Bagnold's (1966) stream power concept, Wu *et al.* (2000b) related the suspended-load transport rate to the rate of energy available in the alluvial system and to the resistance to sediment suspension. The former was expressed as τU, and the latter was accounted for by the settling velocity ω_s and the critical shear stress τ_c. Here, τ is the shear stress on the wetted perimeter of the cross-section: $\tau = \gamma R S_f$. Through dimensional analysis, the independent parameter $(\tau/\tau_c - 1)U/\omega_s$ was derived. By using the laboratory data of non-uniform suspended load measured by Samaga *et al.* (1986b) and two sets of field data in the Yampa River and the Yellow River, the relation between the fractional suspended-load transport rate q_{s*k} and the parameter $(\tau/\tau_{ck} - 1)U/\omega_{sk}$ was established. It is shown in Fig. 3.22 and expressed as

$$\Phi_{sk} = 0.0000262 \left[\left(\frac{\tau}{\tau_{ck}} - 1 \right) \frac{U}{\omega_{sk}} \right]^{1.74} \qquad (3.102)$$

where $\Phi_{sk} = q_{s*k}/[p_{bk}\sqrt{(\gamma_s/\gamma - 1)gd_k^3}]$, with q_{s*k} being the suspended-load transport rate by volume per unit time and width (m^2s^{-1}); and τ_{ck} is determined using Eq. (3.45), which takes into account the hiding and exposure effect in non-uniform sediment transport. The sediment settling velocity ω_{sk} is calculated using the Zhang formula (3.12).

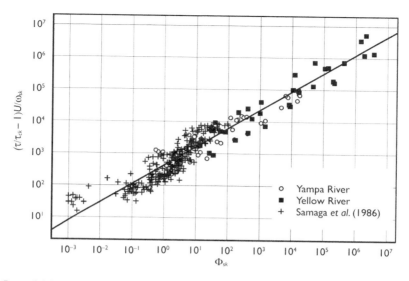

Figure 3.22 Relation of fractional suspended-load transport rate (Wu *et al.*, 2000b).

3.6 BED-MATERIAL LOAD TRANSPORT

Bed-material load is the sum of bed load and suspended load. Therefore, one may either separately calculate the bed-load and suspended-load transport rates, or directly calculate the bed-material load transport rate. Examples of the former approach are the methods of Einstein (1950), van Rijn (1984a & b), and Wu *et al.* (2000b), which are introduced in Sections 3.4 and 3.5. Examples of the latter approach are Laursen's (1958), Engelund and Hansen's (1967), Ackers and White's (1973), Yang's (1973), and Karim's (1998) methods, which are introduced below.

3.6.1 Total transport rate of bed-material load

Laursen formula

The Laursen (1958) formula divides a sediment mixture into size classes and calculates the total average concentration of bed-material load as

$$C_{t*} = 0.01\gamma \sum_{k=1}^{N} p_k \left(\frac{d_k}{h}\right)^{7/6} \left(\frac{\tau_b'}{\tau_{ck}} - 1\right) f\left(\frac{U_*}{\omega_{sk}}\right) \qquad (3.103)$$

where C_{t*} is the sediment concentration by weight per unit volume; p_k is the fraction of the kth size class of available sediment material; N is the total number of size classes; τ_{ck} is the critical shear stress for the incipient motion of sediment size d_k, given by the Shields diagram; and τ_b' is the bed shear stress due to grain roughness, determined using the Manning-Strickler equation:

$$\tau_b' = \frac{\rho U^2}{58} \left(\frac{d_{50}}{h}\right)^{1/3} \tag{3.104}$$

The function $f(U_*/\omega_s)$ in Eq. (3.103) is given as two different curves for bed load and bed-material load, as shown in Fig. 3.23. Therefore, the Laursen formula can be used to determine either bed load or bed-material load.

Note that the Laursen formula can provide the sediment concentration for each size class, but it is generally used to determine only the total sediment concentration. It has good reliability for fine sediments.

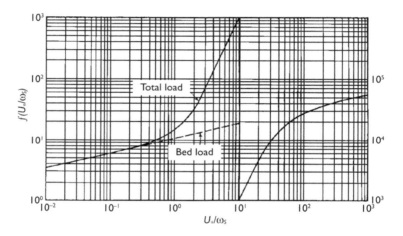

Figure 3.23 Function $f(U_*/\omega_s)$ in the Laursen (1958) formula.

Engelund-Hansen formula

Using Bagnold's stream power concept and the similarity principle, Engelund and Hansen (1967) established the following sediment transport formula:

$$f'\Phi_t = 0.1\Theta^{5/2} \tag{3.105}$$

where f' is the friction factor, defined as $f' = 2gRS_f/U^2$; $\Phi_t = q_{t*}/[\gamma_s\sqrt{(\gamma_s/\gamma - 1)gd^3}]$, with q_{t*} being the bed-material load transport rate by weight per unit time and width; Θ is the Shields number $\tau_b/[(\gamma_s - \gamma)d]$; and d is the median fall diameter of bed material.

Strictly speaking, the Engelund-Hansen formula should be applied to dune-bed streams in accordance with the similarity principle. However, many tests have shown that it can be applied to the upper flow regime with particle size greater than 0.15 mm (Chang, 1988).

Yang formula

Yang (1973, 1984) related the bed-material load transport to the unit stream power as follows:

$$\log C_{t*} = M + N \log \left(\frac{US_f}{\omega_s} - \frac{U_c S_f}{\omega_s} \right) \tag{3.106}$$

where C_{t*} is the sediment concentration in parts per million (ppm) by weight, U_c is determined using Eq. (3.29), and M and N are coefficients. For sand ($d \leq 2$ mm)

$$M = 5.435 - 0.286 \log \frac{\omega_s d}{\nu} - 0.457 \log \frac{U_*}{\omega_s}$$

$$N = 1.799 - 0.409 \log \frac{\omega_s d}{\nu} - 0.314 \log \frac{U_*}{\omega_s} \tag{3.107}$$

and for gravel (2 mm $< d <$ 10 mm)

$$M = 6.681 - 0.633 \log \frac{\omega_s d}{\nu} - 4.816 \log \frac{U_*}{\omega_s}$$

$$N = 2.784 - 0.305 \log \frac{\omega_s d}{\nu} - 0.282 \log \frac{U_*}{\omega_s} \tag{3.108}$$

Ackers-White formula

The transport of coarse sediments, which are mainly in bed load, is attributed to the stream power corresponding to the grain shear stress, $\tau_b' U$, while the transport of fine sediments, which are mainly in suspended load, is related to the turbulence intensity and in turn the total stream power, $\tau_b U$. Based on this concept, Ackers and White (1973) proposed a mobility factor of sediment transport:

$$F_{gr} = \frac{U_*^n}{[(\gamma_s/\gamma - 1)gd]^{1/2}} \left[\frac{U}{\sqrt{32} \log(10h/d)} \right]^{1-n} \tag{3.109}$$

and related the bed-material load to this mobility factor as follows:

$$G_{gr} = \frac{C_{t*}h}{d\gamma_s/\gamma} \left(\frac{U_*}{U} \right)^n = \Lambda \left(\frac{F_{gr}}{A_c} - 1 \right)^m \tag{3.110}$$

where C_{t*} is the sediment concentration by weight, Λ is an empirical coefficient, m is an empirical exponent, n is the transition exponent, and A_c may be interpreted as the critical value of F_{gr} for sediment incipient motion. Coefficients Λ, A_c, m, and n were related to the dimensionless grain diameter $D_* = d[(\rho_s/\rho - 1)g/\nu^2]^{1/3}$, as listed in Table 3.5, based on best-fit curves of laboratory data with sediment sizes greater than 0.04 mm and Froude numbers less than 0.8.

Table 3.5 Coefficients of the Ackers-White formula

$D_* \geq 60$	$1 < D_* < 60$
$n = 0.0$	$n = 1.00 - 0.56 \log D_*$
$A_c = 0.17$	$A_c = 0.23D_*^{-1/2} + 0.14$
$m = 1.50$	$m = 9.66D_*^{-1} + 1.34$
$\Lambda = 0.025$	$\log \Lambda = -3.53 + 2.86 \log D_* - (\log D_*)^2$

Many tests have shown that the Ackers-White formula overpredicts the transport rate for fine sediments (smaller than 0.2 mm).

3.6.2 Fractional transport rate of bed-material load

Modified Ackers-White formula (Day, 1980; Proffitt and Sutherland, 1983)

Day (1980) and Proffitt and Sutherland (1983) extended the Ackers-White (1973) formula to calculate the fractional bed-material load transport rate:

$$G_{gr,k} = \Lambda \left(\frac{F_{gr,k}}{A_c} - 1 \right)^m \tag{3.111}$$

where

$$F_{gr,k} = \eta_k \frac{U_*^n}{[(\gamma_s/\gamma - 1)gd_k]^{1/2}} \left[\frac{U}{\sqrt{32} \log(10h/d_k)} \right]^{1-n}, \quad G_{gr,k} = \frac{C_{t*k}h}{p_{bk}d_k\gamma_s/\gamma} \left(\frac{U_*}{U} \right)^n$$

with C_{t*k} being the sediment concentration by weight of size class k, and η_k the hiding and exposure correction factor. Day's correction factor is

$$\eta_k = \frac{1}{0.4(d_k/d_A)^{-0.5} + 0.6} \tag{3.112}$$

where d_A is the reference diameter, determined by

$$\frac{d_A}{d_{50}} = 1.6 \left(\frac{d_{84}}{d_{16}} \right)^{-0.28} \tag{3.113}$$

Proffitt and Sutherland's correction factor reads

$$\eta_k = \begin{cases} 0.40, & d_k/d_u \leq 0.075 \\ 0.53 \log(d_k/d_u) + 1.0, & 0.075 < d_k/d_u \leq 3.7 \\ 1.30, & d_k/d_u > 3.7 \end{cases} \tag{3.114}$$

where d_u is the reference diameter used by Proffitt and Sutherland (1983).

SEDTRA module (Garbrecht et al., 1995)

The SEDTRA module (Garbrecht *et al.*, 1995) calculates the fractional sediment transport rates using three established transport formulas: the Laursen (1958) formula for size classes from 0.01 to 0.25 mm, the Yang (1973) formula for size classes from 0.25 to 2.0 mm, and the Meyer-Peter-Mueller (1948) formula for size classes from 2.0 to 50.0 mm. The total concentration of sediment C_{t*} is calculated by

$$C_{t*} = \sum_k p_k C_{t*k} \tag{3.115}$$

where C_{t*k} is the sediment concentration of size class k; and p_k is the fraction of the kth size class of available sediment, usually set as the bed-material gradation.

In order to account for the hiding and exposure effect in non-uniform bed material, the sediment size d_{ek}, used to calculate the critical flow strength for the incipient motion of each size class, is adjusted using the following equation (Kuhnle, 1993; Wilcock, 1993; Garbrecht *et al.*, 1995):

$$d_{ek} = d_k \left(\frac{d_k}{d_m} \right)^{-x} \tag{3.116}$$

where d_m is the mean diameter of bed material; and x is an empirical parameter, determined by $x = 1.7/B_m$, with B_m being a bimodality parameter (Wilcock, 1993):

$$B_m = \left(\frac{d_c}{d_f} \right)^{1/2} \sum p_m \tag{3.117}$$

where d_c and d_f are the representative diameters of coarse and fine modes, respectively; and p_m is the portion of the sediment mixture contained in the two modes.

When B_m is less than 1.7, $x = 1$, and for high values of B_m, x approaches zero. Table 3.6 lists the values of x recommended by Kuhnle *et al.* (1996). The mixture names for Wilcock and Southard's (1988) data refer to the standard deviation of bed material, and those for Kuhnle's (1993) data refer to the percentage of gravel in bed material, e.g., SG25 for the mixture with 25% gravel and 75% sand.

The SEDTRA module takes the advantages of the three formulas used and thus performs well in general; however, these formulas may not transit smoothly in the

Table 3.6 Values of x recommended by Kuhnle et al. (1996)

Mixture name	Reference	d_m (mm)	Mixture type	B_m	x
SG10 (lab.)	Kuhnle (1993)	0.616	Bimodal	2.49	0.7
SG25 (lab.)	Kuhnle (1993)	0.927	Bimodal	2.60	0.7
SG45 (lab.)	Kuhnle (1993)	1.454	Bimodal	2.73	0.6
1/2 ψ (lab.)	Wilcock & S. (1988)	1.82	Unimodal	0.67	1.0
ψ (lab.)	Wilcock & S. (1988)	1.85	Unimodal	0.37	1.0
Goodwin Creek	Kuhnle (1993)	1.189	Bimodal	3.10	0.5

case of low sediment transport, because they adopt different criteria for incipient motion.

Karim formula

Karim (1998) related the availability of sediment to the areal fraction of bed material, and established the following formula for the fractional transport rate of bed-material load:

$$\frac{q_{t*k}}{\sqrt{(\gamma_s/\gamma - 1)gd_k^3}} = 0.00139 \left[\frac{U}{\sqrt{(\gamma_s/\gamma - 1)gd_k}} \right]^{2.97} \left(\frac{U_*}{\omega_{sk}} \right)^{1.47} p_{ak}\eta_k$$

(3.118)

where q_{t*k} is in m^2s^{-1}; p_{ak} is the areal fraction of bed material, related to the volumetric fraction of bed material, p_{bk}, by

$$p_{ak} = \frac{p_{bk}}{d_k} \Bigg/ \left(\sum_{k=1}^{N} \frac{p_{bk}}{d_k} \right)$$

(3.119)

and η_k is the hiding and exposure correction factor:

$$\eta_k = C_1 \left(\frac{d_k}{d_{50}} \right)^{C_2}$$

(3.120)

where d_{50} is the median size of bed material; and C_1 and C_2 are coefficients:

$$C_1 = 1.15\omega_{s50}/U_*$$

(3.121)

$$C_2 = 0.60\omega_{s50}/U_*$$

(3.122)

with ω_{s50} being the settling velocity for d_{50}.

Eqs. (3.121) and (3.122) show that C_1 and C_2 increase as ω_{s50}/U_* increases. This suggests that the coarser the sediment mixture, the stronger the hiding and exposure effect. This is physically reasonable. However, the correction factor in Eq. (3.121) is not equal to 1 for uniform sediment, so that Eq. (3.118) may be significantly different from the original uniform sediment transport formula.

3.6.3   Comparison of bed-material load formulas

Many investigators — e.g., Vanoni (1975), Alonso (1980), Brownlie (1981), Shen and Hung (1983), van Rijn (1984b), Nakato (1990), and Woo and Yoo (1991) — have compared the existing formulas for the total and fractional transport rates of bed-material load using extensive flume and field data. Several examples are briefly introduced below. More details can be found in relevant publications.

Alonso (1980) tested eight formulas, including those of Ackers and White (1973), Engelund and Hansen (1967), Laursen (1958), Yang (1973), Bagnold (1966), Meyer-Peter and Mueller (1948), Yalin (1972), and the combination of the Meyer-Peter-Mueller bed-load formula and the Einstein (1950) suspended-load formula (denoted as MPME), using 225 sets of flume data and 40 sets of field data. Alonso limited his comparisons against those field data where bed-material load could be measured by special facilities to avoid the uncertainty of unmeasured load. Table 3.7 shows the discrepancies of the selected formulas. The Yang (1973), Ackers-White (1973), Engelund-Hansen (1967), and Laursen (1958) formulas are more reliable.

Table 3.7 Comparison of sediment transport formulas (Alonso, 1980)

Formula	Ratio of predicted and measured discharges								
	Flume data with $h/d \geq 70$ (177 sets)			Flume data with $h/d < 70$ (48 sets)			Field data (40 sets)		
	Mean	σ	% in 0.5–2	Mean	σ	% in 0.5–2	Mean	σ	% in 0.5–2
Ackers-White	1.34	1.29	73.0	1.12	0.52	89.6	1.27	0.68	87.8
Engelund-H.	0.73	0.68	51.1	0.75	0.50	66.7	1.46	0.56	82.9
Laursen	0.81	0.51	71.4	1.04	0.99	79.2	0.65	0.48	56.1
MPME	3.11	2.75	42.1	1.34	1.04	66.7	0.83	1.02	58.5
Yang	0.99	0.60	79.8	0.90	0.51	85.4	1.01	0.39	92.7
Bagnold	0.85	2.50	20.8	1.53	1.14	45.8	0.39	0.26	32.0
Meyer-Peter-M.	0.40	0.49	18.5	1.03	0.83	72.9	0.24	0.09	0
Yalin	1.62	4.08	32.6	1.92	1.65	64.6	2.59	1.62	46.3

Note: σ = standard deviation, and % in 0.5–2 means percentage of data in error range of 0.5–2.

Brownlie (1981) compared fourteen formulas. The discrepancies resulting from these formulas are shown in Fig. 3.24. The median and 16th and 84th percentile values in the figure are based on the approximation of a log-normal distribution of errors. The Brownlie (1981), Ackers-White (1973), and Engelund-Hansen (1967) formulas provide good results for the data sets used in the comparison.

Woo and Yoo (1991) tested ten sediment transport formulas using the data carefully selected from Brownlie's (1981) compilation. Fig. 3.25 presents the discrepancy ratios of the calculated and measured sediment discharges. The Engelund-Hansen (1967), Ackers-White (1973), and van Rijn (1984a & b) formulas are more reliable than other compared formulas.

The author compared the Engelund-Hansen (1967), Ackers-White (1973), Yang (1973, 1984), and Wu et al. (2000b) formulas as well as the SEDTRA module (Garbrecht et al., 1995) against 1,859 sets of uniform bed-material load data selected from Brownlie's (1981) compilation by limiting the standard deviation of bed material $\sigma < 1.2$ and the Shields number $\Theta > 0.055$. These data cover flow discharges of 0.00094–297 m^{3}s^{-1}, flow depths of 0.01–2.56 m, flow velocities of 0.086–2.88 m·s^{-1}, surface slopes of 0.0000735–0.0367, and sediment sizes of 0.088–28.7 mm. None of them was used to calibrate the Wu et al. formulas (3.80) and (3.102). The discrepancies between the calculated and measured bed-material transport rates are listed in Table 3.8. All these five formulas have comparable reliability for uniform bed-material load transport rate.

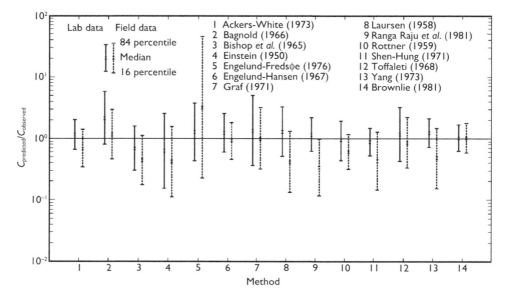

Figure 3.24 Comparison of sediment transport formulas (Brownlie, 1981).

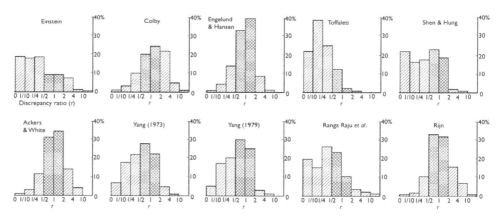

Figure 3.25 Comparison of sediment transport formulas (Woo and Yoo, 1991).

The author tested the Wu *et al.* (2000b) formula, the modified Ackers-White formula (Proffitt and Sutherland, 1983), the modified Zhang formula (Wu and Li, 1992), the Karim (1998) formula, and the SEDTRA module for fractional bed-material load. Because the modified Zhang formula is only for the fractional discharge of suspended load, it was combined with Eq. (3.80) to obtain the fractional discharge of bed-material load. The non-uniform sediment data collected by Toffaletti (1968) were used, including experimental data observed by Nomicos, Einstein-Chien, and Vanoni-Brooks, and field data in the Rio Grande, Middle Loup, Niobrara, and Mississippi Rivers. In order

Table 3.8 Calculated versus measured transport rates of uniform bed-material load for Brownlie's data (Wu and Wang, 2003)

Error range	% of calculated transport rates in error range				
	Ackers- White	Yang	Engelund-Hansen	SEDTRA	Wu et al.
$0.8 \leq r \leq 1.25$	37.3	33.4	33.6	36.6	40.4
$0.667 \leq r \leq 1.5$	57.9	56.6	55.4	59.1	62.7
$0.5 \leq r \leq 2$	82.4	76.6	77.0	78.1	81.3

Note: r = calculation/measurement.

Table 3.9 Comparison of fractional bed-material load formulas

Error range	% of calculated transport rates in error range				
	Modified Ackers-W.	Karim	Modified Zhang	SEDTRA	Wu et al.
$0.5 \leq r \leq 2$	5.6	42.7	48.1	56.9	57.9
$0.33 \leq r \leq 3$	11.1	63.5	67.9	73.1	76.1
$0.25 \leq r \leq 4$	20.8	73.3	80.7	80.9	85.2

to avoid the deficiency in the measurement of suspended load close to river bed, the used field data were selected by limiting the height of the lowest measurement point to be within 0.2 m (0.4 m in some of the Mississippi River data) above the bed. These data cover flow discharges up to 21,600 m³s⁻¹, flow depths up to 17.5 m, and sediment sizes from 0.062 to 1 mm. Table 3.9 lists the discrepancies between the calculated and measured fractional discharges of non-uniform bed-material load. The Wu *et al.* formula and SEDTRA module perform better.

As shown in Tables 3.8 and 3.9, multi-fraction sediment transport formulas usually have larger discrepancies than single-fraction formulas. This is because interactions exist among different size classes in non-uniform bed materials, and it is difficult to ensure all size classes at equilibrium states during measurements. Due to the fact that the possible discrepancy of any existing sediment transport formula may exceed two or three folds, verification using the data measured at the study site or similar sites prior to application is recommended.

3.7 SEDIMENT TRANSPORT OVER STEEP SLOPES

Because channel slopes in most natural rivers are very gentle, the effect of gravity on sediment transport is usually ignored. However, this effect is significant and should be considered if longitudinal and/or transverse slopes are steep. Several methods considering this effect are described below.

Nakagawa et al. formula

Nakagawa *et al.* (1986) established a formula to determine the pickup rate of sediment particles on a steep side slope:

$$p_s\sqrt{\frac{d}{(\rho_s/\rho - 1)g}} = 0.03G_*\Theta\left(1 - \frac{0.7\Omega\Theta_c}{\Theta}\right)^3 \tag{3.123}$$

where p_s is the pickup rate, defined as the probability density per unit time for a sediment particle to be dislodged from the bed; and G_* and Ω are coefficients:

$$G_* = \frac{\sin(\beta_d + \delta_d) + k_L\mu_s}{1 + k_L\mu_s} \tag{3.124}$$

$$\Omega = \frac{\mu_s\cos\theta_n - \sin\theta_n\cos\beta_d}{\mu_s G_*} \tag{3.125}$$

where μ_s is the static friction factor, with a value of about 0.7; k_L is the drag and lift force ratio, which is about 0.85; θ_n is the transverse slope angle; δ_d represents the deflection angle of the flow velocity vector from the longitudinal direction; and β_d is the angle of the sediment movement (resultant force) direction measured from the p-axis defined along the wetted perimeter.

Damgaard et al. formula

Damgaard *et al.* (1997) modified the Meyer-Peter-Mueller (1948) bed-load formula (3.65) to consider the effect of gravity in longitudinal slopes. The modified formula is written as

$$\Phi_b = 8(\Theta - \Theta_{c\varphi})^{3/2}f_{slope} \tag{3.126}$$

where $\Phi_b = q_{b*}/[\gamma_s\sqrt{(\gamma_s/\gamma - 1)gd_{50}^3}]$, Θ is the Shields number $\tau_b/[(\gamma_s - \gamma)d_{50}]$, and $\Theta_{c\varphi}$ is the critical Shields number on sloped beds determined by

$$\frac{\Theta_{c\varphi}}{\Theta_c} = \frac{\sin(\phi_r - \varphi)}{\sin\phi_r} \tag{3.127}$$

where ϕ_r is the repose angle; φ is the bed slope angle, with positive values for downslope beds; and Θ_c is the critical Shields number on the horizontal bed, calculated using the following algebraic representation of the Shields curve suggested by Soulsby (1996):

$$\Theta_c = \frac{0.24}{D_*} + 0.055(1 - e^{-D_*/50}) \tag{3.128}$$

with $D_* = d_{50}[(\rho_s/\rho - 1)g/\nu^2]^{1/3}$.

The parameter f_{slope} is a correction factor, determined by

$$f_{slope} = \begin{cases} 1 & -\phi_r < \varphi \leq 0 \\ 1 + 0.8(\Theta_c/\Theta)^{0.2}(1 - \Theta_{c\varphi}/\Theta_c)^{1.5+\Theta/\Theta_c} & 0 < \varphi < \phi_r \end{cases} \tag{3.129}$$

Wu's method

Many sediment transport formulas, such as those of van Rijn (1984a & b) and Wu *et al.* (2000b), can be expressed as $q_b = f(\tau_b'/\tau_c)$ or $f(\tau_b/\tau_c)$. Two approaches may be used to consider the effect of bed slope in this group of formulas. One is to correct the critical shear stress τ_c using the method of Brooks (1963) or van Rijn (1989). A disadvantage of this approach is that when the bed slope angle is close to the repose angle, the corrected critical shear stress usually goes to zero and, thus, the calculated sediment transport rate perhaps tends to be infinite. This situation should be limited. The other approach is to add the streamwise component of gravity to the grain shear stress τ_b' or the bed shear stress τ_b without modifying τ_c so that the situation of zero critical shear stress can be avoided. The effective tractive force τ_{be} (Wu, 2004) is thus determined by

$$\tau_{be} = \tau_b' + \lambda_s \frac{a\pi}{6}(\rho_s - \rho)gd\sin\varphi \tag{3.130}$$

where a is a coefficient related to the shape and location of sediment particles, and λ_s is a friction factor. Note that τ_b' may be replaced by τ_b in Eq. (3.130), depending on the formula considered.

Because the friction factor λ_s is difficult to determine, Eq. (3.130) is not ready for use. In the case where the bed slope angle φ is equal to the repose angle ϕ_r, sediment particles will start moving ($\tau_{be} = \tau_c$) even without any hydraulic action ($\tau_b' = 0$). Using this condition, one can derive $\lambda_s = \tau_c/[\frac{a\pi}{6}(\rho_s - \rho)gd\sin\phi_r]$. Inserting this relation into Eq. (3.130) yields

$$\tau_{be} = \tau_b' + \tau_c\sin\varphi/\sin\phi_r \tag{3.131}$$

The coefficients λ_s and a are replaced by the critical shear stress τ_c and the repose angle ϕ_r, which are easier to evaluate. However, the test performed by Wu (2004) using the experimental data of Damgaard *et al.* (1997) shows that Eq. (3.131) is adequate for negative (up) slopes, but for positive (down) slopes the following modification is needed:

$$\tau_{be} = \tau_b' + \lambda_0\tau_c\sin\varphi/\sin\phi_r \tag{3.132}$$

where λ_0 is a coefficient. λ_0 may consider the difference in τ_c on horizontal, upslope, and downslope beds; it is related to flow and sediment conditions as well as bed slope. When the above correction is applied to the Wu *et al.* (2000b) bed-load and suspended-load transport formulas (3.80) and (3.102), λ_0 has the following form (Wu, 2004):

$$\lambda_0 = \begin{cases} 1 & \varphi \leq 0 \\ 1 + 0.22(\tau_b'/\tau_c)^{0.15}e^{2\sin\varphi/\sin\phi_r} & \varphi > 0 \end{cases} \tag{3.133}$$

3.8   TEMPORAL LAGS BETWEEN FLOW AND SEDIMENT TRANSPORT

Sediment transport exhibits temporal lags with flow due to flow and sediment velocity difference and bed form development. In particular, such lags become significant for coarse sediment transport under strongly unsteady flow conditions (e.g., Bell, 1980; Tsujimoto *et al.*, 1988; Phillips and Sutherland, 1990; Song and Graf, 1997; de Sutter *et al.*, 2001; Wu *et al.*, 2006).

Lag between flow and suspended-load transport

There is a lag between the local flow and suspended-load velocities. This has been observed experimentally by Muste and Patel (1997) and discussed in detail by Cheng (2004). A two-phase flow model (Wu and Wang, 2000; Greimann and Holly, 2001) can be used to describe this local velocity lag in general situations. However, according to the experimental observations of Muste and Patel (1997), the local streamwise velocity of suspended load with a diameter of 0.23 mm is less than the local flow velocity by as much as 4%; this local velocity difference is negligible in comparison with the depth-averaged flow and suspended-load velocity difference (Wu *et al.*, 2006). Thus, the local velocity lag may be ignored, and only the depth-averaged velocity lag is discussed below.

The concentration-weighted velocity of suspended load can be defined as

$$U_{sed} = \int_\delta^h u_s c\, dz \bigg/ \int_\delta^h c\, dz \qquad (3.134)$$

which is actually the overall velocity of suspended load from a depth-averaging point of view. Therefore, the correction factor β_s defined in Eq. (2.87) also is the ratio of the depth-averaged suspended-load and flow velocities:

$$\beta_s = U_{sed}/U \qquad (3.135)$$

where U is the depth-averaged flow velocity.

Because higher sediment concentration corresponds to smaller flow velocity near the channel bottom while lower sediment concentration corresponds to larger flow velocity in the upper flow layer, β_s normally is less than 1 and $U_{sed} \leq U$. By using the logarithmic distribution of flow velocity, $u = U\{1 + \sqrt{g}[1 + \ln(z/h)]/(C_h\kappa)\}$, and the Rouse distribution of suspended-load concentration introduced in Section 3.5.1 with the reference level set at $0.01h$, Wu *et al.* (2006) obtained the relation of β_s with the Rouse number $\omega_s/(\kappa U_*)$ and the Chezy coefficient C_h, as shown in Fig. 3.26. It can be seen that β_s decreases as the Rouse number increases and the Chezy coefficient decreases. For fine sediments, β_s is close to 1 and the lag between the depth-averaged flow and sediment velocities can be ignored. However, for coarse sediments, this lag can be up to 50% of the flow velocity and should be considered.

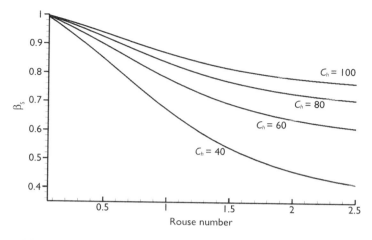

Figure 3.26 Factor β_s as function of the Rouse number and the Chezy coefficient (Wu *et al.*, 2006).

Lag between flow and bed-load transport

Bed load usually moves by rolling, sliding, and saltating, depending on flow and sediment conditions. Saltation is the dominant mode of bed-load transport, while rolling (and to a lesser extent, sliding) occurs only near the threshold of entrainment and between individual saltation jumps (Bridge and Dominic, 1984). Van Rijn (1984a) investigated the characteristics of particle saltation and determined the bed-load velocity as

$$\frac{u_b}{\sqrt{(\rho_s/\rho - 1)gd}} = 1.5T^{0.6} \tag{3.136}$$

where T is the transport stage number defined in Eq. (3.57).

The van Rijn formula (3.136) was verified by Wu *et al.* (2006) using three sets of experimental data measured by Francis (1973), Luque and van Beek (1976), and Lee and Hsu (1994). In the experiments of Francis (1973), the sediment used was 7.5 mm water-worn gravel, and multi-exposure photographs of grains were obtained to show the trajectories and then determine the bed-load velocity. The experiments of Luque and van Beek (1976) were performed in a closed rectangular channel at different surface slopes and using different bed materials. Photographs were taken at regular intervals to measure the mean rate of bed-load transport. The data from these experiments with 0.9 and 1.8 mm sands and 3.3 mm gravel were selected. Lee and Hsu (1994) measured the instantaneous saltation trajectories of sand particles with sizes of 1.36 and 2.47 mm in a slope-adjustable recirculating flume by a real-time flow visualization technique. These three groups of data were used to recalibrate the bed-load velocity formula of van Rijn (1984a). Fig. 3.27 shows the new curve, in which the transport stage parameter T is defined as $T = \tau_b/\tau_c - 1$, with τ_b being the total bed shear stress measured in the three experiments where no significant bed

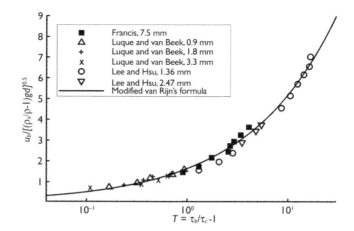

Figure 3.27 Bed-load velocity as function of transport stage parameter (Wu et al., 2006).

forms developed. The new curve can be expressed as Eq. (3.136) with $1.64T^{0.5}$ on the right-hand side.

Fig. 3.27 and Eq. (3.136) show that the lag between flow and bed-load velocities increases as sediment size increases.

Lag between flow and bed form development

It has been recognized that a temporal lag exists between flow and bed form development, but this lag has rarely been investigated experimentally and numerically. A simple empirical impulse response model was suggested by Phillips and Sutherland (1990) to quantify this lag. The interested reader may refer to it. More generally, the development of bed forms and, in turn, the associated temporal lag can be simulated using a vertical 2-D or 3-D model if the selected computational mesh is fine enough (much finer than the lengths and heights of the bed forms simulated). However, this kind of simulation requires a powerful computer and an advanced numerical model.

Chapter 4

Numerical methods

River engineering problems are usually governed by nonlinear differential equations in irregular and movable domains, most of which have to be solved using numerical methods. Introduced in this chapter are the discretization methods for 1-D, 2-D, and 3-D problems on fixed and moving grids, the solution strategies for the Navier-Stokes equations, and the solution methods of algebraic equations. Some of these can be found in Patankar (1980), Hirsch (1988), Fletcher (1991), Ferziger and Peric (1995), Shyy *et al.* (1996), etc.

4.1 CONCEPTS OF NUMERICAL SOLUTION

4.1.1 General procedure of numerical solution

Consider the problem in a domain of $a \leq x \leq b$ shown in Fig. 4.1, governed by a differential equation

$$L(f;x) = S \tag{4.1}$$

with boundary conditions

$$f|_{x=a} = f_a, \quad f|_{x=b} = f_b \tag{4.2}$$

where L is the differential operator, f is the function to be determined, x is the spatial coordinate, and S is the source term.

To acquire a numerical solution, the study domain is first represented by a finite number of points, denoted as $x_1, x_2, \ldots,$ and x_N, which constitute the computational grid (mesh). Here, $x_1 = a$ and $x_N = b$. The distance between two consecutive points, Δx, is the grid size or spacing.

Eq. (4.1) is discretized on the computational grid using a numerical method. A discrete equation L_d is then established to approximate the differential equation at each grid point:

$$L_d(\hat{f};x_i) = S_i \quad (i = 2, \ldots, N-1) \tag{4.3}$$

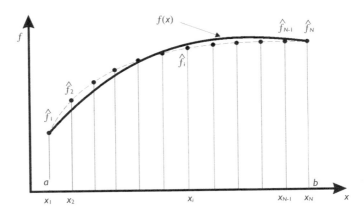

Figure 4.1 Example of numerical solution.

where $\hat{f}$ is the approximate solution of f, which is subject to boundary conditions:

$$\hat{f}_1 = f_a, \quad \hat{f}_N = f_b \qquad (4.4)$$

where $\hat{f}_1$ and $\hat{f}_N$ are the values of $\hat{f}$ at $x = x_1$ and x_N, respectively.

The system of algebraic equations consisting of discrete equations (4.3) and boundary conditions (4.4) is used to determine the approximate solution $(\hat{f}_1, \hat{f}_2, \ldots, \hat{f}_N)$ on the computational grid. A direct or iterative solution method may be adopted to solve the algebraic equations. The obtained approximate solution is a discrete function, which is shown as solid circles in Fig. 4.1.

The quality of the approximate solution usually relies on the computational grid used, the discretization method for the governing equation, and the solution method for the discretized equations.

4.1.2 Properties of numerical solution

The most important properties of numerical solution are accuracy, consistency, stability, and convergence. A brief overview of these terms is given below. Complete descriptions can be found in Hirsch (1988), Fletcher (1991), etc.

Accuracy

Numerical accuracy refers to how well a discretized equation approximates to the differential equation. Eq. (4.3) is said to have an accuracy of mth-order of Δx, if the residual (error) is proportional to Δx^m:

$$R_L = L(f; x_i) - L_d(\hat{f}; x_i) = O(\Delta x^m) \qquad (4.5)$$

The residual term on the right-hand side of Eq. (4.5) can be obtained with the Taylor series expansion method. However, it is usually difficult to judge the overall accuracy for a complex model system, because the governing equations and boundary conditions may be discretized using numerical schemes with different accuracies. An alternative method to analyze the accuracy is through the computation of solution errors on a series of meshes with grid spacings of $\Delta x, 2\Delta x, 3\Delta x$, etc. The root-mean-square error for the solution on each grid is defined as

$$R_f = \left\{ \left[\sum_{i=1}^{N} (f_i - \hat{f}_i)^2 \right] \Big/ N \right\}^{1/2} \tag{4.6}$$

The error R_f is related to the grid spacing Δx, as shown in Fig. 4.2. This relationship can be represented by

$$R_f = a\Delta x^m \tag{4.7}$$

where a is a nearly constant coefficient. The value of m can be determined from the series of R_f and Δx pair values using a regression method.

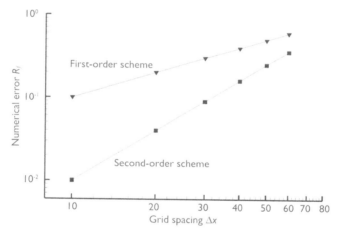

Figure 4.2 Relation between R_f and Δx.

Performing the above numerical accuracy analysis requires that the exact solution be known in advance. This is not feasible for most problems in river engineering. However, the prescribed solution forcing (PSF) method (Dee et al., 1992) can be used instead. The PSF method substitutes the unknown function f in Eq. (4.1) by a known function p. The new equation for p has the form:

$$L(p; x) = S^* \tag{4.8}$$

where S^* is the new source term, which might be different from S because p may not be the exact solution of Eq. (4.1). Note that it is preferable that the function p satisfies boundary conditions (4.2); however, for complex problems, this may be difficult, and one may use the values of p at boundaries as boundary conditions for solving Eq. (4.8).

Because Eqs. (4.1) and (4.8) have the same formulation except for the source terms, the same numerical method can be used to solve Eq. (4.8) and find the approximate solution of p. The above numerical accuracy analysis can then be conducted.

Consistency

The system of discretized equations is considered to be consistent with the original differential equation if it is equivalent to the differential equation at each grid point when the grid spacing reduces to zero.

The consistency analysis can be conducted by expanding all nodal values in the discretized equations as Taylor series about a single point. For consistency, the obtained expression should be made up of the original partial differential equation and a remainder, and the remainder should reduce to zero at each grid point as the grid spacing reduces to zero.

Stability

Numerical stability is concerned with growth or decay in errors introduced at any stage of the computational process. In practice, because of limited computer storage, an infinite decimal number is truncated to a finite number of significant figures, thereby introducing round-off errors. A numerical algorithm is said to be stable if the cumulative effect of the errors produced during its application is negligible.

The von Neumann and matrix methods are commonly used for stability analysis (see Hirsch, 1988; Fletcher, 1991). Both methods can predict whether there will be a growth in numerical errors including the round-off contamination between the true solution of the numerical algorithm and the actually computed solution.

Convergence

A solution of the discretized algebraic equations is said to be convergent if the approximate solution approaches the exact solution of the original differential equation for each dependent variable as the grid spacing reduces to zero. Thus, for problem (4.1), convergence requires $\hat{f} \to f$, as $\Delta x \to 0$.

Proving the convergence of a numerical algorithm is generally very difficult, even for the simplest cases. Nevertheless, for a restricted class of problems, convergence can be established via the Lax equivalent theorem, which was described as follows (Richtmyer and Morton, 1967; Fletcher, 1991):

> Given a properly posed linear initial value problem and a finite difference approximation to it that satisfies the consistency condition, stability is the necessary and sufficient condition for convergence.

The Lax equivalent theorem is very useful to show the convergence through the stability and consistency analyses, which are much easier. However, most of the problems in computational river dynamics are nonlinear, so the Lax equivalent theorem is not always applicable.

In general, the tendency of convergence can be tested by successively refining the computational grid and computing the root-mean-square error of the solution of each dependent variable by Eq. (4.6). If $R_f \to 0$ as $\Delta x \to 0$, the numerical solution is convergent. However, this test shows only the tendency rather than the ultimate convergence ($R_f = 0$), because the mesh cannot be infinitely refined and the round-off errors may increase as the number of grid points increases.

4.1.3 Discretization methods

Widely used discretization methods include finite difference method, finite element method, finite volume method, finite analytical method, and efficient element method. The finite difference method discretizes a differential equation by approximating differential operators with difference operators at each point. The finite analytical method discretizes the differential equation using the analytical solution of its locally linearizd form, and the efficient element method establishes difference operators using interpolation schemes in local elements. Because of their similarity, the finite analytical method and efficient element method are herein grouped with the finite difference method. The finite volume method integrates the differential equation over each control volume, holding the conservation laws of mass, momentum, and energy. In the finite element method, the differential equation is multiplied by a weight function and integrated over the entire domain, and then an approximate solution is constructed using shape functions and optimized by requiring the weighted integral to have a minimum residual.

The algebraic equations resulting from the finite difference and finite volume methods usually have banded and symmetric coefficient matrices that can be handled efficiently, whereas the algebraic equations from the finite element method often have sparse and asymmetric coefficient matrices that require relatively tedious effort for solution. However, the classic finite difference and finite volume methods adopt structured, regular meshes and encounter difficulties in conforming to the irregular domains of river flow, while the finite element method adopts unstructured, irregular meshes and can conveniently handle such irregular domains. Therefore, it has been a trend in recent decades to develop the finite difference and finite volume methods on irregular meshes, which have the grid flexibility of the finite element method and the computational efficiency of the classic finite difference and finite volume methods.

The finite difference method and finite volume method are introduced in this book. The finite element method has also been used in many river models because of its grid flexibility; however, it is absent from this book due to the author's limited expertise. Interested readers are encouraged to consult other references, such as Chung (1978), Fletcher (1991), and Zienkiewicz and Taylor (2000).

One suggestion to new model developers and users is that any numerical method may have its advantages and disadvantages, and subjectivity may prevent you from becoming more successful. You should learn the basic properties — such as accuracy, stability, convergence, and efficiency — of the method that you are going to use and know how to take advantage of its strengths and avoid its weaknesses.

4.2 FINITE DIFFERENCE METHOD

4.2.1 Finite difference method for 1-D problems

4.2.1.1 Taylor-series formulation of finite difference schemes

Fig. 4.3 shows the 1-D computational grid used in the finite difference method. The values of function f at $x = x_{i+1}$ and $x = x_{i-1}$ can be expanded as Taylor series about the point $x = x_i$:

$$f_{i+1} = f_i + \left(\frac{\partial f}{\partial x}\right)_i \Delta x + \frac{1}{2}\left(\frac{\partial^2 f}{\partial x^2}\right)_i \Delta x^2 + \frac{1}{6}\left(\frac{\partial^3 f}{\partial x^3}\right)_i \Delta x^3 + \cdots \qquad (4.9)$$

$$f_{i-1} = f_i - \left(\frac{\partial f}{\partial x}\right)_i \Delta x + \frac{1}{2}\left(\frac{\partial^2 f}{\partial x^2}\right)_i \Delta x^2 - \frac{1}{6}\left(\frac{\partial^3 f}{\partial x^3}\right)_i \Delta x^3 + \cdots \qquad (4.10)$$

where $\Delta x = x_{i+1} - x_i$ or $\Delta x = x_i - x_{i-1}$. Δx is assumed to be uniform on the entire computational grid for convenience in the following analyses.

Figure 4.3 1-D finite difference grid.

Ignoring the high-order terms in Eq. (4.9), the first derivative of function f can be approximated as

$$\left(\frac{\partial f}{\partial x}\right)_i \approx \frac{f_{i+1} - f_i}{\Delta x} \qquad (4.11)$$

Eq. (4.11) is called the forward difference scheme. Similarly, from Eq. (4.10), the backward difference scheme can be obtained as

$$\left(\frac{\partial f}{\partial x}\right)_i \approx \frac{f_i - f_{i-1}}{\Delta x} \qquad (4.12)$$

Subtracting Eqs. (4.9) and (4.10) yields the central difference scheme for the first derivative:

$$\left(\frac{\partial f}{\partial x}\right)_i \approx \frac{f_{i+1} - f_{i-1}}{2\Delta x} \qquad (4.13)$$

Summing Eqs. (4.9) and (4.10) yields the central difference scheme widely used for the second derivative:

$$\left(\frac{\partial^2 f}{\partial x^2}\right)_i \approx \frac{f_{i-1} - 2f_i + f_{i+1}}{\Delta x^2} \qquad (4.14)$$

The forward and backward difference schemes (4.11) and (4.12) are first-order accurate, whereas the central difference schemes (4.13) and (4.14) are second-order accurate. They are bases for many widely used difference schemes.

4.2.1.2 Discretization of 1-D steady problems

Numerical schemes often used in the discretization of 1-D steady problems include the central, upwind, and exponential difference schemes, which are introduced below.

Central and upwind difference schemes

Consider the 1-D steady convection-diffusion equation:

$$u\frac{df}{dx} = \varepsilon_c \frac{d^2 f}{dx^2} + S \tag{4.15}$$

where u is the velocity; and ε_c is the diffusion coefficient, which is positive.

Applying the central difference schemes (4.13) and (4.14) to the convection and diffusion terms in Eq. (4.15), respectively, yields

$$u\frac{f_{i+1} - f_{i-1}}{2\Delta x} = \varepsilon_c \frac{f_{i-1} - 2f_i + f_{i+1}}{\Delta x^2} + S_i \tag{4.16}$$

The central difference scheme (4.14) is adequate for discretizing the diffusion term. However, the use of the central difference scheme (4.13) for the convection term may result in numerical oscillations. Upwind difference schemes are usually preferred for the convection term. The first-order upwind scheme uses the backward or forward difference scheme for the convection term, depending on whether the velocity u is positive or negative, i.e.,

$$u\left(\frac{\partial f}{\partial x}\right)_i = \begin{cases} u\dfrac{f_i - f_{i-1}}{\Delta x} & (u \geq 0) \\[2mm] u\dfrac{f_{i+1} - f_i}{\Delta x} & (u < 0) \end{cases} \tag{4.17}$$

Applying the upwind scheme (4.17) to the convection term and the central difference scheme (4.14) to the diffusion term in Eq. (4.15) yields

$$\begin{cases} u\dfrac{f_i - f_{i-1}}{\Delta x} = \varepsilon_c \dfrac{f_{i-1} - 2f_i + f_{i+1}}{\Delta x^2} + S_i & (u \geq 0) \\[3mm] u\dfrac{f_{i+1} - f_i}{\Delta x} = \varepsilon_c \dfrac{f_{i-1} - 2f_i + f_{i+1}}{\Delta x^2} + S_i & (u < 0) \end{cases} \tag{4.18}$$

Exponential difference scheme

Assuming constant u, ε_c, and S in the segment $x_{i-1} \leq x \leq x_{i+1}$ and imposing boundary conditions $f = f_{i-1}$ at $x = x_{i-1}$ and $f = f_{i+1}$ at $x = x_{i+1}$, one obtains the following analytic solution of Eq. (4.15) in this segment:

$$\frac{f - f_{i-1} - (x - x_{i-1})S_i/u}{f_{i+1} - f_{i-1} - (x_{i+1} - x_{i-1})S_i/u} = \frac{\exp[(x - x_{i-1})u/\varepsilon_c] - 1}{\exp[(x_{i+1} - x_{i-1})u/\varepsilon_c] - 1} \tag{4.19}$$

Imposing $f = f_i$ at $x = x_i$ in Eq. (4.19) yields

$$\frac{f_i - f_{i-1} - S_i\Delta x/u}{f_{i+1} - f_{i-1} - 2S_i\Delta x/u} = \frac{\exp(-P/2)}{\exp(P/2) + \exp(-P/2)} \tag{4.20}$$

where P is the Peclet number, defined as $P = u\Delta x/\varepsilon_c$, which represents the relative importance of convection and diffusion effects.

Eq. (4.20) can be rewritten as

$$a_P f_i = a_W f_{i-1} + a_E f_{i+1} + S_i \tag{4.21}$$

where $a_W = \frac{u}{2\Delta x}\exp(P/2)/\sinh(P/2), a_E = \frac{u}{2\Delta x}\exp(-P/2)/\sinh(P/2)$, and $a_P = a_W + a_E$.

Eq. (4.21) is the exponential difference scheme. It was derived by Lu and Si (1990), who called it a finite analytic scheme. It is similar to Spalding's (1972) exponential scheme based on the finite volume approximation (see Section 4.3.1).

Scheme (4.21) is capable of automatically upwinding and has a diagonally dominant coefficient matrix. It is very stable. It tends to the upwind difference scheme (4.17) for a strong convection problem (large P) and to the central difference scheme (4.14) for a strong diffusion problem (small P).

4.2.1.3   Discretization of 1-D unsteady problems

Time-marching schemes for 1-D unsteady problems include the Euler scheme, leapfrog scheme, Lax scheme, Crank-Nicholson scheme, Preissmann scheme, characteristic difference scheme, and Runge-Kutta method. The former five schemes are discussed below, whereas the others can be found in Abbott (1966), Yeh *et al.* (1995), Fletcher (1991), etc.

Euler scheme

Consider the 1-D unsteady convection equation:

$$\frac{\partial f}{\partial t} + u\frac{\partial f}{\partial x} = S \tag{4.22}$$

The computational grid in the (x, t) plane for solving Eq. (4.22) is shown in Fig. 4.4. The simplest scheme for the temporal term in Eq. (4.22) is the two-level difference

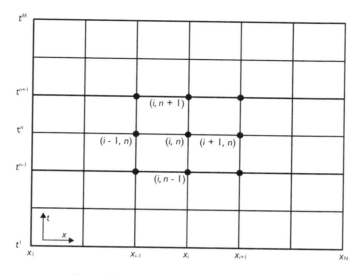

Figure 4.4 Finite difference grid in (x, t) plane.

scheme:

$$\frac{\partial f}{\partial t} = \frac{f_i^{n+1} - f_i^n}{\Delta t} \tag{4.23}$$

where $\Delta t = t^{n+1} - t^n$ is the time step length, and the superscript n is the time step index.

Scheme (4.23) can be the forward or backward difference scheme in time. Applying the forward difference scheme for the time-derivative term in Eq. (4.22) leads to an explicit scheme:

$$\frac{f_i^{n+1} - f_i^n}{\Delta t} + u\frac{f_i^n - f_{i-1}^n}{\Delta x} = S_i^n \quad (u > 0) \tag{4.24}$$

whereas applying the backward difference scheme leads to an implicit scheme:

$$\frac{f_i^{n+1} - f_i^n}{\Delta t} + u\frac{f_i^{n+1} - f_{i-1}^{n+1}}{\Delta x} = S_i^{n+1} \quad (u > 0) \tag{4.25}$$

Note that the convection term in Eqs. (4.24) and (4.25) is discretized using the upwind scheme (4.17) rather than the central difference scheme for better stability, as described in Section 4.2.1.2.

The forward difference scheme, with all other terms evaluated at time level t^n, is also known as the Euler method.

The implicit scheme (4.25) is unconditionally stable, whereas the explicit scheme (4.24) is stable if

$$\Delta t \leq \Delta x / |u| \tag{4.26}$$

which is called the CFL (Courant-Friedrichs-Lewy) condition.

Leapfrog scheme

Using the central difference scheme (4.13) for both the temporal and spatial terms in Eq. (4.22) results in the so-called leapfrog scheme:

$$\frac{f_i^{n+1} - f_i^{n-1}}{2\Delta t} + u\frac{f_{i+1}^n - f_{i-1}^n}{2\Delta x} = S_i^n \tag{4.27}$$

The leapfrog scheme is second-order accurate in time and space. If the CFL condition (4.26) is satisfied, the leapfrog scheme is neutrally stable. However, the leapfrog scheme is a three-level scheme, which requires an alternative method for the first time step.

Lax scheme and Lax-Wendroff scheme

Replacing f_i^n in scheme (4.23) by a weighted average value of f_{i-1}^n, f_i^n, and f_{i+1}^n yields the Lax scheme:

$$\frac{\partial f}{\partial t} = \frac{f_i^{n+1} - \left[\psi f_i^n + \frac{1}{2}(1 - \psi)(f_{i-1}^n + f_{i+1}^n)\right]}{\Delta t} \tag{4.28}$$

where ψ is a spatial weighting coefficient.

Applying the Lax scheme (4.28) for the convection equation (4.22) leads to

$$\frac{f_i^{n+1} - \left[\psi f_i^n + \frac{1}{2}(1 - \psi)(f_{i-1}^n + f_{i+1}^n)\right]}{\Delta t} + u\frac{f_{i+1}^n - f_{i-1}^n}{2\Delta x} = S \tag{4.29}$$

If $\psi = 1 - u^2\Delta t^2/\Delta x^2$ and $S = 0$, the difference equation (4.29) becomes the Lax-Wendroff scheme, which is second-order accurate in time and space.

For the homogeneous convection-diffusion equation

$$\frac{\partial f}{\partial t} + u\frac{\partial f}{\partial x} = \varepsilon_c\frac{\partial^2 f}{\partial x^2} \tag{4.30}$$

the Lax-Wendroff scheme is

$$\frac{f_i^{n+1} - f_i^n}{\Delta t} + u\frac{f_{i+1}^n - f_{i-1}^n}{2\Delta x} = \left(\varepsilon_c + \frac{1}{2}u^2\Delta t\right)\frac{f_{i-1}^n - 2f_i^n + f_{i+1}^n}{\Delta x^2} \tag{4.31}$$

The truncation error of the Lax-Wendroff scheme (4.31) is $O(\Delta t^2, \Delta x^2)$ as well. Its stability condition is $u^2 \Delta t^2 + 2\varepsilon_c \Delta t \leq \Delta x^2$ (see Fletcher, 1991).

Crank-Nicholson scheme

Applying Eq. (4.23) to the time derivative and a weighted average of the central difference scheme (4.14) between time levels n and $n + 1$ to the diffusion term in the 1-D diffusion equation

$$\frac{\partial f}{\partial t} = \varepsilon_c \frac{\partial^2 f}{\partial x^2} + S \tag{4.32}$$

yields

$$\frac{f_i^{n+1} - f_i^n}{\Delta t} = \theta \left[\varepsilon_c \frac{f_{i-1}^{n+1} - 2f_i^{n+1} + f_{i+1}^{n+1}}{\Delta x^2} + S_i^{n+1} \right]$$
$$+ (1 - \theta) \left[\varepsilon_c \frac{f_{i-1}^n - 2f_i^n + f_{i+1}^n}{\Delta x^2} + S_i^n \right] \tag{4.33}$$

where θ is a temporal weighting factor. When $\theta = 0$, Eq. (4.33) is an explicit scheme, and when $\theta > 0$, Eq. (4.33) is an implicit scheme. For $\theta = 1/2$, Eq. (4.33) is called the Crank-Nicholson scheme, which is second-order accurate in time and space.

Preissmann scheme

Preissmann (1961) proposed an implicit scheme based on two levels in time and two points in space, as shown in Fig. 4.5. This scheme replaces the continuous function f and its time and space derivatives by

$$f = \theta[\psi f_{i+1}^{n+1} + (1 - \psi)f_i^{n+1}] + (1 - \theta)[\psi f_{i+1}^n + (1 - \psi)f_i^n] \tag{4.34}$$

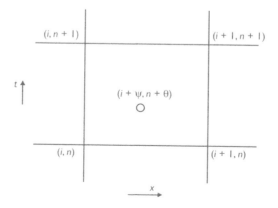

Figure 4.5 Computational element in the Preissmann scheme.

$$\frac{\partial f}{\partial t} = \psi \frac{f_{i+1}^{n+1} - f_{i+1}^{n}}{\Delta t} + (1 - \psi) \frac{f_{i}^{n+1} - f_{i}^{n}}{\Delta t} \tag{4.35}$$

$$\frac{\partial f}{\partial x} = \theta \frac{f_{i+1}^{n+1} - f_{i}^{n+1}}{\Delta x} + (1 - \theta) \frac{f_{i+1}^{n} - f_{i}^{n}}{\Delta x} \tag{4.36}$$

where θ and ψ are the weighting factors in time and space, respectively. The original Preissmann scheme adopts $\psi = 1/2$; thus, Eqs. (4.34)–(4.36) are the generalized version.

Application of the Preissmann scheme in the 1-D simulation of unsteady open-channel flows is discussed in detail in Section 5.2.2.

4.2.1.4   High-order difference schemes

The backward and forward difference schemes (4.11) and (4.12) based on two grid points are the simplest asymmetric difference schemes, and the central difference schemes (4.13) and (4.14) based on three grid points are the simplest symmetric difference schemes. To enhance the accuracy of numerical discretization, one may use more grid points in the difference formulation. For example, the following three-point and four-point asymmetric difference schemes for the first derivative are derived using the Taylor series expansion:

$$\left(\frac{\partial f}{\partial x} \right)_i = \frac{f_{i-2} - 4f_{i-1} + 3f_i}{2\Delta x} + O(\Delta x^2) \tag{4.37}$$

$$\left(\frac{\partial f}{\partial x} \right)_i = \frac{f_{i-2} - 6f_{i-1} + 3f_i + 2f_{i+1}}{6\Delta x} + O(\Delta x^3) \tag{4.38}$$

and the five-point symmetric difference schemes for the first and second derivatives are

$$\left(\frac{\partial f}{\partial x} \right)_i = \frac{f_{i-2} - 8f_{i-1} + 8f_{i+1} - f_{i+2}}{12\Delta x} + O(\Delta x^4) \tag{4.39}$$

$$\left(\frac{\partial^2 f}{\partial x^2} \right)_i = \frac{-f_{i-2} + 16f_{i-1} - 30f_i + 16f_{i+1} - f_{i+2}}{12\Delta x^2} + O(\Delta x^4) \tag{4.40}$$

By using schemes (4.37) and (4.38), the second-order and third-order upwind schemes for the convection terms in Eqs. (4.15) and (4.22) can be established as follows:

$$u \left(\frac{\partial f}{\partial x} \right)_i = \begin{cases} u \dfrac{f_{i-2} - 4f_{i-1} + 3f_i}{2\Delta x} & u \geq 0 \\[2mm] u \dfrac{-3f_i + 4f_{i+1} - f_{i+2}}{2\Delta x} & u < 0 \end{cases} \tag{4.41}$$

$$u \left(\frac{\partial f}{\partial x} \right)_i = \begin{cases} u \dfrac{f_{i-2} - 6f_{i-1} + 3f_i + 2f_{i+1}}{6\Delta x} & u \geq 0 \\[2mm] u \dfrac{-2f_{i-1} - 3f_i + 6f_{i+1} - f_{i+2}}{6\Delta x} & u < 0 \end{cases} \tag{4.42}$$

Upwind schemes (4.41) and (4.42) have better accuracy and less numerical diffusion than the first-order upwind scheme (4.17). However, they may produce numerical oscillations where the function f varies rapidly.

Using the asymmetric difference scheme (4.37) for the time derivative yields the three-level implicit scheme:

$$\left.\frac{\partial f}{\partial t}\right|_i^{n+1} = \frac{3f_i^{n+1} - 4f_i^n + f_i^{n-1}}{2\Delta t} \tag{4.43}$$

which is second-order accurate in time.

In addition, one may establish high-order schemes based on only two or three grid points by including the first and second derivatives of the function in difference formulations (Yang and Cunge, 1989; Wu, 1993). One approach is based on the expansion of f^{n+1} as a Taylor series about t^n:

$$f_i^{n+1} = f_i^n + \sum_{k=1}^m \frac{\Delta t^k}{k!}\left(\frac{\partial^k f}{\partial t^k}\right)_i^n + O(\Delta t^{m+1}) \tag{4.44}$$

For the homogeneous convection equation (4.22) with $S = 0$ and a constant velocity u, one can derive $\partial^k f/\partial t^k = (-u)^k \partial^k f/\partial x^k$. Substituting this relation into Eq. (4.44) yields

$$f_i^{n+1} = f_i^n + \sum_{k=1}^m \frac{\Delta t^k}{k!}(-u)^k \left(\frac{\partial^k f}{\partial x^k}\right)_i^n + O(\Delta t^{m+1}) \tag{4.45}$$

If the first and second derivatives in Eq. (4.45) are evaluated using the central difference schemes (4.13) and (4.14), Eq. (4.45) is exactly the Lax-Wendroff scheme. A fourth-order accurate scheme can be obtained by using the five-point schemes (4.39) and (4.40) for the first and second derivatives and constructing two fourth-order seven-point schemes for the third and fourth derivatives in Eq. (4.45). However, to limit the number of grid points involved, Wu (1993) suggested the following three-point schemes for these derivatives:

$$\left(\frac{\partial f}{\partial x}\right)_{i-1}^n + 4\left(\frac{\partial f}{\partial x}\right)_i^n + \left(\frac{\partial f}{\partial x}\right)_{i+1}^n = 3\frac{f_{i+1}^n - f_{i-1}^n}{\Delta x} + O(\Delta x^4) \tag{4.46}$$

$$\left(\frac{\partial^2 f}{\partial x^2}\right)_i^n = 2\frac{f_{i-1}^n - 2f_i^n + f_{i+1}^n}{\Delta x^2} - \frac{(\partial f/\partial x)_{i+1}^n - (\partial f/\partial x)_{i-1}^n}{2\Delta x} + O(\Delta x^4) \tag{4.47}$$

$$\left(\frac{\partial^3 f}{\partial x^3}\right)_i^n = 15\frac{f_{i+1}^n - f_{i-1}^n}{2\Delta x^3} - 3\frac{(\partial f/\partial x)_{i-1}^n + 8(\partial f/\partial x)_i^n + (\partial f/\partial x)_{i+1}^n}{2\Delta x^2} + O(\Delta x^4)$$

$$\tag{4.48}$$

$$\left(\frac{\partial^4 f}{\partial x^4}\right)_i^n = 36\frac{f_{i-1}^n - 2f_i^n + f_{i+1}^n}{\Delta x^4} - 6\frac{(\partial f/\partial x)_{i+1}^n - (\partial f/\partial x)_{i-1}^n}{\Delta x^3}$$

$$- \frac{24}{\Delta x^2}\left(\frac{\partial^2 f}{\partial x^2}\right)_i^n + O(\Delta x^4) \tag{4.49}$$

Eqs. (4.46)–(4.49) can be derived using the Taylor series expansion. Note that Eq. (4.46) needs to be solved using a direct or iterative method to compute the first derivative at each point.

Higher-order (up to eighth-order) difference schemes can be derived by adding the first and second derivatives at points $i - 1$ and $i + 1$ in Eq. (4.45). The approach described above can also be applied in the derivation of high-order schemes for the diffusion and convection-diffusion equations by substituting relevant relations for $\partial^k f/\partial t^k$ into Eq. (4.44) (Wu, 1993).

However, the above high-order difference schemes must be treated specially at boundary points because external points or boundary values for the first and/or second derivatives are involved. Furthermore, they usually need a uniform mesh that is difficult to conform to the irregular and movable boundaries of river flow. Therefore, the numerical schemes of higher than fourth-order accuracy are rarely used in computational river dynamics.

4.2.2 Finite difference method for multidimensional problems on regular grids

4.2.2.1 Discretization of multidimensional steady problems

It is straightforward to extend the aforementioned 1-D finite difference schemes to the discretization of 2-D and 3-D differential equations on regular grids. For example, on the rectangular grid shown in Fig. 4.6, applying the upwind difference scheme (4.17) for the convection terms and the central difference scheme (4.14) for the diffusion terms in the 2-D steady convection-diffusion equation

$$u_x\frac{\partial f}{\partial x} + u_y\frac{\partial f}{\partial y} = \varepsilon_c\left(\frac{\partial^2 f}{\partial x^2} + \frac{\partial^2 f}{\partial y^2}\right) + S \tag{4.50}$$

yields

$$G_{xi,j} + G_{yi,j} = \varepsilon_c\left(\frac{f_{i-1,j} - 2f_{i,j} + f_{i+1,j}}{\Delta x^2} + \frac{f_{i,j-1} - 2f_{i,j} + f_{i,j+1}}{\Delta y^2}\right) + S_{i,j} \tag{4.51}$$

where Δx and Δy are the grid spacings in the x- and y-directions, respectively; $G_{xi,j}$ is set as $u_x(f_{i,j} - f_{i-1,j})/\Delta x$ when $u_x \geq 0$ and $u_x(f_{i+1,j} - f_{i,j})/\Delta x$ when $u_x < 0$; and $G_{yi,j}$ is $u_y(f_{i,j} - f_{i,j-1})/\Delta y$ when $u_y \geq 0$ and $u_y(f_{i,j+1} - f_{i,j})/\Delta y$ when $u_y < 0$.

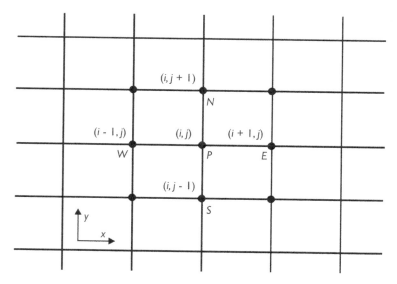

Figure 4.6 2-D finite difference grid.

To apply the exponential difference scheme (4.21), one can rearrange Eq. (4.50) as

$$\left(u_x\frac{\partial f}{\partial x} - \varepsilon_c\frac{\partial^2 f}{\partial x^2}\right) + \left(u_y\frac{\partial f}{\partial y} - \varepsilon_c\frac{\partial^2 f}{\partial y^2}\right) = S \qquad (4.52)$$

Discretizing the convection and diffusion terms with the exponential difference scheme (4.21) in the x- and y-directions, respectively, yields

$$a_P f_{i,j} = a_W f_{i-1,j} + a_E f_{i+1,j} + a_S f_{i,j-1} + a_N f_{i,j+1} + S_{i,j} \qquad (4.53)$$

where

$$a_W = \frac{u_x}{2\Delta x}\exp(P_x/2)/\sinh(P_x/2), \quad a_E = \frac{u_x}{2\Delta x}\exp(-P_x/2)/\sinh(P_x/2),$$

$$a_S = \frac{u_y}{2\Delta y}\exp(P_y/2)/\sinh(P_y/2), \quad a_N = \frac{u_y}{2\Delta y}\exp(-P_y/2)/\sinh(P_y/2), \quad (4.54)$$

$$a_P = a_W + a_E + a_S + a_N$$

with $P_x = u_x\Delta x/\varepsilon_c$ and $P_y = u_y\Delta y/\varepsilon_c$.

Scheme (4.53) is also called the five-point hybrid finite analytic scheme (Li and Yang, 1990; Lu and Si, 1990). Chen and Li (1980) derived the analytic solution for Eq. (4.50) with constant velocity, diffusivity, and source term at the nine-point cluster shown in Fig. 4.6 and established a nine-point finite analytic scheme. The nine-point analytic scheme also has the capability of automatically upwinding and is very stable. The details can be found in Chen and Li (1980).

4.2.2.2   Discretization of multidimensional unsteady problems

The important issue for discretizing 2-D and 3-D unsteady problems is how to arrange spatial difference operators in different directions or fractional steps. Widely used approaches include the full-domain implicit (or explicit) method, alternating direction implicit method, and operator splitting method.

Full-domain implicit (or explicit) method

The full-domain implicit (or explicit) method discretizes all spatial derivatives in 2-D and 3-D differential equations at the same time level. For example, extending Eq. (4.33) to the 2-D diffusion equation

$$\frac{\partial f}{\partial t} = \varepsilon_c \left(\frac{\partial^2 f}{\partial x^2} + \frac{\partial^2 f}{\partial y^2} \right) + S \tag{4.55}$$

yields

$$\frac{f_{i,j}^{n+1} - f_{i,j}^{n}}{\Delta t} = \varepsilon_c \theta \left(\frac{f_{i-1,j}^{n+1} - 2f_{i,j}^{n+1} + f_{i+1,j}^{n+1}}{\Delta x^2} + \frac{f_{i,j-1}^{n+1} - 2f_{i,j}^{n+1} + f_{i,j+1}^{n+1}}{\Delta y^2} \right)$$

$$+ \varepsilon_c (1 - \theta) \left(\frac{f_{i-1,j}^{n} - 2f_{i,j}^{n} + f_{i+1,j}^{n}}{\Delta x^2} + \frac{f_{i,j-1}^{n} - 2f_{i,j}^{n} + f_{i,j+1}^{n}}{\Delta y^2} \right) + S_{i,j}^{n+\theta} \tag{4.56}$$

When $\theta = 0$, Eq. (4.56) is explicit in both x- and y-directions; its sufficient and necessary stability condition is $r = \varepsilon_c \Delta t / h^2 \leq 1/4$, if $\Delta x = \Delta y = h$. When $\theta = 1$, Eq. (4.56) is implicit in both x- and y-directions and unconditionally stable; however, the discretized equation at each grid point involves five unknowns and usually needs to be solved by an iteration method.

Alternating direction implicit method

The alternating direction implicit (ADI) method was proposed by Peaceman and Rachford (1955). It usually divides the computation into two or three steps and discretizes the spatial derivatives implicitly in only one direction at each step.

Consider a 2-D partial differential equation:

$$\frac{\partial f}{\partial t} = L_x f + L_y f \tag{4.57}$$

where L_x and L_y are differential operators in the x- and y-directions. The corresponding two-step ADI difference equations can be written as

$$\begin{cases} \dfrac{f^{n+1/2} - f^n}{\Delta t/2} = \Delta_x f^{n+1/2} + \Delta_y f^n \\[2ex] \dfrac{f^{n+1} - f^{n+1/2}}{\Delta t/2} = \Delta_x f^{n+1/2} + \Delta_y f^{n+1} \end{cases} \tag{4.58}$$

where Δ_x and Δ_y are the spatial difference operators of L_x and L_y, respectively. In the first step, the operator L_x is approximated implicitly, while L_y is approximated explicitly. In the second step, L_x is treated explicitly, while L_y is treated implicitly.

For example, the ADI scheme for the 2-D diffusion equation (4.55) is

$$\begin{cases} \dfrac{f_{i,j}^{n+1/2} - f_{i,j}^n}{\Delta t/2} = \varepsilon_c(\delta_{xx}f_{i,j}^{n+1/2} + \delta_{yy}f_{i,j}^n) + S_{i,j}^{n+1/2} \\ \dfrac{f_{i,j}^{n+1} - f_{i,j}^{n+1/2}}{\Delta t/2} = \varepsilon_c(\delta_{xx}f_{i,j}^{n+1/2} + \delta_{yy}f_{i,j}^{n+1}) + S_{i,j}^{n+1} \end{cases} \tag{4.59}$$

where δ_{xx} and δ_{yy} are the central difference operators corresponding to the second derivatives $\partial^2/\partial x^2$ and $\partial^2/\partial y^2$, respectively.

The ADI scheme (4.59) is unconditionally stable. Because only a 1-D difference equation with three unknowns needs to be solved at each step, it is simpler than the full-domain implicit difference equation (4.56).

Operator splitting method

The operator splitting method was proposed by Yanenko (1971) and others. It splits the differential equation into several operators and then treats each operator separately.

Consider a differential equation:

$$\frac{\partial f}{\partial t} = L_1 f + L_2 f \tag{4.60}$$

where L_1 and L_2 are spatial operators. The corresponding difference equation is written as

$$\frac{f^{n+1} - f^n}{\Delta t} = \Delta_1 f^{n+1} + \Delta_2 f^{n+1} \tag{4.61}$$

where Δ_1 and Δ_2 are the difference operators of L_1 and L_2, respectively.

Eq. (4.61) can be split as

$$\begin{cases} \dfrac{f^{n+1/2} - f^n}{\Delta t} = \Delta_1 f^{n+1/2} \\ \dfrac{f^{n+1} - f^{n+1/2}}{\Delta t} = \Delta_2 f^{n+1} \end{cases} \tag{4.62}$$

Note that the operator splitting method can be used for 1-D, 2-D, and 3-D problems. In other words, operators L_1 and L_2 in Eq. (4.60) can be one-, two-, or three-dimensional.

The consistency of the operator splitting method for linear differential equations has been proven, but not yet for nonlinear differential equations. However, extensive numerical tests have shown that it can provide satisfactory results for many practical nonlinear problems.

The advantage of the operator splitting method is that each operator can be handled with an appropriate method specific to that operator. However, boundary conditions may be difficult to implement, and the overall accuracy is hard to judge even though high-order schemes might be used for every operator.

4.2.3 Finite difference method for multidimensional problems on curvilinear grids

River flow problems usually have irregular and even movable domains. When the classic finite difference method on regular grids is used to solve these problems, difficulties may arise near boundaries. However, boundary conditions are essential to the properties of the solution. Therefore, the finite difference method on fixed and moving curvilinear grids has been established in the past decades via coordinate transformation and interpolation, as described below.

4.2.3.1 Governing equations in generalized coordinate system

In general, the unsteady coordinate transformation from the Cartesian coordinate system (x_i, t) to a moving, curvilinear coordinate system (ξ_m, τ) can be written as

$$\begin{cases} x_i = x_i(\xi_m, \tau) & (i = 1, 2, 3; m = 1, 2, 3) \\ t = \tau \end{cases} \tag{4.63}$$

where ξ_1, ξ_2, and $\xi_3 (= \xi, \eta, \zeta)$ are the coordinates, and τ is the time in the curvilinear system.

Coordinate transformation (4.63) includes time and hence can be applied to both fixed and movable grids (Wu, 1996a; Shyy $et\ al.$, 1996). Its Jacobian matrix is

$$B = \begin{bmatrix} \dfrac{\partial x_1}{\partial \xi_1} & \dfrac{\partial x_2}{\partial \xi_1} & \dfrac{\partial x_3}{\partial \xi_1} & 0 \\ \dfrac{\partial x_1}{\partial \xi_2} & \dfrac{\partial x_2}{\partial \xi_2} & \dfrac{\partial x_3}{\partial \xi_2} & 0 \\ \dfrac{\partial x_1}{\partial \xi_3} & \dfrac{\partial x_2}{\partial \xi_3} & \dfrac{\partial x_3}{\partial \xi_3} & 0 \\ \dfrac{\partial x_1}{\partial \tau} & \dfrac{\partial x_2}{\partial \tau} & \dfrac{\partial x_3}{\partial \tau} & 1 \end{bmatrix} \tag{4.64}$$

and Jacobian determinant is $J = |B|$. For a monotonic and reversible coordinate transformation, the Jacobian determinant should be non-zero and have finite bounds, i.e., $0 < J < +\infty$.

Denote $\alpha_i^m = \partial \xi_m / \partial x_i$ and $\beta_m^i = \partial x_i / \partial \xi_m$. Then α_i^m and $\partial \xi_m / \partial t$ can be determined by

$$\alpha_i^m = \frac{M_m^i}{J}, \quad \frac{\partial \xi_m}{\partial t} = \frac{M_m^4}{J} \tag{4.65}$$

where M_m^i $(i = 1, 2, 3)$ and M_m^4 are the cofactors of β_m^i and $\partial t / \partial \xi_m$ in the Jacobian matrix B, respectively.

Under coordinate transformation (4.63), the first and second derivatives of function f are given by

$$\frac{\partial f}{\partial t} = \frac{\partial f}{\partial \tau} + \frac{\partial \xi_m}{\partial t} \frac{\partial f}{\partial \xi_m} \tag{4.66}$$

$$\frac{\partial f}{\partial x_i} = \alpha_i^m \frac{\partial f}{\partial \xi_m} \tag{4.67}$$

$$\frac{\partial^2 f}{\partial x_i \partial x_j} = \alpha_i^m \frac{\partial}{\partial \xi_m} \left(\alpha_j^n \frac{\partial f}{\partial \xi_n} \right) \tag{4.68}$$

and the substantial derivative is

$$\frac{Df}{Dt} = \frac{\partial f}{\partial t} + u_i \frac{\partial f}{\partial x_i} = \frac{\partial f}{\partial \tau} + \hat{u}_m \frac{\partial f}{\partial \xi_m} \tag{4.69}$$

where u_i and $\hat{u}_m$ are the velocities in the (x_i, t) and (ξ_m, τ) coordinate systems, respectively. They are related as follows:

$$\hat{u}_m = \frac{\partial \xi_m}{\partial t} + \alpha_i^m u_i, \quad u_i = \frac{\partial x_i}{\partial \tau} + \beta_m^i \hat{u}_m \tag{4.70}$$

Note that like the Cartesian coordinate index i, the curvilinear coordinate index m is also subject to Einstein's summation convention.

In the (ξ_m, τ) coordinate system, the continuity and Navier-Stokes equations of incompressible flows are

$$\frac{\partial J}{\partial \tau} + \frac{\partial (J \hat{u}_m)}{\partial \xi_m} = 0 \tag{4.71}$$

$$\frac{\partial u_i}{\partial \tau} + \hat{u}_m \frac{\partial u_i}{\partial \xi_m} = F_i - \frac{1}{\rho} \alpha_i^m \frac{\partial p}{\partial \xi_m} + \frac{1}{\rho} \alpha_j^m \frac{\partial \tau_{ij}}{\partial \xi_m} \tag{4.72}$$

and the scalar transport equation is

$$\frac{\partial c}{\partial \tau} + \hat{u}_m \frac{\partial c}{\partial \xi_m} = \alpha_j^m \frac{\partial}{\partial \xi_m} \left(\alpha_j^n \varepsilon_c \frac{\partial c}{\partial \xi_n} \right) + S \tag{4.73}$$

where c is a scalar quantity, such as mass concentration and temperature.

4.2.3.2 Typical coordinate transformations

Boundary-fitted coordinate transformation

A boundary-fitted coordinate transformation was adopted by Thompson *et al.* (1985) to simulate flows around physical bodies. In the 2-D case shown in Fig. 4.7,

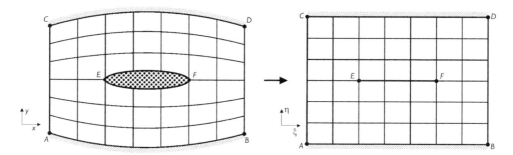

Figure 4.7 Boundary-fitted coordinate transformation.

the coordinate transformation between physical domain (x, y) and logical domain (ξ, η) is governed by the Poisson equations:

$$\begin{cases} \dfrac{\partial^2 \xi}{\partial x^2} + \dfrac{\partial^2 \xi}{\partial y^2} = P(\xi, \eta) \\[2mm] \dfrac{\partial^2 \eta}{\partial x^2} + \dfrac{\partial^2 \eta}{\partial y^2} = Q(\xi, \eta) \end{cases} \qquad (4.74)$$

where P and Q are source terms, which essentially determine the grid density and smoothness.

Because the grid in the physical domain is unknown, it is inconvenient to solve the equation set (4.74). Exchanging the independent and dependent parameters yields the corresponding equations for x and y with respect to ξ and η:

$$\begin{cases} A\dfrac{\partial^2 x}{\partial \xi^2} - 2B\dfrac{\partial^2 x}{\partial \xi \partial \eta} + C\dfrac{\partial^2 x}{\partial \eta^2} + J^2\left(P\dfrac{\partial x}{\partial \xi} + Q\dfrac{\partial x}{\partial \eta}\right) = 0 \\[2mm] A\dfrac{\partial^2 y}{\partial \xi^2} - 2B\dfrac{\partial^2 y}{\partial \xi \partial \eta} + C\dfrac{\partial^2 y}{\partial \eta^2} + J^2\left(P\dfrac{\partial y}{\partial \xi} + Q\dfrac{\partial y}{\partial \eta}\right) = 0 \end{cases} \qquad (4.75)$$

where $A = (\partial x/\partial \eta)^2 + (\partial y/\partial \eta)^2$, $B = \partial x/\partial \xi \cdot \partial x/\partial \eta + \partial y/\partial \xi \cdot \partial y/\partial \eta$, and $C = (\partial x/\partial \xi)^2 + (\partial y/\partial \xi)^2$. Because the grid in the logical domain is prescribed, the two equations in (4.75) can be solved conveniently.

The Jacobian determinant J of the coordinate transformation is

$$J = \frac{\partial x}{\partial \xi}\frac{\partial y}{\partial \eta} - \frac{\partial x}{\partial \eta}\frac{\partial y}{\partial \xi} \qquad (4.76)$$

Under coordinate transformation (4.74), the first derivatives of f are

$$\frac{\partial f}{\partial x} = \alpha_1^1 \frac{\partial f}{\partial \xi} + \alpha_1^2 \frac{\partial f}{\partial \eta} \qquad (4.77)$$

$$\frac{\partial f}{\partial y} = \alpha_2^1 \frac{\partial f}{\partial \xi} + \alpha_2^2 \frac{\partial f}{\partial \eta} \qquad (4.78)$$

where

$$\alpha_1^1 = J^{-1}\frac{\partial y}{\partial \eta}, \quad \alpha_1^2 = -J^{-1}\frac{\partial y}{\partial \xi}, \quad \alpha_2^1 = -J^{-1}\frac{\partial x}{\partial \eta}, \quad \alpha_2^2 = J^{-1}\frac{\partial x}{\partial \xi} \qquad (4.79)$$

The second derivatives are

$$\frac{\partial^2 f}{\partial x^2} = \alpha_1^1\alpha_1^1\frac{\partial^2 f}{\partial \xi^2} + 2\alpha_1^1\alpha_1^2\frac{\partial^2 f}{\partial \xi \partial \eta} + \alpha_1^2\alpha_1^2\frac{\partial^2 f}{\partial \eta^2} + \left(\alpha_1^1\frac{\partial \alpha_1^1}{\partial \xi} + \alpha_1^2\frac{\partial \alpha_1^1}{\partial \eta}\right)\frac{\partial f}{\partial \xi}$$

$$+ \left(\alpha_1^1\frac{\partial \alpha_1^2}{\partial \xi} + \alpha_1^2\frac{\partial \alpha_1^2}{\partial \eta}\right)\frac{\partial f}{\partial \eta} \qquad (4.80)$$

$$\frac{\partial^2 f}{\partial x \partial y} = \alpha_1^1\alpha_2^1\frac{\partial^2 f}{\partial \xi^2} + (\alpha_1^1\alpha_2^2 + \alpha_1^2\alpha_2^1)\frac{\partial^2 f}{\partial \xi \partial \eta} + \alpha_1^2\alpha_2^2\frac{\partial^2 f}{\partial \eta^2}$$

$$+ \left(\alpha_1^1\frac{\partial \alpha_2^1}{\partial \xi} + \alpha_1^2\frac{\partial \alpha_2^1}{\partial \eta}\right)\frac{\partial f}{\partial \xi} + \left(\alpha_1^1\frac{\partial \alpha_2^2}{\partial \xi} + \alpha_1^2\frac{\partial \alpha_2^2}{\partial \eta}\right)\frac{\partial f}{\partial \eta} \qquad (4.81)$$

$$\frac{\partial^2 f}{\partial y^2} = \alpha_2^1\alpha_2^1\frac{\partial^2 f}{\partial \xi^2} + 2\alpha_2^1\alpha_2^2\frac{\partial^2 f}{\partial \xi \partial \eta} + \alpha_2^2\alpha_2^2\frac{\partial^2 f}{\partial \eta^2} + \left(\alpha_2^1\frac{\partial \alpha_2^1}{\partial \xi} + \alpha_2^2\frac{\partial \alpha_2^1}{\partial \eta}\right)\frac{\partial f}{\partial \xi}$$

$$+ \left(\alpha_2^1\frac{\partial \alpha_2^2}{\partial \xi} + \alpha_2^2\frac{\partial \alpha_2^2}{\partial \eta}\right)\frac{\partial f}{\partial \eta} \qquad (4.82)$$

Local coordinate transformation on fixed grids

The previous boundary-fitted coordinate transformation can provide high-quality numerical grids with global properties, such as orthogonality and smoothness. However, two partial differential equations need to be solved in the entire domain. A simpler method for handling irregular boundary problems is the local coordinate transformation that is based on only individual elements.

Suppose a 2-D physical domain is represented by a quadrilateral grid, and the nine-point quadrilateral isoparametric element shown in Fig. 4.8 is used as the basic element (Wu and Li, 1992; Wang and Hu, 1993). At each element all nine points are numbered 1 through 9 and the 5th point is the central point. This irregular element is converted into a rectangle by the following coordinate transformation:

$$x = \sum_{k=1}^{9} x_k \varphi_k(\xi, \eta), \quad y = \sum_{k=1}^{9} y_k \varphi_k(\xi, \eta) \qquad (4.83)$$

where x_k and y_k are the coordinate values of the kth point in the (x, y) coordinate system; and $\varphi_k(k = 1, 2, \ldots, 9)$ are the interpolation functions, which are quadratic

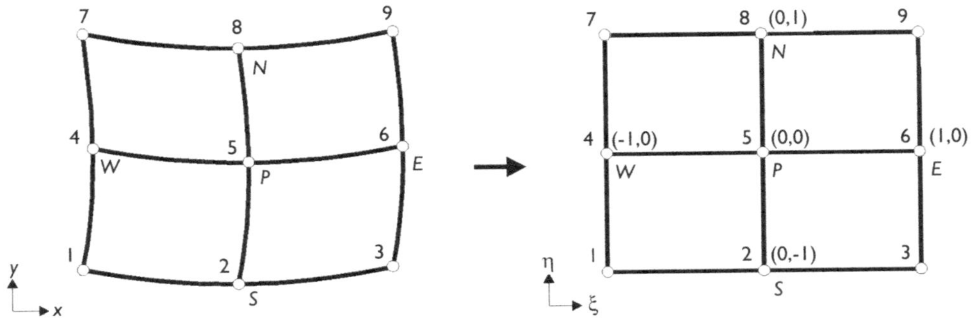

Figure 4.8 2-D local coordinate transformation.

and satisfy $\varphi_k(\xi_j, \eta_j) = \delta_j^k$ and $\sum_{k=1}^{9} \varphi_k = 1$. They are written as

$$
\varphi_k = \begin{cases}
(\xi\xi_k + \xi^2)(\eta\eta_k + \eta^2)/4 & k = 1,3,7,9 \\
(\xi\xi_k + \xi^2)(1 - \eta^2)/2 & k = 4,6 \\
(1 - \xi^2)(\eta\eta_k + \eta^2)/2 & k = 2,8 \\
(1 - \xi^2)(1 - \eta^2) & k = 5
\end{cases} \tag{4.84}
$$

Differentiating coordinate transformation (4.83) with respect to ξ and η leads to

$$
\frac{\partial x}{\partial \xi} = \sum_{k=1}^{9} x_k \frac{\partial \varphi_k}{\partial \xi}, \quad \frac{\partial x}{\partial \eta} = \sum_{k=1}^{9} x_k \frac{\partial \varphi_k}{\partial \eta},
$$

$$
\frac{\partial y}{\partial \xi} = \sum_{k=1}^{9} y_k \frac{\partial \varphi_k}{\partial \xi}, \quad \frac{\partial y}{\partial \eta} = \sum_{k=1}^{9} y_k \frac{\partial \varphi_k}{\partial \eta}, \tag{4.85}
$$

and

$$
\frac{\partial^2 x}{\partial \xi^2} = \sum_{k=1}^{9} x_k \frac{\partial^2 \varphi_k}{\partial \xi^2}, \quad \frac{\partial^2 x}{\partial \xi \partial \eta} = \sum_{k=1}^{9} x_k \frac{\partial^2 \varphi_k}{\partial \xi \partial \eta}, \quad \frac{\partial^2 x}{\partial \eta^2} = \sum_{k=1}^{9} x_k \frac{\partial^2 \varphi_k}{\partial \eta^2},
$$

$$
\frac{\partial^2 y}{\partial \xi^2} = \sum_{k=1}^{9} y_k \frac{\partial^2 \varphi_k}{\partial \xi^2}, \quad \frac{\partial^2 y}{\partial \xi \partial \eta} = \sum_{k=1}^{9} y_k \frac{\partial^2 \varphi_k}{\partial \xi \partial \eta}, \quad \frac{\partial^2 y}{\partial \eta^2} = \sum_{k=1}^{9} y_k \frac{\partial^2 \varphi_k}{\partial \eta^2}. \tag{4.86}
$$

The Jacobian determinant is Eq. (4.76). The first and second derivatives are given by Eqs. (4.77), (4.78), and (4.80)–(4.82).

For a 3-D problem, the volume formed by twenty-seven points shown in Fig. 4.9 is adopted as the basic element. This irregular element is turned into a cube by the following coordinate transformation between the physical domain (x, y, z) and the logical domain (ξ, η, ζ):

$$
x = \sum_{k=1}^{27} x_k \varphi_k(\xi, \eta, \zeta), \quad y = \sum_{k=1}^{27} y_k \varphi_k(\xi, \eta, \zeta), \quad z = \sum_{k=1}^{27} z_k \varphi_k(\xi, \eta, \zeta) \tag{4.87}
$$

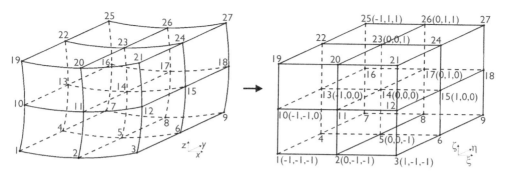

Figure 4.9 3-D local coordinate transformation.

where x_k, y_k, and z_k are the coordinate values of the kth point in the (x, y, z) coordinate system, and $\varphi_k (k = 1, 2, \ldots, 27)$ are interpolation functions (Wu, 1996b):

$$
\varphi_k = \begin{cases}
(\xi\xi_k + \xi^2)(\eta\eta_k + \eta^2)(\zeta\zeta_k + \zeta^2)/8 & k = 1, 3, 7, 9, 19, 21, 25, 27 \\
(1 - \xi^2)(\eta\eta_k + \eta^2)(\zeta\zeta_k + \zeta^2)/4 & k = 2, 8, 20, 26 \\
(\xi\xi_k + \xi^2)(1 - \eta^2)(\zeta\zeta_k + \zeta^2)/4 & k = 4, 6, 22, 24 \\
(\xi\xi_k + \xi^2)(\eta\eta_k + \eta^2)(1 - \zeta^2)/4 & k = 10, 12, 16, 18 \\
(1 - \xi^2)(1 - \eta^2)(\zeta\zeta_k + \zeta^2)/2 & k = 5, 23 \\
(1 - \xi^2)(\eta\eta_k + \eta^2)(1 - \zeta^2)/2 & k = 11, 17 \\
(\xi\xi_k + \xi^2)(1 - \eta^2)(1 - \zeta^2)/2 & k = 13, 15 \\
(1 - \xi^2)(1 - \eta^2)(1 - \zeta^2) & k = 14
\end{cases} \tag{4.88}
$$

Note that the local coordinate transformations (4.83) and (4.87) do not specify how to generate the computational grid. The grid can be generated by either the boundary-fitted coordinate method or another more arbitrary method. However, to ensure a monotonic coordinate transformation, the angles between ξ, η, and ζ grid lines should be away from $0°$ and $180°$ in the physical space. It is preferable that the angles are between $45°$ and $135°$.

Local coordinate transformation on moving grids

The local coordinate transformations (4.83) and (4.87) on fixed grids can be extended to moving grids. For a 2-D case, for example, the physical domain is represented by a boundary-fitted quadrilateral grid at each time or iteration step. Because the grid adapts to the changing boundaries, the coordinate values of each grid point are functions of time, i.e., $x_k = x_k(\tau)$ and $y_k = y_k(\tau)$, as shown in Fig. 4.10. Therefore, the local coordinate transformation at each element reads (Wu, 1996a)

$$
x = \sum_{k=1}^{9} x_k(\tau)\varphi_k(\xi, \eta), \quad y = \sum_{k=1}^{9} y_k(\tau)\varphi_k(\xi, \eta), \quad t = \tau \tag{4.89}
$$

where $\varphi_k (k = 1, 2, \ldots, 9)$ are the interpolation functions expressed in Eq. (4.84).

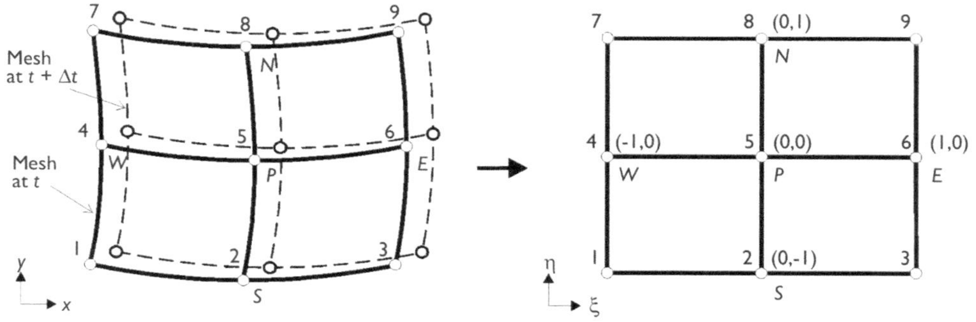

Figure 4.10 Local coordinate transformation on moving grid.

Because coordinate transformation (4.89) is time-dependent, it has relations (4.85) and (4.86) as well as the following:

$$\frac{\partial x}{\partial \tau} = \sum_{k=1}^{9} \frac{\partial x_k}{\partial \tau} \varphi_k, \quad \frac{\partial y}{\partial \tau} = \sum_{k=1}^{9} \frac{\partial y_k}{\partial \tau} \varphi_k, \quad \frac{\partial t}{\partial \xi} = 0, \quad \frac{\partial t}{\partial \eta} = 0, \quad \frac{\partial t}{\partial \tau} = 1 \quad (4.90)$$

The local coordinate transformation is convenient for complex movable boundary problems. Because at each time (or iteration) step the used grid conforms to the physical domain, the complex irregular and movable boundaries can be resolved effectively.

Stretching coordinate transformation

The stretching coordinate transformation, which is also called the σ-coordinate transformation, is an algebraic example of the unsteady coordinate transformation introduced in Section 4.2.3.1. If the boundaries are simple and vary gradually, the physical domain can be expanded or contracted along one or more directions by the stretching coordinate transformation, so that a fixed, regular logical domain is obtained. For example, for 2-D gradually varied open-channel flows, the stretching coordinate transformation shown in Fig. 4.11 is often used, which is expressed as

$$\begin{cases} \xi = x \\ \zeta = \dfrac{z - z_b}{h} H \\ \tau = t \end{cases} \quad (4.91)$$

where h is the width of the physical domain, either the flow depth or channel width; H is the width of the logical domain; and z_b is the distance from the lower boundary to the x axis. For the vertical 2-D case, z_b and h are the bed elevation and flow depth and vary with x and t.

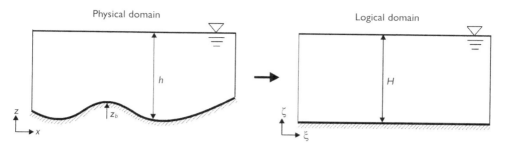

Figure 4.11 Stretching coordinate transformation.

Stretching coordinate transformation (4.91) has the following analytic relations:

$$\frac{\partial x}{\partial \xi} = 1, \qquad \frac{\partial x}{\partial \zeta} = 0, \qquad \frac{\partial x}{\partial \tau} = 0,$$

$$\frac{\partial z}{\partial \xi} = \frac{\partial z_b}{\partial \xi} + \frac{\zeta}{H}\frac{\partial h}{\partial \xi}, \qquad \frac{\partial z}{\partial \zeta} = \frac{h}{H}, \qquad \frac{\partial z}{\partial \tau} = \frac{\partial z_b}{\partial \tau} + \frac{\zeta}{H}\frac{\partial h}{\partial \tau}, \qquad (4.92)$$

$$\frac{\partial t}{\partial \xi} = 0, \qquad \frac{\partial t}{\partial \zeta} = 0, \qquad \frac{\partial t}{\partial \tau} = 1,$$

and

$$\frac{\partial \xi}{\partial x} = 1, \qquad \frac{\partial \xi}{\partial z} = 0, \qquad \frac{\partial \xi}{\partial t} = 0,$$

$$\frac{\partial \zeta}{\partial x} = -\frac{H}{h}\frac{\partial z_b}{\partial x} - \frac{\zeta}{h}\frac{\partial h}{\partial x}, \qquad \frac{\partial \zeta}{\partial z} = \frac{H}{h}, \qquad \frac{\partial \zeta}{\partial t} = -\frac{H}{h}\frac{\partial z_b}{\partial t} - \frac{\zeta}{h}\frac{\partial h}{\partial t}, \qquad (4.93)$$

$$\frac{\partial \tau}{\partial x} = 0, \qquad \frac{\partial \tau}{\partial z} = 0, \qquad \frac{\partial \tau}{\partial t} = 1.$$

The 2-D stretching coordinate transformation (4.91) can be easily extended to the 3-D case by adding one stretching function in the third direction. Because analytic transformation relations exist in the entire domain, it is very convenient to solve the transformed governing equations in the fixed, regular logical domain. However, this stretching coordinate transformation is inconvenient for the complex boundary problems that do not have analytic transformation relations.

4.2.3.3 *Discretization of the transformed equations*

As mentioned above, the irregular (and/or moving) physical domain or element is converted to a regular logical domain or element under coordinate transformations (4.74), (4.83), (4.87), (4.89), and (4.91). Therefore, many classic finite difference schemes based on regular grids can be used to solve the transformed equation on the regular logical domain or element. For example, the convection-diffusion equation (4.50) in the (x, y) coordinate system can be converted to the following form in the

(ξ, η) coordinate system by these coordinate transformations:

$$\hat{u}_\xi \frac{\partial f}{\partial \xi} + \hat{u}_\eta \frac{\partial f}{\partial \eta} = \varepsilon_{\xi\xi} \frac{\partial^2 f}{\partial \xi^2} + \varepsilon_{\eta\eta} \frac{\partial^2 f}{\partial \eta^2} + S^* \tag{4.94}$$

where $\hat{u}_\xi = \alpha_1^1 u_x + \alpha_2^1 u_y$, $\hat{u}_\eta = \alpha_1^2 u_x + \alpha_2^2 u_y$, $\varepsilon_{\xi\xi} = \varepsilon_c(\alpha_1^1 \alpha_1^1 + \alpha_2^1 \alpha_2^1)$, $\varepsilon_{\eta\eta} = \varepsilon_c(\alpha_1^2 \alpha_1^2 + \alpha_2^2 \alpha_2^2)$, and

$$S^* = S + 2\varepsilon_c(\alpha_1^1 \alpha_1^2 + \alpha_2^1 \alpha_2^2) \frac{\partial^2 f}{\partial \xi \partial \eta} + \varepsilon_c \left(\alpha_1^1 \frac{\partial \alpha_1^1}{\partial \xi} + \alpha_1^2 \frac{\partial \alpha_1^1}{\partial \eta} + \alpha_2^1 \frac{\partial \alpha_2^1}{\partial \xi} + \alpha_2^2 \frac{\partial \alpha_2^1}{\partial \eta} \right) \frac{\partial f}{\partial \xi}$$

$$+ \varepsilon_c \left(\alpha_1^1 \frac{\partial \alpha_1^2}{\partial \xi} + \alpha_1^2 \frac{\partial \alpha_1^2}{\partial \eta} + \alpha_2^1 \frac{\partial \alpha_2^2}{\partial \xi} + \alpha_2^2 \frac{\partial \alpha_2^2}{\partial \eta} \right) \frac{\partial f}{\partial \eta}.$$

The transformed equation (4.94) is still a convection-diffusion equation, which can be easily discretized using the upwind difference scheme (4.17), exponential difference scheme (4.21), or another scheme on the rectangular logical domain or element. For example, using the exponential difference scheme (4.21) for $\hat{u}_\xi \partial f / \partial \xi - \varepsilon_{\xi\xi} \partial^2 f / \partial \xi^2$ and $\hat{u}_\eta \partial f / \partial \eta - \varepsilon_{\eta\eta} \partial^2 f / \partial \eta^2$ in Eq. (4.94) yields Eq. (4.53) with coefficients (Wu, 1996b):

$$a_W = \frac{\hat{u}_\xi}{2\Delta\xi} \exp(P_\xi/2)/\sinh(P_\xi/2), \quad a_E = \frac{\hat{u}_\xi}{2\Delta\xi} \exp(-P_\xi/2)/\sinh(P_\xi/2),$$

$$a_S = \frac{\hat{u}_\eta}{2\Delta\eta} \exp(P_\eta/2)/\sinh(P_\eta/2), \quad a_N = \frac{\hat{u}_\eta}{2\Delta\eta} \exp(-P_\eta/2)/\sinh(P_\eta/2), \tag{4.95}$$

$$a_P = a_W + a_E + a_S + a_N,$$

and the source term replaced by S^*. Here, $P_\xi = \hat{u}_\xi \Delta\xi / \varepsilon_{\xi\xi}$ and $P_\eta = \hat{u}_\eta \Delta\eta / \varepsilon_{\eta\eta}$, with $\Delta\xi$ and $\Delta\eta$ being the grid lengths in the (ξ, η) system. For the local element shown in Fig. 4.8, $\Delta\xi = \Delta\eta = 1$.

4.2.4 Interpolation method

4.2.4.1 Isoparametric interpolation method on fixed grids

At the nine-point isoparametric element shown in Fig. 4.8, the function f can be approximated by interpolation:

$$f = \sum_{k=1}^{9} f_k \varphi_k \tag{4.96}$$

where f_k are the values of f on grid points, and φ_k are the interpolation functions given in Eq. (4.84).

The following difference schemes for the first and second derivatives of f can be derived from Eq. (4.96) (Wu, 1996b):

$$\frac{\partial f}{\partial x} = \sum_{k=1}^{9} a_k f_k, \quad \frac{\partial f}{\partial y} = \sum_{k=1}^{9} b_k f_k \tag{4.97}$$

$$\frac{\partial^2 f}{\partial x^2} = \sum_{k=1}^{9} c_k f_k, \quad \frac{\partial^2 f}{\partial y^2} = \sum_{k=1}^{9} d_k f_k \tag{4.98}$$

where a_k, b_k, c_k, and d_k are coefficients:

$$a_k = \alpha_1^1 \frac{\partial \varphi_k}{\partial \xi} + \alpha_1^2 \frac{\partial \varphi_k}{\partial \eta}, \quad b_k = \alpha_2^1 \frac{\partial \varphi_k}{\partial \xi} + \alpha_2^2 \frac{\partial \varphi_k}{\partial \eta},$$

$$c_k = \alpha_1^1 \alpha_1^1 \frac{\partial^2 \varphi_k}{\partial \xi^2} + 2\alpha_1^1 \alpha_1^2 \frac{\partial^2 \varphi_k}{\partial \xi \partial \eta} + \alpha_1^2 \alpha_1^2 \frac{\partial^2 \varphi_k}{\partial \eta^2} + \left(\alpha_1^1 \frac{\partial \alpha_1^1}{\partial \xi} + \alpha_1^2 \frac{\partial \alpha_1^1}{\partial \eta} \right) \frac{\partial \varphi_k}{\partial \xi}$$

$$+ \left(\alpha_1^1 \frac{\partial \alpha_1^2}{\partial \xi} + \alpha_1^2 \frac{\partial \alpha_1^2}{\partial \eta} \right) \frac{\partial \varphi_k}{\partial \eta},$$

and

$$d_k = \alpha_2^1 \alpha_2^1 \frac{\partial^2 \varphi_k}{\partial \xi^2} + 2\alpha_2^1 \alpha_2^2 \frac{\partial^2 \varphi_k}{\partial \xi \partial \eta} + \alpha_2^2 \alpha_2^2 \frac{\partial^2 \varphi_k}{\partial \eta^2} + \left(\alpha_2^1 \frac{\partial \alpha_2^1}{\partial \xi} + \alpha_2^2 \frac{\partial \alpha_2^1}{\partial \eta} \right) \frac{\partial \varphi_k}{\partial \xi}$$

$$+ \left(\alpha_2^1 \frac{\partial \alpha_2^2}{\partial \xi} + \alpha_2^2 \frac{\partial \alpha_2^2}{\partial \eta} \right) \frac{\partial \varphi_k}{\partial \eta}.$$

Eqs. (4.97) and (4.98) can be applied to any point in the local element. For example, for the central point 5, one can obtain a_k, b_k, c_k, and d_k by specifying $\xi = \eta = 0$.

The isoparametric interpolation formula (4.96) can be extended to the 3-D case using the interpolation functions (4.88) based on the 27-point element shown in Fig. 4.9, and similar difference schemes for the 3-D first and second derivatives can be easily derived.

Note that the difference schemes (4.97) are similar to the central difference scheme (4.13); thus, they are not as adequate for strong convection problems as the exponential difference scheme (4.95) and the upwind interpolation method introduced in the next subsection.

4.2.4.2 Upwind interpolation method on fixed grids

Wang and Hu (1993) analytically solved the following convection-diffusion equation with constant velocity, diffusivity, and source term in the 1-D local element shown in Fig. 4.12:

$$\hat{u} \frac{df}{d\xi} = \varepsilon_\xi \frac{d^2 f}{d\xi^2} + S^* \tag{4.99}$$

and derived the upwind interpolation functions:

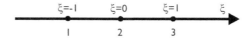

$$\xi=-1 \quad \xi=0 \quad \xi=1 \quad \xi$$

1   2   3

Figure 4.12 1-D local element.

$$c_1 = \frac{1}{2T}\{(2e^{P\xi} - e^{-P} - e^{P}) + T[1 - \xi(R+1)]\}$$

$$c_3 = \frac{1}{2T}\{(2e^{P\xi} - e^{-P} - e^{P}) + T[1 - \xi(R-1)]\} \qquad (4.100)$$

$$c_2 = 1 - c_1 - c_3$$

where $T = e^{-P} + e^{P} - 2$, $R = (e^{P} - e^{-P})/T$, P is the Peclet number defined as $P = \hat{u}/\varepsilon_\xi$, $\hat{u}$ is the local velocity, and ε_ξ is the diffusion coefficient.

Fig. 4.13 shows the behavior of the upwind interpolation functions at various Peclet numbers. It can be seen that these functions become more asymmetric as the Peclet number increases, i.e., when convection becomes more dominant. This upwind feature stabilizes this interpolation method in the simulation of strong convection problems.

The upwind interpolation functions in 2-D and 3-D cases can be obtained by applying Eq. (4.100) in every direction. For example, the upwind interpolation functions for the 2-D element shown in Fig. 4.8 are constructed by

$$\varphi_k = c_i(\xi)c_j(\eta) \qquad (4.101)$$

where k is corresponding to the pair of i and j according to Table 4.1.

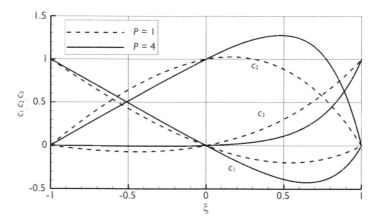

Figure 4.13 Upwind interpolation functions.

This upwind interpolation method is called the efficient element method by Wang and Hu (1993). It has been used in hydrodynamic modeling by Jia and Wang (1999).

Table 4.1 Relation between k and (i, j)

k	1	2	3	4	5	6	7	8	9
(i, j)	(1, 1)	(2, 1)	(3, 1)	(1, 2)	(2, 2)	(3, 2)	(1, 3)	(2, 3)	(3, 3)

4.2.4.3 Interpolation method on moving grids

For a moving grid, Eq. (4.96) can still be used to interpolate the function f at the element shown in Fig. 4.10 at each time step. Consequently the spatial derivatives of f are discretized by schemes (4.97) and (4.98), while the time derivative is discretized as (Wu, 1996a)

$$\left(\frac{\partial f}{\partial t}\right)_5 = \frac{f_5^{n+1} - f_5^n}{\Delta \tau} + \sum_{k=1}^{9} e_k f_k^{n+\theta} \qquad (4.102)$$

where e_k is the difference coefficient, defined as $e_k = (\frac{\partial \xi}{\partial t}\frac{\partial \varphi_k}{\partial \xi} + \frac{\partial \eta}{\partial t}\frac{\partial \varphi_k}{\partial \eta})_5$; $\Delta \tau$ is the time step length; and θ is an index: $= 0$ for explicit schemes and $= 1$ for implicit schemes.

The second term on the right-hand side of scheme (4.102) appears due to grid movement.

4.3 FINITE VOLUME METHOD

4.3.1 Finite volume method for 1-D problems

4.3.1.1 Discretization of 1-D steady problems

Consider the 1-D steady, homogeneous convection-diffusion equation, which is written in conservative form as

$$\frac{d}{dx}(\rho u \phi) = \frac{d}{dx}\left(\Gamma \frac{d\phi}{dx}\right) \qquad (4.103)$$

where ϕ is the quantity to be determined, and Γ is the diffusion coefficient. Γ is related to ε_c in Eq. (4.15) by $\Gamma = \rho \varepsilon_c$. Note that the flow density ρ is included in Eq. (4.103) to consider its possible changes due to sediment, temperature, salinity, etc.

Fig. 4.14 shows the commonly used 1-D finite volume grid. For a grid point P, the point on its west side or in the negative x direction is denoted as W, and the point on its east side or in the positive x direction is denoted as E. The further west and east points are WW and EE, respectively. The control volume (cell) for point P is embraced by two faces w and e, which are located midway (not absolutely necessary) between W and P and between P and E, respectively.

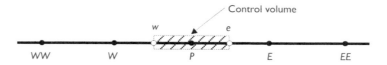

Figure 4.14 1-D finite volume grid.

Integrating Eq. (4.103) over the control volume centered at point P shown in Fig. 4.14 yields

$$(\rho u \phi)_e - (\rho u \phi)_w = \left(\Gamma \frac{d\phi}{dx}\right)_e - \left(\Gamma \frac{d\phi}{dx}\right)_w \qquad (4.104)$$

To complete the discretization, the convection flux $\rho u \phi$ and diffusion flux $\Gamma d\phi/dx$ at faces w and e are determined using the schemes described below.

Central scheme

The central scheme adopts a piecewise linear profie for ϕ, as shown in Fig. 4.15. Thus, the values of ϕ at cell faces are given as the average of two neighboring nodal values:

$$\phi_w = \frac{1}{2}(\phi_P + \phi_W), \quad \phi_e = \frac{1}{2}(\phi_E + \phi_P) \qquad (4.105)$$

and the diffusion fluxes are determined by

$$\left(\Gamma \frac{d\phi}{dx}\right)_w = \frac{\Gamma_w(\phi_P - \phi_W)}{\Delta x_w}, \quad \left(\Gamma \frac{d\phi}{dx}\right)_e = \frac{\Gamma_e(\phi_E - \phi_P)}{\Delta x_e}, \qquad (4.106)$$

where Δx_w and Δx_e are the distances from W to P and from P to E, respectively.

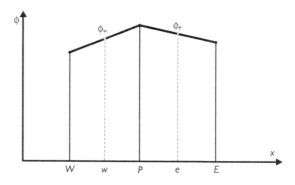

Figure 4.15 Piecewise linear profie in central scheme.

Substituting Eqs. (4.105) and (4.106) into Eq. (4.104) yields

$$\frac{1}{2}(\rho u)_e(\phi_E + \phi_P) - \frac{1}{2}(\rho u)_w(\phi_P + \phi_W) = \frac{\Gamma_e(\phi_E - \phi_P)}{\Delta x_e} - \frac{\Gamma_w(\phi_P - \phi_W)}{\Delta x_w}$$

(4.107)

The values of ρ, u, and Γ at faces w and e can be obtained by interpolation of their values at points W, P, and E. This is to be discussed in the end of Section 4.3.2 for general situations. The discretized equation (4.107) is reformulated as

$$a_P\phi_P = a_W\phi_W + a_E\phi_E$$

(4.108)

where a_W, a_P, and a_E are coefficients:

$$a_W = D_w + F_w/2$$
$$a_E = D_e - F_e/2$$
$$a_P = a_W + a_E + (F_e - F_w)$$

(4.109)

with $F = \rho u$ and $D = \Gamma/\Delta x$.

Integrating the 1-D continuity equation over the control volume shown in Fig. 4.14 leads to $F_e = F_w$, which is not introduced here in detail. Therefore, $F_e - F_w$ can be eliminated from the expression of a_P in Eq. (4.109).

Because the coefficients in Eq. (4.109) could become negative and $|a_P| < |a_E| + |a_W|$ when $|F| > 2D$ (or $|P| > 2$), the central scheme may result in unrealistic solutions; see also Section 4.2.1.2. Here, P is the Peclet number, defined in Eq. (4.20) or as F/D. The numerical oscillations for the central scheme at large Peclet numbers are due to the assumption that the convected property of ϕ at a cell face is given the average of its values at two neighboring points. Schemes that overcome this problem are upwind scheme (Courant *et al.*, 1952), exponential scheme (Spalding, 1972), hybrid upwind/central scheme (Spalding, 1972), QUICK scheme (Leonard, 1979), SOUCUP (Zhu and Rodi, 1991), HLPA scheme (Zhu, 1991), etc., as discussed below.

Upwind scheme

In the upwind scheme, the formulation of the diffusion flux remains unchanged. For the convection flux the value of ϕ at face w is set as its value at the upwind adjacent grid point, as shown in Fig. 4.16, thus yielding

$$\phi_w = \begin{cases} \phi_W, & \text{if} \quad F_w \geq 0 \\ \phi_P, & \text{if} \quad F_w < 0 \end{cases}$$

(4.110)

which can be rewritten as

$$F_w\phi_w = \phi_W \max(F_w, 0) - \phi_P \max(-F_w, 0)$$

(4.111)

An expression similar to Eq. (4.111) can be derived for the convection flux at face e. When Eq. (4.105) is replaced by this concept, the coefficients of the discretized

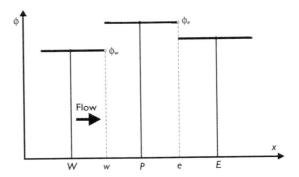

Figure 4.16 Stepwise profie in upwind scheme.

equation (4.108) become

$$a_W = D_w + \max(F_w, 0)$$
$$a_E = D_e + \max(-F_e, 0) \qquad (4.112)$$
$$a_P = a_W + a_E + (F_e - F_w)$$

It is evident that no negative coefficients would arise in Eq. (4.112). Thus, the solution will always be physically realistic. However, the upwind scheme is only first-order accurate and has strong numerical diffusion.

Hybrid scheme

As mentioned above, the central scheme is second-order accurate, but it may encounter difficulties when $|F| > 2D$; while the upwind scheme can solve these difficulties although it is only first-order accurate. Combining these two schemes leads to a hybrid scheme, which has the advantages of both schemes. The concept is that when $|F| \leq 2D$, the central scheme is used, and when $|F| > 2D$, the upwind scheme is used. Thus, the coefficient a_W for the hybrid scheme is

$$a_W = \begin{cases} F_w, & \text{if} \quad P_w > 2 \\ D_w + F_w/2, & \text{if} \quad -2 \leq P_w \leq 2 \\ 0 & \text{if} \quad P_w < -2 \end{cases} \qquad (4.113)$$

where P_w is the Peclet number at face w. The resulting discretized equation can then be written as Eq. (4.108) with coefficients:

$$a_W = \max(F_w, D_w + F_w/2, 0)$$
$$a_E = \max(-F_e, D_e - F_e/2, 0) \qquad (4.114)$$
$$a_P = a_W + a_E + (F_e - F_w)$$

Exponential scheme

As discussed in Section 4.2.1.2, if ρu and Γ are assumed to be constant, Eq. (4.103) has an exact solution. If a domain $0 \leq x \leq L$ is considered, with boundary conditions $\phi = \phi_0$ at $x = 0$ and $\phi = \phi_L$ at $x = L$, the solution of Eq. (4.103) is

$$\frac{\phi - \phi_0}{\phi_L - \phi_0} = \frac{\exp(\rho u x / \Gamma) - 1}{\exp(\rho u L / \Gamma) - 1} \tag{4.115}$$

Define the total flux $I = \rho u \phi - \Gamma d\phi/dx$. Using the exact solution (4.115) as a profie between points P and E, as shown in Fig. 4.17, yields the expression for I_w:

$$I_w = F_w \left[\phi_W + \frac{\phi_W - \phi_P}{\exp(P_w) - 1} \right] \tag{4.116}$$

Substituting Eq. (4.116) and a similar expression for I_e into Eq. (4.104) leads to

$$F_e \left[\phi_P + \frac{\phi_P - \phi_E}{\exp(P_e) - 1} \right] - F_w \left[\phi_W + \frac{\phi_W - \phi_P}{\exp(P_w) - 1} \right] = 0 \tag{4.117}$$

which can be written as Eq. (4.108) with coefficients:

$$a_W = \frac{F_w \exp(F_w/D_w)}{\exp(F_w/D_w) - 1}$$

$$a_E = \frac{F_e}{\exp(F_e/D_e) - 1} \tag{4.118}$$

$$a_P = a_W + a_E + (F_e - F_w)$$

The exponential scheme (4.117) is based on the formulation first presented by Spalding (1972). It is similar to the exponential difference scheme introduced in Section 4.2.1.2.

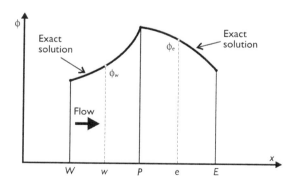

Figure 4.17 Sketch of exponential scheme.

QUICK scheme

As shown in Fig. 4.18, QUICK (Quadratic Upwind Interpolation for Convective Kinematics) scheme, proposed by Leonard (1979), approximates the face value ϕ_w by fitting a parabolic curve through the values of ϕ at points WW, W, and P when $F_w \geq 0$, and at W, P, and E when $F_w < 0$:

$$
\phi_w = \begin{cases}
-\dfrac{1}{8}\phi_{WW} + \dfrac{3}{4}\phi_W + \dfrac{3}{8}\phi_P & F_w \geq 0 \\[2mm]
\dfrac{3}{8}\phi_W + \dfrac{3}{4}\phi_P - \dfrac{1}{8}\phi_E & F_w < 0
\end{cases}
\tag{4.119}
$$

A similar expression can be derived for the value of ϕ at face e. The interpolation scheme (4.119) has a third-order truncation error.

The QUICK scheme is widely applied, but it may have numerical oscillations where the function ϕ changes sharply.

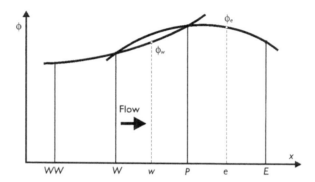

Figure 4.18 Quadratic profie in QUICK scheme.

SOUCUP scheme

SOUCUP (Composite Second-Order Upwind/Central Difference/First-Order Upwind) scheme was proposed by Zhu and Rodi (1991). When $F_w \geq 0$, the SOUCUP scheme approximates the face value ϕ_w as

$$
\phi_w = \begin{cases}
1.5\phi_W - 0.5\phi_{WW} & 0 \leq \hat{\phi}_W \leq 0.5 \\[1mm]
0.5(\phi_P + \phi_W) & 0.5 < \hat{\phi}_W \leq 1 \\[1mm]
\phi_W & \text{otherwise}
\end{cases}
\tag{4.120}
$$

where $\hat{\phi}_W = (\phi_W - \phi_{WW})/(\phi_P - \phi_{WW})$.

Fig. 4.19 shows the relation between $\hat{\phi}_W$ and $\hat{\phi}_w$ for the SOUCUP scheme. When $0 \leq \hat{\phi}_W \leq 0.5, \phi_w$ is approximated by the second-order upwind scheme. When $0.5 < \hat{\phi}_W \leq 1, \phi_w$ is determined by the central scheme. When $\hat{\phi}_W < 0$ or $\hat{\phi}_W > 1, \phi_w$ is approximated by the first-order upwind scheme.

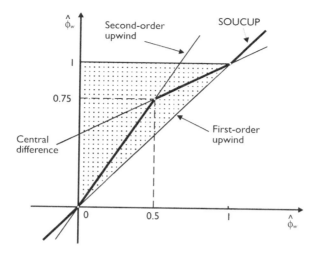

Figure 4.19 Relation between $\hat{\phi}_W$ and $\hat{\phi}_w$ for SOUCUP scheme.

HLPA scheme

HLPA (Hybrid Linear/Parabolic Approximation) scheme was proposed by Zhu (1991). When $F_w \geq 0$, the HLPA scheme approximates the face value ϕ_w as

$$\phi_w = \phi_W + \gamma_w(\phi_P - \phi_W)\frac{\phi_W - \phi_{WW}}{\phi_P - \phi_{WW}} \tag{4.121}$$

where $\gamma_w = 1$ if $0 \leq \hat{\phi}_W \leq 1$; otherwise, $\gamma_w = 0$. $\hat{\phi}_W$ is defined in Eq. (4.120).

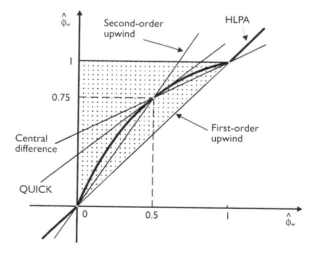

Figure 4.20 Relation between $\hat{\phi}_W$ and $\hat{\phi}_w$ for HLPA scheme.

Fig. 4.20 shows the relation between $\hat{\phi}_W$ and $\hat{\phi}_w$ for the HLPA scheme. When $0 \le \hat{\phi}_W \le 1, \phi_w$ is approximated by the parabolic function through three points $(0, 0)$, $(0.5, 0.75)$, and $(1, 1)$. When $\hat{\phi}_W < 0$ or $\hat{\phi}_W > 1, \phi_w$ is approximated by the first-order upwind scheme. Note that the point $(0.5, 0.75)$ is the intersect point of the QUICK scheme, central scheme, and second-order upwind scheme in the $(\hat{\phi}_W, \hat{\phi}_w)$ plane.

The SOUCUP and HLPA schemes have good performance for convection-dominated problems.

4.3.1.2 Discretization of 1-D unsteady problems

Integrating the 1-D unsteady, heterogeneous convection-diffusion equation

$$\frac{\partial(\rho\phi)}{\partial t} + \frac{\partial}{\partial x}(\rho u \phi) = \frac{\partial}{\partial x}\left(\Gamma \frac{\partial\phi}{\partial x}\right) + S \tag{4.122}$$

over the control volume centered at point P shown in Fig. 4.14 yields

$$\frac{\partial(\rho\phi)}{\partial t}\Delta x_P + (\rho u \phi)_e - (\rho u \phi)_w = \left(\Gamma \frac{\partial\phi}{\partial x}\right)_e - \left(\Gamma \frac{\partial\phi}{\partial x}\right)_w + S\Delta x_P \tag{4.123}$$

where Δx_P is the length of the control volume.

Applying the backward difference scheme (4.23) for the time-derivative term, one of the numerical schemes introduced in Section 4.3.1.1 for the convection fluxes, and the central difference scheme for the diffusion fluxes in Eq. (4.123) yields

$$\frac{(\rho\phi)_P^{n+1} - (\rho\phi)_P^n}{\Delta t}\Delta x_P = a_W\phi_W^{n+1} + a_E\phi_E^{n+1} - a_P\phi_P^{n+1} + S\Delta x_P \tag{4.124}$$

The source term in Eq. (4.124) can be linearized as

$$S\Delta x_P = S_U + S_P\phi_P^{n+1} \tag{4.125}$$

where S_U and S_P are coefficients. The linearization formulation (4.125) should be a good representation of the $S \sim \phi$ relationship, and S_P must be nonpositive (Patankar, 1980).

The final form of the discretized equation is

$$a_P'\phi_P^{n+1} = a_W\phi_W^{n+1} + a_E\phi_E^{n+1} + S_U' \tag{4.126}$$

where

$$a_P' = a_P + \rho_P^{n+1}\Delta x_P/\Delta t - S_P, \quad S_U' = S_U + \rho_P^n\phi_P^n\Delta x_P/\Delta t \tag{4.127}$$

Similarly, one can apply the three-level implicit scheme (4.43) for the time-derivative term in Eq. (4.123). The resulting discretized equation is Eq. (4.126) with

$$a_P' = a_P + 1.5\rho_P^{n+1}\Delta x_P/\Delta t - S_P, \quad S_U' = S_U + (2\rho_P^n\phi_P^n - 0.5\rho_P^{n-1}\phi_P^{n-1})\Delta x_P/\Delta t$$

$$(4.128)$$

4.3.2 Finite volume method for multidimensional problems on fixed grids

Discretization of 2-D transport equation

The 2-D transport equation in the fixed, curvilinear coordinate system is written in conservative form as

$$\frac{\partial}{\partial t}(\rho J\phi) + \frac{\partial}{\partial \xi}\left(\rho J\hat{u}_\xi\phi - \Gamma J\alpha_j^1\alpha_j^1\frac{\partial\phi}{\partial\xi}\right) + \frac{\partial}{\partial\eta}\left(\rho J\hat{u}_\eta\phi - \Gamma J\alpha_j^2\alpha_j^2\frac{\partial\phi}{\partial\eta}\right) = JS$$

$$(4.129)$$

where $\hat{u}_\xi$ and $\hat{u}_\eta$ are the components of flow velocity in the ξ- and η-directions, which are related to the velocity components u_x and u_y in the Cartesian coordinate system by $\hat{u}_\xi = \alpha_1^1 u_x + \alpha_2^1 u_y$ and $\hat{u}_\eta = \alpha_1^2 u_x + \alpha_2^2 u_y$. Note that the source term S includes the cross-derivative terms but excludes the second derivatives of coordinates (curvature terms) that are very sensitive to grid smoothness.

The computational domain is discretized into a finite number of control volumes (cells) by a computational grid. The grid may be the body-fitted grid generated by Eq. (4.74) or another more arbitrary grid. One of the commonly used methods for the control volume setup is shown in Fig. 4.21. The grid lines are identified as cell faces, and the computational point is placed at the geometric center of each control volume. The control volume centered at point P is embraced by four faces w, s, e, and n, which are the linear segments between cell corners nw, sw, se, and ne. It is connected with four adjacent control volumes centered at points W, E, S, and N. Here, W denotes the west or the negative ξ direction, E the east or the positive ξ direction, S the south or the negative η direction, and N the north or the positive η direction.

Integrating the transport equation (4.129) over the control volume shown in Fig. 4.21 yields

$$\frac{\rho_P^{n+1}\phi_P^{n+1} - \rho_P^n\phi_P^n}{\Delta t}(J\Delta\xi\Delta\eta)_P + \left(\rho J\hat{u}_\xi\phi - \Gamma J\alpha_j^1\alpha_j^1\frac{\partial\phi}{\partial\xi}\right)_e^{n+1}\Delta\eta_e$$

$$-\left(\rho J\hat{u}_\xi\phi - \Gamma J\alpha_j^1\alpha_j^1\frac{\partial\phi}{\partial\xi}\right)_w^{n+1}\Delta\eta_w + \left(\rho J\hat{u}_\eta\phi - \Gamma J\alpha_j^2\alpha_j^2\frac{\partial\phi}{\partial\eta}\right)_n^{n+1}\Delta\xi_n$$

$$-\left(\rho J\hat{u}_\eta\phi - \Gamma J\alpha_j^2\alpha_j^2\frac{\partial\phi}{\partial\eta}\right)_s^{n+1}\Delta\xi_s = S(J\Delta\xi\Delta\eta)_P \qquad (4.130)$$

where $\Delta\eta_w$, $\Delta\eta_e$, $\Delta\xi_s$, and $\Delta\xi_n$ are the widths of faces w, e, s, and n in the (ξ,η) coordinate system; and $\Delta\xi_P$ and $\Delta\eta_P$ are the lengths of the control volume centered at P in the ξ- and η-directions.

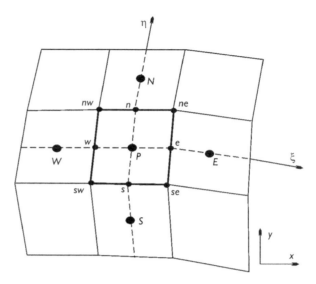

Figure 4.21 Typical 2-D control volume.

The numerical schemes previously introduced for the 1-D case can be extended to determine the convection and diffusion fluxes at faces w, e, s, and n. For example, inserting the exponential scheme (4.116) into Eq. (4.130) leads to

$$\frac{\rho_P^{n+1}\phi_P^{n+1} - \rho_P^n\phi_P^n}{\Delta t}\Delta A_P + F_e\left[\phi_P^{n+1} + \frac{\phi_P^{n+1} - \phi_E^{n+1}}{\exp(F_e/D_e) - 1}\right]$$

$$- F_w\left[\phi_W^{n+1} + \frac{\phi_W^{n+1} - \phi_P^{n+1}}{\exp(F_w/D_w) - 1}\right] + F_n\left[\phi_P^{n+1} + \frac{\phi_P^{n+1} - \phi_N^{n+1}}{\exp(F_n/D_n) - 1}\right]$$

$$- F_s\left[\phi_S^{n+1} + \frac{\phi_S^{n+1} - \phi_P^{n+1}}{\exp(F_s/D_s) - 1}\right] = S\Delta A_P \qquad (4.131)$$

where $\Delta A_P = (J\Delta\xi\Delta\eta)_P$ is the area of the control volume at point P; F_w, F_e, F_s, and F_n are the convection fluxes at cell faces w, e, s, and n, respectively, approximated by the midpoint integral rule as follows:

$$F_w = \rho_w^{n+1}(J\Delta\eta)_w\hat{u}_{\xi,w}^{n+1}, \quad F_e = \rho_e^{n+1}(J\Delta\eta)_e\hat{u}_{\xi,e}^{n+1},$$

$$F_s = \rho_s^{n+1}(J\Delta\xi)_s\hat{u}_{\eta,s}^{n+1}, \quad F_n = \rho_n^{n+1}(J\Delta\xi)_n\hat{u}_{\eta,n}^{n+1}; \qquad (4.132)$$

and D_w, D_e, D_s, and D_n are the diffusion parameters:

$$D_w = (\Gamma J\alpha_j^1\alpha_j^1\Delta\eta)_w/\Delta\xi_w, \quad D_e = (\Gamma J\alpha_j^1\alpha_j^1\Delta\eta)_e/\Delta\xi_e,$$

$$D_s = (\Gamma J\alpha_j^2\alpha_j^2\Delta\xi)_s/\Delta\eta_s, \quad D_n = (\Gamma J\alpha_j^2\alpha_j^2\Delta\xi)_n/\Delta\eta_n. \qquad (4.133)$$

The final discretized transport equation is written as

$$\frac{\rho_P^{n+1}\phi_P^{n+1} - \rho_P^n\phi_P^n}{\Delta t}\Delta A_P = a_W\phi_W^{n+1} + a_E\phi_E^{n+1} + a_S\phi_S^{n+1} + a_N\phi_N^{n+1} - a_P\phi_P^{n+1} + b$$

(4.134)

where

$$a_W = \frac{F_w\exp(F_w/D_w)}{\exp(F_w/D_w) - 1}, \quad a_E = \frac{F_e}{\exp(F_e/D_e) - 1},$$

$$a_S = \frac{F_s\exp(F_s/D_s)}{\exp(F_s/D_s) - 1}, \quad a_N = \frac{F_n}{\exp(F_n/D_n) - 1},$$

$$a_P = a_W + a_E + a_S + a_N + (F_e - F_w + F_n - F_s), \text{ and } b = S\Delta A_P.$$

(4.135)

The term $F_e - F_w + F_n - F_s$ in the coefficient a_P can be treated by using the discretized continuity equation introduced in Section 4.4.

In fact, ΔA_P and the quantities F and D at cell faces in Eqs. (4.132) and (4.133) can be evaluated using only the parameters in the Cartesian coordinate system without involving the increments $\Delta\xi$ and $\Delta\eta$ (Peric, 1985; Zhu, 1992a). The area of the control volume is calculated by

$$\Delta A_P = \frac{1}{2}|(x_{ne} - x_{sw})(y_{nw} - y_{se}) - (x_{nw} - x_{se})(y_{ne} - y_{sw})|$$

(4.136)

The convection fluxes at faces w and s are determined by

$$F_w = \rho_w^{n+1}(J\Delta\eta)_w\hat{u}_{\xi,w}^{n+1} = \rho_w^{n+1}(b_1^1 u_x + b_2^1 u_y)_w^{n+1}$$

$$F_s = \rho_s^{n+1}(J\Delta\xi)_s\hat{u}_{\eta,s}^{n+1} = \rho_s^{n+1}(b_1^2 u_x + b_2^2 u_y)_s^{n+1}$$

(4.137)

where $b_1^1 = J\alpha_1^1\Delta\eta \approx (\partial y/\partial\eta)\Delta\eta$, $b_2^1 = J\alpha_2^1\Delta\eta \approx -(\partial x/\partial\eta)\Delta\eta$, $b_1^2 = J\alpha_1^2\Delta\xi \approx -(\partial y/\partial\xi)\Delta\xi$, and $b_2^2 = J\alpha_2^2\Delta\xi \approx (\partial x/\partial\xi)\Delta\xi$ according to Eq. (4.79). The difference equations for b_i^m at center P and faces w and s of the control volume shown in Fig. 4.21 are:

$$b_{1P}^1 = y_n - y_s, \quad b_{1w}^1 = y_{nw} - y_{sw}, \quad b_{1s}^1 = y_P - y_S,$$

$$b_{2P}^1 = x_s - x_n, \quad b_{2w}^1 = x_{sw} - x_{nw}, \quad b_{2s}^1 = x_S - x_P,$$

$$b_{1P}^2 = y_w - y_e, \quad b_{1w}^2 = y_W - y_P, \quad b_{1s}^2 = y_{sw} - y_{se},$$

$$b_{2P}^2 = x_e - x_w, \quad b_{2w}^2 = x_P - x_W, \quad b_{2s}^2 = x_{se} - x_{sw}.$$

(4.138)

The diffusion parameters at faces w and s are computed by

$$D_w = (\Gamma J\alpha_j^1\alpha_j^1\Delta\eta)_w/\Delta\xi_w = \Gamma_w(b_1^1 b_1^1 + b_2^1 b_2^1)_w/\Delta A_w$$

$$D_s = (\Gamma J\alpha_j^2\alpha_j^2\Delta\xi)_s/\Delta\eta_s = \Gamma_s(b_1^2 b_1^2 + b_2^2 b_2^2)_s/\Delta A_s$$

(4.139)

where $\Delta A_w = (J\Delta\xi\Delta\eta)_w = (\Delta A_W + \Delta A_P)/2$, and $\Delta A_s = (J\Delta\xi\Delta\eta)_s = (\Delta A_S + \Delta A_P)/2$.

Note that only formulations for the quantities at faces w and s are given in Eqs. (4.137)–(4.139). The reason is that the face e of each cell is the face w of the next cell on the east side, and the face n of each cell is the face s of the next cell on the north side. The quantities at each cell face need to be calculated only once. This ensures the quantities across cell faces to be consistent.

Discretization of 3-D transport equation

The 3-D transport equation in the fixed, curvilinear coordinate system is written in conservative form as

$$\frac{\partial}{\partial t}(\rho J\phi) + \frac{\partial}{\partial \xi}\left(\rho J\hat{u}_\xi\phi - \Gamma J\alpha_j^1\alpha_j^1\frac{\partial\phi}{\partial\xi}\right) + \frac{\partial}{\partial \eta}\left(\rho J\hat{u}_\eta\phi - \Gamma J\alpha_j^2\alpha_j^2\frac{\partial\phi}{\partial\eta}\right)$$

$$+ \frac{\partial}{\partial \zeta}\left(\rho J\hat{u}_\zeta\phi - \Gamma J\alpha_j^3\alpha_j^3\frac{\partial\phi}{\partial\zeta}\right) = JS \tag{4.140}$$

where $\hat{u}_\xi$, $\hat{u}_\eta$, and $\hat{u}_\zeta$ are the components of flow velocity in the ξ-, η-, and ζ-directions, related to the velocity components u_x, u_y, and u_z in the Cartesian coordinate system by $\hat{u}_\xi = \alpha_1^1 u_x + \alpha_2^1 u_y + \alpha_3^1 u_z$, $\hat{u}_\eta = \alpha_1^2 u_x + \alpha_2^2 u_y + \alpha_3^2 u_z$, and $\hat{u}_\zeta = \alpha_1^3 u_x + \alpha_2^3 u_y + \alpha_3^3 u_z$.

Fig. 4.22 shows the 3-D control volume centered at point P, which is embraced by six faces w, e, s, n, b, and t. The cell faces are identified by the grid lines, and the point P is placed at the geometric center of the cell. Compared to the 2-D case, point P is connected to two more points B (bottom) and T (top) in the ζ direction. Integrating Eq. (4.140) in this control volume leads to

$$\frac{\rho_P^{n+1}\phi_P^{n+1} - \rho_P^n\phi_P^n}{\Delta t}(J\Delta\xi\Delta\eta\Delta\zeta)_P + \left(\rho J\hat{u}_\xi\phi - \Gamma J\alpha_j^1\alpha_j^1\frac{\partial\phi}{\partial\xi}\right)_e^{n+1}\Delta\eta_e\Delta\zeta_e$$

$$- \left(\rho J\hat{u}_\xi\phi - \Gamma J\alpha_j^1\alpha_j^1\frac{\partial\phi}{\partial\xi}\right)_w^{n+1}\Delta\eta_w\Delta\zeta_w + \left(\rho J\hat{u}_\eta\phi - \Gamma J\alpha_j^2\alpha_j^2\frac{\partial\phi}{\partial\eta}\right)_n^{n+1}\Delta\xi_n\Delta\zeta_n$$

$$- \left(\rho J\hat{u}_\eta\phi - \Gamma J\alpha_j^2\alpha_j^2\frac{\partial\phi}{\partial\eta}\right)_s^{n+1}\Delta\xi_s\Delta\zeta_s + \left(\rho J\hat{u}_\zeta\phi - \Gamma J\alpha_j^3\alpha_j^3\frac{\partial\phi}{\partial\zeta}\right)_t^{n+1}\Delta\xi_t\Delta\eta_t$$

$$- \left(\rho J\hat{u}_\zeta\phi - \Gamma J\alpha_j^3\alpha_j^3\frac{\partial\phi}{\partial\zeta}\right)_b^{n+1}\Delta\xi_b\Delta\eta_b = S(J\Delta\xi\Delta\eta\Delta\zeta)_P \tag{4.141}$$

The backward difference scheme (4.23) is used to discretize the time-derivative term, and the numerical schemes introduced in Section 4.3.1 are employed for the convection and diffusion fluxes, thus yielding

$$\frac{\rho_P^{n+1}\phi_P^{n+1} - \rho_P^n\phi_P^n}{\Delta t}\Delta V_P = a_W\phi_W^{n+1} + a_E\phi_E^{n+1} + a_S\phi_S^{n+1} + a_N\phi_N^{n+1}$$

$$+ a_B\phi_B^{n+1} + a_T\phi_T^{n+1} - a_P\phi_P^{n+1} + b \tag{4.142}$$

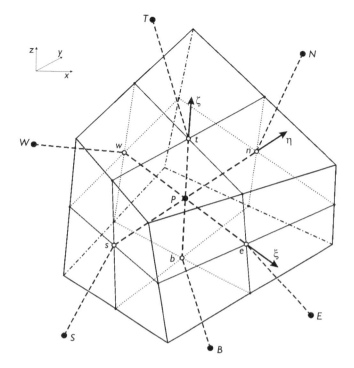

Figure 4.22 Typical 3-D control volume.

where ΔV_P is the volume of the cell centered at point P, defined as $\Delta V_P = (J\Delta\xi\Delta\eta\Delta\zeta)_P$.

If the exponential scheme is used, the coefficients in Eq. (4.142) are

$$a_W = \frac{F_w\exp(F_w/D_w)}{\exp(F_w/D_w)-1}, \quad a_E = \frac{F_e}{\exp(F_e/D_e)-1},$$

$$a_S = \frac{F_s\exp(F_s/D_s)}{\exp(F_s/D_s)-1}, \quad a_N = \frac{F_n}{\exp(F_n/D_n)-1}, \quad (4.143)$$

$$a_B = \frac{F_b\exp(F_b/D_b)}{\exp(F_b/D_b)-1}, \quad a_T = \frac{F_t}{\exp(F_t/D_t)-1},$$

$$a_P = a_W + a_E + a_S + a_N + a_B + a_T$$

$$+ (F_e - F_w + F_n - F_s + F_t - F_b), \text{ and } b = S\Delta V_P$$

where

$$F_w = \rho_w^{n+1}(J\Delta\eta\Delta\zeta)_w\hat{u}_{\xi,w}^{n+1}, \quad F_e = \rho_e^{n+1}(J\Delta\eta\Delta\zeta)_e\hat{u}_{\xi,e}^{n+1},$$

$$F_s = \rho_s^{n+1}(J\Delta\xi\Delta\zeta)_s\hat{u}_{\eta,s}^{n+1}, \quad F_n = \rho_n^{n+1}(J\Delta\xi\Delta\zeta)_n\hat{u}_{\eta,n}^{n+1},$$

$$F_b = \rho_b^{n+1}(J\Delta\xi\Delta\eta)_b \hat{u}_{\zeta,b}^{n+1}, \quad F_t = \rho_t^{n+1}(J\Delta\xi\Delta\eta)_t \hat{u}_{\zeta,t}^{n+1}, \tag{4.144}$$

and

$$D_w = (\Gamma J\alpha_j^1\alpha_j^1\Delta\eta\Delta\zeta)_w/\Delta\xi_w, \quad D_e = (\Gamma J\alpha_j^1\alpha_j^1\Delta\eta\Delta\zeta)_e/\Delta\xi_e,$$

$$D_s = (\Gamma J\alpha_j^2\alpha_j^2\Delta\xi\Delta\zeta)_s/\Delta\eta_s, \quad D_n = (\Gamma J\alpha_j^2\alpha_j^2\Delta\xi\Delta\zeta)_n/\Delta\eta_n, \tag{4.145}$$

$$D_b = (\Gamma J\alpha_j^3\alpha_j^3\Delta\xi\Delta\eta)_b/\Delta\zeta_b, \quad D_t = (\Gamma J\alpha_j^3\alpha_j^3\Delta\xi\Delta\eta)_t/\Delta\zeta_t.$$

Like the 2-D case, ΔV_P and the quantities F and D at cell faces in Eqs. (4.144) and (4.145) can be calculated using only the parameters in the Cartesian coordinate system, and the final discretized equation does not involve the increments $\Delta\xi$, $\Delta\eta$, and $\Delta\zeta$ (Peric, 1985; Zhu, 1992b). This is demonstrated below.

Kordulla and Vinokur (1983) suggested a method to calculate the volume of a 3-D cell. As shown in Fig. 4.23, the cell with points $A, B, \ldots$, and H as its eight vertices is decomposed into six tetrahedra, all containing the same diagonal joining points A and H. The volume of the tetrahedron with vertices A, E, G, and H can be calculated as

$$\Delta V_1 = \frac{1}{6}|(\overrightarrow{AE} \times \overrightarrow{AG}) \cdot \overrightarrow{AH}| \tag{4.146}$$

Thus, the volume of the total cell is

$$\Delta V = \frac{1}{6}\left|\left[\left(\overrightarrow{AE} \times \overrightarrow{AG}\right) + \left(\overrightarrow{AF} \times \overrightarrow{AE}\right) + \left(\overrightarrow{AG} \times \overrightarrow{AC}\right)\right.\right.$$
$$\left.\left. + \left(\overrightarrow{AB} \times \overrightarrow{AF}\right) + \left(\overrightarrow{AC} \times \overrightarrow{AD}\right) + \left(\overrightarrow{AD} \times \overrightarrow{AB}\right)\right] \cdot \overrightarrow{AH}\right| \tag{4.147}$$

It is of interest to note that the above method of calculating cell volumes ensures the conservation of space, i.e., the sum of all cell volumes gives exactly the total volume of the solution domain. This is a necessary condition for guaranteeing the true conservation of transported quantities (Zhu, 1992b).

The convection fluxes at faces w, s, and b are determined by

$$F_w = \rho_w^{n+1}(J\Delta\eta\Delta\zeta)_w \hat{u}_{\xi,w}^{n+1} = \rho_w^{n+1}(b_1^1 u_x + b_2^1 u_y + b_3^1 u_z)_w^{n+1}$$

$$F_s = \rho_s^{n+1}(J\Delta\xi\Delta\zeta)_s \hat{u}_{\eta,s}^{n+1} = \rho_s^{n+1}(b_1^2 u_x + b_2^2 u_y + b_3^2 u_z)_s^{n+1} \tag{4.148}$$

$$F_b = \rho_b^{n+1}(J\Delta\xi\Delta\eta)_b \hat{u}_{\zeta,b}^{n+1} = \rho_b^{n+1}(b_1^3 u_x + b_2^3 u_y + b_3^3 u_z)_b^{n+1}$$

where $b_i^1 = J\alpha_i^1\Delta\eta\Delta\zeta$, $b_i^2 = J\alpha_i^2\Delta\xi\Delta\zeta$, and $b_i^3 = J\alpha_i^3\Delta\xi\Delta\eta$ ($i = 1, 2, 3$). The coefficients α_i^m and, in turn, b_i^m can be calculated using Eq. (4.65) and the cofactors of β_m^i in the Jacobian matrix B in Eq. (4.64). For example, using $b_1^1 = $

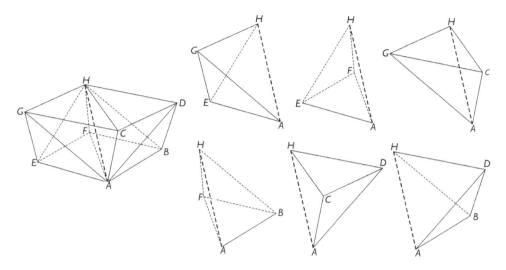

Figure 4.23 Volume of a 3-D cell.

$(\partial y/\partial \eta \; \partial z/\partial \zeta - \partial y/\partial \zeta \; \partial z/\partial \eta)\Delta\eta\Delta\zeta$ and discretizing the coordinate derivatives at cell center P yields

$$b_{1P}^1 = (y_n - y_s)(z_t - z_b) - (y_t - y_b)(z_n - z_s) \qquad (4.149)$$

Following this procedure, one can derive the discretized equations for all b_i^m at cell center P and faces w, s, and b, which are not introduced here.

The diffusion parameters are

$$D_w = (\Gamma J \alpha_j^1 \alpha_j^1 \Delta\eta\Delta\zeta)_w/\Delta\xi_w = \Gamma_w(b_1^1 b_1^1 + b_2^1 b_2^1 + + b_3^1 b_3^1)_w/\Delta V_w$$
$$D_s = (\Gamma J \alpha_j^2 \alpha_j^2 \Delta\xi\Delta\zeta)_s/\Delta\eta_s = \Gamma_s(b_1^2 b_1^2 + b_2^2 b_2^2 + b_3^2 b_3^2)_s/\Delta V_s \qquad (4.150)$$
$$D_b = (\Gamma J \alpha_j^3 \alpha_j^3 \Delta\xi\Delta\eta)_b/\Delta\zeta_b = \Gamma_b(b_1^3 b_1^3 + b_2^3 b_2^3 + b_3^3 b_3^3)_b/\Delta V_b$$

where $\Delta V_w = (J\Delta\xi\Delta\eta\Delta\zeta)_w = (\Delta V_W + \Delta V_P)/2$, $\Delta V_s = (J\Delta\xi\Delta\eta\Delta\zeta)_s = (\Delta V_S + \Delta V_P)/2$, and $\Delta V_b = (J\Delta\xi\Delta\eta\Delta\zeta)_b = (\Delta V_B + \Delta V_P)/2$.

In addition, the values of parameters, such as velocity, density, and diffusivity, at cell faces need to be interpolated from their values at adjacent cell centers. The often used method is linear interpolation. For example, the quantities of ϕ at faces w, s, and b are computed by linear interpolation between the values at two adjacent cell centers of each face as follows:

$$\phi_w = f_{x,P}\phi_P + (1 - f_{x,P})\phi_W$$
$$\phi_s = f_{y,P}\phi_P + (1 - f_{y,P})\phi_S \qquad (4.151)$$
$$\phi_b = f_{z,P}\phi_P + (1 - f_{z,P})\phi_B$$

where $f_{x,P}$, $f_{y,P}$, and $f_{z,P}$ are the interpolation factors, defined as $f_{x,P} = \Delta l_{wW}/(\Delta l_{Pw} + \Delta l_{wW})$, $f_{y,P} = \Delta l_{sS}/(\Delta l_{Ps} + \Delta l_{sS})$, and $f_{z,P} = \Delta l_{bB}/(\Delta l_{Pb} + \Delta l_{bB})$, with Δl_{wW}, Δl_{Pw}, Δl_{sS}, Δl_{Ps}, Δl_{bB}, and Δl_{Pb} being the lengths of linear segments wW, Pw, sS, Ps, bB, and Pb, respectively.

4.3.3 Finite volume method for multidimensional problems on moving grids

The transport equation in the moving, curvilinear coordinate system (ξ_m, τ) can be written in conservative form as

$$\frac{\partial}{\partial \tau}(\rho J \phi) + \frac{\partial}{\partial \xi_m}\left(\rho J \hat{u}_m \phi - \Gamma J \alpha_j^m \alpha_j^m \frac{\partial \phi}{\partial \xi_m}\right) = JS \qquad (4.152)$$

where $\hat{u}_m$ ($m = 1, 2$ or $1, 2, 3$) are the velocity components in the (ξ_m, τ) system, defined in Eq. (4.70).

Compared with Eq. (4.129) or (4.140) on fixed grids, Eq. (4.152) has additional terms related to the moving grid. In particular, $\hat{u}_m$ include the term $\partial \xi_m / \partial t$ that is related to the grid moving velocity. These terms can be eliminated for a steady problem, but they should be considered for an unsteady problem.

Because the grid is moving, it needs to be generated repeatedly. Like the discretization of governing equations, the grid generation can be treated explicitly or implicitly. In the explicit treatment, the grid is generated before the solution of governing equations at every time step, whereas in the implicit treatment, the grid generation is coupled with the solution of governing equations.

The control volume in the moving grid system can still be arranged as Fig. 4.21 or 4.22 at every time step. For a 2-D case, integrating Eq. (4.152) over the control volume, moving the cross-derivative diffusion terms into the source term, and then using the numerical schemes described in Section 4.3.1.1 to determine the convection and normal-derivative diffusion terms on cell faces yields the following discretized equation (Wu, 1996a):

$$\frac{\rho_P^{n+1} \Delta A_P^{n+1} \phi_P^{n+1} - \rho_P^n \Delta A_P^n \phi_P^n}{\Delta \tau} = a_W \phi_W^{n+1} + a_E \phi_E^{n+1} + a_S \phi_S^{n+1}$$
$$+ a_N \phi_N^{n+1} - a_P \phi_P^{n+1} + b \qquad (4.153)$$

Because of grid movement, the control volume area ΔA_P varies with time. The coefficients in Eq. (4.153) are evaluated in the same way as for the fixed grid in Section 4.3.2.

4.4 NUMERICAL SOLUTION OF NAVIER-STOKES EQUATIONS

For incompressible flows, the momentum (Navier-Stokes) equations link the velocity to the pressure gradient, while the continuity equation is just an additional constraint on the velocity field without directly linking to the pressure. Because of such a weak linkage, the convergence and stability of a numerical solution of the Navier-Stokes equations depend largely on how the pressure gradient and velocity are evaluated. Storing the variables at the geometric center of the control volume coupled with the use of linear interpolation for internodal variation usually leads to non-physical node-to-node (checkerboard) oscillations. One approach for eliminating such oscillations is to use the staggered grid, as adopted in Harlow and Welch's (1965) MAC (Marker and Cell) method, Chorin's (1968) projection method, and Patankar and Spalding's (1972) SIMPLE (Semi-Implicit Method for Pressure-Linked Equations) algorithm. The other approach is to use the momentum interpolation technique proposed by Rhie and Chow (1983) based on the non-staggered grid. In addition, the stream function and vorticity approach is also useful for solving the 2-D Navier-Stokes equations.

4.4.1 Primitive variables: MAC formulation on staggered grid

The 2-D Navier-Stokes equations (2.34) and (2.35) with a constant flow density are written in the Cartesian coordinate system as

$$\frac{\partial u_x}{\partial x} + \frac{\partial u_y}{\partial y} = 0 \tag{4.154}$$

$$\frac{\partial u_x}{\partial t} + u_x \frac{\partial u_x}{\partial x} + u_y \frac{\partial u_x}{\partial y} = \frac{1}{\rho} F_x - \frac{1}{\rho} \frac{\partial p}{\partial x} + \frac{1}{\rho} \frac{\partial \tau_{xx}}{\partial x} + \frac{1}{\rho} \frac{\partial \tau_{xy}}{\partial y} \tag{4.155}$$

$$\frac{\partial u_y}{\partial t} + u_x \frac{\partial u_y}{\partial x} + u_y \frac{\partial u_y}{\partial y} = \frac{1}{\rho} F_y - \frac{1}{\rho} \frac{\partial p}{\partial y} + \frac{1}{\rho} \frac{\partial \tau_{yx}}{\partial x} + \frac{1}{\rho} \frac{\partial \tau_{yy}}{\partial y} \tag{4.156}$$

The MAC method first proposed by Harlow and Welch (1965) solves the Navier-Stokes equations on the staggered rectangular grid, which stores the variables u_x, u_y, and p at different grid points, as shown in Fig. 4.24. The continuity equation (4.154) is discretized as

$$D_{i,j}^{n+1} = \frac{u_{x,i+1/2,j}^{n+1} - u_{x,i-1/2,j}^{n+1}}{\Delta x} + \frac{u_{y,i,j+1/2}^{n+1} - u_{y,i,j-1/2}^{n+1}}{\Delta y} = 0 \tag{4.157}$$

where $D_{i,j}^{n+1}$ is the dilatation of the cell (i,j).

The momentum equations (4.155) and (4.156) are discretized as

$$u_{x,i+1/2,j}^{n+1} = u_{x,i+1/2,j}^{n} + F_{i+1/2,j}^{n} - \frac{\Delta t}{\rho \Delta x}(p_{i+1,j}^{n+1} - p_{i,j}^{n+1}) \tag{4.158}$$

$$u_{y,i,j+1/2}^{n+1} = u_{y,i,j+1/2}^{n} + G_{i,j+1/2}^{n} - \frac{\Delta t}{\rho \Delta y}(p_{i,j+1}^{n+1} - p_{i,j}^{n+1}) \tag{4.159}$$

where $F_{i+1/2,j}^{n}$ and $G_{i,j+1/2}^{n}$ include the convection and diffusion terms in the momentum equations (4.155) and (4.156) discretized by the finite difference schemes introduced in Section 4.2.

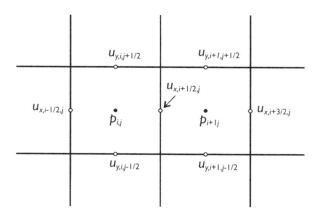

Figure 4.24 Staggered grid in MAC method.

Substituting Eqs. (4.158) and (4.159) into (4.157) leads to the discretized Poisson equation for the pressure:

$$\frac{p_{i+1,j}^{n+1} - 2p_{i,j}^{n+1} + p_{i-1,j}^{n+1}}{\Delta x^2} + \frac{p_{i,j+1}^{n+1} - 2p_{i,j}^{n+1} + p_{i,j-1}^{n+1}}{\Delta y^2}$$

$$= \frac{\rho D_{i,j}^n}{\Delta t} + \frac{\rho}{\Delta t}\left(\frac{F_{i+1/2,j}^n - F_{i-1/2,j}^n}{\Delta x} + \frac{G_{i,j+1/2}^n - G_{i,j-1/2}^n}{\Delta y}\right) \qquad (4.160)$$

In Eq. (4.160) $D_{i,j}^n/\Delta t$ may be interpreted as a discretization of $-(\partial D/\partial t)_{i,j}$ with $D_{j,k}^{n+1} = 0$. Thus, the pressure solution resulting from Eq. (4.160) is such as to allow the discretized continuity equation (4.157) to be satisfied at time level $n+1$.

Eq. (4.160) can be solved by using an iterative or direct method. Once it is solved, substituting the obtained p^{n+1} into Eqs. (4.158) and (4.159) permits u_x^{n+1} and u_y^{n+1} to be calculated. Because Eqs. (4.158) and (4.159) are explicit algorithms, the maximum time step for a stable solution is restricted by (Peyret and Taylor, 1983)

$$0.25(|u_x| + |u_y|)^2 \Delta t \, \text{Re} \, \leq 1 \quad \text{and} \quad \Delta t/(\text{Re}\Delta x^2) \leq 0.25 \qquad (4.161)$$

with the assumption of $\Delta x = \Delta y$. Re is the Reynolds number.

4.4.2 Primitive variables: projection formulation on staggered grid

The projection method first proposed by Chorin (1968) solves the transport equations to predict intermediate velocities and then project these velocities onto a space of divergence-free field. Write the Navier-Stokes equations as

$$\frac{\partial \mathbf{u}}{\partial t} + (\mathbf{u} \cdot \nabla)\mathbf{u} = \mathbf{F} - \frac{1}{\rho}\nabla p + \nu \Delta \mathbf{u} \tag{4.162}$$

$$\nabla \cdot \mathbf{u} = 0 \tag{4.163}$$

where $\mathbf{u}$ is the velocity vector; $\mathbf{F}$ is the external force; ∇ is the divergence or gradient operator, defined as $\nabla = \mathbf{i}_x \partial/\partial x + \mathbf{i}_y \partial/\partial y + \mathbf{i}_z \partial/\partial z$, with $\mathbf{i}_x$, $\mathbf{i}_y$, and $\mathbf{i}_z$ being unit vectors in the x-, y-, and z-axes in the Cartesian coordinate system; and Δ is the Laplace operator, defined as $\Delta = \partial^2/\partial x^2 + \partial^2/\partial y^2 + \partial^2/\partial z^2$.

The projection method consists of two steps in time. The first step computes the intermediate velocity field $\mathbf{u}^*$ by omitting the pressure term from the momentum equation:

$$\frac{\mathbf{u}^* - \mathbf{u}^n}{\Delta t} + (\mathbf{u}^n \cdot \nabla)\mathbf{u}^n = \mathbf{F} + \nu \Delta \mathbf{u}^* \tag{4.164}$$

The second step projects $\mathbf{u}^*$ to the space of divergence-free field to obtain $\mathbf{u}^{n+1}$:

$$\begin{cases} \mathbf{u}^{n+1} = \mathbf{u}^* - \dfrac{\Delta t}{\rho}\nabla p^{n+1} \\ \nabla \cdot \mathbf{u}^{n+1} = 0 \end{cases} \tag{4.165}$$

Substituting the first equation into the second equation in (4.165) yields a Poisson equation for the pressure:

$$\frac{\Delta t}{\rho}\Delta p^{n+1} = \nabla \cdot \mathbf{u}^* \tag{4.166}$$

To solve Eq. (4.166), the following boundary condition is often applied:

$$\frac{\partial p^{n+1}}{\partial \mathbf{n}} = 0 \tag{4.167}$$

where $\mathbf{n}$ denotes the direction normal to the boundary.

In the projection method, Eqs. (4.162) and (4.163) are usually discretized on a staggered grid (such as the MAC grid in Fig. 4.24). The convection terms are commonly discretized using an upwind scheme, and the diffusion terms can be discretized using the central difference scheme.

Various variants of the projection method have been proposed in the literature to solve the shallow water equations and the Navier-Stokes equations. Some of them are introduced in Chapters 6 and 7.

4.4.3   Primitive variables: SIMPLE(C) formulation on staggered grid

SIMPLE algorithm

In the finite volume method, the 2-D Navier-Stokes equations are usually written in conservative form as

$$\frac{\partial \rho}{\partial t} + \frac{\partial(\rho u_x)}{\partial x} + \frac{\partial(\rho u_y)}{\partial y} = 0 \tag{4.168}$$

$$\frac{\partial(\rho u_x)}{\partial t} + \frac{\partial(\rho u_x^2)}{\partial x} + \frac{\partial(\rho u_y u_x)}{\partial y} = F_x - \frac{\partial p}{\partial x} + \frac{\partial \tau_{xx}}{\partial x} + \frac{\partial \tau_{xy}}{\partial y} \tag{4.169}$$

$$\frac{\partial(\rho u_y)}{\partial t} + \frac{\partial(\rho u_x u_y)}{\partial x} + \frac{\partial(\rho u_y^2)}{\partial y} = F_y - \frac{\partial p}{\partial y} + \frac{\partial \tau_{yx}}{\partial x} + \frac{\partial \tau_{yy}}{\partial y} \tag{4.170}$$

Note that the flow density ρ may vary with sediment concentration, temperature, salinity, etc.

Fig. 4.25(a) shows the staggered grid used in the SIMPLE algorithm of Patankar and Spalding (1972). For simplicity, a rectangular grid is used here. The control volume for the x-momentum equation is shown in Fig. 4.25(b). Applying the finite volume discretization introduced in Section 4.3.2 to Eq. (4.169) in this control volume leads to the following discretized equation for $u_{x,e}$:

$$a_e^u u_{x,e}^{n+1} = \sum_l a_l^u u_{x,l}^{n+1} + S_u + A_e(p_P^{n+1} - p_E^{n+1}) \tag{4.171}$$

where A_e is the width of face e, i.e., Δy_e. Note that the index l sweeps over all four u_x neighbors outside the control volume in Fig. 4.25(b).

As explained in Eq. (4.126), in the case of unsteady flow the discretized time-derivative term is split and added to the source term S_{ui} and the coefficient a_e^u. Therefore, Eq. (4.171) can be used for both steady and unsteady flows.

The control volume for the y-momentum equation is shown in Fig. 4.25(c). The discretized equation for $u_{y,n}$ can be written as

$$a_n^v u_{y,n}^{n+1} = \sum_l a_l^v u_{y,l}^{n+1} + S_v + A_n(p_P^{n+1} - p_N^{n+1}) \tag{4.172}$$

where $A_n = \Delta x_n$.

Once the pressure field is given, the discretized momentum equations (4.171) and (4.172) can be solved. However, the pressure field is still to be determined. In an iterative solution process, a pressure field p^* is first guessed and then an approximate velocity field is obtained using the following equations:

$$a_e^u u_{x,e}^* = \sum_l a_l^u u_{x,l}^* + S_u + A_e(p_P^* - p_E^*) \tag{4.173}$$

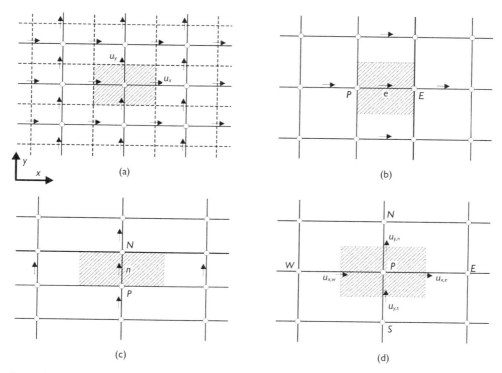

Figure 4.25 Staggered grid in SIMPLE algorithm: (a) sketch of global grid, (b) control volume for u_x, (c) control volume for u_y, and (d) control volume for p.

$$a_n^v u_{y,n}^* = \sum_l a_l^v u_{y,l}^* + S_v + A_n(p_P^* - p_N^*) \tag{4.174}$$

The approximate velocities u_x^* and u_y^* do not satisfy the continuity equation. Hence, the pressure correction p' and velocity corrections u_x' and u_y' are defined as

$$p' = p^{n+1} - p^* \tag{4.175}$$

$$u_x' = u_x^{n+1} - u_x^*, \quad u_y' = u_y^{n+1} - u_y^* \tag{4.176}$$

Subtracting Eq. (4.173) from Eq. (4.171) and neglecting the term $\sum_l a_l^u u_{x,l}'$ yields the u_x'-equation:

$$u_{x,e}' = d_e(p_P' - p_E') \tag{4.177}$$

where $d_e = A_e/a_e^u$.

In a similar manner, the u'_y-equation is derived by subtracting Eqs. (4.172) and (4.174) as

$$u'_{y,n} = d_n(p'_P - p'_N) \qquad (4.178)$$

where $d_n = A_n/a_n^v$.

The control volume for the pressure is shown in Fig. 4.25(d), over which the continuity equation (4.168) can be integrated as

$$\frac{\rho_P^{n+1} - \rho_P^n}{\Delta t}\Delta x\Delta y + [(\rho u_x)_e^{n+1} - (\rho u_x)_w^{n+1}]\Delta y + [(\rho u_y)_n^{n+1} - (\rho u_y)_s^{n+1}]\Delta x = 0 \qquad (4.179)$$

Inserting Eqs. (4.176)–(4.178) into Eq. (4.179) leads to the discrete equation for p':

$$a_P^p p'_P = a_W^p p'_W + a_E^p p'_E + a_S^p p'_S + a_N^p p'_N + b_p \qquad (4.180)$$

where $a_W^p = \rho_w^{n+1} d_w \Delta y$, $a_E^p = \rho_e^{n+1} d_e \Delta y$, $a_S^p = \rho_s^{n+1} d_s \Delta x$, $a_N^p = \rho_n^{n+1} d_n \Delta x$, $a_P^p = a_W^p + a_E^p + a_S^p + a_N^p$, and $b_p = -(\rho_P^{n+1} - \rho_P^n)\Delta x\Delta y/\Delta t - (\rho_e^{n+1}u_{x,e}^* - \rho_w^{n+1}u_{x,w}^*)\Delta y - (\rho_n^{n+1}u_{y,n}^* - \rho_s^{n+1}u_{y,s}^*)\Delta x$.

The computation is performed in the following sequence:

(1) Guess the pressure field p^*;
(2) Solve the momentum equations (4.173) and (4.174) to obtain u_x^* and u_y^*;
(3) Calculate p' using (4.180);
(4) Calculate p using Eq. (4.175);
(5) Calculate u_x^{n+1} and u_y^{n+1} using Eqs. (4.176)–(4.178);
(6) Treat the corrected pressure p as a new guessed pressure p^*, and repeat the procedure from step 2 to 6 until a converged solution is obtained, and
(7) Conduct the calculation of the next time step if unsteady flow is concerned.

SIMPLEC algorithm

Because the term $\sum_l a_l^u u'_{x,l}$ is neglected in the derivation of Eq. (4.177), the pressure is not exactly solved in the aforementioned SIMPLE algorithm. Several algorithms, such as SIMPLER (SIMPLE Revised, Patankar, 1980), PISO (Issa, 1982), and SIMPLEC (SIMPLE Consistence, van Doormaal and Raithby, 1984), have been proposed to improve this. Van Doormaal and Raithby (1984) have reported that significant savings on computation time can be achieved by the SIMPLEC algorithm in several applications, as compared to the SIMPLE and SIMPLER algorithms. Therefore, the SIMPLEC algorithm is introduced below.

The full velocity correction equation reads

$$a_e^u u'_{x,e} = \sum_l a_l^u u'_{x,l} + A_e(p'_P - p'_E) \qquad (4.181)$$

To introduce a consistent approximation, the term $\sum_l a_l^u u_{x,e}'$ is subtracted from both sides of Eq. (4.181), yielding

$$\left(a_e^u - \sum_l a_l^u\right) u_{x,e}' = \sum_l a_l^u (u_{x,l}' - u_{x,e}') + A_e(p_P' - p_E') \qquad (4.182)$$

Assuming that all $u_{x,k}'$ are of the same order of magnitude as $u_{x,e}'$, the first term on the right-hand side of Eq. (4.182) can be omitted, thus yielding the new u_x'-equation at face e as

$$u_{x,e}' = \tilde{d}_e(p_P' - p_E') \qquad (4.183)$$

where $\tilde{d}_e = A_e/(a_e^u - \sum_l a_l^u)$.

In a similar manner, the new u_y'-equation at face n is derived as

$$u_{y,n}' = \tilde{d}_n(p_P' - p_N') \qquad (4.184)$$

where $\tilde{d}_n = A_n/(a_n^v - \sum_l a_l^v)$.

Substituting Eqs. (4.176), (4.183), and (4.184) into Eq. (4.179) leads to the pressure correction equation (4.180) with d_e, d_w, d_n, and d_s replaced by $\tilde{d}_e, \tilde{d}_w, \tilde{d}_n$, and $\tilde{d}_s$, respectively.

The SIMPLEC and SIMPLE algorithms have almost the same computational sequence. Therefore, they can be implemented easily together in a computer code.

Note that both SIMPLE and SIMPLEC algorithms ignore terms in the derivation of the pressure correction equation. This is not essential, because the pressure correction equation is only an intermediate algorithm that leads to the correct pressure field, without directly affecting the final solution. As long as a converged solution is obtained ($p' \to 0$), all formulations of the p' equation will give the same final solution (Pantakar, 1980, p.128).

4.4.4 Primitive variables: SIMPLE(C) formulation on non-staggered grid

The non-staggered grid, also called the collocated grid, stores all variables on the same set of grid points. Many terms in the discretized equations on the non-staggered grid are identical, and the number of coefficients that must be computed and stored is minimized. Thus, the non-staggered grid approach has simpler computer codes and handles complex domains more easily than the staggered grid approach, especially in 3-D situations. However, the non-staggered grid encountered difficulties in the coupling of pressure and velocity, as well as numerical oscillations in the pressure field; thus, it had rarely been used for the computation of incompressible flows, until Rhie and Chow (1983) proposed the momentum interpolation technique. This interpolation technique improved the pressure-velocity coupling on the non-staggered grid.

The works of Peric (1985), Majumdar (1988), and Ferziger and Peric (1995) further popularized the non-staggered grid. The SIMPLE and SIMPLEC algorithms based on the non-staggered grid with the momentum interpolation technique are introduced below.

SIMPLE algorithm

The conservative form of the 2-D Navier-Stokes equations in the curvilinear grid system reads

$$\frac{\partial}{\partial \tau}(\rho J) + \frac{\partial}{\partial \xi_m}(\rho J \hat{u}_m) = 0 \tag{4.185}$$

$$\frac{\partial}{\partial \tau}(\rho J u_i) + \frac{\partial}{\partial \xi_m}\left(\rho J \hat{u}_m u_i - \Gamma J \alpha_j^m \alpha_j^m \frac{\partial u_i}{\partial \xi_m}\right) = -\frac{\partial}{\partial \xi_m}(J \alpha_i^m p) + JS \tag{4.186}$$

where S includes the cross-derivative diffusion terms and the external forces.

Discretizating the momentum equation (4.186) in the control volume shown in Fig. 4.21 yields the following equation for velocities $u_{i,P}$ ($i = 1, 2$):

$$u_{i,P}^{n+1} = \frac{1}{a_P^u}\left(\sum_{l=W,E,S,N} a_l^u u_{i,l}^{n+1} + S_{ui}\right) + D_i^1(p_w^{n+1} - p_e^{n+1}) + D_i^2(p_s^{n+1} - p_n^{n+1}) \tag{4.187}$$

where $D_i^1 = (J\alpha_i^1 \Delta\eta)_P/a_P^u$, and $D_i^2 = (J\alpha_i^2 \Delta\xi)_P/a_P^u$. Note that the values of p at faces $w, e, s,$ and n are calculated by linear interpolation between two adjacent points, as expressed in Eq. (4.151).

In analogy to Eq. (4.126), Eq. (4.187) can be used for both steady and unsteady flows.

After an under-relaxation is introduced to stabilize the iterative solution process, Eq. (4.187) is rewritten as (Majumdar, 1988)

$$u_{i,P}^{n+1} = \alpha_u[H_{i,P} + D_i^1(p_w^{n+1} - p_e^{n+1}) + D_i^2(p_s^{n+1} - p_n^{n+1})] + (1 - \alpha_u)u_{i,P}^o \tag{4.188}$$

where $H_{i,P} = (\sum_{l=W,E,S,N} a_l^u u_{i,l}^{n+1} + S_{ui})/a_P^u$, and $u_{i,P}^o$ are the old values of $u_{i,P}^{n+1}$ in the previous iteration step.

Because the pressure is unknown, a pressure field p^* is guessed, and then the approximate values of the velocities are obtained by

$$u_{i,P}^* = \alpha_u[H_{i,P}^* + D_i^1(p_w^* - p_e^*) + D_i^2(p_s^* - p_n^*)] + (1 - \alpha_u)u_{i,P}^o \tag{4.189}$$

Subtracting Eq. (4.189) from Eq. (4.188) and neglecting the terms $H_{i,P} - H_{i,P}^*$ leads to the relation of velocity and pressure corrections at cell center:

$$u_{i,P}^{n+1} = u_{i,P}^* + \alpha_u[D_i^1(p_w' - p_e') + D_i^2(p_s' - p_n')] \tag{4.190}$$

and the pressure correction has the following relation:

$$p^{n+1} = p^* + p' \tag{4.191}$$

The momentum interpolation technique proposed by Rhie and Chow (1983) calculates the values of u_i at face w as

$$u_{i,w}^* = \alpha_u[(1 - f_{x,P})G_{i,PW}^{1*} + f_{x,P}G_{i,P}^{1*}] + \alpha_u[(1 - f_{x,P})/a_{PW}^u + f_{x,P}/a_P^u]$$
$$\times (J\alpha_i^1 \Delta \eta)_w(p_W^* - p_P^*) + (1 - \alpha_u)[(1 - f_{x,P})u_{i,W}^o + f_{x,P}u_{i,P}^o] \tag{4.192}$$

where $G_{i,P}^{1*} = H_{i,P}^* + D_i^2(p_s^* - p_n^*)$, and $G_{i,PW}^{1*}$ and a_{PW}^u are the values of $G_{i,P}^{1*}$ and a_P^u for the neighboring control volume centered at point W.

Similarly, the values of u_i at face s are calculated by

$$u_{i,s}^* = \alpha_u[(1 - f_{y,P})G_{i,PS}^{2*} + f_{y,P}G_{i,P}^{2*}] + \alpha_u[(1 - f_{y,P})/a_{PS}^u + f_{y,P}/a_P^u]$$
$$\times (J\alpha_i^2 \Delta \xi)_s(p_S^* - p_P^*) + (1 - \alpha_u)[(1 - f_{y,P})u_{i,S}^o + f_{y,P}u_{i,P}^o] \tag{4.193}$$

where $G_{i,P}^{2*} = H_{i,P}^* + D_i^1(p_w^* - p_e^*)$, and $G_{i,PS}^{2*}$ and a_{PS}^u are the values of $G_{i,P}^{2*}$ and a_P^u for the neighboring control volume centered at point S.

Subtracting Eqs. (4.192) and (4.193) from their counterparts for $u_{i,w}^{n+1}$ and $u_{i,s}^{n+1}$ under the pressure field p^{n+1} and neglecting the terms $G_{i,P}^{1*} - G_{i,P}^1$, $G_{i,P}^{2*} - G_{i,P}^2$, etc., leads to

$$u_{i,w}^{n+1} = u_{i,w}^* + \alpha_u Q_{i,w}^1(p_W' - p_P') \tag{4.194}$$
$$u_{i,s}^{n+1} = u_{i,s}^* + \alpha_u Q_{i,s}^2(p_S' - p_P') \tag{4.195}$$

where $Q_{i,w}^1 = [(1 - f_{x,P})/a_{PW}^u + f_{x,P}/a_P^u](J\alpha_i^1 \Delta \eta)_w$, and $Q_{i,s}^2 = [(1 - f_{y,P})/a_{PS}^u + f_{y,P}/a_P^u] (J\alpha_i^2 \Delta \xi)_s$.

Using the definition (4.132) of the fluxes at cell faces yields

$$F_w = F_w^* + a_W^p(p_W' - p_P') \tag{4.196}$$
$$F_s = F_s^* + a_S^p(p_S' - p_P') \tag{4.197}$$

where $a_W^p = \alpha_u \rho_w^{n+1}(J\alpha_i^1 \Delta \eta)_w Q_{i,w}^1$, $a_S^p = \alpha_u \rho_s^{n+1}(J\alpha_i^2 \Delta \xi)_s Q_{i,s}^2$, and F_w^* and F_s^* are the fluxes determined using Eq. (4.132) in terms of the approximate velocities $u_{i,w}^*$ abd $u_{i,s}^*$.

Integrating the continuity equation (4.185) over the control volume shown in Fig. 4.21 and discretizing the time-derivative term with the backward difference scheme yields

$$\frac{\rho_P^{n+1} - \rho_P^n}{\Delta\tau}\Delta A_P + F_e - F_w + F_n - F_s = 0 \tag{4.198}$$

Substituting Eqs. (4.196) and (4.197) and two similar equations for the fluxes at faces e and n into Eq. (4.198) leads to the equation for pressure correction:

$$a_P^p p'_P = a_W^p p'_W + a_E^p p'_E + a_S^p p'_S + a_N^p p'_N + S_p \tag{4.199}$$

where $a_P^p = a_W^p + a_E^p + a_S^p + a_N^p$, and $S_p = -(\rho_P^{n+1} - \rho_P^n)\Delta A_P/\Delta\tau - (F_e^* - F_w^* + F_n^* - F_s^*)$.

The computation procedure of the SIMPLE algorithm on the non-staggered grid is similar to that on the staggered grid, as introduced in Section 4.4.3.

SIMPLEC algorithm

Following Van Doormaal and Raithby (1984), the term $\sum_k a_k^u u'_{i,k}$ is kept in the derivation of Eq. (4.190), thus yielding

$$a_P^u u'_{i,P} = \alpha_u \sum_{l=E,W,N,S} a_l^u u'_{i,l} + \alpha_u a_P^u [D_i^1(p'_w - p'_e) + D_i^2(p'_s - p'_n)] \tag{4.200}$$

The term $\alpha_u \sum a_l^u u'_{i,P}$ is then subtracted from both sides of Eq. (4.200), yielding

$$\left(a_P^u - \alpha_u \sum_{l=E,W,N,S} a_l^u\right) u'_{i,P} = \alpha_u \sum_{l=E,W,N,S} a_l^u (u'_{i,l} - u'_{i,P})$$
$$+ \alpha_u a_P^u [D_i^1(p'_w - p'_e) + D_i^2(p'_s - p'_n)] \tag{4.201}$$

Assuming that all $u'_{i,k}$ are of about the same order as $u'_{i,P}$ and neglecting the first term on the right-hand side of Eq. (4.201) leads to

$$u_{i,P}^{n+1} = u_{i,P}^* + \alpha_u [\tilde{D}_i^1(p'_w - p'_e) + \tilde{D}_i^2(p'_s - p'_n)] \tag{4.202}$$

where $\tilde{D}_i^m = D_i^m/(1 - \alpha_u \sum_{l=E,W,N,S} a_l^u/a_P^u), m = 1, 2$.

Using the momentum interpolation technique introduced above, the velocity corrections at cell faces w and s are derived as

$$u_{i,w}^{n+1} = u_{i,w}^* + \alpha_u \tilde{Q}_{i,w}^1(p'_W - p'_P) \tag{4.203}$$

$$u_{i,s}^{n+1} = u_{i,s}^* + \alpha_u \tilde{Q}_{i,s}^2(p'_S - p'_P) \tag{4.204}$$

where $\tilde{Q}_{i,w}^1 = Q_{i,w}^1/[1-\alpha_u(1-f_{x,P})(\sum_{l=E,W,N,S} a_l^u/a_P^u)_W - \alpha_u f_{x,P}(\sum_{l=E,W,N,S} a_l^u/a_P^u)_P]$, and $\tilde{Q}_{i,s}^2 = Q_{i,s}^2/[1-\alpha_u(1-f_{y,P})(\sum_{l=E,W,N,S} a_l^u/a_P^u)_S - \alpha_u f_{y,P}(\sum_{l=E,W,N,S} a_l^u/a_P^u)_P]$.

The fluxes at cell faces are still determined by Eqs. (4.196) and (4.197), and the pressure correction equation is still written as Eq. (4.199), with $Q_{i,w}^1$ and $Q_{i,s}^2$ replaced by $\tilde{Q}_{i,w}^1$ and $\tilde{Q}_{i,s}^2$.

4.4.5 Stream function and vorticity formulation

In the 2-D case, it is possible to avoid explicit appearance of the pressure in the Navier-Stokes equations by introducing stream function and vorticity as dependent variables. The voticity Ω is defined as

$$\Omega = \frac{\partial u_y}{\partial x} - \frac{\partial u_x}{\partial y} \tag{4.205}$$

Cross-differentiating the u_x and u_y momentum equations (4.155) and (4.156) with respect to y and x and then subtracting them yields the transport equation of vorticity:

$$\frac{\partial \Omega}{\partial t} + \frac{\partial(u_x\Omega)}{\partial x} + \frac{\partial(u_y\Omega)}{\partial y} = v\left(\frac{\partial^2\Omega}{\partial x^2} + \frac{\partial^2\Omega}{\partial y^2}\right) \tag{4.206}$$

Eq. (4.206) is for laminar flows. A similar equation can be derived for turbulent flows.

The stream function ψ is defined by

$$u_x = \frac{\partial \psi}{\partial y}, \quad u_y = -\frac{\partial \psi}{\partial x} \tag{4.207}$$

Substituting Eq. (4.207) into the continuity equation (4.154) leads to the following Poisson equation for stream function:

$$\frac{\partial^2\psi}{\partial x^2} + \frac{\partial^2\psi}{\partial y^2} = -\Omega \tag{4.208}$$

Eqs. (4.206) and (4.208) replace the continuity and Navier-Stokes equations (4.154)–(4.156) and constitute the new governing equations. They can be solved conveniently using the finite difference method, finite volume method, or finite element method.

Since the pressure does not appear in Eqs. (4.206) and (4.208) and the continuity equation (4.154) is automatically satisfied, the stream function and vorticity method is convenient in the 2-D case. However, extension of this method to the 3-D case is not straightforward and loses the merits of the 2-D version.

4.5 SOLUTION OF ALGEBRAIC EQUATIONS

After a partial differential equation is discretized using one of the previously introduced numerical methods, the next task is to solve the resulting algebraic equations. If an explicit scheme is used for an unsteady problem, only one unknown appears at each time step, so the calculation can be easily performed step by step. If an implicit scheme is used for an unsteady problem or a numerical scheme involving more than two grid points is used for a steady problem, multiple unknowns appear in the algebraic equations that must be solved together. The implicit scheme is usually more stable and allows for larger time steps than the explicit scheme, yet its overall efficiency depends on the method used to solve the algebraic equations.

The algebraic equations can be solved directly or iteratively. Direct methods, such as the Gaussian elimination, are often used to solve linear algebraic equations; iteration methods are usually used for nonlinear equations, because the coefficients have to be updated and the equations have to be solved repeatedly. The methods often used for solving algebraic equations in computational river dynamics are introduced below.

4.5.1 Thomas algorithm

The Thomas algorithm, also called the double sweep algorithm, is often used to solve the set of algebraic equations resulting from the use of a three-point implicit finite difference or finite volume method for a 1-D second-order differential equation. The algebraic equations at internal points are

$$a_{P,i}\phi_i = a_{W,i}\phi_{i-1} + a_{E,i}\phi_{i+1} + b_i \quad (i = 2, 3, \ldots, m - 1) \tag{4.209}$$

and boundary conditions are

$$a_{P,1}\phi_1 = a_{E,1}\phi_2 + b_1 \tag{4.210}$$
$$a_{P,m}\phi_m = a_{W,m}\phi_{m-1} + b_m \tag{4.211}$$

where m is the total number of grid points.

The set of equations (4.209)–(4.211) can be written in matrix form as

$$
\begin{bmatrix}
-a_{P,1} & a_{E,1} & & & & & \\
a_{W,2} & -a_{P,2} & a_{E,2} & & & & \\
 & \cdot & \cdot & \cdot & & & \\
 & & a_{W,i} & -a_{P,i} & a_{E,i} & & \\
 & & & \cdot & \cdot & \cdot & \\
 & & & & a_{W,m-1} & -a_{P,m-1} & a_{E,m-1} \\
 & & & & & a_{W,m} & -a_{P,m}
\end{bmatrix}
\begin{bmatrix}
\phi_1 \\
\phi_2 \\
\cdot \\
\phi_i \\
\cdot \\
\phi_{m-1} \\
\phi_m
\end{bmatrix}
=
\begin{bmatrix}
b_1 \\
b_2 \\
\cdot \\
b_i \\
\cdot \\
b_{m-1} \\
b_m
\end{bmatrix}
$$

$$\tag{4.212}$$

which has a tridiagonal coefficient matrix.

Assuming that ϕ_{i-1} and ϕ_i are related by

$$\phi_{i-1} = c_i\phi_i + d_i \tag{4.213}$$

and substituting Eq. (4.213) into Eq. (4.209) leads to

$$\phi_i = c_{i+1}\phi_{i+1} + d_{i+1} \tag{4.214}$$

where

$$c_{i+1} = a_{E,i}/(a_{P,i} - c_i a_{W,i}), \quad d_{i+1} = (b_i + d_i a_{W,i})/(a_{P,i} - c_i a_{W,i}) \tag{4.215}$$

Comparing the boundary condition (4.210) with Eq. (4.214) at the first point yields the coefficients c_2 and d_2:

$$c_2 = a_{E,1}/a_{P,1}, \quad d_2 = b_1/a_{P,1} \tag{4.216}$$

and then the coefficients c_{i+1} and d_{i+1} are determined by Eq. (4.215) in the order of increasing i from 2 to $m - 1$. This is the forward sweep.

At the last grid point, substituting $\phi_{m-1} = c_m\phi_m + d_m$ into the boundary condition (4.211) yields

$$\phi_m = (b_m + d_m a_{W,m})/(a_{P,m} - c_m a_{W,m}) \tag{4.217}$$

Now all ϕ_i can be obtained using Eq. (4.213) in the order of decreasing i from m to 2. This is the backward sweep.

The Thomas algorithm is a direct solution method; it is particularly economical and requires only $5m - 4$ operations (multiplications and divisions) for linear problems. For non-linear problems, the coefficients and source term in Eq. (4.209) are related to the solution of ϕ, so an iteration procedure is needed. At each iteration step, an initial guess is given to ϕ at each point, the coefficients and source term are evaluated using the guessed ϕ, and then the double sweep calculations are performed to obtain the new value of ϕ at each point. This procedure is repeated until a convergent solution is reached. However, to obtain the convergent solution, it is necessary that

$$|a_{P,i}| > |a_{E,i}| + |a_{W,i}| \tag{4.218}$$

4.5.2 Jacobi and Gauss-Seidel iteration methods

Jacobi and Gauss-Seidel methods solve the algebraic equations point by point in a certain order. They can be used in the solution of 1-D, 2-D, and 3-D problems. Consider the following algebraic equation resulting from a 2-D second-order differential equation:

$$a_P\phi_{i,j} = a_W\phi_{i-1,j} + a_E\phi_{i+1,j} + a_S\phi_{i,j-1} + a_N\phi_{i,j+1} + b \qquad (4.219)$$

Assume that $a_P \neq 0$. If an initial approximation $\phi^{(0)}$ to the solution is chosen, the Jacobi iteration method gives a new approximation by

$$\phi_{i,j}^{(1)} = (a_W\phi_{i-1,j}^{(0)} + a_E\phi_{i+1,j}^{(0)} + a_S\phi_{i,j-1}^{(0)} + a_N\phi_{i,j+1}^{(0)} + b)/a_P \qquad (4.220)$$

If the solution is calculated in the order of increasing i and j, two points $(i-1, j)$ and $(i, j-1)$ have been visited before the solution at point (i, j) is calculated, as shown in Fig. 4.26. Therefore, the latest values at these two visited points are not used in Eq. (4.220) in the current iteration step. For this reason, the Jacobi method is not efficient. Improvement can be made using the Gauss-Seidel iteration method, which replaces $\phi_{i-1,j}^{(0)}$ and $\phi_{i,j-1}^{(0)}$ in Eq. (4.220) by the latest values:

$$\phi_{i,j}^{(1)} = (a_W\phi_{i-1,j}^{(1)} + a_S\phi_{i,j-1}^{(1)} + a_E\phi_{i+1,j}^{(0)} + a_N\phi_{i,j+1}^{(0)} + b)/a_P \qquad (4.221)$$

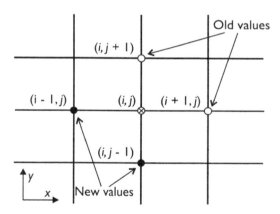

Figure 4.26 Calculation sequence in Jacobi and Gauss-Seidel methods.

4.5.3 ADI iteration method

Alternating Direction Implicit (ADI) iteration method splits or factorizes the 2-D or 3-D algebraic equation in different directions, and then solves the resulting equations using the TDMA method line by line. The ADI iteration method has many variants (Hageman and Young, 1981). As an example, a simple 2-D ADI method is presented here. For the algebraic equation (4.219), the following ADI iteration method, which has two fractional steps along i and j lines as shown in Fig. 4.27, is often used:

$$-a_W\phi_{i-1,j}^{(1/2)} + a_P\phi_{i,j}^{(1/2)} - a_E\phi_{i+1,j}^{(1/2)} = a_S\phi_{i,j-1}^{(0)} + a_N\phi_{i,j+1}^{(0)} + b \qquad (4.222)$$

$$-a_S\phi_{i,j-1}^{(1)} + a_P\phi_{i,j}^{(1)} - a_N\phi_{i,j+1}^{(1)} = a_W\phi_{i-1,j}^{(1/2)} + a_E\phi_{i+1,j}^{(1/2)} + b \qquad (4.223)$$

Eqs. (4.222) and (4.223) are implicit in single directions and can be directly solved using the Thomas algorithm described above. Because the boundary-condition information from the two ends of the grid line is transmitted at once to the interior of the domain, the ADI method converges faster than the Jacobi and Gauss-Seidel iteration methods.

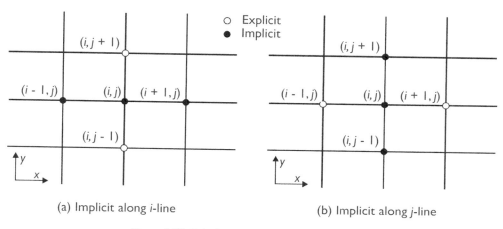

Figure 4.27 Calculation sequence in ADI method.

Efficiency of the ADI method can be improved by using the value at the new time level for one of the variables on the right-hand sides of Eqs. (4.222) and (4.223). For example, in Eq. (4.222) $\phi_{i,j-1}^{(0)}$ can be replaced by its latest value $\phi_{i,j-1}^{(1/2)}$, which has been calculated in the solution of Eq. (4.222) along the $j-1$ line, if the sweep is done by the order of increasing j. The same approach can be applied to Eq. (4.223) along the i line, in which $\phi_{i-1,j}^{(1/2)}$ can be replaced by $\phi_{i-1,j}^{(1)}$ if the sweep is carried out in the order of increasing i.

4.5.4 SIP iteration method

Consider a 2-D problem discretized by a five-point numerical scheme, the algebraic equations of which are Eq. (4.219). One may write the set of algebraic equations in matrix form:

$$A\Phi = b \qquad (4.224)$$

where A is the coefficient matrix, Φ is the vector of the unknowns, and b is the vector of the source terms. If Φ is numbered in the order of increasing j and then increasing i,

A can be assembled to be a penta-diagonal matrix, as shown in Fig. 4.28, in which non-zero entries are shaded, and each horizontal set of boxes corresponds to one grid line.

For the correction vector $\Delta\Phi = \Phi^{(1)} - \Phi^{(0)}$, one obtains

$$A\Delta\Phi = R \tag{4.225}$$

where R is the residual matrix defined as $R = b - A\Phi^{(0)}$.

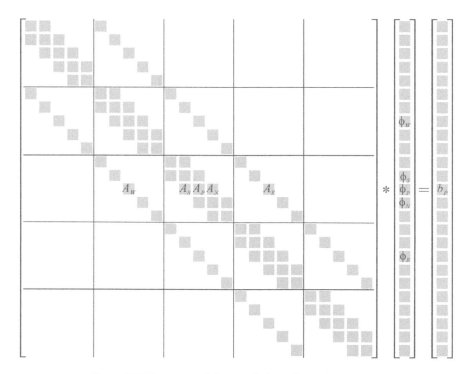

Figure 4.28 Structure of the matrix for a five-point scheme.

If the matrix A can be completely factorized as the product of a lower triangular matrix and an upper triangular matrix, Eq. (4.225) can be easily solved. Unfortunately, this complete LU decomposition usually is not feasible for the penta-diagonal matrix shown in Fig. 4.28. Nevertheless, this observation leads to the idea of approximating the matrix A by a matrix M that is the product of a lower triangular matrix L and an upper triangular matrix U. The LU decomposition of matrix M is shown in Fig. 4.29. The rules of matrix multiplication give that the product matrix $M = LU$ should be a seven-diagonal matrix. Two non-zero diagonals in M that correspond to the zero diagonals of A are shown by dashed lines in Fig. 4.29. The accurate relation should be

$$M = LU = A + C \tag{4.226}$$

where C is the remaining matrix of A after the factorization.

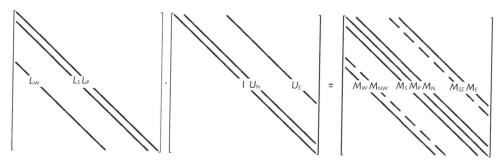

Figure 4.29 Sketch of LU decomposition of matrix M.

In order to make up the difference between A and M, one choice is to let the matrix C contain just the two non-zero diagonals of M that correspond to the zero diagonals of A. This is the standard incomplete LU decomposition method. However, this method converges slowly. A better choice, which was proposed by Stone (1968), is to allow C to have non-zero elements on the diagonals corresponding to all seven non-zero diagonals of LU. C must contain the 'two extra' diagonals of M, and the elements on the remaining diagonals of C are chosen to ensure $C\Delta\Phi \approx 0$. The resulting equations relating L and U with A are (see Ferziger and Peric, 1995)

$$\begin{cases} L_{W,l} = A_{W,l}/(1 + \alpha_s U_{N,l-m_j}) \\ L_{S,l} = A_{S,l}/(1 + \alpha_s U_{E,l-1}) \\ L_{P,l} = A_{P,l} + \alpha_s (L_{W,l} U_{N,l-m_j} + L_{S,l} U_{E,l-1}) - L_{W,l} U_{E,l-m_j} - L_{S,l} U_{N,l-1} \\ U_{N,l} = (A_{N,l} - \alpha_s L_{W,l} U_{N,l-m_j})/L_{P,l} \\ U_{E,l} = (A_{E,l} - \alpha_s L_{S,l} U_{E,l-1})/L_{P,l} \end{cases}$$

(4.227)

where $L_{W,l}$, $L_{S,l}$, and $L_{P,l}$ are the coefficients of matrix L; $U_{N,l}$ and $U_{E,l}$ are the coefficients of matrix U (the coefficients in the main diagonal are set to be 1 to get the unique solution in LU decomposition); α_s is a coefficient less than 1; l is the index of points in the matrix A, related to i and j by $l = j + (i - 1)m_j$; and m_j is the number of points on the j line.

The set of equations (4.227) can be solved in a sequential order beginning at the southeast corner of the grid. For points next to boundaries, any matrix element that carries the index of a boundary point is set to be zero.

After L and U are determined, the approximate formulation for Eq. (4.225) is obtained as

$$LU\Delta\Phi = R \tag{4.228}$$

By defining $U\Delta\Phi = \Psi$, Eq. (4.228) can be split as

$$\begin{cases} L\Psi = R \\ U\Delta\Phi = \Psi \end{cases} \tag{4.229}$$

The two equations in (4.229) can be solved separately by a direct method. After $\Delta\Phi$ is calculated, the new values of Φ can be obtained conveniently by

$$\Phi^{(1)} = \Phi^{(0)} + \Delta\Phi \qquad (4.230)$$

The above five-point SIP method has been extended to solve the algebraic equations related to seven-point (Leister and Peric, 1994) and nine-point numerical schemes (Schneider and Zedan, 1981). Because the coefficient matrices of the algebraic equations resulting from many numerical schemes are or can be approximated as five-, seven-, or nine-diagonal matrices, the SIP method is often used in computational river dynamics.

It should be noted that because the approximation used in the SIP method is related to the discretization of partial differential equations, the SIP method makes little sense for generic algebraic equations.

4.5.5 Over-relaxation and under-relaxation

After the correction vector $\Delta\Phi$ is calculated, the new values $\Phi^{(1)}$ can also be obtained by the relaxation method:

$$\Phi^{(1)} = \Phi^{(0)} + \alpha_\phi \Delta\Phi \qquad (4.231)$$

where α_ϕ is the relaxation factor. $\alpha_\phi > 1$ for over-relaxation, and $\alpha_\phi < 1$ for under-relaxation.

The over- and under-relaxation methods can accelerate or decelerate the convergence speed. Over-relaxation is often used in conjunction with the Gauss-Seidel method, yielding the Successive Over-Relaxation (SOR) method. Under-relaxation is very useful for nonlinear problems and can avoid divergence.

Faster convergence can be achieved when using an optimum value of α_ϕ. However, because the optimum α_ϕ value depends on many factors, such as the nature of the problem, number of grid points, grid spacing, and iteration procedure, there are no general rules to determine it. Usually, a suitable value of α_ϕ can be found by experience and from exploratory computation for the problem under consideration.

Chapter 5

1-D numerical models

1-D models simulate the flow and sediment transport in the streamwise direction of a channel without solving the details over the cross-section. They are often applied in the study of long-term sedimentation problems in rivers, reservoirs, estuaries, etc. Described in this chapter are approaches and issues regarding 1-D models, such as channel network routing, decoupled and coupled flow and sediment calculations, non-uniform total-load transport, equilibrium and non-equilibrium sediment transport, lateral allocation of bed change, bank erosion, data requirements, and parameter sensitivity.

5.1 FORMULATION OF 1-D DECOUPLED FLOW AND SEDIMENT TRANSPORT MODEL

As discussed in Section 2.2.3, in the case of low sediment concentration, the influence of sediment on the flow field is negligible, and thus the simulation of the water and sediment two-phase flow can be simplified as a problem of solving the clear water flow with sediment transport. Moreover, because the bed usually changes at a much lower rate than the flow (especially when bed load is the main transport mode), the bed elevation can be assumed to be "fixed" at each time step, and the flow can be calculated based on the channel geometry estimated at the previous time step. With these simplifications, the flow and sediment calculations can be performed in a decoupled manner. Such decoupled calculations are introduced in Sections 5.1–5.3.

5.1.1 Formulation of 1-D clear water flow model

5.1.1.1 1-D hydrodynamic equations

Dynamic wave model

In the 1-D dynamic wave model, open-channel flows are governed by Eqs. (2.102) and (2.104), which are called the St. Venant equations. In the case with side flows (inflow and/or outflow), these equations are written as

$$\frac{\partial A}{\partial t} + \frac{\partial Q}{\partial x} = q_l \tag{5.1}$$

$$\frac{\partial Q}{\partial t} + \frac{\partial}{\partial x}\left(\frac{\beta Q^2}{A}\right) + gA\frac{\partial z_s}{\partial x} + gAS_f = q_l v_x \tag{5.2}$$

where x is the spatial coordinate representing the streamwise distance; A is the flow area; Q is the flow discharge, defined as $Q = AU$, with U being the flow velocity averaged over the cross-section; z_s is the water stage; β is the correction factor for momentum due to the non-uniformity of streamwise velocity over the cross-section; q_l is the side flow discharge per unit channel length; v_x is the velocity of side flows in the direction of the x-coordinate; and S_f is the friction slope:

$$S_f = \frac{Q|Q|}{K^2} \tag{5.3}$$

where K is the conveyance. For a simple cross-section, $K = AR^{2/3}/n$, with R being the hydraulic radius and n the Manning roughness coefficient of the channel. For a compound cross-section, determining K or n is introduced in Section 5.1.1.4.

Note that "$\wedge$", representing the section-averaged quantities in Eqs. (2.102)–(2.111), is omitted hereafter, for simplicity.

Diffusion wave model

The diffusion wave model assumes that the local and convective accelerations in the momentum equation (5.2) are negligible, thus yielding

$$gA\frac{\partial z_s}{\partial x} + gAS_f = q_l v_x \tag{5.4}$$

The continuity equation (5.1) is still used in the diffusion wave model.

The diffusion wave model is more stable than the dynamic wave model, but the latter is more accurate and can be applied in a wider range of flow conditions. Wu and Vieira (2002) investigated the errors of the diffusion wave assumption in various cases. One example was steady flow through a channel contraction, as shown in Fig. 5.1. The diffusion wave model exhibits errors in the computed water surface profile in the transition region near the contraction, whereas the two models give identical results in the upstream and downstream regions with uniform flow. Normally, the relative errors are less than 10%, if the Froude number is less than 0.5.

Kinematic wave model

For the kinematic wave, the variations in flow velocity and depth are negligible in comparison with the variation in channel bed elevation, and thus, the momentum equation (5.2) can be simplified considerably as follows:

$$S_f = S_0 \tag{5.5}$$

where S_0 is the channel slope in the longitudinal direction.

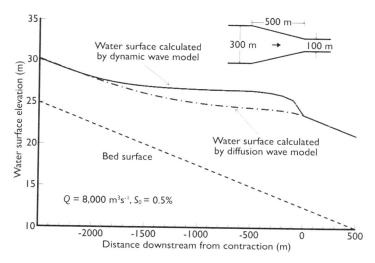

Figure 5.1 Comparison of dynamic and diffusion wave models at channel contraction.

Using the Manning equation, Eq. (5.5) can be rewritten as

$$Q = \frac{A}{n} R^{2/3} S_0^{1/2} \tag{5.6}$$

The continuity equation (5.1) is still used in the kinematic wave model.
The kinematic wave assumption is generally applicable, if (Dingman, 1984)

$$g\Lambda S_0/U^2 > 10 \tag{5.7}$$

where Λ represents the length of the channel under study, and U is the average velocity of uniform flow. Eq. (5.7) implies that the kinematic wave model is valid in steep channels.

5.1.1.2   Imposition of boundary and initial conditions of flow

To establish a well-posed problem, boundary and initial conditions should be provided for Eqs. (5.1) and (5.2). These dynamic wave equations constitute a hyperbolic system that has two characteristics:

$$\frac{dx}{dt} = C^+ = U + \sqrt{gh} \tag{5.8}$$

$$\frac{dx}{dt} = C^- = U - \sqrt{gh} \tag{5.9}$$

where U and h are the flow velocity and depth averaged over the cross-section, respectively.

For subcritical flow ($Fr = U/\sqrt{gh} < 1$), the characteristics $C^+ > 0$ and $C^- < 0$. As shown in Fig. 5.2(a), a C^+ characteristic curve enters from outside to the solution domain through the inlet, and a C^- characteristic curve enters through the outlet. To determine the flow properties at the inlet and outlet, information from each characteristic curve entering from outside has to be provided by a boundary condition. Therefore, a boundary condition should be specified at each of the two boundaries. Usually, a time series of flow discharge is specified at the inlet, and a time series of water stage or a stage-discharge rating curve is imposed at the outlet.

For supercritical flow ($Fr > 1$), two characteristics are positive: $C^+ > 0$ and $C^- > 0$. As shown in Fig. 5.2(b), both C^+ and C^- characteristic curves enter from outside through the inlet, and no characteristic curve enters through the outlet. Therefore, two boundary conditions should be imposed at the inlet, and none is required at the outlet.

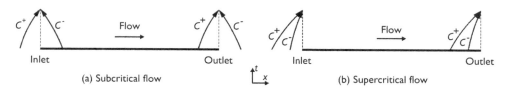

Figure 5.2 Characteristic curves of dynamic wave model at inlet and outlet boundaries.

Similarly, it can be derived that the diffusive wave model requires two boundary conditions, which are specified at the inlet and outlet, respectively. The kinematic wave model requires only one boundary condition, which is often specified at the inlet.

In addition, the initial water stage and flow discharge in the solution domain should be given for an unsteady flow simulation.

5.1.1.3 Manning roughness coefficient

The Manning roughness coefficient n accounts for the effect of bed roughness on the flow field, and its determination is essential to the accuracy of the calculated flow, sediment transport, and bed change. For a movable bed with sediment grains and bed forms as roughness elements, the Manning n can be evaluated using one of the empirical formulas introduced in Section 3.3.3. However, the Manning n generally depends on a number of factors, including channel size, cross-section shape, channel alignment, channel meandering and curvature, surface roughness, bed forms, obstructions, vegetation, sediment transport, temperature, and seasonal changes. Therefore, it is suggested that the Manning n should be calibrated, if gauged water surface profiles and high water marks are available; otherwise, the Manning n values in similar stream conditions should be used as guides. There are several references available for determining the Manning n, e.g., Chow (1959), Fasken (1963), Barnes (1967), and Hicks and Mason (1991).

In the case of reservoir sedimentation, because the water stage is raised significantly, bank roughness becomes important and should be considered in the flow calculation. Usually, the Manning n values on banks and bed are different. In addition, due to significant deposition, the bed material in the reservoir gradually becomes finer, and thus the Manning n of the channel bed decreases with time. This can be described using the movable bed roughness formulas introduced in Section 3.3.3, or using the following relation proposed by Han *et al.* (1986):

$$n^{3/2} = n_e^{3/2} + (n_0^{3/2} - n_e^{3/2})(1 - a/a_e)^{1/4} \tag{5.10}$$

where n_0, n, and n_e are the Manning roughness coefficients in the beginning, transitional period, and equilibrium state of reservoir deposition, respectively; a is the deposition area accumulated with time at a cross-section; and a_e is the final deposition area when the reservoir reaches equilibrium. The values of n_e can be determined by referring to those in the downstream alluvial channels with flow and sediment conditions similar to the equilibrium state of the reservoir.

5.1.1.4 Composite hydraulic properties

If hydraulic properties, such as roughness and conveyance, are non-uniform across the channel, their composite values need to be computed. The often used methods include the alpha method, hydraulic radius division method, energy slope division method, and conveyance method, which are described below.

Alpha method

In the alpha method, the cross-section is divided into panels between coordinate points (stations), as shown in Fig. 5.3. The divisions between the panels are assumed to be vertical. The cross-section is not distinguished between the main channel and overbanks in this method.

The flow area A_j, wetted perimeter χ_j, hydraulic radius R_j, and conveyance K_j of panel j are calculated by

$$A_j = [z_s - 0.5(z_{b,j} + z_{b,j+1})]\Delta y_j \tag{5.11}$$

$$\chi_j = \sqrt{(z_{b,j} - z_{b,j+1})^2 + \Delta y_j^2} \tag{5.12}$$

$$R_j = A_j/\chi_j \tag{5.13}$$

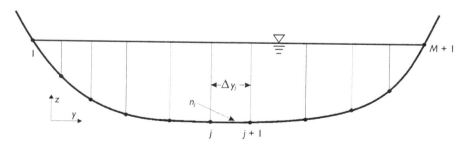

Figure 5.3 Representation of cross-section in alpha method.

$$K_j = A_j R_j^{2/3} / n_j \qquad (5.14)$$

where $z_{b,j}$ is the bed elevation at station j, Δy_j is the distance between stations j and $j+1$, and n_j is the Manning roughness coefficient in panel j.

The composite cross-sectional area of flow is defined as the sum of all panel subareas and is the true area. The composite velocity is defined as the total discharge divided by the cross-sectional area, conserving continuity. The composite hydraulic radius is conveyance-weighted as

$$R = \sum_{j=1}^{M} R_j K_j \bigg/ \sum_{j=1}^{M} K_j \qquad (5.15)$$

where M is the number of the wetted panels.

Because the alpha method ignores the effect of vertical walls, it is not adequate in situations where vertical sidewalls or steep bank slopes exist.

Division of hydraulic radius

Einstein (1950) proposed a more adequate method for determining the composite hydraulic properties for the cross-section with rough vertical sidewalls or steep bank slopes, based on the division of hydraulic radius. This method assumes equal velocity in all panels, and calculates all hydraulic variables in the normal way, except for the composite Manning roughness coefficient.

The total shear stress τ in the cross-section can be computed as

$$\chi \tau = \sum_{j=1}^{M} \chi_j \tau_j \qquad (5.16)$$

where χ is the total wetted perimeter, i.e., $\chi = \sum_{j=1}^{M} \chi_j$, and τ_j is the shear stress in panel j.

Einstein's method determines

$$\tau = \gamma R S_f \qquad (5.17)$$

$$\tau_j = \gamma R_j S_f \qquad (5.18)$$

Applying the equal velocity assumption and the Manning equation in the entire cross-section and each panel yields

$$R = (nU/S_f^{1/2})^{3/2}, \quad R_j = (n_j U/S_f^{1/2})^{3/2} \qquad (5.19)$$

Inserting Eqs. (5.17)–(5.19) into Eq. (5.16) yields

$$n = \left(\sum_{j=1}^{M} \chi_j n_j^{3/2} / \chi \right)^{2/3}$$
(5.20)

Division of energy slope

The method based on the division of energy slope originated from Engelund (1966) is another option for determining the composite hydraulic properties for the cross-section with rough vertical sidewalls or steep bank slopes. This method gives

$$\tau_j = \gamma R S_{f,j}$$
(5.21)

and applies the equal velocity assumption and the Manning equation in the entire cross-section and each panel:

$$S_f = (nU/R^{2/3})^2, \quad S_{f,j} = (n_j U/R^{2/3})^2$$
(5.22)

Inserting Eqs. (5.17), (5.21), and (5.22) into Eq. (5.16) yields the following equation for the composite Manning n:

$$n = \left(\sum_{j=1}^{M} \chi_j n_j^2 / \chi \right)^{1/2}$$
(5.23)

Conveyance method

The assumption of equal velocity used in the previous methods, based on the division of either hydraulic radius or energy slope, is only applicable in simple channels. For compound channels with floodplains, the flow velocities in the main channel and floodplains may be significantly different. A more adequate method for determining the composite hydraulic properties in compound channels is the conveyance method.

The conveyance method divides the cross-section into subsections in such a way that the equal velocity assumption can be approximately valid in each subsection. Each subsection can be further divided into panels. The flow area, wetted perimeter, and conveyance of each subsection can be calculated in the normal way. The conveyances of all subsections are then summed to provide the total conveyance for the entire cross-section. For example, the compound cross-section shown in Fig. 5.4 can be divided into three subsections: main channel, left floodplain, and right floodplain, and the total conveyance is determined by

$$K = \frac{A_{LF}^{5/3}}{n_{LF}\chi_{LF}^{2/3}} + \frac{A_{MC}^{5/3}}{n_{MC}\chi_{MC}^{2/3}} + \frac{A_{RF}^{5/3}}{n_{RF}\chi_{RF}^{2/3}}$$
(5.24)

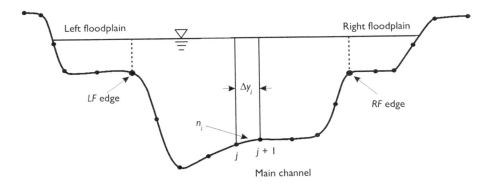

Figure 5.4 Representation of compound cross-section with floodplains.

where the subscript *LF* denotes the left floodplain, *MC* the main channel, and *RF* the right floodplain.

For each subsection, the Manning *n* can be determined using the hydraulic radius or energy slope division method. For example, the Manning *n* in the main channel is determined using these two methods as follows:

$$n_{MC} = \left(\sum_{j=LCB}^{j=RCB} \chi_j n_j^{3/2} / \chi_{MC} \right)^{2/3}, \quad n_{MC} = \left(\sum_{j=LCB}^{j=RCB} \chi_j n_j^2 / \chi_{MC} \right)^{1/2} \quad (5.25)$$

where *LCB* and *RCB* represent the main-channel panels adjacent to the left and right floodplain edges (denoted as *LF* and *RF* in Fig. 5.4), respectively.

5.1.1.5 *Momentum correction factor*

The correction factor β for momentum in Eq. (5.2) is close to 1 for a simple cross-section. For the compound cross-section shown in Fig. 5.4, β is determined by

$$\beta = \frac{1}{QU} \iint_A u^2 dA = \frac{1}{QU}(Q_{LF}U_{LF} + Q_{MC}U_{MC} + Q_{RF}U_{RF})$$

$$= \frac{A}{K^2} \left(\frac{K_{LF}^2}{A_{LF}} + \frac{K_{MC}^2}{A_{MC}} + \frac{K_{RF}^2}{A_{RF}} \right) \quad (5.26)$$

5.1.2 Formulation of 1-D sediment transport model

5.1.2.1 *1-D non-equilibrium sediment transport equations*

For general applications, the transport of non-uniform total load is considered here. Because the total load can be divided into bed load and suspended load, or into bed-material load and wash load, as shown in Fig. 2.3, two total-load modeling approaches are usually adopted. One is to simulate bed load and suspended load separately, while the other is to compute bed-material (total) load directly. Both approaches have advantages and disadvantages.

1-D bed-load and suspended-load transport model

As described in Section 2.7.1, the non-uniform sediment mixture is divided into a suitable number of size classes (N). In the case of low sediment concentration, interactions among different size classes are usually ignored and, thus, the transport of each size class is simulated separately. For each size class, the moving sediment is further divided into suspended load and bed load. Introducing Eq. (2.132) into Eq. (2.108) and considering the lateral exchange with banks and tributaries yields the 1-D transport equation of the kth size class of suspended load:

$$\frac{\partial}{\partial t}\left(\frac{AC_k}{\beta_{sk}}\right) + \frac{\partial(AUC_k)}{\partial x} = \alpha\omega_{sk}B(C_{*k} - C_k) + q_{slk} \quad (k = 1, 2, \ldots, N)$$

(5.27)

where C_k and C_{*k} are the actual and equilibrium (capacity) average concentrations of the kth size class of suspended load, respectively; α is the adaptation coefficient of suspended load; q_{slk} is the suspended-load side discharge per unit channel length due to the lateral exchange with banks and tributaries; and β_{sk} is the correction coefficient, which is determined using Eq. (3.135) in general, but may be set to 1 in the simulation of long-term sedimentation processes.

In analogy to Eq. (2.158), the 1-D bed-load transport equation is

$$\frac{\partial}{\partial t}\left(\frac{Q_{bk}}{U_{bk}}\right) + \frac{\partial Q_{bk}}{\partial x} = \frac{1}{L}(Q_{b*k} - Q_{bk}) + q_{blk}$$

(5.28)

where Q_{bk} and Q_{b*k} are the actual and equilibrium (capacity) transport rates of the kth size class of bed load, respectively; L is the adaptation length of sediment, defined in Section 2.6.2; and q_{blk} is the bed-load side discharge per unit channel length.

The bed-load velocity U_{bk} needs to be determined using one of the empirical formulas described in Section 3.8. However, the storage term, the first term on the left-hand side of Eq. (5.28), is often ignored in the simulation of long-term sedimentation processes.

The equilibrium suspended-load concentration and bed-load transport rate can be determined using the existing formulas described in Sections 3.4 and 3.5. For convenience, these formulas are written in general forms:

$$C_{*k} = p_{bk}C_k^*, \quad Q_{b*k} = p_{bk}Q_{bk}^*$$

(5.29)

where p_{bk} is the sediment availability factor, usually set as the fraction of size class k in the mixing layer of bed material; C_k^* is the potential equilibrium concentration for the kth size class of suspended load; and Q_{bk}^* is the potential equilibrium transport rate for the kth size class of bed load. C_k^* and Q_{bk}^* can be interpreted as the equilibrium suspended-load concentration and bed-load transport rate of uniform sediment with the same size as d_k, taking into consideration, however, the hiding and exposure effects in non-uniform bed material.

In analogy to Eq. (2.159), the 1-D fractional bed change equation is

$$(1 - p'_m)\left(\frac{\partial A_b}{\partial t}\right)_k = \alpha \omega_s B(C_k - C_{*k}) + \frac{1}{L}(Q_{bk} - Q_{b*k}) \qquad (5.30)$$

where $(\partial A_b/\partial t)_k$ is the rate of change in bed area due to size class k.

The total rate of change in bed area, $\partial A_b/\partial t$, is determined by

$$\frac{\partial A_b}{\partial t} = \sum_{k=1}^{N}\left(\frac{\partial A_b}{\partial t}\right)_k \qquad (5.31)$$

As described in Section 2.7.2, the bed material is divided into layers. The temporal variation of the mixing-layer bed-material gradation p_{bk} is determined by Eq. (2.161), which is rewritten in the 1-D model as follows:

$$\frac{\partial(A_m p_{bk})}{\partial t} = \left(\frac{\partial A_b}{\partial t}\right)_k + p_{bk}^*\left(\frac{\partial A_m}{\partial t} - \frac{\partial A_b}{\partial t}\right) \qquad (5.32)$$

where A_m is the cross-sectional area of the mixing layer, and p_{bk}^* is p_{bk} when $\partial A_b/\partial t - \partial A_m/\partial t \geq 0$ and the fraction of size class k in the second layer of bed material when $\partial A_b/\partial t - \partial A_m/\partial t < 0$. Accordingly, the bed material sorting equation (2.162) in the second layer is rewritten as

$$\frac{\partial(A_{sub} p_{sbk})}{\partial t} = -p_{bk}^*\left(\frac{\partial A_m}{\partial t} - \frac{\partial A_b}{\partial t}\right) \qquad (5.33)$$

where p_{sbk} is the fraction of size class k in the second layer of bed material, and A_{sub} is the cross-sectional area of the second layer. Note that Eq. (5.33) assumes no exchange between the second and third layers.

Eqs. (5.27)–(5.33) constitute the governing equations of the total-load transport model that discerns bed load and suspended load. This model provides the ratio of bed load and suspended load. However, many reliable bed-material load transport capacity formulas, such as the Ackers-White (1973), Engelund-Hansen (1967), and Yang (1973) formulas, cannot be used directly in this approach.

1-D bed-material load transport model

When the bed-material (total) load transport is simulated without separating bed load and suspended load, introducing Eq. (2.149) into Eq. (2.111) and considering the lateral exchange with banks and tributaries yields the bed-material load transport equation:

$$\frac{\partial}{\partial t}\left(\frac{Q_{tk}}{\beta_{tk} U}\right) + \frac{\partial Q_{tk}}{\partial x} = \frac{1}{L_t}(Q_{t*k} - Q_{tk}) + q_{tlk} \qquad (k = 1, 2, \ldots, N) \qquad (5.34)$$

where Q_{tk} and Q_{t*k} are the actual and equilibrium (capacity) transport rates of the kth size class of bed-material load, respectively; L_t is the adaptation length of bed-material load; q_{tlk} is the side discharge of bed-material load per unit channel length; and β_{tk} is the correction factor, which is determined in analogy to Eq. (2.92) but may be set to 1 in the simulation of long-term sedimentation processes.

The sediment transport capacity can be written in the general form:

$$Q_{t*k} = p_{bk}Q_{tk}^* \qquad (5.35)$$

where Q_{tk}^* is the potential equilibrium transport rate for the kth size class of bed-material load.

Extending Eq. (2.149) to the 1-D model yields the following equation for the fractional change in bed area:

$$(1 - p_m')\left(\frac{\partial A_b}{\partial t}\right)_k = \frac{1}{L_t}(Q_{tk} - Q_{t*k}) \qquad (5.36)$$

The total change in bed area is calculated using Eq. (5.31), while the bed material sorting is determined using Eqs. (5.32) and (5.33).

Eqs. (5.31)–(5.36) constitute the governing equations of the total-load transport model that directly computes bed-material load. This model has N less transport equations than the previous bed-load and suspended-load transport model. Not only can those aforementioned reliable bed-material load transport capacity formulas be used, but also many bed-load and suspended-load transport capacity formulas, such as the Wu *et al.* (2000b) formulas, can be applied jointly in this approach. However, it does not provide the ratio of bed load and suspended load.

Note that if $L = L_t$ and $\alpha = Uh/(L_t\omega_s)$, the bed-load and suspended-load model and the bed-material load model give the same results for total sediment discharge, bed change, and bed-material gradation. This explains why L is found to be approximately equal to L_t, as stated in Section 2.6.2. Because normally $L_s \geq L_b$, the condition $\alpha = Uh/(L_t\omega_s)$ can usually be satisfied using Eqs. (2.154) and (2.155). For very coarse sediments, this condition may be violated, but because such sediments move mainly in bed load, the difference between the two models is small and the bed-material load model is preferable.

In addition, both models can simulate the transport of wash load by setting the adaptation coefficient α in Eq. (5.27) to zero and the adaptation lengths in Eqs. (5.28) and (5.34) to be infinitely large. The wash-load size range can be defined using the bed-material diameter d_{10} or the Rouse number $\omega_{sk}/(\kappa U_*) < 0.06$, as discussed in Section 3.5.1. The latter method is more convenient for numerical modeling.

Both models can also simulate sediment transport over non-erodible channel beds. This is often called the hard-bottom problem. On the non-erodible cross-sections, the sediment transport capacity Q_{t*k} in Eqs. (5.34) and (5.36) is replaced by $\min(Q_{t*k}, Q_{tk})$, or the sediment transport capacities Q_{b*k} and C_{*k} in Eqs. (5.27), (5.28), and (5.30) are replaced by $\min(Q_{b*k}, Q_{bk})$ and $\min(C_{*k}, C_k)$, respectively. This method allows only deposition on the hard-bottom points. It can be easily extended to 2-D and 3-D models.

5.1.2.2 1-D equilibrium sediment transport equations

The assumption of local equilibrium transport described in Section 2.6.1 ignores the temporal and spatial lags of sediment transport and sets the actual sediment transport rate to be equal to the equilibrium (capacity) one at each cross-section:

$$Q_{tk} = Q_{t*k}(U, h, \tau, B, d_k, p_{bk}, \gamma_s, \gamma \dots) \quad (k = 1, 2, \dots, N) \tag{5.37}$$

In the equilibrium transport model, the change in bed area due to size class k is calculated by

$$(1 - p'_m)\frac{\partial A_{bk}}{\partial t} + \frac{\partial Q_{tk}}{\partial x} = 0 \tag{5.38}$$

The total change in bed area is calculated by Eq. (5.31) and the bed-material gradations by Eqs. (5.32) and (5.33).

It should be noted that the local equilibrium assumption does not mean that the sediment transport in the entire channel is at equilibrium. Conversely, the sediment transport capacities at two consecutive cross-sections may be different under varying flow and sediment conditions, and thus the channel bed between these two cross-sections may change according to Eq. (5.38).

5.1.2.3 Characteristics of equilibrium and non-equilibrium transport models

The equilibrium sediment transport model is simple but may lead to a numerical difficulty near the inlet with constrained sediment loading. Fig. 5.5 shows the sediment discharge profiles determined by the equilibrium transport model on a finite difference mesh in cases of erosion ($Q_{t0} = 0$) and deposition ($Q_{t0} = 2Q_{t*}$). Here, Q_{t0} is the sediment discharge loaded at the inlet ($x = 0$), and Q_{t*} is assumed constant in the entire channel. The sediment discharge at cross-section 1 is specified by the boundary condition (constraint), while the sediment discharge at cross-section 2 is determined using Eq. (5.37). If the sediment is strongly over- or under-loaded, the sediment discharges at these two cross-sections will be significantly different, and strong deposition or erosion will be computed in the first reach. The smaller the grid spacing, the larger the deposition or erosion rate calculated in this reach. This is physically unreasonable, and may cause numerical instability. Therefore, the application of the equilibrium sediment transport model should be limited to situations with near-equilibrium loading at the inlet.

For uniform sediment under steady flow conditions, the non-equilibrium transport equation (5.34) with constant L_t and Q_{t*} and without side discharge has an analytical solution:

$$Q_t = Q_{t*} + (Q_{t0} - Q_{t*})\exp\left(-\frac{x}{L_t}\right) \tag{5.39}$$

Fig. 5.6 illustrates the sediment discharge profiles determined by Eq. (5.39) for the

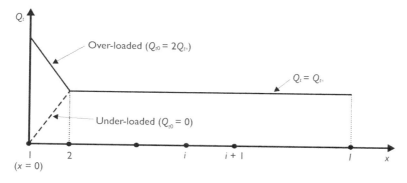

Figure 5.5 Sediment discharge profiles in equilibrium transport model.

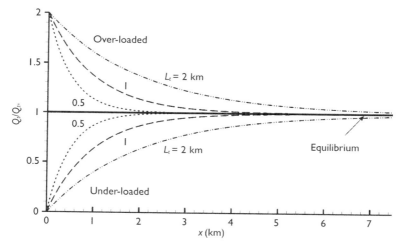

Figure 5.6 Sediment discharge profiles in non-equilibrium transport model.

same cases of erosion ($Q_{t0} = 0$) and deposition ($Q_{t0} = 2Q_{t*}$) shown in Fig. 5.5. The actual sediment discharge does not adjust to the equilibrium one immediately near the inlet, but after a certain distance downstream. In addition, Fig. 5.6 also shows that the adaptation length L_t is an important parameter in the non-equilibrium transport model. It essentially determines the sediment discharge profile. At a distance equal to one adaptation length ($x = L_t$), $(Q_t - Q_{t*})/(Q_{t0} - Q_{t*}) \approx 0.3679$.

A comparison of Figs. 5.5 and 5.6 shows that the non-equilibrium transport model is physically more realistic and can handle the constrained sediment loading more easily than the equilibrium transport model. In addition, as $L_t \rightarrow 0$, the exchange term in Eq. (5.34) becomes dominant; thus, Eq. (5.34) reduces to Eq. (5.37) and the sediment discharge profiles in Fig. 5.6 become those in Fig. 5.5. This implies that the non-equilibrium transport model is more general and includes the equilibrium transport model as a special case.

5.1.2.4   Boundary and initial conditions of sediment

The fractional sediment discharges for all size classes must be imposed at the inflow boundary in each time step, but no sediment boundary condition is required at the outflow boundary in the 1-D model. The initial sediment discharge, channel topography, and bed-material gradation should be provided for the simulation of unsteady sediment transport and channel morphological evolution.

5.2   1-D CALCULATION OF OPEN-CHANNEL FLOW

5.2.1   1-D steady flow calculation

5.2.1.1   Discretization of steady flow equations

For steady open-channel flow without side inflow or outflow, Eq. (5.1) reduces to $\partial Q/\partial x = 0$ and leads to a constant flow discharge along the study reach, while Eq. (5.2) can be rewritten as the energy equation:

$$\frac{\partial}{\partial x}\left(\frac{\beta' Q^2}{2A^2}\right) + g\frac{\partial z_s}{\partial x} + g\frac{Q|Q|}{K^2} = 0 \qquad (5.40)$$

where β' is the correction factor for kinetic energy due to the non-uniformity of streamwise velocity over the cross-section. For the compound cross-section shown in Fig. 5.4, β' can be determined using the discharge-weighted average kinetic energy:

$$\begin{aligned}\beta' &= \frac{1}{QU^2}(Q_{LF}U_{LF}^2 + Q_{MC}U_{MC}^2 + Q_{RF}U_{RF}^2)\\ &= \frac{A^2}{K^3}\left(\frac{K_{LF}^3}{A_{LF}^2} + \frac{K_{MC}^3}{A_{MC}^2} + \frac{K_{RF}^3}{A_{RF}^2}\right)\end{aligned} \qquad (5.41)$$

where all parameters are the same as those in Eq. (5.26).

Suppose that the computational domain of a single channel is divided into $I - 1$ reaches by I cross-sections (computational points), as shown in Fig. 5.7. The cross-sections are numbered 1 through I in the downstream direction. Each cross-section is represented by an adequate number of points (stations), as shown in Fig. 5.4, with each point characterized by a pair of values of the distance to the left bank and the bed elevation. In the longitudinal direction, each reach is characterized by its length. For a simple channel, the reach length measures the path of the main flow or channel thalweg. For a compound channel, the flow paths in the main channel and floodplains may be significantly different, and an average, such as the discharge-weighted average, of their lengths should be used as the reach length.

Applying the standard step method to discretize Eq. (5.40) yields

$$\frac{\beta_i' Q_i^2}{2gA_i^2} + z_{s,i} = \frac{\beta_{i+1}' Q_{i+1}^2}{2gA_{i+1}^2} + z_{s,i+1} + \frac{\Delta x_{i+1/2}}{2}\left(\frac{Q_{i+1}|Q_{i+1}|}{K_{i+1}^2} + \frac{Q_i|Q_i|}{K_i^2}\right) \qquad (5.42)$$

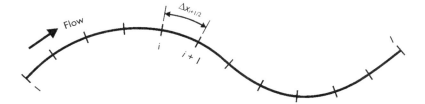

Figure 5.7 Finite difference grid in 1-D channel model.

where $\Delta x_{i+1/2}$ represents the length of the reach between cross-sections i and $i+1$.

In Eq. (5.42), the friction slope is represented by the arithmetic mean between cross-sections i and $i+1$. It can also be represented by the harmonic mean

$$S_{f,i+1/2} = 2 \bigg/ \left(\frac{K_{i+1}^2}{Q_{i+1}|Q_{i+1}|} + \frac{K_i^2}{Q_i|Q_i|} \right) \tag{5.43}$$

the geometric mean

$$S_{f,i+1/2} = \left(\frac{Q_{i+1}|Q_{i+1}|}{K_{i+1}^2} \frac{Q_i|Q_i|}{K_i^2} \right)^{1/2} \tag{5.44}$$

or the conveyance mean

$$S_{f,i+1/2} = \left(\frac{Q_{i+1} + Q_i}{K_{i+1} + K_i} \right)^2 \tag{5.45}$$

If the channel cross-section is suddenly expanded or contracted, a local head loss should be considered and Eq. (5.42) is replaced by

$$\frac{\beta_i' Q_i^2}{2gA_i^2} + z_{s,i} = \frac{\beta_{i+1}' Q_{i+1}^2}{2gA_{i+1}^2} + z_{s,i+1} + \frac{\Delta x_{i+1/2}}{2} \left(\frac{Q_{i+1}|Q_{i+1}|}{K_{i+1}^2} + \frac{Q_i|Q_i|}{K_i^2} \right)$$
$$+ \lambda_{i+1/2} \left| \frac{\beta_{i+1}' Q_{i+1}^2}{2gA_{i+1}^2} - \frac{\beta_i' Q_i^2}{2gA_i^2} \right| \tag{5.46}$$

where $\lambda_{i+1/2}$ is the coefficient of local head loss due to channel expansion or contraction in the reach between cross-sections i and $i+1$.

5.2.1.2   Solution of discretized steady flow equations

The solution procedure for Eq. (5.42) differs in cases of subcritical and supercritical flows. For subcritical flow, a flow discharge is usually specified at the inlet and a water stage is specified at the outlet. Therefore, the flow discharge in the solution domain can be calculated easily in a forewater sweep by applying mass continuity, and the water stage can then be determined by backwater calculation using Eq. (5.42). Because of its nonlinearity, Eq. (5.42) needs to be solved iteratively.

Define the following function:

$$F = \frac{\beta'_{i+1}Q^2_{i+1}}{2gA^2_{i+1}} - \frac{\beta'_i Q^2_i}{2gA^2_i} + z_{s,i+1} - z_{s,i} + \frac{\Delta x_{i+1/2}}{2}\left(\frac{Q_{i+1}|Q_{i+1}|}{K^2_{i+1}} + \frac{Q_i|Q_i|}{K^2_i}\right)$$

(5.47)

Because $z_{s,i+1}$ and the corresponding A_{i+1} and K_{i+1} at cross-section $i+1$ have been obtained from the previous calculation in the reach between cross-sections $i+1$ and $i+2$, or from the given water stage at the outlet, now the problem is determining $z_{s,i}$ and the corresponding A_i and K_i by ensuring $F = 0$. The following bisection method is often used:

(1) Find a segment $[Z_{lower}, Z_{upper}]$ in which the solution of $z_{s,i}$ exists, i.e., $F_{upper}F_{lower} < 0$, with F_{upper} and F_{lower} being the values of F corresponding to Z_{upper} and Z_{lower}, respectively;
(2) Set $Z_{middle} = (Z_{upper} + Z_{lower})/2$ and calculate F_{middle}, the value of F corresponding to Z_{middle};
(3) If $F_{middle} = 0$ (or less than a certain tolerance), Z_{middle} is the solution of $z_{s,i}$ and then stop iteration; otherwise, if $F_{middle}F_{lower} < 0$, then set $Z_{upper} = Z_{middle}$, and if $F_{upper}F_{middle} < 0$, then set $Z_{lower} = Z_{middle}$;
(4) If $Z_{upper} - Z_{lower}$ is less than a reasonable tolerance, then set $(Z_{upper} + Z_{lower})/2$ to be the solution of $z_{s,i}$ and stop iteration; otherwise, repeat from step (2) until the convergent solution is obtained.

Note that the search in step (1) for the lower and upper bounds Z_{lower} and Z_{upper} of the initial segment where the solution exists can start from either the channel thalweg elevation or $z_{s,i+1}$. The search starting from the thalweg is upward only, whereas the search starting from $z_{s,i+1}$ must be conducted upward and downward. The former search is simpler and can guarantee the solution.

For supercritical flow, both flow discharge and water stage are usually specified at the inlet. Therefore, the water stage in the solution domain can be determined by forewater calculation using Eq. (5.42). Similarly, Eq. (5.42) must be solved using an iteration method, such as the bisection method. The difference is only that $z_{s,i}$ and the corresponding A_i and K_i are known while $z_{s,i+1}$ and the corresponding A_{i+1} and K_{i+1} are unknown.

For flow in mixed regimes, the entire computational domain is divided into subdomains according to the flow regimes, and then the previous methods are used to solve the subcritical and supercritical flows in all subdomains individually. Usually, internal boundary conditions should be applied in the transition regions between subdomains. Because the energy equation (5.40) may not be applicable in regions with hydraulic jumps, internal boundary conditions should be derived from the momentum equation instead, which may be found in Chow (1959) and HEC (1997).

5.2.1.3   Treatments for flow at channel confluences and splits

Fig. 5.8 shows a typical network of channels connected with confluences, splits, and hydraulic structures. To compute the steady flow in such a channel network, external boundary conditions at inlets and outlets and internal boundary conditions at channel confluences, splits, and hydraulic structures have to be imposed. The imposition of external boundary conditions of flow is introduced in Section 5.1.1.2, and the handling of hydraulic structures is discussed in Section 5.2.2.4. Treating channel confluences and splits is discussed here.

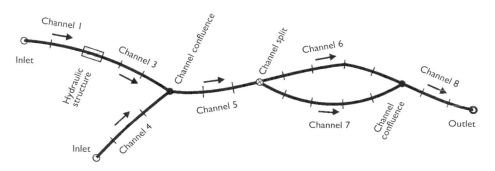

Figure 5.8 Sketch of a channel network.

Channel confluences

A confluence of two channels is depicted in Fig. 5.9, in which cross-sections 1 and 2 are placed at the ends of the upstream channels (denoted as 1 and 2), and cross-section 3 is at the beginning of the downstream channel (denoted as 3). The flow discharges at cross-sections 1, 2, and 3 are denoted as Q_1, Q_2, and Q_3, respectively. The continuity equation at the confluence reads

$$Q_3 = Q_1 + Q_2 \tag{5.48}$$

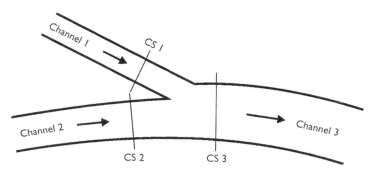

Figure 5.9 Configuration of channel confluence.

Applying Eq. (5.46) in the reaches from cross-sections 1 and 2 to 3 yields

$$\frac{\beta_1' Q_1^2}{2gA_1^2} + z_{s1} = \frac{\beta_3' Q_3^2}{2gA_3^2} + z_{s3} + \frac{\Delta x_{13}}{2}\left(\frac{Q_3|Q_3|}{K_3^2} + \frac{Q_1|Q_1|}{K_1^2}\right) + \lambda_{13}\left|\frac{\beta_3' Q_3^2}{2gA_3^2} - \frac{\beta_1' Q_1^2}{2gA_1^2}\right|$$

(5.49)

$$\frac{\beta_2' Q_2^2}{2gA_2^2} + z_{s2} = \frac{\beta_3' Q_3^2}{2gA_3^2} + z_{s3} + \frac{\Delta x_{23}}{2}\left(\frac{Q_3|Q_3|}{K_3^2} + \frac{Q_2|Q_2|}{K_2^2}\right) + \lambda_{23}\left|\frac{\beta_3' Q_3^2}{2gA_3^2} - \frac{\beta_2' Q_2^2}{2gA_2^2}\right|$$

(5.50)

where Δx_{13} and Δx_{23} represent the distances from cross-sections 1 and 2 to 3, respectively.

If the flow is subcritical, the water stage z_{s3} at cross-section 3 is obtained first by backwater calculation in channel 3. The water stages z_{s1} and z_{s2} at cross-sections 1 and 2 can then be obtained by solving Eqs. (5.49) and (5.50), following the procedure introduced in Section 5.2.1.2.

As a simplified case, if the distances Δx_{13} and Δx_{23} are very small, the water stages or energy heads of the three cross-sections at the confluence can be assumed to be identical. Thus, the calculated water stage at cross-section 3 is specified to cross-sections 1 and 2 if the flow is subcritical.

If the flow is supercritical, the forewater calculations are carried out in channels 1 and 2 down to cross-sections 1 and 2. The reach controlling the flow at the confluence has a larger specific force $A\bar{z}_s + \beta QU/g$ (Chow, 1959). Here, $\bar{z}_s$ is the depth from the water surface to the centroid of the flow area. The forewater calculation is made from the controlling upstream cross-section down to cross-section 3.

Channel splits

A split of one channel to two channel branches is depicted in Fig. 5.10, in which cross-section 1 is placed at the end of the upstream channel (denoted as 1), and cross-sections 2 and 3 are at the beginnings of the downstream channels (denoted as 2 and 3). The continuity equation at the channel split reads

$$Q_2 + Q_3 = Q_1 \qquad (5.51)$$

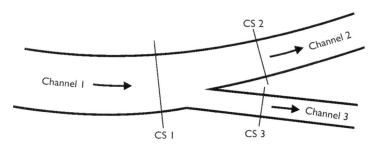

Figure 5.10 Configuration of channel split.

Applying Eq. (5.46) in the reaches from cross-section 1 to cross-sections 2 and 3 yields

$$\frac{\beta_1' Q_1^2}{2gA_1^2} + z_{s1} = \frac{\beta_2' Q_2^2}{2gA_2^2} + z_{s2} + \frac{\Delta x_{12}}{2}\left(\frac{Q_2|Q_2|}{K_2^2} + \frac{Q_1|Q_1|}{K_1^2}\right) + \lambda_{12}\left|\frac{\beta_2' Q_2^2}{2gA_2^2} - \frac{\beta_1' Q_1^2}{2gA_1^2}\right|$$

(5.52)

$$\frac{\beta_1' Q_1^2}{2gA_1^2} + z_{s1} = \frac{\beta_3' Q_3^2}{2gA_3^2} + z_{s3} + \frac{\Delta x_{13}}{2}\left(\frac{Q_3|Q_3|}{K_3^2} + \frac{Q_1|Q_1|}{K_1^2}\right) + \lambda_{13}\left|\frac{\beta_3' Q_3^2}{2gA_3^2} - \frac{\beta_1' Q_1^2}{2gA_1^2}\right|$$

(5.53)

where Δx_{12} and Δx_{13} represent the lengths of the two reaches.

If the flow is subcritical, the water stages z_{s2} and z_{s3} at cross-sections 2 and 3 are determined first by backwater calculations in channels 2 and 3. The water stage z_{s1} at cross-section 1 can then be obtained by solving Eqs. (5.52) and (5.53). However, because the ratio of flow discharges Q_2 and Q_3 is unknown, the following iteration procedure is needed:

(1) Assume the flow discharges Q_2 and Q_3 that satisfy the continuity equation (5.51);
(2) Determine the water stages z_{s2} and z_{s3} at cross-sections 2 and 3 through backwater calculations in channels 2 and 3, respectively;
(3) Calculate the water stage at cross-section 1, denoted as z_{s1}', from cross-section 2 using Eq. (5.52), according to the procedure in Section 5.2.1.2;
(4) Calculate the water stage at cross-section 1, denoted as z_{s1}'', from cross-section 3 using Eq. (5.53), according to the procedure in Section 5.2.1.2;
(5) If $|z_{s1}' - z_{s1}''|$ is less than a reasonable tolerance, then set $(z_{s1}' + z_{s1}'')/2$ as the water stage z_{s1} at cross-section 1, and stop iteration; otherwise, repeat from step (2) by reducing Q_2 and increasing Q_3 if $z_{s1}' > z_{s1}''$, or increasing Q_2 and reducing Q_3 if $z_{s1}' < z_{s1}''$, until a convergent solution is obtained.

If the lengths Δx_{12} and Δx_{13} are very small, the water stages of the three cross-sections at the split can be set the same. However, an iteration method similar to the one described above is still needed to determine the ratio of flow discharges in the two downstream channels.

If the flow is supercritical, the forewater calculation is carried out in channel 1 down to cross-section 1. The water stages at cross-sections 2 and 3 are calculated by performing forewater calculations from cross-section 1 to cross-sections 2 and 3 using Eqs. (5.52) and (5.53), respectively. In order to determine the ratio of flow discharges Q_2 and Q_3, an additional equation is needed. The momentum balance equation at the split is usually used. More details can be found in HEC (1997).

5.2.2 1-D unsteady flow calculation

5.2.2.1 Discretization of unsteady flow equations

The governing equations for unsteady open-channel flows are the St. Venant equations (5.1) and (5.2). However, a variety of forms of these equations have been used in the literature. For example, Wu and Vieira (2002) simplified Eq. (5.2) to the following form, by ignoring the momentum contribution from side flows and dividing by the flow area A:

$$\frac{\partial}{\partial t}\left(\frac{Q}{A}\right) + \frac{\partial}{\partial x}\left(\frac{\beta' Q^2}{2A^2}\right) + g\frac{\partial z_s}{\partial x} + gS_f = 0 \qquad (5.54)$$

where the correction factor β' is not exactly the same as the momentum correction factor β in Eq. (5.2). By comparing Eqs. (5.40) and (5.54), it is recognized that β' is similar in these two equations and thus can be determined using Eq. (5.41).

The numerical solution of the St. Venant equations is a classic problem in computational river dynamics. Mahmood and Yevjevich (1975) and Cunge *et al.* (1980) described some of the historical developments. One of the most widely used schemes is the Preissmann (1961) scheme, based on two computational points in two time levels, as presented in Eqs. (4.34)–(4.36). Its application to Eqs. (5.1) and (5.54) yields

$$\frac{\psi}{\Delta t}(A_{i+1}^{n+1} - A_{i+1}^n) + \frac{1-\psi}{\Delta t}(A_i^{n+1} - A_i^n) + \frac{\theta}{\Delta x}(Q_{i+1}^{n+1} - Q_i^{n+1})$$

$$+ \frac{1-\theta}{\Delta x}(Q_{i+1}^n - Q_i^n) - \theta[\psi q_{l,i+1}^{n+1} + (1-\psi)q_{l,i}^{n+1}]$$

$$- (1-\theta)[\psi q_{l,i+1}^n + (1-\psi)q_{l,i}^n] = 0 \qquad (5.55)$$

$$\frac{\psi}{\Delta t}\left(\frac{Q_{i+1}^{n+1}}{A_{i+1}^{n+1}} - \frac{Q_{i+1}^n}{A_{i+1}^n}\right) + \frac{1-\psi}{\Delta t}\left(\frac{Q_i^{n+1}}{A_i^{n+1}} - \frac{Q_i^n}{A_i^n}\right)$$

$$+ \frac{\theta}{\Delta x}\left[\frac{\beta_{i+1}'^{n+1}}{2}\left(\frac{Q_{i+1}^{n+1}}{A_{i+1}^{n+1}}\right)^2 - \frac{\beta_i'^{n+1}}{2}\left(\frac{Q_i^{n+1}}{A_i^{n+1}}\right)^2\right]$$

$$+ \frac{1-\theta}{\Delta x}\left[\frac{\beta_{i+1}'^n}{2}\left(\frac{Q_{i+1}^n}{A_{i+1}^n}\right)^2 - \frac{\beta_i'^n}{2}\left(\frac{Q_i^n}{A_i^n}\right)^2\right] + \frac{\theta g}{\Delta x}(z_{s,i+1}^{n+1} - z_{s,i}^{n+1})$$

$$+ \frac{(1-\theta)g}{\Delta x}(z_{s,i+1}^n - z_{s,i}^n) + \theta g[\psi_R S_{f,i+1}^{n+1} + (1-\psi_R)S_{f,i}^{n+1}]$$

$$+ (1-\theta)g[\psi_R S_{f,i+1}^n + (1-\psi_R)S_{f,i}^n] = 0 \qquad (5.56)$$

where θ and ψ are the temporal and spatial weighting factors in the Preissmann scheme, as shown in Fig. 4.5; and ψ_R is the spatial weighting factor for friction slope in the case of low flow depth, as described in Section 5.2.2.6.

5.2.2.2 *Local linearization of discretized unsteady flow equations*

Eqs. (5.55) and (5.56) constitute a nonlinear system that needs to be solved iteratively. During the iteration process, the following relations are expected:

$$A_i^{n+1} = A_i^* + B_i^* \delta h_i \qquad (5.57)$$

$$Q_i^{n+1} = Q_i^* + \delta Q_i \qquad (5.58)$$

where $*$ denotes the estimates at the last iteration step, δh is the water stage (flow depth) increment, δQ is the flow discharge increment, and B is the channel width at the water surface.

Substituting Eqs. (5.57) and (5.58) into Eq. (5.55) yields

$$\frac{\psi}{\Delta t} B_{i+1}^* \delta h_{i+1} + \frac{1-\psi}{\Delta t} B_i^* \delta h_i + \frac{\theta}{\Delta x} \delta Q_{i+1} - \frac{\theta}{\Delta x} \delta Q_i$$
$$= -\frac{\psi}{\Delta t}(A_{i+1}^* - A_{i+1}^n) - \frac{1-\psi}{\Delta t}(A_i^* - A_i^n) - \frac{\theta}{\Delta x}(Q_{i+1}^* - Q_i^*)$$
$$- \frac{1-\theta}{\Delta x}(Q_{i+1}^n - Q_i^n) + \theta[\psi q_{l,i+1}^{n+1} + (1-\psi)q_{l,i}^{n+1}]$$
$$+ (1-\theta)[\psi q_{l,i+1}^n + (1-\psi)q_{l,i}^n] \qquad (5.59)$$

Eq. (5.59) can be written as

$$a_i \delta h_i + b_i \delta Q_i + c_i \delta h_{i+1} + d_i \delta Q_{i+1} = p_i \qquad (5.60)$$

where $a_i = (1-\psi)B_i^*/\Delta t$, $b_i = -\theta/\Delta x$, $c_i = \psi B_{i+1}^*/\Delta t$, $d_i = \theta/\Delta x$, and

$$p_i = -\frac{\psi}{\Delta t}(A_{i+1}^* - A_{i+1}^n) - \frac{1-\psi}{\Delta t}(A_i^* - A_i^n) - \frac{\theta}{\Delta x}(Q_{i+1}^* - Q_i^*)$$
$$- \frac{1-\theta}{\Delta x}(Q_{i+1}^n - Q_i^n) + \theta[\psi q_{l,i+1}^{n+1} + (1-\psi)q_{l,i}^{n+1}]$$
$$+ (1-\theta)[\psi q_{l,i+1}^n + (1-\psi)q_{l,i}^n]$$

To linearize the discretized momentum equation, the following relations based on the first-order Taylor series expansion in terms of δh and δQ are used:

$$z_{s,i}^{n+1} = z_{s,i}^* + \delta h_i \qquad (5.61)$$

$$(Q_i^{n+1})^2 = (Q_i^*)^2 + 2Q_i^* \delta Q_i \qquad (5.62)$$

$$\frac{1}{(K_i^{n+1})^2} = \frac{1}{(K_i^*)^2} - \frac{2}{(K_i^*)^3}\left(\frac{\partial K}{\partial z_s}\right)_i^* \delta h_i \qquad (5.63)$$

$$S_{f,i}^{n+1} = S_{f,i}^* + \frac{2|Q_i^*|}{(K_i^*)^2}\delta Q_i - \frac{2S_{f,i}^*}{K_i^*}\left(\frac{\partial K}{\partial z_s}\right)_i^* \delta h_i \qquad (5.64)$$

$$\frac{Q_i^{n+1}}{A_i^{n+1}} = \frac{Q_i^*}{A_i^*} + \frac{1}{A_i^*}\delta Q_i - \frac{Q_i^* B_i^*}{(A_i^*)^2}\delta h_i \tag{5.65}$$

$$\left(\frac{Q_i^{n+1}}{A_i^{n+1}}\right)^2 = \left(\frac{Q_i^*}{A_i^*}\right)^2 + \frac{2Q_i^*}{(A_i^*)^2}\delta Q_i - \frac{2(Q_i^*)^2 B_i^*}{(A_i^*)^3}\delta h_i \tag{5.66}$$

$$\beta_i'^{n+1} = \beta_i'^* \tag{5.67}$$

Substituting Eqs. (5.61)–(5.67) into the discretized momentum equation (5.56) yields the locally linearized form:

$$a_i'\delta h_i + b_i'\delta Q_i + c_i'\delta h_{i+1} + d_i'\delta Q_{i+1} = p_i' \tag{5.68}$$

where

$$a_i' = -\frac{1-\psi}{\Delta t}\frac{Q_i^* B_i^*}{(A_i^*)^2} + \frac{\theta}{\Delta x}\frac{\beta_i'^*(Q_i^*)^2 B_i^*}{(A_i^*)^3} - \frac{\theta g}{\Delta x} - 2\theta(1-\psi_R)g\frac{S_{f,i}^*}{K_i^*}\left(\frac{\partial K}{\partial z_s}\right)_i^*;$$

$$b_i' = \frac{1-\psi}{\Delta t}\frac{1}{A_i^*} - \frac{\theta}{\Delta x}\frac{\beta_i'^* Q_i^*}{(A_i^*)^2} + 2\theta(1-\psi_R)g\frac{|Q_i^*|}{(K_i^*)^2};$$

$$c_i' = -\frac{\psi}{\Delta t}\frac{Q_{i+1}^* B_{i+1}^*}{(A_{i+1}^*)^2} - \frac{\theta}{\Delta x}\frac{\beta_{i+1}'^*(Q_{i+1}^*)^2 B_{i+1}^*}{(A_{i+1}^*)^3} + \frac{\theta g}{\Delta x} - 2\theta\psi_R g\frac{S_{f,i+1}^*}{K_{i+1}^*}\left(\frac{\partial K}{\partial z_s}\right)_{i+1}^*;$$

$$d_i' = \frac{\psi}{\Delta t}\frac{1}{A_{i+1}^*} + \frac{\theta}{\Delta x}\frac{\beta_{i+1}'^* Q_{i+1}^*}{(A_{i+1}^*)^2} + 2\theta\psi_R g\frac{|Q_{i+1}^*|}{(K_{i+1}^*)^2}; \text{ and}$$

$$p_i' = -\frac{\psi}{\Delta t}\left(\frac{Q_{i+1}^*}{A_{i+1}^*} - \frac{Q_{i+1}^n}{A_{i+1}^n}\right) - \frac{1-\psi}{\Delta t}\left(\frac{Q_i^*}{A_i^*} - \frac{Q_i^n}{A_i^n}\right)$$

$$- \frac{\theta}{\Delta x}\left[\frac{\beta_{i+1}'^*}{2}\left(\frac{Q_{i+1}^*}{A_{i+1}^*}\right)^2 - \frac{\beta_i'^*}{2}\left(\frac{Q_i^*}{A_i^*}\right)^2\right]$$

$$- \frac{1-\theta}{\Delta x}\left[\frac{\beta_{i+1}'^n}{2}\left(\frac{Q_{i+1}^n}{A_{i+1}^n}\right)^2 - \frac{\beta_i'^n}{2}\left(\frac{Q_i^n}{A_i^n}\right)^2\right]$$

$$- \frac{\theta g}{\Delta x}(z_{s,i+1}^* - z_{s,i}^*) - \frac{(1-\theta)g}{\Delta x}(z_{s,i+1}^n - z_{s,i}^n)$$

$$- \theta g[\psi_R S_{f,i+1}^* + (1-\psi_R)S_{f,i}^*]$$

$$- (1-\theta)g[\psi_R S_{f,i+1}^n + (1-\psi_R)S_{f,i}^n]$$

Note that the momentum equation (5.2) in the dynamic wave model and its simplification Eq. (5.4) in the diffusion wave model can also be discretized using the Preissmann implicit scheme and linearized locally using Eqs. (5.61)–(5.67), and the resulting equations can be written as Eq. (5.68) with different coefficients. The detailed derivation is left to interested readers.

5.2.2.3 Solution of discretized unsteady flow equations

Algorithm for a single channel

As shown in Fig. 5.7, a single channel is segmented to $I - 1$ reaches with I cross-sections. The pentadiagonal matrix of Eqs. (5.60) and (5.68) is solved by successively applying a double sweep algorithm, which is often called the Thomas algorithm.

A linear relationship between the unknowns δh_i and δQ_i is assumed to be of the type:

$$\delta Q_i = S_i \delta h_i + T_i \tag{5.69}$$

Substituting Eq. (5.69) into Eqs. (5.60) and (5.68) and eliminating δh_i yields

$$\delta Q_{i+1} = S_{i+1} \delta h_{i+1} + T_{i+1} \tag{5.70}$$

where S_{i+1} and T_{i+1} are recurrence coefficients:

$$S_{i+1} = -\frac{(a_i + b_i S_i)c'_i - (a'_i + b'_i S_i)c_i}{(a_i + b_i S_i)d'_i - (a'_i + b'_i S_i)d_i} \tag{5.71}$$

$$T_{i+1} = \frac{(a_i + b_i S_i)(p'_i - b'_i T_i) - (a'_i + b'_i S_i)(p_i - b_i T_i)}{(a_i + b_i S_i)d'_i - (a'_i + b'_i S_i)d_i} \tag{5.72}$$

In the first (forward) sweep, Eqs. (5.71) and (5.72) are applied recursively, with i varying from 1 to $I - 1$. To perform this sweep, S_1 and T_1 at cross-section 1 (inlet) are derived from the upstream boundary condition. For simplicity, the case of subcritical flow is considered here. Therefore, Q_1^{n+1} is known by the given discharge hydrograph at the inlet, and the recurrence coefficients S_1 and T_1 read

$$S_1 = 0, \quad T_1 = Q_1^{n+1} - Q_1^* \tag{5.73}$$

Substituting Eq. (5.69) into Eq. (5.60) yields

$$\delta h_i = \frac{(p_i - b_i T_i) - (c_i \delta h_{i+1} + d_i \delta Q_{i+1})}{a_i + b_i S_i} \tag{5.74}$$

Therefore, in the second (return) sweep, δh_i and δQ_i can be calculated using Eqs. (5.74) and (5.69) recursively, with i from $I - 1$ to 1.

To perform the second sweep, the values of δh_I and δQ_I at cross-section I (outlet) are derived from the downstream boundary condition, which can be a time series of water stage, a stage-discharge rating curve, etc. If the water-stage time series is specified, $z_{s,I}^{n+1}$ is known and the stage increment at cross-section I is

$$\delta h_I = z_{s,I}^{n+1} - z_{s,I}^* \qquad (5.75)$$

and the discharge increment δQ_I can then be determined using Eq. (5.69).

If a measured stage-discharge rating curve, $Q = f(z_s)$, is specified at the outlet, a discretized equation can be obtained by applying the first-order Taylor series expansion:

$$\delta Q_I - \frac{df}{dz_s}\delta h_I = f^* - Q_I^* \qquad (5.76)$$

and the stage increment at point I can then be derived from Eqs. (5.69) and (5.76) as

$$\delta h_I = \frac{T_I + Q_I^* - f^*}{df/dz_s - S_I} \qquad (5.77)$$

In the outlet located in a nearly prismatic channel with a positive slope (downslope), the flow can be assumed to be uniform; thus, a relation of $Q = K(z_s)\sqrt{S_0}$ exists, in which K is the conveyance and S_0 is the channel slope. In analogy to Eq. (5.77), the following equation for the stage increment at point I can be derived:

$$\delta h_I = \frac{T_I + Q_I^* - K^*\sqrt{S_0}}{\sqrt{S_0}dK^*/dz_s - S_I} \qquad (5.78)$$

If the outlet is controlled by an in-stream structure, such as spillway or weir, a free overfall flow exists; thus, a stage-discharge rating curve $Q = f(z_s)$ can be obtained using the critical flow condition near the brinkpoint, and then Eq. (5.77) can be applied.

If a flood or tidal wave propagation is concerned, the outflow boundary must also be non-reflective and able to damp out the waves. This type of outflow boundary condition may be found in Hinatsu (1992) and others.

Therefore, the aforementioned two sweeps constitute an iteration step, yielding δh and δQ. z_s^* and Q^* are then updated by $z_s^* + \delta h$ and $Q^* + \delta Q$. The iteration is stopped when the solutions for z_s^* and Q^* have converged ($\delta h \to 0$ and $\delta Q \to 0$). The converged z_s^* and Q^* are eventually given to z_s^{n+1} and Q^{n+1}.

Algorithm for a dendritic channel network

A dendritic (or tree-like) channel network includes tributaries and/or distributaries without any loop. The previous double sweep algorithm can still be applied in the solution of unsteady flows in this type of channel network, provided that a certain computational order is respected.

A dendritic network of three channels, shown in Fig. 5.11, is used as illustration. Suppose that the forward sweep starts from point 1 of channel A, at which a boundary condition, such as the time series of flow discharge or water stage, is given. The recurrence coefficients are calculated along channel A using Eqs. (5.71) and (5.72). At the last point of channel A, the following relation is obtained:

$$\delta Q_{A,M} = S_{A,M}\delta h_{A,M} + T_{A,M} \tag{5.79}$$

where the subscript A denotes channel A, and M denotes the last point in channel A.

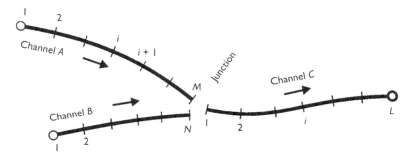

Figure 5.11 Dendritic network with three channels.

The forward sweep in channel B is also carried out from the first to the last point. A boundary condition should be given at the first point, while the following relation is obtained at the last point, N:

$$\delta Q_{B,N} = S_{B,N}\delta h_{B,N} + T_{B,N} \tag{5.80}$$

where the coefficients $S_{B,N}$ and $T_{B,N}$ are determined using Eqs. (5.71) and (5.72).

Now, let us consider how to handle the junction. For convenience, the three cross-sections at the junction are located very close together. Therefore, it can be assumed that the water stages at the three cross-sections are equal, and the flow discharge at the downstream cross-section is equal to the sum of those at the two upstream cross-sections:

$$z_{sA,M}^{n+1} = z_{sB,N}^{n+1} = z_{sC,1}^{n+1} \tag{5.81}$$

$$Q_{C,1}^{n+1} = Q_{A,M}^{n+1} + Q_{B,N}^{n+1} \tag{5.82}$$

Eqs. (5.81) and (5.82) are the compatibility conditions at the junction. Substituting Eqs. (5.58) and (5.61) into Eqs. (5.81) and (5.82) yields

$$\delta h_{A,M} - \delta h_{C,1} = z_{sC,1}^* - z_{sA,M}^* \tag{5.83}$$

$$\delta h_{B,N} - \delta h_{C,1} = z_{sC,1}^* - z_{sB,N}^* \tag{5.84}$$

$$\delta Q_{C,1} - \delta Q_{A,M} - \delta Q_{B,N} = Q_{A,M}^* + Q_{B,N}^* - Q_{C,1}^* \tag{5.85}$$

Substituting Eqs. (5.79) and (5.80) into Eq. (5.85) and then using the expressions for $\delta h_{A,M}$ and $\delta h_{B,N}$ obtained from Eqs. (5.83) and (5.84) yields

$$\delta Q_{C,1} = S_{C,1}\delta h_{C,1} + T_{C,1} \tag{5.86}$$

where $S_{C,1} = S_{A,M} + S_{B,N}$, and $T_{C,1} = Q_{A,M}^* + Q_{B,N}^* - Q_{C,1}^* + S_{A,M}(z_{sC,1}^* - z_{sA,M}^*) + S_{B,N}(z_{sC,1}^* - z_{sB,N}^*) + T_{A,M} + T_{B,N}$.

The forward sweep can then be carried out from the first to the last point in channel C using Eqs. (5.71) and (5.72).

The return sweep starts from the last point of channel C, at which a boundary condition is specified. The stage and discharge increments at the last point are determined using Eqs. (5.75)–(5.78), and at the intermediate points using Eqs. (5.69) and (5.74). At the end of the return sweep in channel C back to the junction, the stage increment $\delta h_{C,1}$ is calculated using Eq. (5.74), and the discharge increment $\delta Q_{C,1}$ is determined using Eq. (5.86). Next, the stage increments $\delta h_{A,M}$ and $\delta h_{B,N}$ are determined using Eqs. (5.83) and (5.84), and the discharge increments $\delta Q_{A,M}$ and $\delta Q_{B,N}$ are computed using Eqs. (5.79) and (5.80). Finally, the return sweep can be carried out along both channels A and B.

It should be noted that the equal water stage condition (5.81) may be replaced with the equal energy level condition at the junction:

$$z_{sA,M}^{n+1} + \frac{1}{2g}\left(\frac{Q_{A,M}^{n+1}}{A_{A,M}^{n+1}}\right)^2 = z_{sB,N}^{n+1} + \frac{1}{2g}\left(\frac{Q_{B,N}^{n+1}}{A_{B,N}^{n+1}}\right)^2 = z_{sC,1}^{n+1} + \frac{1}{2g}\left(\frac{Q_{C,1}^{n+1}}{A_{C,1}^{n+1}}\right)^2$$

$$\tag{5.87}$$

which can be expanded in terms of δh and δQ and used to substitute Eqs. (5.83) and (5.84).

Algorithm for a looped channel network

A "looped" channel network is shown in Fig. 5.12. The difference between "dendritic" and "looped" channel networks is that there is only one possible flow path from a given point to another in a dentritic network, while there are usually several such flow paths in a looped network. The previous double sweep algorithm cannot be applied directly to the solution of unsteady flows in looped channel networks. The looped solution algorithm described by Cunge et al. (1980) is often used instead. In this algorithm, the term "node" is used to represent the junction of several flow paths that originate from either other nodes or boundary points. For example, the nodes in the channel network shown in Fig. 5.12 are A, B, C, and D. The points (cross-sections) between two nodes in each channel are defined as intermediate points.

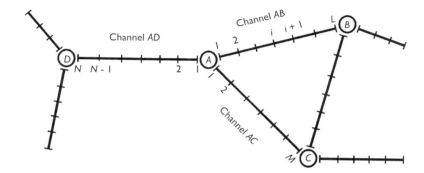

Figure 5.12 Looped channel network.

Consider the channels AB, AC, and AD, which are connected to node A in Fig. 5.12. Suppose that there are L computational points along channel AB, M points along channel AC, and N points along channel AD. For the reach between points i and $i+1$ in a channel, e.g., AB, Eqs. (5.60) and (5.68) can be obtained. Eliminating δQ_{i+1} from them yields

$$\delta h_{i+1} = C_{i+1}\delta h_i + D_{i+1}\delta Q_i + E_{i+1} \qquad (5.88)$$

where $C_{i+1} = -(d_i'a_i - d_i a_i')/(d_i'c_i - d_i c_i')$, $D_{i+1} = -(d_i'b_i - d_i b_i')/(d_i'c_i - d_i c_i')$, and $E_{i+1} = (d_i'p_i - d_i p_i')/(d_i'c_i - d_i c_i')$.

Suppose that the following relation exists at point $i+1$ of channel AB:

$$\delta Q_{i+1} = F_{i+1}\delta h_{i+1} + G_{i+1} + H_{i+1}\delta h_L \qquad (5.89)$$

Substituting Eq. (5.89) into Eqs. (5.60) and (5.68), and eliminating Δh_{i+1} yields

$$\delta Q_i = F_i\delta h_i + G_i + H_i\delta h_L \qquad (5.90)$$

with

$$F_i = -[a_i(c_i' + d_i'F_{i+1}) - a_i'(c_i + d_iF_{i+1})]/$$
$$[b_i(c_i' + d_i'F_{i+1}) - b_i'(c_i + d_iF_{i+1})] \qquad (5.91)$$

$$G_i = [(p_i - d_iG_{i+1})(c_i' + d_i'F_{i+1}) - (p_i' - d_i'G_{i+1})(c_i + d_iF_{i+1})]/$$
$$[b_i(c_i' + d_i'F_{i+1}) - b_i'(c_i + d_iF_{i+1})] \qquad (5.92)$$

$$H_i = -H_{i+1}[d_i(c_i' + d_i'F_{i+1}) - d_i'(c_i + d_iF_{i+1})]/$$
$$[b_i(c_i' + d_i'F_{i+1}) - b_i'(c_i + d_iF_{i+1})] \qquad (5.93)$$

The recurrence coefficients F_{L-1}, G_{L-1}, and H_{L-1} are obtained by eliminating δQ_L from Eqs. (5.60) and (5.68) in the reach between points $L-1$ and L. Therefore, the coefficients F, G, and H can be computed using Eqs. (5.91)–(5.93) in the first sweep from node B to node A, consequently yielding

$$\delta Q_{1,AB} = F_{1,AB}\delta h_{1,AB} + G_{1,AB} + H_{1,AB}\delta h_L \tag{5.94}$$

where $\delta Q_{1,AB}$ denotes the discharge increment at point 1 of channel AB; and $F_{1,AB}$, $G_{1,AB}$, and $H_{1,AB}$ are recurrence coefficients known from Eqs. (5.91)–(5.93).

Similarly, a sweep from node C to node A along channel AC gives

$$\delta Q_{1,AC} = F_{1,AC}\delta h_{1,AC} + G_{1,AC} + H_{1,AC}\delta h_M \tag{5.95}$$

and a sweep from node D to node A along channel AD gives

$$\delta Q_{1,AD} = F_{1,AD}\delta h_{1,AD} + G_{1,AD} + H_{1,AD}\delta h_N \tag{5.96}$$

The compatibility conditions of discharge continuity and equal water stages at node A are written as follows:

$$q_1(t^{n+1}) + \sum_{j=1}^{J} Q_{1,j}^{n+1} = 0 \tag{5.97}$$

$$z_{s1,1}^{n+1} = z_{s1,2}^{n+1} = \ldots = z_{s1,j}^{n+1} = \ldots = z_{s1,J}^{n+1} \tag{5.98}$$

where j is the index of the channels that emanate from node A, J is the total number of such channels and $J = 3$ for node A in Fig. 5.12, and $q_1(t^{n+1})$ is the external inflow (or outflow) to node A at time t^{n+1}.

Applying the Taylor series expansion to Eqs. (5.97) and (5.98) yields

$$q_1(t^{n+1}) + \sum_{j=1}^{J} Q_{1,j}^* + \sum_{j=1}^{J} \delta Q_{1,j} = 0 \tag{5.99}$$

$$\delta h_{1,1} = \delta h_{1,2} = \ldots = \delta h_{1,j} = \ldots = \delta h_{1,J} \tag{5.100}$$

Substituting Eqs. (5.94)–(5.96) and (5.100) into Eq. (5.99) yields a linear algebraic equation in terms of $J + 1$ unknowns:

$$f(\delta h_A, \delta h_L, \delta h_M, \delta h_N) = 0 \tag{5.101}$$

Eq. (5.101) is derived based on node A. Performing the above procedure for all nodes in the network eventually leads to a system of linear equations for the stage increments δh at all nodes as unknowns:

$$[S]\{\delta h\} = \{b\} \qquad (5.102)$$

where $[S]$ is the coefficient matrix with $m \times m$ elements, $\{\delta h\}$ is the vector of m unknowns, and $\{b\}$ is the vector with m elements holding all free terms. Here, m is the total number of nodes.

The system represented by Eq. (5.102) can be solved using a matrix inversion technique. Once the stage increments are solved at the nodes, Eqs. (5.94)–(5.96) are used to determine the discharge increments at the ends of each channel, and Eqs. (5.88) and (5.89) are used to compute δh_{i+1} and δQ_{i+1} for all intermediate points in a generalized return sweep.

In principle, Eq. (5.102) can be solved using any matrix inversion technique. However, because the matrix may be quite large, a direct inversion computation can be expensive. An iterative inverse computation may also have trouble when the matrix loses diagonal dominance. The block tri-diagonal matrix solution technique suggested by Mahmood and Yevjevich (1975) has been found to be very efficient in the solution of equation system (5.102). The details can be found in that reference.

5.2.2.4 *Treatment of hydraulic structures as internal boundaries*

Because of their complexity, it is almost impossible to simulate the detailed flow patterns around in-stream hydraulic structures, such as culverts, bridge crossings, drop structures, weirs, sluice gates, spillways, and measuring flumes, using a 1-D model. Simplifications must be made to obtain a feasible solution. The storage effect of the flow at a hydraulic structure is usually neglected, so the same flow discharge is imposed at its upstream and downstream ends:

$$Q_{up} = Q_{down} \qquad (5.103)$$

which can be expanded as

$$\delta Q_{up} - \delta Q_{down} = Q^*_{down} - Q^*_{up} \qquad (5.104)$$

The water stage at the hydraulic structure is often determined using a stage-discharge relation, which is related to whether the flow is upstream or downstream controlled. The upstream control flow is treated as a free overfall flow that is critical, while the downstream control flow is treated as an orifice-like flow. For the upstream control flow, the critical flow condition implies

$$Q = A_c \sqrt{g \frac{A_c}{B_c}} \qquad (5.105)$$

where A_c and B_c are the area and top width of flow at the structure, respectively. Both are functions of flow depth. Thus, the following general stage-discharge relation can be established:

$$Q = f(z_{s,up}) \tag{5.106}$$

The first-order Taylor series expansion of Eq. (5.106) reads

$$\delta Q - \frac{\partial f}{\partial z_{s,up}} \delta h_{up} = f^* - Q^* \tag{5.107}$$

For the downstream control flow, the following relation of orifice-like flow is usually used:

$$Q = A\sqrt{\frac{2g(z_{s,up} - z_{s,down})}{K_L}} \tag{5.108}$$

where K_L is the coefficient of energy loss at the hydraulic structure.

Because it cannot handle the situation of $z_{s,up} \leq z_{s,down}$, Eq. (5.108) is reformulated as

$$z_{s,up} - z_{s,down} = \frac{K_L}{2g} \frac{Q|Q|}{A^2} \tag{5.109}$$

which is then expanded as

$$\delta h_{up} - \delta h_{down} = -z^*_{s,up} + z^*_{s,down} + \frac{K_L}{2g} \frac{Q^*|Q^*|}{A^{*2}} \tag{5.110}$$

Note that the stage and discharge increments originated from the term on the right-hand side of Eq. (5.109) are ignored in Eq. (5.110). They may be included for the sake of completion.

A dam structure may have various flow passage facilities, such as spillways, sluice gates, and power generators. The flows through these facilities may be free overflow and/or under control. Thus, the stage-discharge rating relation for a dam structure may be Eqs. (5.106), (5.108), or a combination of them.

In addition, the water stage or flow discharge measured at a dam and other structures can be used as the internal condition. If a time series of the water stage is known:

$$z_{s,up} = z_s(t) \tag{5.111}$$

the stage increment at the upstream point is determined by

$$\delta h_{up} = z^{n+1}_{s,up} - z^*_{s,up} \tag{5.112}$$

If a time series of the flow discharge is known:

$$Q_{up} = Q(t) \tag{5.113}$$

the discharge increment at the upstream point is

$$\delta Q_{up} = Q^{n+1} - Q^* \qquad (5.114)$$

Eq. (5.104) and one of Eqs. (5.107), (5.110), (5.112), and (5.114) are used to determine the flow at a hydraulic structure. Eq. (5.104) can be written in the form of Eq. (5.60), with the coefficients being $a_i = 0$, $b_i = 1$, $c_i = 0$, $d_i = -1$, and $p_i = Q^*_{down} - Q^*_{up}$. Eqs. (5.107), (5.110), (5.112), and (5.114) can be written as Eq. (5.68). The coefficients are: $a'_i = -\partial f / \partial z_{s,up}$, $b'_i = 1$, $c'_i = 0$, $d'_i = 0$, and $p'_i = f^* - Q^*$ for Eq. (5.107); $a'_i = 1$, $b'_i = 0$, $c'_i = -1$, $d'_i = 0$, and $p'_i = -z^*_{s,up} + z^*_{s,down} + K_L Q^* |Q^*| / (2gA^{*2})$ for Eq. (5.110); $a'_i = 1$, $b'_i = 0$, $c'_i = 0$, $d'_i = 0$, and $p'_i = z^{n+1}_{s,up} - z^*_{s,up}$ for Eq. (5.112); and $a'_i = 0$, $b'_i = 1$, $c'_i = 0$, $d'_i = 0$, and $p'_i = Q^{n+1} - Q^*$ for Eq. (5.114). Thus, Eqs. (5.104), (5.107), (5.110), (5.112), and (5.114) can be intrinsically incorporated into the solution algorithm.

It should be noted that described above are the general methods for considering hydraulic structures in a 1-D channel network model. For specific hydraulic structures, empirical stage-discharge relations may be used (see Wu and Vieira, 2002).

5.2.2.5 Stability of Preissmann scheme for unsteady flow equations

The numerical stability of the Preissmann scheme for the St. Venant equations was studied by Lyn and Goodwin (1987) and Venutelli (2002). Lyn and Goodwin's findings are introduced below.

Eqs. (5.1) and (5.2) are written as

$$\frac{\partial F}{\partial t} + M \frac{\partial F}{\partial x} = b \qquad (5.115)$$

where $F = (u, h)$, M is the coefficient matrix, and b is the vector of inhomogeneous terms.

Appling the Preissmann scheme (4.34)–(4.36) to discretize Eq. (5.115) and linearizing the discretized equation locally yields

$$\psi(F^{n+1}_{i+1} - F^n_{i+1}) + (1 - \psi)(F^{n+1}_i - F^n_i) + rM_0[\theta(F^{n+1}_{i+1} - F^{n+1}_i)$$
$$+ (1 - \theta)(F^n_{i+1} - F^n_i)] = b\Delta t \qquad (5.116)$$

where $r = \Delta t / \Delta x$, and M_0 is the coefficient matrix of M at locally uniform state.

The Fourier component, $\delta = \delta_* e^{-i(\omega \Delta t - \sigma \Delta x)}$, corresponding to F is governed by

$$\psi(\delta^{n+1}_{i+1} - \delta^n_{i+1}) + (1 - \psi)(\delta^{n+1}_i - \delta^n_i) + rM_0[\theta(\delta^{n+1}_{i+1} - \delta^{n+1}_i)$$
$$+ (1 - \theta)(\delta^n_{i+1} - \delta^n_i)] = 0 \qquad (5.117)$$

The growth factor of δ in Eq. (5.117) is

$$\xi = 1 - \frac{rc_k}{\psi + \frac{1}{\eta-1} + rc_k\theta} \qquad (5.118)$$

where $\xi = e^{-i\omega\Delta t}$, $\eta = e^{i\sigma\Delta x}$, and c_k is a characteristic wave speed of the system.

For the homogeneous problem, the von Neumann condition for numerical stability is $|\xi| \leq 1$, which implies

$$\left(\psi - \frac{1}{2}\right)\frac{1}{C_r} + \left(\theta - \frac{1}{2}\right) \geq 0 \qquad (5.119)$$

where C_r is the Courant number, defined as $C_r = rc_k$.

When $C_r > 0$, the stability condition (5.119) is schematically shown in Fig. 5.13. The Preissmann scheme is unconditionally stable in the quarter of $\theta \geq 1/2$ and $\psi \geq 1/2$, and unconditionally unstable in the quarter of $\theta < 1/2$ and $\psi < 1/2$. In the other two quarters, the stability depends on the Courant number. Usually, $\psi = 1/2$ is used for better accuracy regarding space. This leads to the conclusion that for unconditional stability — i.e., stable for all Courant numbers — it is necessary that $\theta \geq 1/2$. If $\psi \neq 1/2$, the stability will depend on the sign of C_r, or equivalently, on the direction of travel of a characteristic wave. Because of this, $\psi \neq 1/2$ should be used with caution in situations where characteristics travel in both directions, particularly where characteristic directions may change. As described by Meselhe and Holly (1993) and Kutija and Hewett (2002), the Preissmann scheme may encounter numerical instability in the transition between supercritical and subcritical flow regimes. This may be avoided by using some newly developed schemes, such as that proposed by Kutija and Hewett (2002). Some schemes for dam-break flow simulation introduced in Section 9.1 may also be used in the simulation of mixed-regime flows.

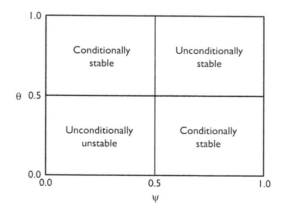

Figure 5.13 Regions of stability in $\psi - \theta$ plane when $C_r > 0$.

5.2.2.6   Auxiliary treatments for unsteady flow calculation

Representation of friction slope

The friction slope in Eq. (5.2) may be represented in various ways, such as arithmetic mean, harmonic mean, geometric mean, and conveyance mean (French, 1985), as expressed in Eqs. (5.42)–(5.45) for a steady flow model. For an unsteady flow model, the arithmetic mean friction slope is introduced in Eq. (5.56), and the conveyance mean friction slope is given as

$$S_f = \left\{ \frac{\theta[\psi_R Q_{i+1}^{n+1} + (1 - \psi_R)Q_i^{n+1}] + (1 - \theta)[\psi_R Q_{i+1}^n + (1 - \psi_R)Q_i^n]}{\theta[\psi_R K_{i+1}^{n+1} + (1 - \psi_R)K_i^{n+1}] + (1 - \theta)[\psi_R K_{i+1}^n + (1 - \psi_R)K_i^n]} \right\}^2$$

$$(5.120)$$

The harmonic and geometric mean friction slopes are left to interested readers.

Small flow depth

Computational difficulties arise when the flow depth becomes small. As the flow depth approaches zero, the conveyance and flow discharge go to zero, and thus the friction slope becomes indeterminate. This was explained well by Cunge *et al.* (1980). Meselhe and Holly (1993) showed that the characteristic curves are vertical and do not intersect when the flow depth is zero; consequently, a solution does not exist.

Cunge *et al.* (1980) proposed and Meselhe and Holly (1993) developed further an approach for handling the dry-bed problem. The basic idea is to switch the weighting for friction slope from central ($\psi_R = 0.5$, highest accuracy) to upstream ($0 \leq \psi_R \leq 0.5$). In a diffusive wave model, Langendoen (1996) related the weighting factor to the flow depth as

$$\psi_R = \min(0.5, ab^b) \qquad (5.121)$$

where the coefficient $a \approx 0.7$ and the exponent $b \approx 0.35$.

In the author's experience, Eq. (5.121) may fail and $\psi_R = 0$ is occasionally necessary to ensure stable solutions when using the dynamic wave model. One of the best choices is to try several values of ψ_R and find the value closest to 0.5 that allows a stable solution for a specific case.

Storage effect of still water zones

Still waters, or very slow flows, exist in sudden expansions, appendix channels, ponds, or small lakes that are connected to the main stream. These still water zones do not have significant momentum exchange with the main stream, but their storages may affect the main flow. To consider the storage effect, the continuity equation (5.1) is substituted by

$$\frac{\partial(A + A_0)}{\partial t} + \frac{\partial Q}{\partial x} = q_l \qquad (5.122)$$

where A_0 is the cross-sectional area of still water zones, and A is the cross-sectional area of main flow. The momentum equation (5.2) or (5.54) is not changed, in which only the main flow area A is used.

5.3 I-D CALCULATION OF SEDIMENT TRANSPORT

5.3.1 I-D equilibrium sediment transport model

The sediment continuity equation (5.38) is used to determine bed change in the equilibrium sediment transport model. For uniform sediment in a rectangular channel, it becomes

$$(1 - p'_m)\frac{\partial z_b}{\partial t} + \frac{\partial q_t}{\partial x} = 0 \tag{5.123}$$

where z_b is the bed elevation, and q_t is the sediment transport rate per unit channel width. As demonstrated in Eq. (5.37), q_t is determined using a sediment transport capacity formula.

Many numerical schemes have been used to discretize Eq. (5.123). Saiedi (1997) summarized some of them. For example, applying the Preissmann scheme expressed as Eqs. (4.35) and (4.36) to Eq. (5.123) yields

$$(1 - p'_m)\frac{\psi \Delta z_{b,i+1} + (1 - \psi)\Delta z_{b,i}}{\Delta t} + \frac{1}{\Delta x}[\theta(q_{t,i+1}^{n+1} - q_{t,i}^{n+1})$$
$$+ (1 - \theta)(q_{t,i+1}^n - q_{t,i}^n)] = 0 \tag{5.124}$$

where $\Delta z_{b,i}$ is the change in bed elevation at cross-section i in time step Δt, i.e., $\Delta z_{b,i} = z_{b,i}^{n+1} - z_{b,i}^n$. The spatial and temporal weighting factors in Eq. (5.124) were given various values, e.g., $\psi = 0.5$ by Cunge and Perdreau (1973).

De Vries (1981) adopted a Lax-type scheme for the bed change term and an explicit central difference scheme for the gradient of sediment discharge:

$$(1 - p'_m)\frac{1}{\Delta t}\left\{z_{b,i}^{n+1} - \left[(1 - \psi_b)z_{b,i}^n + \frac{\psi_b}{2}(z_{b,i-1}^n + z_{b,i+1}^n)\right]\right\}$$
$$+ \frac{1}{2\Delta x}(q_{t,i+1}^n - q_{t,i-1}^n) = 0 \tag{5.125}$$

where ψ_b is a weighting factor, which can enhance numerical stability but may introduce numerical diffusion. A small value should be used for ψ_b.

Gessler (1971) and Thomas (1982) used the forward difference scheme for the bed change term and the central difference scheme for the gradient of sediment discharge:

$$(1 - p'_m)\frac{\Delta z_{b,i}}{\Delta t} + \frac{1}{2\Delta x}(q_{t,i+1}^n - q_{t,i-1}^n) = 0 \tag{5.126}$$

Because of its special link between bed change and sediment discharge, Eq. (5.123) can be easily solved using the finite volume method on a staggered grid, in which sediment discharge is stored at cell faces and bed change is stored at cell centers. Integrating Eq. (5.123) over the control volume in Fig. 4.14 yields

$$(1 - p'_m)\frac{\Delta z_{b,P}}{\Delta t}\Delta x_P + q_{t,e} - q_{t,w} = 0 \qquad (5.127)$$

where $q_{t,e}$ and $q_{t,w}$ are the sediment discharges at faces e and w and can be determined using the first-order upwind scheme or the QUICK scheme introduced in Section 4.3.1.

The calculations at each time step are executed as follows: (a) compute flow using the steady or unsteady flow model introduced in Section 5.2; (b) determine sediment discharge using an empirical sediment transport formula; (c) calculate bed change using one of Eqs. (5.124)–(5.127); and (d) update channel geometry. In addition, bed material sorting is also calculated for non-uniform sediment transport. This is introduced in Sections 5.3.2 and 5.3.3.

5.3.2 1-D quasi-steady non-equilibrium sediment transport model

5.3.2.1 Representation of hydrographs

Denote the characteristic length, time, and velocity of fluvial processes in an open channel as Λ, T, and U. If $T \gg \Lambda/U$, the time-derivative terms in the St. Venant equations (5.1) and (5.2) and sediment transport equations (5.27), (5.28), and (5.34) can be omitted. Therefore, the fluvial processes can be simulated using a step-wise quasi-steady model. As demonstrated in Fig. 5.14, the continuous time series of flow discharge, water stage, and sediment discharge are represented by step functions that are constructed with the corresponding representative quantities over a suitable number of

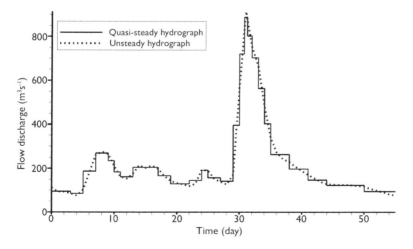

Figure 5.14 Representation of hydrographs in quasi-steady model.

time intervals. At each time interval, the flow and sediment transport are assumed to be steady, but the bed change and bed material sorting are still calculated, and thus the temporal evolution of channel morphology is simulated.

The time intervals are generally used as time steps in the calculation of bed change and bed material sorting, and should thus be restricted by the stability criteria of the sediment transport model. In addition, the time intervals should be defined in such a way that the temporal variations of flow discharge, water stage, and sediment discharge are well represented. This means that shorter intervals should be used for high flow periods and longer intervals may be used for low flow periods. Normally, the time intervals can be hours or days.

5.3.2.2 Discretization of quasi-steady sediment transport equations

As discussed in Section 5.1.2.1, two approaches may be used in the simulation of total-load transport. The approach that computes bed load and suspended load separately is adopted here.

Under steady flow conditions, the suspended-load and bed-load transport equations (5.27) and (5.28) without side sediment discharges are written as

$$\frac{d(QC_k)}{dx} = \alpha \omega_{sk} B(C_{*k} - C_k) \qquad (5.128)$$

$$\frac{dQ_{bk}}{dx} = \frac{1}{L}(Q_{b*k} - Q_{bk}) \qquad (5.129)$$

Eqs. (5.128) and (5.129) are first-order ordinary differential equations. They can be discretized using many numerical schemes, such as the Euler scheme, central difference scheme, and the Runge-Kutta method. Han (1980) established the following exponential difference scheme for Eq. (5.128), based on its analytical solution:

$$C_{k,i+1} = C_{*k,i+1} + (C_{k,i} - C_{*k,i}) \exp\left(-\frac{\alpha \omega_{sk} B_{i+1/2} \Delta x_{i+1/2}}{Q_{i+1/2}}\right)$$

$$+ (C_{*k,i} - C_{*k,i+1}) \frac{Q_{i+1/2}}{\alpha \omega_{sk} B_{i+1/2} \Delta x_{i+1/2}}$$

$$\times \left[1 - \exp\left(-\frac{\alpha \omega_{sk} B_{i+1/2} \Delta x_{i+1/2}}{Q_{i+1/2}}\right)\right] \qquad (5.130)$$

The Han scheme (5.130) is very stable, but it is not strictly conservative. However, many tests have shown that it has good accuracy.

Similarly, Eq. (5.129) can be discretized using the following exponential difference scheme:

$$Q_{bk,i+1} = Q_{b*k,i+1} + (Q_{bk,i} - Q_{b*k,i}) \exp\left(-\frac{\Delta x_{i+1/2}}{L}\right)$$

$$+ (Q_{b*k,i} - Q_{b*k,i+1}) \frac{L}{\Delta x_{i+1/2}} \left[1 - \exp\left(-\frac{\Delta x_{i+1/2}}{L}\right)\right] \qquad (5.131)$$

Eqs. (5.130) and (5.131) do not involve time, but they are applied at time level $n+1$ in the computation of channel morphological evolution.

The bed change equation (5.30) is discretized as

$$(1 - p'_m)\frac{\Delta A_{bk,i+1}}{\Delta t} = \alpha \omega_{sk} B_{i+1}(C_{k,i+1} - C_{*k,i+1}) + \frac{Q_{bk,i+1} - Q_{b*k,i+1}}{L} \tag{5.132}$$

where $\Delta A_{bk,i+1}$ is the change in bed area due to size class k at time step Δt.

The total change in bed area, $\Delta A_{b,i+1}$, is calculated by

$$\Delta A_{b,i+1} = \sum_{k=1}^{N} \Delta A_{bk,i+1} \tag{5.133}$$

and the bed elevation is updated by

$$z_{b,i+1,j}^{n+1} = z_{b,i+1,j}^{n} + \Delta z_{b,i+1,j} \tag{5.134}$$

where j is the point index in the cross-section, and $\Delta z_{b,i+1,j}$ is the local change in bed elevation obtained by allocating the bed area change $\Delta A_{b,i+1}$ along the cross-section. For a uniform allocation, $\Delta z_{b,i+1,j} = \Delta A_{b,i+1}/B_{i+1}$. More allocation options are discussed in Section 5.3.5.

The bed material sorting equations (5.32) and (5.33) in the mixing and second layers are discretized as

$$p_{bk,i+1}^{n+1} = \frac{\Delta A_{bk,i+1} + A_{m,i+1}^{n} p_{bk,i+1}^{n} + p_{bk,i+1}^{*n}(A_{m,i+1}^{n+1} - A_{m,i+1}^{n} - \Delta A_{b,i+1})}{A_{m,i+1}^{n+1}} \tag{5.135}$$

$$p_{sbk,i+1}^{n+1} = \frac{A_{sub,i+1}^{n} p_{sbk,i+1}^{n} - p_{bk,i+1}^{*n}(A_{m,i+1}^{n+1} - A_{m,i+1}^{n} - \Delta A_{b,i+1})}{A_{sub,i+1}^{n+1}} \tag{5.136}$$

where $p_{bk,i+1}^{*n}$ is $p_{bk,i+1}^{n}$ if $\Delta A_{b,i+1} + A_{m,i+1}^{n} \geq A_{m,i+1}^{n+1}$ and $p_{sbk,i+1}^{n}$ if $\Delta A_{b,i+1} + A_{m,i+1}^{n} < A_{m,i+1}^{n+1}$.

The bed-material gradations in other subsurface layers are calculated according to mass conservation, if there is exchange between them.

The bed-material gradation p_{bk} in Eq. (5.29) can be treated explicitly or implicitly. If the implicit scheme is used, the discretized sediment equations (5.130)–(5.136) can be solved in a coupled form, using the direct method proposed by Wu (1991). This coupled solution procedure is not presented here, because a similar one is introduced in the next subsection.

5.3.3 1-D unsteady non-equilibrium sediment transport model

5.3.3.1 Discretization of unsteady sediment transport equations

For simplicity, the bed-material load transport model introduced in Section 5.1.2.1 is adopted here. Applying the Preissmann scheme to discretize the total-load transport equation (5.34) yields (Wu et al., 2004a)

$$
\begin{aligned}
&\frac{\psi}{\Delta t}\left(\frac{Q_{tk,i+1}^{n+1}}{\beta_{tk,i+1}^{n+1}U_{i+1}^{n+1}} - \frac{Q_{tk,i+1}^{n}}{\beta_{tk,i+1}^{n}U_{i+1}^{n}}\right) + \frac{1-\psi}{\Delta t}\left(\frac{Q_{tk,i}^{n+1}}{\beta_{tk,i}^{n+1}U_{i}^{n+1}} - \frac{Q_{tk,i}^{n}}{\beta_{tk,i}^{n}U_{i}^{n}}\right) \\
&+ \frac{\theta}{\Delta x}(Q_{tk,i+1}^{n+1} - Q_{tk,i}^{n+1}) + \frac{1-\theta}{\Delta x}(Q_{tk,i+1}^{n} - Q_{tk,i}^{n}) \\
&+ \theta\left[\psi\frac{Q_{tk,i+1}^{n+1} - Q_{t*k,i+1}^{n+1}}{L_{t,i+1}^{n+1}} + (1-\psi)\frac{Q_{tk,i}^{n+1} - Q_{t*k,i}^{n+1}}{L_{t,i}^{n+1}}\right] \\
&+ (1-\theta)\left[\psi\frac{Q_{tk,i+1}^{n} - Q_{t*k,i+1}^{n}}{L_{t,i+1}^{n}} + (1-\psi)\frac{Q_{tk,i}^{n} - Q_{t*k,i}^{n}}{L_{t,i}^{n}}\right] \\
&= \theta[\psi q_{tlk,i+1}^{n+1} + (1-\psi)q_{tlk,i}^{n+1}] + (1-\theta)[\psi q_{tlk,i+1}^{n} + (1-\psi)q_{tlk,i}^{n}]
\end{aligned}
$$

(5.137)

which can be written as

$$
c_1 Q_{tk,i+1}^{n+1} = c_2 Q_{tk,i}^{n+1} + c_3 Q_{tk,i+1}^{n} + c_4 Q_{tk,i}^{n} + c_{0k}
\tag{5.138}
$$

where

$$
c_1 = \frac{\psi}{\beta_{tk,i+1}^{n+1}U_{i+1}^{n+1}\Delta t} + \frac{\theta}{\Delta x} + \frac{\theta\psi}{L_{t,i+1}^{n+1}}
$$

$$
c_2 = -\frac{1-\psi}{\beta_{tk,i}^{n+1}U_i^{n+1}\Delta t} + \frac{\theta}{\Delta x} - \frac{\theta(1-\psi)}{L_{t,i}^{n+1}}
$$

$$
c_3 = \frac{\psi}{\beta_{tk,i+1}^{n}U_{i+1}^{n}\Delta t} - \frac{1-\theta}{\Delta x} - \frac{(1-\theta)\psi}{L_{t,i+1}^{n}}
$$

$$
c_4 = \frac{1-\psi}{\beta_{tk,i}^{n}U_i^{n}\Delta t} + \frac{1-\theta}{\Delta x} - \frac{(1-\theta)(1-\psi)}{L_{t,i}^{n}}
$$

$$
\begin{aligned}
c_{0k} = &\ \theta\psi\frac{Q_{t*k,i+1}^{n+1}}{L_{t,i+1}^{n+1}} + \theta(1-\psi)\frac{Q_{t*k,i}^{n+1}}{L_{t,i}^{n+1}} + (1-\theta)\psi\frac{Q_{t*k,i+1}^{n}}{L_{t,i+1}^{n}} + (1-\theta)(1-\psi)\frac{Q_{t*k,i}^{n}}{L_{t,i}^{n}} \\
&+ \theta\psi q_{tlk,i+1}^{n+1} + \theta(1-\psi)q_{tlk,i}^{n+1} + (1-\theta)\psi q_{tlk,i+1}^{n} + (1-\theta)(1-\psi)q_{tlk,i}^{n}
\end{aligned}
$$

In order to satisfy sediment continuity, the sediment exchange terms in Eqs. (5.34) and (5.36) should be discretized using the same scheme. Thus, the bed change equation (5.36) is discretized as

$$(1 - p'_m)\frac{\Delta A_{bk,i+1}}{\Delta t} = \theta\frac{Q^{n+1}_{tk,i+1} - Q^{n+1}_{t*k,i+1}}{L^{n+1}_{t,i+1}} + (1 - \theta)\frac{Q^n_{tk,i+1} - Q^n_{t*k,i+1}}{L^n_{t,i+1}}$$

(5.139)

The total change in bed area is calculated using Eq. (5.133), the bed elevation is updated using Eq. (5.134), and the bed-material gradations are determined using Eqs. (5.135) and (5.136).

5.3.3.2 Solution of discretized unsteady sediment transport equations

Decoupled sediment calculation

A decoupled procedure for solving the discretized sediment transport, bed change, and bed material sorting equations can be established, if the bed-material gradation p_{bk} in Eq. (5.35) is treated explicitly:

$$Q^{n+1}_{t*k,i+1} = p^n_{bk,i+1}Q^{*n+1}_{tk,i+1}$$

(5.140)

The sediment quantities at cross-section $i + 1$ are then obtained in the following sequence:

(1) Compute $Q^{n+1}_{t*k,i+1}$ using Eq. (5.140) with the known $p^n_{bk,i+1}$;
(2) Calculate $Q^{n+1}_{tk,i+1}$ using Eq. (5.138);
(3) Compute $\Delta A_{bk,i+1}$ using Eq. (5.139);
(4) Calculate $\Delta A_{b,i+1}$ using Eq. (5.133);
(5) Compute $p^{n+1}_{bk,i+1}$ using Eq. (5.135), and
(6) Update the cross-section topography using Eq. (5.134), and calculate the bed-material gradations in the subsurface layers.

Once the sediment discharges at the inlet have been determined using boundary conditions, the forewater calculation of sediment transport can be performed cross-section by cross-section, following the procedure laid out above. This decoupled procedure is very simple but may be subject to non-physical phenomena, such as numerical oscillation and negative bed-material gradation.

The decoupled sediment calculation is usually decoupled from the flow calculation. Therefore, the entire flow and sediment calculations are fully decoupled.

Coupled sediment calculation

A coupled procedure for solving the discretized sediment equations described above can be established, if the bed-material gradation p_{bk} in Eq. (5.35) is treated implicitly:

$$Q^{n+1}_{t*k,i+1} = p^{n+1}_{bk,i+1}Q^{*n+1}_{tk,i+1}$$

(5.141)

This coupled solution procedure can eliminate those non-physical phenomena that exist in the decoupled procedure. However, the set of discretized sediment equations should be solved simultaneously. An iteration method is normally needed, and then the computational effort will be significantly increased. To avoid this, the direct solution method proposed by Wu (1991) can be used (Wu *et al.*, 2004a), as described below.

For convenience, Eq. (5.138) is written as

$$Q_{tk,i+1}^{n+1} = e_k Q_{t*k,i+1}^{n+1} + e_{0k} \tag{5.142}$$

where $e_k = \theta\psi/(c_1 L_{t,i+1}^{n+1})$, $e_{0k} = (c_2 Q_{tk,i}^{n+1} + c_3 Q_{tk,i+1}^{n} + c_4 Q_{tk,i}^{n} + c'_{0k})/c_1$, and c'_{0k} is c_{0k} without $\theta\psi Q_{t*k,i+1}^{n+1}/L_{t,i+1}^{n+1}$.

Eq. (5.139) is written as

$$\Delta A_{bk,i+1} = f_1 Q_{tk,i+1}^{n+1} - f_2 Q_{t*k,i+1}^{n+1} + f_{0k} \tag{5.143}$$

where $f_1 = f_2 = \theta\Delta t/[(1 - p'_m)L_{t,i+1}^{n+1}]$, and $f_{0k} = (1 - \theta)\Delta t(Q_{tk,i+1}^{n} - Q_{t*k,i+1}^{n})/[(1 - p'_m)L_{t,i+1}^{n}]$.

Inserting Eqs. (5.141) and (5.142) into Eq. (5.143) yields

$$\Delta A_{bk,i+1} = (f_1 e_k - f_2)p_{bk,i+1}^{n+1} Q_{tk,i+1}^{*n+1} + (f_1 e_{0k} + f_{0k}) \tag{5.144}$$

and then substituting Eq. (5.135) into Eq. (5.144) leads to

$$\Delta A_{bk,i+1} = \Delta A_{b,i+1}\frac{(f_2 - f_1 e_k)Q_{tk,i+1}^{*n+1} p_{bk,i+1}^{*n}}{A_{m,i+1}^{n+1} + (f_2 - f_1 e_k)Q_{tk,i+1}^{*n+1}} + \frac{(f_1 e_{0k} + f_{0k})A_{m,i+1}^{n+1}}{A_{m,i+1}^{n+1} + (f_2 - f_1 e_k)Q_{tk,i+1}^{*n+1}}$$
$$- \frac{(f_2 - f_1 e_k)Q_{tk,i+1}^{*n+1}[p_{bk,i+1}^{n}A_{m,i+1}^{n} + p_{bk,i+1}^{*n}(A_{m,i+1}^{n+1} - A_{m,i+1}^{n})]}{A_{m,i+1}^{n+1} + (f_2 - f_1 e_k)Q_{tk,i+1}^{*n+1}} \tag{5.145}$$

Summing Eq. (5.145) over all size classes and using Eq. (5.133) yields the following equation for the total change in bed area:

$$\Delta A_{b,i+1} = \left\{ -\sum_{k=1}^{N} \frac{(f_2 - f_1 e_k)Q_{tk,i+1}^{*n+1}[p_{bk,i+1}^{n}A_{m,i+1}^{n} + p_{bk,i+1}^{*n}(A_{m,i+1}^{n+1} - A_{m,i+1}^{n})]}{A_{m,i+1}^{n+1} + (f_2 - f_1 e_k)Q_{tk,i+1}^{*n+1}} \right.$$
$$\left. + \sum_{k=1}^{N} \frac{(f_1 e_{0k} + f_{0k})A_{m,i+1}^{n+1}}{A_{m,i+1}^{n+1} + (f_2 - f_1 e_k)Q_{tk,i+1}^{*n+1}} \right\} \Bigg/$$
$$\left[1 - \sum_{k=1}^{N} \frac{(f_2 - f_1 e_k)Q_{tk,i+1}^{*n+1}p_{bk,i+1}^{*n}}{A_{m,i+1}^{n+1} + (f_2 - f_1 e_k)Q_{tk,i+1}^{*n+1}} \right] \tag{5.146}$$

The discretized sediment equations are then directly solved in the following sequence:

(1) Compute $\Delta A_{b,i+1}$ using Eq. (5.146);
(2) Calculate $\Delta A_{bk,i+1}$ using Eq. (5.145);
(3) Compute $p_{bk,i+1}^{n+1}$ using Eq. (5.135);
(4) Calculate $Q_{t*k,i+1}^{n+1}$ using Eq. (5.141);
(5) Compute $Q_{tk,i+1}^{n+1}$ using Eq. (5.142), and
(6) Update the cross-section topography using Eq. (5.134), and calculate the bed-material gradations in the subsurface layers.

However, the coupled sediment calculation is still decoupled from the flow calculation so that the entire flow and sediment calculation procedure is in a semi-coupled form.

The above direct solution method can also be used in the bed-load and suspended-load transport model in Section 5.3.2, by writing Eqs. (5.130)–(5.132) as Eqs. (5.142) and (5.143) with $C_{k,i+1}$, $C_{*k,i+1}$, $Q_{bk,i+1}$, and $Q_{b*k,i+1}$ as unknowns and deriving an equation similar to Eq. (5.146) to compute the total bed change $\Delta A_{b,i+1}$ directly.

5.3.3.3 Stability of Preissmann scheme for sediment transport equation

Neglecting the influence of the source term, the error in the sediment transport rate determined using Eq. (5.138) is governed by

$$c_1 \delta_{i+1}^{n+1} = c_2 \delta_i^{n+1} + c_3 \delta_{i+1}^n + c_4 \delta_i^n \qquad (5.147)$$

where δ_i^n is the Fourier component of the error at point i and time level n, defined as $\delta_i^n = V^n e^{i\sigma x_i}$, with V^n and σ being its amplitude and wave number, respectively. Inserting this definition expression into Eq. (5.147) yields the growth factor:

$$r = \frac{V^{n+1}}{V^n} = \frac{c_3 e^{i\sigma \Delta x} + c_4}{c_1 e^{i\sigma \Delta x} - c_2} \qquad (5.148)$$

The coefficients of Eq. (5.138) satisfy that $c_1 \geq 0$ and $c_2 + c_3 + c_4 \leq c_1$ at the locally uniform state. Supposing c_2, c_3, and $c_4 \geq 0$ yields

$$|r| = \left| \frac{c_3 e^{i\sigma \Delta x} + c_4}{c_1 e^{i\sigma \Delta x} - c_2} \right| = \left| \frac{c_3 + c_4 e^{-i\sigma \Delta x}}{c_1 - c_2 e^{-i\sigma \Delta x}} \right| \leq \frac{c_3 + c_4}{c_1 - c_2} \leq 1 \qquad (5.149)$$

which means that the von Neumann stability condition is satisfied. At the locally uniform state, in which U and L_t are constant in each element, the constraints c_2, c_3, and $c_4 \geq 0$ imply

$$\max\left\{1 - \frac{\psi}{C_r + \psi D_r}, \frac{1 - \psi}{C_r - (1 - \psi)D_r}\right\} \leq \theta \leq 1 \quad \text{with}$$

$$\max\left\{0, 1 - \frac{C_r}{D_r}\right\} \leq \psi \leq 1 \tag{5.150}$$

where C_r is the Courant number $U\Delta t/\Delta x$; and D_r is a scale factor of non-equilibrium sediment transport, defined as $D_r = U\Delta t/L_t$. Note that β_{tk} is set to 1 here.

Condition (5.150) is sufficient but not necessary for the numerical stability of Eq. (5.138). If $L_t \ll \Delta x$, $D_r \gg C_r$, then ψ and θ should be given values close to 1.

5.3.3.4 Advantages of the coupled sediment calculation procedure

Stabilities of explicit and implicit schemes for bed-material gradation

The decoupled and coupled sediment calculation procedures are compared by analyzing the stabilities of the explicit and implicit schemes for the bed-material gradation in Eq. (5.35). For convenience, Eqs. (5.140) and (5.141) are replaced by

$$Q_{t*k,i+1}^{n+1} = [\theta_p p_{bk,i+1}^{n+1} + (1 - \theta_p)p_{bk,i+1}^n]Q_{tk,i+1}^{*n+1} \tag{5.151}$$

where θ_p is the temporal weighting factor for bed-material gradation: $= 1$ for the implicit scheme (coupled calculation procedure), and 0 for the explicit scheme (decoupled calculation procedure).

Inserting Eqs. (5.142), (5.143), and (5.151) into Eq. (5.135) yields the equation for the bed-material gradation in the mixing layer:

$$\begin{aligned}
p_{bk,i+1}^{n+1} = &\frac{(A_{m,i+1}^{n+1} - A_{m,i+1}^n - \Delta A_{b,i+1})p_{bk,i+1}^{*n}}{A_{m,i+1}^{n+1} + (f_2 - f_1 e_k)\theta_p Q_{tk,i+1}^{*n+1}} \\
&+ \frac{[A_{m,i+1}^n - (f_2 - f_1 e_k)(1 - \theta_p)Q_{tk,i+1}^{*n+1}]p_{bk,i+1}^n}{A_{m,i+1}^{n+1} + (f_2 - f_1 e_k)\theta_p Q_{tk,i+1}^{*n+1}} \\
&+ \frac{f_1 e_{0k} + f_{0k}}{A_{m,i+1}^{n+1} + (f_2 - f_1 e_k)\theta_p Q_{tk,i+1}^{*n+1}}
\end{aligned} \tag{5.152}$$

To simplify the analysis, it is assumed that $A_m^{n+1} \approx A_m^n$. For deposition, usually $\Delta A_{b,i+1} + A_{m,i+1}^n \geq A_{m,i+1}^{n+1}$, then $p_{bk,i+1}^{*n} = p_{bk,i+1}^n$, and the bed-material gradation error, δ, is governed by

$$\delta^{n+1} = \delta^n \frac{A_{m,i+1}^{n+1} - \Delta A_{b,i+1} - (f_2 - f_1 e_k)(1 - \theta_p)Q_{tk,i+1}^{*n+1}}{A_{m,i+1}^{n+1} + (f_2 - f_1 e_k)\theta_p Q_{tk,i+1}^{*n+1}} \tag{5.153}$$

Numerical stability requires $r = |\delta^{n+1}/\delta^n \le 1|$, which implies that for the implicit scheme,

$$\frac{\theta(1 - e_k)Q_{tk,i+1}^{*n+1}\Delta t}{(1 - p_m')L_{t,i+1}^{n+1}} \ge \Delta A_{b,i+1} - 2A_{m,i+1}^{n+1} \tag{5.154}$$

and for the explicit scheme,

$$\frac{\theta(1 - e_k)Q_{tk,i+1}^{*n+1}\Delta t}{(1 - p_m')L_{t,i+1}^{n+1}} \le 2A_{m,i+1}^{n+1} - \Delta A_{b,i+1} \tag{5.155}$$

For erosion, usually $\Delta A_{b,i+1} + A_{m,i+1}^n < A_{m,i+1}^{n+1}$, then $p_{bk,i+1}^{*n}$ is the bed-material gradation in the second layer, the influence of which is assumed to be negligible on the numerical stability of the bed-material gradation in the mixing layer. Thus, the bed-material gradation error is governed by

$$\delta^{n+1} = \delta^n \frac{A_{m,i+1}^n - (f_2 - f_1 e_k)(1 - \theta_p)Q_{tk,i+1}^{*n+1}}{A_{m,i+1}^{n+1} + (f_2 - f_1 e_k)\theta_p Q_{tk,i+1}^{*n+1}} \tag{5.156}$$

from which it is known that the implicit scheme is unconditionally stable, and the stability condition for the explicit scheme is

$$\frac{\theta(1 - e_k)Q_{tk,i+1}^{*n+1}\Delta t}{(1 - p_m')L_{t,i+1}^{n+1}} \le A_{m,i+1}^n + A_{m,i+1}^{n+1} \tag{5.157}$$

By definition, the mixing layer should be thicker than the change in bed elevation, i.e., $A_m^{n+1} \ge |\Delta A_b|$. Because $e_k < 1$, the stability condition (5.154) for the implicit scheme is automatically satisfied; conditions (5.155) and (5.157) require upper limits for the time step Δt in the explicit scheme. It is evident that the implicit scheme is much more stable than the explicit scheme.

Requirement of non-negative bed-material gradation

In calculating bed-material gradation, negative values may occur under certain conditions. Of course, this is a non-physical phenomenon and must be eliminated. The condition $p_{bk,i+1}^{n+1} \ge 0$ for Eq. (5.152) implies that

$$f_1 e_{0k} + f_{0k} + [A_{m,i+1}^n - (f_2 - f_1 e_k)(1 - \theta_p)Q_{tk,i+1}^{*n+1}]p_{bk,i+1}^n$$
$$+ (A_{m,i+1}^{n+1} - A_{m,i+1}^n - \Delta A_{b,i+1})p_{bk,i+1}^{*n} \ge 0 \tag{5.158}$$

and then

$$\Delta A_{b,i+1} \leq A_{m,i+1}^{n+1} + \frac{f_1 e_{0k} + f_{0k} + A_{m,i+1}^n (p_{bk,i+1}^n - p_{bk,i+1}^{*n})}{p_{bk,i+1}^{*n}}$$

$$- \frac{\theta(1 - e_k)(1 - \theta_p)Q_{tk,i+1}^{*n+1}\Delta t}{(1 - p_m')L_{t,i+1}^{n+1}} \frac{p_{bk,i+1}^n}{p_{bk,i+1}^{*n}} \qquad (5.159)$$

Because the last term on the right-hand side of inequality (5.159) is negative but vanishes when $\theta_p = 1$, the implicit scheme allows for larger time steps than the explicit scheme. After considering the stability conditions of the Preissmann scheme for sediment transport equation, condition (5.159) for the implicit scheme can be easily satisfied. One of the safest treatments is to impose $\theta = 1$, $|\Delta A_b| \leq A_m^{n+1}$ and $A_m^{n+1} \approx A_m^n$, which is a sufficient but not necessary condition.

Sensitivity of bed-material gradation to mixing layer thickness

Assuming $A_{m,i+1}^{n+1} = A_{m,i+1}^n = A_{m,i+1}$ in Eq. (5.152) and differentiating $p_{bk,i+1}^{n+1}$ with respect to $A_{m,i+1}$ yields

$$\left(\frac{\partial p_{bk,i+1}^{n+1}}{\partial A_{m,i+1}}\right)_{\theta_p=1} \bigg/ \left(\frac{\partial p_{bk,i+1}^{n+1}}{\partial A_{m,i+1}}\right)_{\theta_p=0} = \left[\frac{A_{m,i+1}}{A_{m,i+1} + f_1(1 - e_k)Q_{tk,i+1}^{*n+1}}\right]^2 \leq 1$$

$$(5.160)$$

The gradient $\partial p_{bk,i+1}^{n+1}/\partial A_{m,i+1}$ represents the change in bed-material gradation per unit change in mixing layer thickness. Eq. (5.160) shows that the implicit scheme has smaller $\partial p_{bk,i+1}^{n+1}/\partial A_{m,i+1}$ and is thus less sensitive to $A_{m,i+1}$ than the explicit scheme.

5.3.4   Treatments for sediment transport in channel networks

If a channel network is concerned, the sediment transport at channel confluences and splits, as well as hydraulic structures, needs to be treated specially. At hydraulic structures such as culverts, drop structures, weirs, and measuring flumes, erosion is not allowed; thus, the beds are fixed and the sediment discharges are constant through them. For bridge crossings, 1-D models are able to simulate the bed change due to channel contraction, but not the local scour due to 3-D flow features. However, the maximum local scour depth and volume can be estimated using empirical functions.

In analogy to the flow calculation described in Section 5.2.1.3, the sediment transport at a channel confluence or split can generally be computed by applying Eqs. (5.130) and (5.131) or Eq. (5.137). For this computation, the downstream cross-section at the confluence or the upstream cross-section at the split needs to be divided into two parts. This approach was successfully used by Wu (1991) in a quasi-steady model. However, a simpler approach, which is described below, may be used if the three cross-sections at the confluence or split are located very close together.

For the confluence shown in Fig. 5.9, the suspended-load concentration $C_{k,3}$ and bed-load transport rate $Q_{bk,3}$ at cross-section 3 can be calculated using the following mass balance equations:

$$C_{k,3} = (Q_1 C_{k,1} + Q_2 C_{k,2})/Q_3 \qquad (5.161)$$

$$Q_{bk,3} = Q_{bk,1} + Q_{bk,2} \qquad (5.162)$$

where $C_{k,1}$ and $C_{k,2}$ are the suspended-load concentrations and $Q_{bk,1}$ and $Q_{bk,2}$ are the bed-load transport rates at cross-sections 1 and 2, respectively, which are known from the previous calculations in channels 1 and 2.

For the split shown in Fig. 5.10, the following mass balance equations exist:

$$Q_2 C_{k,2} + Q_3 C_{k,3} = Q_1 C_{k,1} \qquad (5.163)$$

$$Q_{bk,2} + Q_{bk,3} = Q_{bk,1} \qquad (5.164)$$

which, however, cannot uniquely determine $C_{k,2}$, $C_{k,3}$, $Q_{bk,2}$, and $Q_{bk,3}$ without additional relations. For suspended load, the ratio of $C_{k,2}$ and $C_{k,3}$ can be determined using Ding and Qiu's (1981) method. Fig. 5.15 depicts the vertical distribution of suspended-load concentrations entering cross-sections 2 and 3. The bed elevations of these two cross-sections are denoted as z_{b2} and z_{b3}, and their water stages are assumed to be z_s. Suppose that cross-section 2 is in the main branch channel, i.e., $z_{b2} < z_{b3}$. It is assumed that the sediment concentration at cross-section 3 corresponds to the upper layer above z_{b3} in the distribution curve of cross-section 2. Therefore, the ratio of $C_{k,2}$ and $C_{k,3}$ can be approximated as

$$\frac{C_{k,2}}{C_{k,3}} = \frac{\int_{z_{b2}+\delta}^{z_s} f_k(z)dz}{\int_{z_{b3}}^{z_s} f_k(z)dz} \qquad (5.165)$$

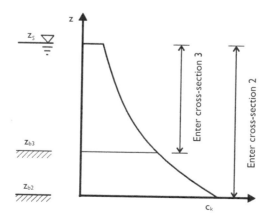

Figure 5.15 Sediment concentrations at channel split.

where $f_k(z)$ is the Rouse distribution of suspended-load concentration, and δ is the thickness of the bed-load zone at cross-section 2.

Then, $C_{k,2}$ and $C_{k,3}$ can be determined using Eqs. (5.163) and (5.165).

The determination of bed-load transport rates $Q_{bk,2}$ and $Q_{bk,3}$ has not been well investigated, due to the complexity. Their ratio may be assumed to be approximately equal to that of bed-load transport capacities $Q_{b*k,2}$ and $Q_{b*k,3}$ at cross-sections 2 and 3:

$$\frac{Q_{bk,2}}{Q_{bk,3}} \approx \frac{Q_{b*k,2}}{Q_{b*k,3}} \tag{5.166}$$

and $Q_{bk,2}$ and $Q_{bk,3}$ can then be determined using Eqs. (5.164) and (5.166).

Note that Eqs. (5.161)–(5.166) can be applied in both quasi-steady and unsteady sediment transport models.

5.3.5   Lateral allocation of bed change in 1-D model

A 1-D model provides only the lumped change in bed area, ΔA_b, at a cross-section. In order to acquire a reasonable prediction for long-term river morphological evolution, ΔA_b must be allocated appropriately to the local change in bed elevation, Δz_b, along the cross-section at each time step. The obtained Δz_b is used to update the cross-section geometry, as expressed in Eq. (5.134).

The simplest method is the uniform distribution of bed change along the cross-section, except for water edges, where the bed change may be zero. Wu (1991) suggested a slight modification of this method, assuming uniform deposition and erosion for wide channels and horizontal deposition and uniform erosion for narrow channels, as shown in Fig. 5.16. This method is more adequate for suspended load (fine sediments) than for bed load (coarse sediments), because the suspended-load concentration tends to be relatively uniform along the cross-section while the bed load usually moves in strips.

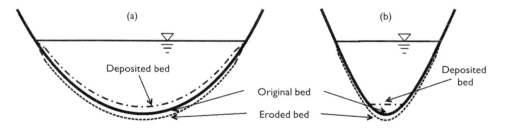

Figure 5.16 Allocation of bed change in cross-section: (a) wide channel and (b) narrow channel.

A more general method, used by Chang (1988), allocates deposition and erosion along the cross-section by a power function of excess shear stress:

$$\Delta z_b = \frac{(\tau_b - \tau_c)^m}{\sum_B (\tau_b - \tau_c)^m \Delta y} \Delta A_b \tag{5.167}$$

where τ_b is the local bed shear stress, determined by $\tau_b = \gamma h S$, with h being the local flow depth; τ_c is the critical shear stress, which is given zero in the case of deposition; m is an exponent; y is the cross-stream coordinate; and B is the channel width at the water surface.

The value of m is generally between 0 and 1; it essentially affects the pattern of bed change distribution. A small value means a fairly uniform distribution of Δz_b along the cross-section, while a larger value gives a less uniform distribution of Δz_b. In Chang's model, the value of m is determined at each time step, such that a correction in the channel bed profile will result in the most rapid movement toward uniformity in power expenditure, or linear water surface profile, along the channel.

Eq. (5.167) is only applied in straight channels. For curved channels, the following curvature-weighting relation is used to adjust the cross-section:

$$\Delta z_b = \frac{(\tau_b - \tau_c)^m / r}{\sum_B (\tau_b - \tau_c)^m \Delta y / r} \Delta A_b \qquad (5.168)$$

where r is the coordinate along the radius of channel bend.

Similar relations can be obtained by replacing the excess shear stress $\tau_b - \tau_c$ in Eqs. (5.167) and (5.168) with the excess velocity $U - U_c$.

A simplification can be made by setting $\tau_c = 0$ and S to be constant along the cross-section. Thus, Eq. (5.167) becomes

$$\Delta z_b = \frac{h^m}{\sum_B h^m \Delta y} \Delta A_b \qquad (5.169)$$

A more complicated method for lateral allocation of bed change is the stream tube model proposed by Yang et al. (1998). The entire cross-section is divided into several stream tubes, and a 1-D model is adopted to simulate the flow, sediment transport, and bed change in each stream tube. This technique is more like a quasi-two-dimensional approach. The shape of the cross-section is adjusted according to the assumption of minimum stream power.

In addition, the change in bed elevation due to the consolidation of cohesive bed material needs to be considered. This is discussed in Section 11.1.6.

5.3.6 I-D simulation of bank erosion and channel meandering

5.3.6.1 I-D bank erosion model

Stream bank erosion occurs due to channel degradation, toe erosion, mass failure, seepage flow, weathering, etc. Channel bed degradation increases bank heights, and lateral erosion undercuts bank toes. Both processes make banks steeper and more unstable. Seepage flow and weathering may aggravate these processes. Once the stability criterion is exceeded, a bank mass failure event occurs and the bank top retreats. The failed bank material is first piled on the bed near the bank toe and then washed away by flow. Thus, bank erosion can significantly affect sediment balance and channel morphology in rivers.

Fluvial erosion at bank toes

The fluvial (particle-by-particle) erosion at bank toes directly influences channel bed width and bank angle, and causes bank instability with respect to mass failure under gravity. Arulanandan *et al.* (1980) proposed an empirical formula to compute the fluvial erosion of cohesive bank material:

$$\frac{dw}{dt} = \frac{r}{\gamma_s}\left(\frac{\tau - \tau_{ce}}{\tau_{ce}}\right) \tag{5.170}$$

where dw/dt is the lateral erosion rate near the bank toe $(\text{m} \cdot \text{min}^{-1})$; τ is the flow shear stress $(\text{dynes} \cdot \text{cm}^{-2})$ applied on the bank toe, determined by $\tau = \gamma RS$; τ_{ce} is the critical shear stress $(\text{dynes} \cdot \text{cm}^{-2})$ for bank toe erosion, related to water and soil properties; γ_s is the unit weight of the soil $(\text{kN} \cdot \text{m}^{-3})$; and r is the initial rate of soil erosion $(\text{g} \cdot \text{cm}^{-2}\text{min}^{-1})$, given by $r = 0.0223\tau_{ce} \exp(-0.13\tau_{ce})$.

The eroded bank material is treated as side inflow in sediment transport equations (5.27), (5.28), and (5.34).

Bank mass failure

Depending on bank geometry, water table, surface runoff, seepage, vegetation, and soil properties, channel banks may fail by various mechanisms, which may be planar (e.g., Osman and Thorne, 1988; Simon *et al.*, 2000), rotational (Osman, 1985), cantilever (Thorne and Tovey, 1981), or piping- or sapping-type (Hagerty, 1991). Planar and rotational failures usually occur on the homogeneous, non-layered banks, whereas cantilever failures usually happen on the layered banks. Piping- or sapping-type failures most likely occur on the heterogeneous banks, where seepage flow is often observed. A stability analysis of planar failures is introduced below, while those for other failure types can be found in relevant references.

Osman and Thorne (1988) analyzed the planar failure shown in Fig. 5.17. It is assumed that the failure plane intersects the bank toe. The factor of safety is defined as

$$f_s = \frac{F_r}{F_d} \tag{5.171}$$

where F_d and F_r are the driving and resisting forces, respectively:

$$F_d = W_t \sin\beta = \frac{\gamma_s}{2}\left(\frac{H^2 - y_d^2}{\tan\beta} - \frac{H'^2}{\tan\alpha}\right)\sin\beta \tag{5.172}$$

$$F_r = \frac{(H - y_d)C}{\sin\beta} + \frac{\gamma_s}{2}\left(\frac{H^2 - y_d^2}{\tan\beta} - \frac{H'^2}{\tan\alpha}\right)\cos\beta\tan\phi \tag{5.173}$$

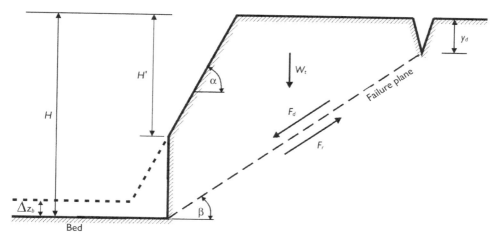

Figure 5.17 Sketch of planar bank failure.

where W_t is the weight of failure block, C is the soil cohesion (kPa), ϕ is the soil friction angle (degrees), α is the angle of bank slope, y_d is the depth of tension crack (m), and β is the angle of failure plane. The failure angle is determined by

$$\beta = \frac{1}{2}\left\{\tan^{-1}\left[\frac{H}{H'}(1 - K_{tc}^2)\tan\alpha\right] + \phi\right\} \tag{5.174}$$

with K_{tc} being the ratio of the observed tension crack depth to the bank height.

Once a mass failure is predicted ($f_s < 1$), the retreat distance of the bank top, Λ, and the volume of the failure block, V_f, are determined by

$$\Lambda = \frac{H - y_d}{\tan\beta} - \frac{H'}{\tan\alpha} \tag{5.175}$$

$$V_f = \frac{1}{2}\left(\frac{H^2 - y_d^2}{\tan\beta} - \frac{H'^2}{\tan\alpha}\right) \tag{5.176}$$

Simon *et al.* (2000) proposed a more sophisticated bank stability and toe erosion model, which considers wedge-shaped bank failures with distinct bank material layers and user-defined bank geometry. Their model is able to incorporate the root reinforcement and surcharge effects of six vegetation species, including willows, grasses, and large trees, and simulate saturated and unsaturated soil strengths, taking into consideration the effect of pore-water pressure. The details can be found in Simon *et al.* (2000).

5.3.6.2 1-D channel meandering model

Simulating channel meandering processes using a 1-D numerical model is quite difficult, because of the strongly three-dimensional features of flow and sediment transport in meandering channels. However, ignoring the channel meandering processes may induce significant errors in the simulation of long-term flow and sediment transport in alluvial channels. Therefore, many investigators have studied this problem and proposed empirical or semi-empirical methods to account for the influence of channel meandering on sediment transport.

From a computational point of view, one needs to know the flow path and its potential change during the simulation using a 1-D model. One approach is to adopt empirical relations to calculate the flow path of a meandering channel under given flow and sediment conditions. Some channel regime theories may be used.

The other approach is to use meandering migration models, which may be kinematic or dynamic. The kinematic migration models relate the migration rate to channel width, curvature, etc. Examples are Ferguson (1983) and Howard and Knutson (1984). The dynamic migration models solve the simplified dynamic equation of flow to estimate the flow properties in the meandering channel, and relate the bank erosion rate to the excess velocity or shear stress at the outer bank. Examples can be found in Johannesson and Parker (1989) and Odgaard (1989).

Because of the truly three-dimensional flows and highly complex soil properties in meandering channels, most of the current 1-D channel meandering models are only applicable in simple cases. Further study on this problem is needed.

5.3.7 Overall procedure for 1-D decoupled flow and sediment calculations

The individual modeling components in the fully decoupled and semi-coupled models, which decouple flow and sediment calculations, have been introduced in previous sections. The overall calculation procedure for the 1-D decoupled unsteady model consists of the following steps:

(1) Calculate the unsteady flow using the Preissmann scheme and the double sweep method based on initial channel geometry;
(2) Calculate sediment transport, bed change, and bed material sorting from upstream to downstream, using the known flow conditions;
(3) Determine the bed change due to bed material consolidation, if needed;
(4) Correct channel geometry by allocating the bed change along the cross-section;
(5) Calculate bank erosion and mass failure, if needed, and
(6) Return to step (1) and conduct the calculations for the next time step, based on the new channel geometry, until all time steps are finished.

The calculation procedure for the 1-D decoupled quasi-steady model is almost the same as the above procedure, except that the standard step method is used for the quasi-steady flow calculation in step (1) and the time interval (in hours or days) used in the quasi-steady model is usually longer than the time step (in minutes) used in the unsteady model.

5.4 1-D COUPLED CALCULATION OF FLOW AND SEDIMENT TRANSPORT

Fluvial processes in a river consist of simultaneous motions of water and sediment. Any change in flow conditions may be associated with a variation in sediment transport and channel topography, and vice versa. Thus, the decoupled flow and sediment transport models introduced in Sections 5.1–5.3 have limitations and are only applicable in the case of weak or mild sediment transport. In the case of strong sediment transport, a coupled model should be used, to take into account the interactions between flow and sediment transport. How to couple the flow and sediment calculations is introduced below.

5.4.1 1-D coupled flow and sediment transport equations

In general, 1-D unsteady sediment-laden flows are described by Eqs. (2.126) and (2.127), which are rewritten below to consider side flows:

$$\frac{\partial(\rho A)}{\partial t} + \frac{\partial(\rho Q)}{\partial x} + \rho_b \frac{\partial A_b}{\partial t} = \rho_0 q_l \tag{5.177}$$

$$\frac{\partial(\rho Q)}{\partial t} + \frac{\partial}{\partial x}\left(\frac{\rho \beta Q^2}{A}\right) + \rho g A \frac{\partial z_s}{\partial x} + \frac{1}{2} g A h_p \frac{\partial \rho}{\partial x} + \rho g A S_f = \rho_0 q_l v_x \tag{5.178}$$

where ρ is the density of the water and sediment mixture in the water column, determined by $\rho = \rho_f(1 - C_t) + \rho_s C_t$, with C_t being the volumetric concentration of sediment; and ρ_0 is the density of the water and sediment mixture from tributaries and banks.

The effect of alluvial bed roughness is accounted for through the dependence of the Manning roughness coefficient on flow and sediment conditions:

$$n = f(U, B, h, \tau_b, d_{50}, \dots) \tag{5.179}$$

which can be one of the formulas introduced in Section 3.3.3.

Sediment transport, bed change, and bed material sorting equations are the same as those introduced in Section 5.1.2.1. For simplicity, the bed-material load transport model is presented here. The governing equations include the total-load transport equation (5.34), bed change equations (5.31) and (5.36), mixing-layer bed material sorting equation (5.32), and sediment transport capacity (5.35). Note that Eq. (5.33) should also be included, but it is not listed here because the bed material sorting in subsurface layers can be computed separately. In addition, the sediment settling velocity is related to sediment concentration, but it can be set as an intermediate variable.

The system described above has $4N + 4$ equations that are used to determine $4N + 4$ unknowns: A, Q, n, $\partial A_b/\partial t$, $(\partial A_b/\partial t)_k$, Q_{tk}, Q_{t*k}, and p_{bk} ($k = 1, 2, \dots, N$).

5.4.2 Discretization of coupled flow and sediment transport equations

Like Eqs. (5.1) and (5.2), Eqs. (5.177) and (5.178) can be solved numerically, using the Preissmann scheme for common flows, as described below, or using shock-capturing schemes for dam-break and overtopping flows, as discussed in Section 9.2.

Applying the Preissmann scheme to Eqs. (5.177) and (5.178) yields

$$
\begin{aligned}
&\frac{\psi}{\Delta t}(\rho_{i+1}^{n+1}A_{i+1}^{n+1} - \rho_{i+1}^{n}A_{i+1}^{n}) + \frac{1-\psi}{\Delta t}(\rho_{i}^{n+1}A_{i}^{n+1} - \rho_{i}^{n}A_{i}^{n}) \\
&+ \frac{\theta}{\Delta x}(\rho_{i+1}^{n+1}Q_{i+1}^{n+1} - \rho_{i}^{n+1}Q_{i}^{n+1}) + \frac{1-\theta}{\Delta x}(\rho_{i+1}^{n}Q_{i+1}^{n} - \rho_{i}^{n}Q_{i}^{n}) \\
&+ \frac{\psi}{\Delta t}\rho_{b,i+1}\Delta A_{b,i+1} + \frac{1-\psi}{\Delta t}\rho_{b,i}\Delta A_{b,i} \\
&- \theta[\psi\rho_{0,i+1}^{n+1}q_{l,i+1}^{n+1} + (1-\psi)\rho_{0,i}^{n+1}q_{l,i}^{n+1}] \\
&- (1-\theta)[\psi\rho_{0,i+1}^{n}q_{l,i+1}^{n} + (1-\psi)\rho_{0,i}^{n}q_{l,i}^{n}] = 0
\end{aligned} \tag{5.180}
$$

$$
\begin{aligned}
&\frac{\psi}{\Delta t}(\rho_{i+1}^{n+1}Q_{i+1}^{n+1} - \rho_{i+1}^{n}Q_{i+1}^{n}) + \frac{1-\psi}{\Delta t}(\rho_{i}^{n+1}Q_{i}^{n+1} - \rho_{i}^{n}Q_{i}^{n}) \\
&+ \frac{\theta}{\Delta x}\left[\frac{\rho_{i+1}^{n+1}\beta_{i+1}^{n+1}(Q_{i+1}^{n+1})^2}{A_{i+1}^{n+1}} - \frac{\rho_{i}^{n+1}\beta_{i}^{n+1}(Q_{i}^{n+1})^2}{A_{i}^{n+1}}\right] \\
&+ \frac{1-\theta}{\Delta x}\left[\frac{\rho_{i+1}^{n}\beta_{i+1}^{n}(Q_{i+1}^{n})^2}{A_{i+1}^{n}} - \frac{\rho_{i}^{n}\beta_{i}^{n}(Q_{i}^{n})^2}{A_{i}^{n}}\right] \\
&+ g(\rho A)_{i+\psi}^{n+\theta}\left[\frac{\theta}{\Delta x}(z_{s,i+1}^{n+1} - z_{s,i}^{n+1}) + \frac{1-\theta}{\Delta x}(z_{s,i+1}^{n} - z_{s,i}^{n})\right] \\
&+ \frac{g}{2}(Ah_p)_{i+\psi}^{n+\theta}\left[\frac{\theta}{\Delta x}(\rho_{i+1}^{n+1} - \rho_{i}^{n+1}) + \frac{1-\theta}{\Delta x}(\rho_{i+1}^{n} - \rho_{i}^{n})\right] \\
&+ g(\rho A)_{i+\psi}^{n+\theta}\left\{\theta[\psi_R S_{f,i+1}^{n+1} + (1-\psi_R)S_{f,i}^{n+1}]\right. \\
&\left. + (1-\theta)[\psi_R S_{f,i+1}^{n} + (1-\psi_R)S_{f,i}^{n}]\right\} = (\rho_0 q_l v_x)_{i+\psi}^{n+\theta}
\end{aligned} \tag{5.181}
$$

where $(\rho A)_{i+\psi}^{n+\theta} = \theta[\psi\rho_{i+1}^{n+1}A_{i+1}^{n+1} + (1-\psi)\rho_{i}^{n+1}A_{i}^{n+1}] + (1-\theta)[\psi\rho_{i+1}^{n}A_{i+1}^{n} + (1-\psi)\rho_{i}^{n}A_{i}^{n}]$, $(Ah_p)_{i+\psi}^{n+\theta} = \theta[\psi A_{i+1}^{n+1}h_{p,i+1}^{n+1} + (1-\psi)A_{i}^{n+1}h_{p,i}^{n+1}] + (1-\theta)[\psi A_{i+1}^{n}h_{p,i+1}^{n} + (1-\psi)A_{i}^{n}h_{p,i}^{n}]$, and $(\rho_0 q_l v_x)_{i+\psi}^{n+\theta} = \theta[\psi\rho_{0,i+1}^{n+1}q_{l,i+1}^{n+1}v_{x,i+1}^{n+1} + (1-\psi)\rho_{0,i}^{n+1}q_{l,i}^{n+1}v_{x,i}^{n+1}] + (1-\theta)[\psi\rho_{0,i+1}^{n}q_{l,i+1}^{n}v_{x,i+1}^{n} + (1-\psi)\rho_{0,i}^{n}q_{l,i}^{n}v_{x,i}^{n}]$.

The Manning roughness coefficient relation (5.179) is treated as

$$
n_{i+1}^{n+1} = f(U_{i+1}^{n+1}, B_{i+1}^{n+1}, h_{i+1}^{n+1}, \tau_{b,i+1}^{n+1}, d_{50,i+1}^{n+1}, \ldots) \tag{5.182}
$$

The discretizations of sediment transport, bed change, and bed material sorting equations remain unchanged. The resulting equations include the dicretized sediment transport equation (5.137), the discretized fractional bed change equation (5.139), the discretized total bed change equation (5.133), the discretized bed material sorting equation (5.135), and the implicitly treated sediment transport capacity equation (5.141).

Flow equations (5.180)–(5.182) and sediment equations (5.133), (5.135), (5.137), (5.139), and (5.141) can be solved in either iteratively or fully coupled form, as described in the next subsection.

5.4.3 Solution of discretized coupled flow and sediment transport equations

Iteratively coupled solution procedure

Iteratively coupled models can be found in Holly *et al.* (1990). In such models, all the flow and sediment calculations are divided into two loops. The first loop is the "flow loop," which solves flow equations (5.180) and (5.181) using the latest estimates of sediment discharge, bed elevation, and bed-material gradation. Eqs. (5.180) and (5.181) are locally linearized by using Δh and ΔQ as unknowns. The Manning coefficient n^{n+1}, mixture density ρ_{i+1}^{n+1}, and bed change $\Delta A_{b,i+1}$ in these two equations are set as intermediate variables and replaced with the latest estimates n^*, ρ_{i+1}^*, and $\Delta A_{b,i+1}^*$, which are determined using Eq. (5.182), related to sediment concentration, and determined by the sediment model, respectively. The locally linearized equations can be written as Eqs. (5.60) and (5.68), and solved using the algorithms described in Section 5.2.2.

The second loop is the "sediment loop," which solves Eqs. (5.133), (5.135), (5.137), (5.139), and (5.141) to estimate sediment discharge, bed change, and bed-material gradation. These equations can be solved using the direct solution method described in Section 5.3.3.

To obtain a simultaneous solution of flow and sediment transport, these two loops are coupled through the following iteration procedure:

(1) Load the imposed boundary conditions, such as mainstream and tributary water and sediment inflows, and downstream water stage;
(2) Calculate water stage, flow discharge, and other flow parameters, using the latest estimates of Manning n, flow density, and bed elevation;
(3) Compute sediment discharge, bed change, and bed-material gradation, using the calculated flow conditions;
(4) Estimate new Manning n, flow density, and bed elevation, using the computed flow and sediment quantities, and
(5) Repeat steps (2)–(4) iteratively until the successive estimates of bed elevation, Manning n, etc., no longer change.

This iteratively coupled solution procedure can take into account interactions between flow and sediment transport, indirectly giving a simultaneous solution. As a simplified case, the flow and sediment models described in Sections 5.2.2 and 5.3.3

can be treated as "flow loop" and "sediment loop," respectively, and then coupled using this iteration procedure.

Fully coupled solution procedure

Fully coupled models can be found in Lyn and Goodwin (1987), Holly and Rahuel (1990), Correia et al. (1992), and Yeh et al. (1995). In particular, Holly and Rahuel (1990) proposed a fully coupled procedure for calculating non-uniform sediment transport, which is herein extended to solve Eqs. (5.133), (5.135), (5.137), (5.139) (5.141), and (5.180)–(5.182). To reduce the effort of matrix inversion required in the fully coupled solution procedure, these equations are re-organized as follows.

Inserting Eqs. (5.135) and (5.141) into Eq. (5.139) yields

$$
\begin{aligned}
\Delta A_{bk,i+1} = {} & \Delta A_{b,i+1} \frac{\theta \Delta t p_{bk,i+1}^{*n} Q_{tk,i+1}^{*n+1}}{\theta \Delta t Q_{tk,i+1}^{*n+1} + (1 - p_m') L_{t,i+1}^{n+1} A_{m,i+1}^{n+1}} \\
& + Q_{tk,i+1}^{n+1} \frac{\theta \Delta t A_{m,i+1}^{n+1}}{\theta \Delta t Q_{tk,i+1}^{*n+1} + (1 - p_m') L_{t,i+1}^{n+1} A_{m,i+1}^{n+1}} \\
& - \frac{\theta \Delta t Q_{tk,i+1}^{*n+1} [A_{m,i+1}^n p_{bk,i+1}^n + p_{bk,i+1}^{*n} (A_{m,i+1}^{n+1} - A_{m,i+1}^n)]}{\theta \Delta t Q_{tk,i+1}^{*n+1} + (1 - p_m') L_{t,i+1}^{n+1} A_{m,i+1}^{n+1}} \\
& + \frac{(1 - \theta) \Delta t L_{t,i+1}^{n+1} A_{m,i+1}^{n+1} (Q_{tk,i+1}^n - Q_{t*k,i+1}^n)/L_{t,i+1}^n}{\theta \Delta t Q_{tk,i+1}^{*n+1} + (1 - p_m') L_{t,i+1}^{n+1} A_{m,i+1}^{n+1}}
\end{aligned}
\tag{5.183}
$$

and then summing Eq. (5.183) over all size classes and using Eq. (5.133) yields

$$
\Delta A_{b,i+1} = \sum_{k=1}^{N} r_{k,i+1} Q_{tk,i+1}^{n+1} + s_{i+1}
\tag{5.184}
$$

where

$$
r_{k,i+1} = \frac{1}{\lambda} \frac{\theta \Delta t A_{m,i+1}^{n+1}}{\theta \Delta t Q_{tk,i+1}^{*n+1} + (1 - p_m') L_{t,i+1}^{n+1} A_{m,i+1}^{n+1}}, \text{ and}
$$

$$
\begin{aligned}
s_{i+1} = {} & -\frac{1}{\lambda} \sum_{k=1}^{N} \frac{\theta \Delta t Q_{tk,i+1}^{*n+1} [A_{m,i+1}^n p_{bk,i+1}^n + p_{bk,i+1}^{*n} (A_{m,i+1}^{n+1} - A_{m,i+1}^n)]}{\theta \Delta t Q_{tk,i+1}^{*n+1} + (1 - p_m') L_{t,i+1}^{n+1} A_{m,i+1}^{n+1}} \\
& + \frac{1}{\lambda} \sum_{k=1}^{N} \frac{(1 - \theta) \Delta t L_{t,i+1}^{n+1} A_{m,i+1}^{n+1}}{\theta \Delta t Q_{tk,i+1}^{*n+1} + (1 - p_m') L_{t,i+1}^{n+1} A_{m,i+1}^{n+1}} \frac{Q_{tk,i+1}^n - Q_{t*k,i+1}^n}{L_{t,i+1}^n}
\end{aligned}
$$

$$
\text{with } \lambda = 1 - \sum_{k=1}^{N} \frac{\theta \Delta t p_{bk,i+1}^{*n} Q_{tk,i+1}^{*n+1}}{\theta \Delta t Q_{tk,i+1}^{*n+1} + (1 - p_m') L_{t,i+1}^{n+1} A_{m,i+1}^{n+1}}.
$$

Inserting Eqs. (5.139) and (5.141) into Eq. (5.135) yields

$$p_{bk,i+1}^{n+1} = e_{k,i+1} Q_{tk,i+1}^{n+1} + f_{k,i+1} \Delta A_{b,i+1} + g_{k,i+1} \qquad (5.185)$$

where

$$e_{k,i+1} = \frac{\theta \Delta t}{\theta \Delta t Q_{tk,i+1}^{*n+1} + (1 - p_m') L_{t,i+1}^{n+1} A_{m,i+1}^{n+1}},$$

$$f_{k,i+1} = -\frac{(1 - p_m') L_{t,i+1}^{n+1} p_{bk,i+1}^{*n}}{\theta \Delta t Q_{tk,i+1}^{*n+1} + (1 - p_m') L_{t,i+1}^{n+1} A_{m,i+1}^{n+1}}, \text{ and}$$

$$g_{k,i+1} = \frac{(1 - p_m') L_{t,i+1}^{n+1} [A_{m,i+1}^n p_{bk,i+1}^n + p_{bk,i+1}^{*n}(A_{m,i+1}^{n+1} - A_{m,i+1}^n)]}{\theta \Delta t Q_{tk,i+1}^{*n+1} + (1 - p_m') L_{t,i+1}^{n+1} A_{m,i+1}^{n+1}}$$

$$+ \frac{(1 - \theta) \Delta t L_{t,i+1}^{n+1}(Q_{tk,i+1}^n - Q_{t*k,i+1}^n)/L_{t,i+1}^n}{\theta \Delta t Q_{tk,i+1}^{*n+1} + (1 - p_m') L_{t,i+1}^{n+1} A_{m,i+1}^{n+1}}.$$

Substituting Eq. (5.184) into the discretized continuity equation (5.180) yields

$$F_1 = \frac{\psi}{\Delta t}(\rho_{i+1}^{n+1} A_{i+1}^{n+1} - \rho_{i+1}^n A_{i+1}^n) + \frac{1 - \psi}{\Delta t}(\rho_i^{n+1} A_i^{n+1} - \rho_i^n A_i^n)$$

$$+ \frac{\theta}{\Delta x}(\rho_{i+1}^{n+1} Q_{i+1}^{n+1} - \rho_i^{n+1} Q_i^{n+1}) + \frac{1 - \theta}{\Delta x}(\rho_{i+1}^n Q_{i+1}^n - \rho_i^n Q_i^n)$$

$$+ \frac{\psi}{\Delta t}\rho_{b,i+1}\left(\sum_{k=1}^{N} r_{k,i+1} Q_{tk,i+1}^{n+1} + s_{i+1}\right) + \frac{1 - \psi}{\Delta t}\rho_{b,i}\left(\sum_{k=1}^{N} r_{k,i} Q_{tk,i}^{n+1} + s_i\right)$$

$$- \theta[\psi \rho_{0,i+1}^{n+1} q_{l,i+1}^{n+1} + (1 - \psi)\rho_{0,i}^{n+1} q_{l,i}^{n+1}]$$

$$- (1 - \theta)[\psi \rho_{0,i+1}^n q_{l,i+1}^n + (1 - \psi)\rho_{0,i}^n q_{l,i}^n] = 0 \qquad (5.186)$$

Eq. (5.181) is rewritten as

$$F_2 = \frac{\psi}{\Delta t}(\rho_{i+1}^{n+1} Q_{i+1}^{n+1} - \rho_{i+1}^n Q_{i+1}^n) + \frac{1 - \psi}{\Delta t}(\rho_i^{n+1} Q_i^{n+1} - \rho_i^n Q_i^n)$$

$$+ \frac{\theta}{\Delta x}\left[\frac{\rho_{i+1}^{n+1} \beta_{i+1}^{n+1}(Q_{i+1}^{n+1})^2}{A_{i+1}^{n+1}} - \frac{\rho_i^{n+1} \beta_i^{n+1}(Q_i^{n+1})^2}{A_i^{n+1}}\right]$$

$$+ \frac{1 - \theta}{\Delta x}\left[\frac{\rho_{i+1}^n \beta_{i+1}^n(Q_{i+1}^n)^2}{A_{i+1}^n} - \frac{\rho_i^n \beta_i^n(Q_i^n)^2}{A_i^n}\right]$$

$$+ g(\rho A)_{i+\psi}^{n+\theta}\left[\frac{\theta}{\Delta x}(z_{s,i+1}^{n+1} - z_{s,i}^{n+1}) + \frac{1 - \theta}{\Delta x}(z_{s,i+1}^n - z_{s,i}^n)\right]$$

$$+ \frac{g}{2}(Ab_p)_{i+\psi}^{n+\theta}\left[\frac{\theta}{\Delta x}(\rho_{i+1}^{n+1} - \rho_i^{n+1}) + \frac{1 - \theta}{\Delta x}(\rho_{i+1}^n - \rho_i^n)\right]$$

$$+ g(\rho A)_{i+\psi}^{n+\theta}\{\theta[\psi_R S_{f,i+1}^{n+1} + (1 - \psi_R)S_{f,i}^{n+1}]$$

$$+ (1 - \theta)[\psi_R S_{f,i+1}^n + (1 - \psi_R)S_{f,i}^n]\} - (\rho_0 q_l v_x)_{i+\psi}^{n+\theta} = 0 \qquad (5.187)$$

Substituting Eqs. (5.141), (5.184), and (5.185) into the discretized sediment transport equation (5.137) yields

$$
F_{3k} = \frac{\psi}{\Delta t}\left(\frac{Q_{tk,i+1}^{n+1}}{\beta_{tk,i+1}^{n+1} U_{i+1}^{n+1}} - \frac{Q_{tk,i+1}^{n}}{\beta_{tk,i+1}^{n} U_{i+1}^{n}}\right) + \frac{1-\psi}{\Delta t}\left(\frac{Q_{tk,i}^{n+1}}{\beta_{tk,i}^{n+1} U_{i}^{n+1}} - \frac{Q_{tk,i}^{n}}{\beta_{tk,i}^{n} U_{i}^{n}}\right)
$$

$$
+ \frac{\theta}{\Delta x}(Q_{tk,i+1}^{n+1} - Q_{tk,i}^{n+1}) + \frac{1-\theta}{\Delta x}(Q_{tk,i+1}^{n} - Q_{tk,i}^{n}) + \frac{\theta\psi}{L_{t,i+1}^{n+1}}\left\{Q_{tk,i+1}^{n+1}\right.
$$

$$
- \left[e_{k,i+1}Q_{tk,i+1}^{n+1} + f_{k,i+1}\left(\sum_{k=1}^{N} r_{k,i+1}Q_{tk,i+1}^{n+1} + s_{i+1}\right) + g_{k,i+1}\right]Q_{tk,i+1}^{*n+1}\right\}
$$

$$
+ \frac{\theta(1-\psi)}{L_{t,i}^{n+1}}\left\{Q_{tk,i}^{n+1} - \left[e_{k,i}Q_{tk,i}^{n+1} + f_{k,i}\left(\sum_{k=1}^{N} r_{k,i}Q_{tk,i}^{n+1} + s_i\right) + g_{k,i}\right]Q_{tk,i}^{*n+1}\right\}
$$

$$
+ (1-\theta)\left[\psi\frac{Q_{tk,i+1}^{n} - Q_{t*k,i+1}^{n}}{L_{t,i+1}^{n}} + (1-\psi)\frac{Q_{tk,i}^{n} - Q_{t*k,i}^{n}}{L_{t,i}^{n}}\right]
$$

$$
- \theta[\psi q_{tlk,i+1}^{n+1} + (1-\psi)q_{tlk,i}^{n+1}] - (1-\theta)[\psi q_{tlk,i+1}^{n} + (1-\psi)q_{tlk,i}^{n}] = 0
$$

$$
(k = 1, 2, \dots, N) \qquad (5.188)
$$

For the channel with $I - 1$ reaches shown in Fig. 5.7, the system of equations (5.186)–(5.188) has $(2 + N)(I - 1)$ equations, which are used to determine $(2 + N)I$ unknowns: A, Q, and Q_{tk} ($k = 1, 2, \dots, N$). The system is closed by imposing $2 + N$ boundary conditions. For simplicity, the Manning roughness coefficient and flow density are treated as intermediate variables. An alternative treatment for flow density may be to remove it from the left-hand sides of Eqs. (5.177) and (5.178), as described in Section 9.2.

Eqs. (5.186)–(5.188) can be solved by many methods. The following Newton-Raphson solution procedure is given as an example, which is almost the same as that used by Holly and Rahuel (1990).

The Newton-Raphson correction equations for each reach are written in the following matrix form:

$$
[L_i]\{\delta W_i\} + [R_i]\{\delta W_{i+1}\} + \{S_i\} = 0 \qquad (5.189)
$$

where $\{\delta W_i\}$ is the vector of unknown corrections to the $2 + N$ primary variables: δA_i, δQ_i, $\delta Q_{t1,i}$, $\delta Q_{t2,i}$, $\dots$, and $\delta Q_{tN,i}$; $\{S_i\}$ is the known vector of functions F_1, F_2, and F_{3k} defined in Eqs. (5.186)–(5.188); and $[L_i]$ and $[R_i]$ are the matrices of Jacobian derivatives with $(2 + N) \times (2 + N)$ elements, e.g.,

$$
[L_i] = \begin{bmatrix}
\partial F_1/\partial A_i & \partial F_1/\partial Q_i & \partial F_1/\partial Q_{t1,i} & \cdots & \partial F_1/\partial Q_{tN,i} \\
\partial F_2/\partial A_i & \partial F_2/\partial Q_i & \partial F_2/\partial Q_{t1,i} & \cdots & \partial F_2/\partial Q_{tN,i} \\
\partial F_{31}/\partial A_i & \partial F_{31}/\partial Q_i & \partial F_{31}/\partial Q_{t1,i} & \cdots & \partial F_{31}/\partial Q_{tN,i} \\
\vdots & \vdots & \vdots & \cdots & \vdots \\
\partial F_{3N}/\partial A_i & \partial F_{3N}/\partial Q_i & \partial F_{3N}/\partial Q_{t1,i} & \cdots & \partial F_{3N}/\partial Q_{tN,i}
\end{bmatrix} \qquad (5.190)
$$

$[L_i]$, $[R_i]$, and $\{S_i\}$ are evaluated using the latest estimates of primary variables. To determine the matrices $[L_i]$ and $[R_i]$, the derivatives of each function (F_1, F_2, F_{3k}) with respect to each primary dependent variable (A, Q, Q_{tk}) at two ends of each reach are required. The derivatives with respect to auxiliary variables are transformed into the primary-variable derivatives through chain-rule expansion. The dependence of the Manning n and flow density on primary variables should be considered.

The equation system (5.189) for all reaches and the associated boundary conditions are solved using a block-bidiagonal algorithm. Suppose that there is a relation of the form:

$$[E_i]\{\delta W_1\} + [H_i]\{\delta W_i\} + \{G_i\} = 0 \qquad (5.191)$$

Then deriving $\{\delta W_i\}$ from Eq. (5.189) and substituting it into Eq. (5.191) yields

$$[E_{i+1}]\{\delta W_1\} + [H_{i+1}]\{\delta W_{i+1}\} + \{G_{i+1}\} = 0 \qquad (5.192)$$

where the coefficient matrices are

$$[E_{i+1}] = [E_i] \qquad (5.193)$$
$$[H_{i+1}] = -[H_i][L_i]^{-1}[R_i] \qquad (5.194)$$
$$\{G_{i+1}\} = -[H_i][L_i]^{-1}\{S_i\} + \{G_i\} \qquad (5.195)$$

Comparing Eqs. (5.189) and (5.192) at $i = 1$ results in $[E_2] = [L_1]$, $[H_2] = [R_1]$, and $\{G_2\} = \{S_1\}$. The forward sweep can then be carried out using Eqs. (5.193)–(5.195) from $i = 2, 4, \ldots, I$. At the end of the forward sweep, the following relation is obtained:

$$[E_I]\{\delta W_1\} + [H_I]\{\delta W_I\} + \{G_I\} = 0 \qquad (5.196)$$

The boundary conditions at upstream and downstream points can be linearized locally and written in the vector form:

$$[\alpha]\{\delta W_1\} + [\beta]\{\delta W_I\} + \{\gamma\} = 0 \qquad (5.197)$$

Eliminating $\{\delta W_1\}$ from Eqs. (5.196) and (5.197) yields

$$\{\delta W_I\} = [-[\alpha][E_I]^{-1}[H_I] + [\beta]]^{-1}\{[\alpha][E_I]^{-1}\{G_I\} - \{\gamma\}\} \qquad (5.198)$$

Once $\{\delta W_I\}$ is determined, the remaining unknown vectors can be obtained by the "return-sweep" inversion of Eq. (5.189):

$$\{\delta W_i\} = -[L_i]^{-1}[R_i]\{\delta W_{i+1}\} - [L_i]^{-1}\{S_i\} \qquad (5.199)$$

It can be seen from Eqs. (5.193)–(5.195), (5.198), and (5.199) that this algorithm requires $(I - 1)$ inversions of a $(2 + N) \times (2 + N)$ matrix for each iteration step.

After A, Q, and Q_{tk} are solved, the sediment transport capacity, Manning n, bed change, and bed-material gradation can be calculated using Eqs. (5.141), (5.182), (5.184), and (5.185), respectively.

As compared to the iteratively coupled procedure, the fully coupled procedure is much more complicated. In particular, if channel network routing, bank erosion, and bed material consolidation need to be considered, the fully coupled procedure becomes cumbersome and tedious.

5.4.4　Justification of decoupled and coupled models

Decoupled flow and sediment transport models have been widely used in the solution of many real-life engineering problems. They are relatively easy to implement, and their results may be justified due to different time scales in flow and sediment transport and the use of empirical formulas for bed roughness and sediment transport capacity. Most of the criticisms against the decoupled models are related to the equilibrium sediment transport model. The application of the non-equilibrium transport model and Wu's (1991) coupling procedure for non-uniform sediment transport simulation has significantly enhanced the numerical stability of the decoupled flow and sediment transport models. However, it is true that the applicability of the decoupled models is restricted due to the assumption of low sediment concentration and small bed change at each time step.

The coupled models take into account the physical coupling of water and sediment phases, so that they should be more reasonable and could be applied in a wider range of flow and sediment conditions. The coupled models are usually more stable and can use larger time steps than the decoupled models (Saiedi, 1997; Cao *et al.*, 2002). However, the implementation of the coupled models, especially the fully coupled models for non-uniform sediment transport in looped channel networks, is very complicated. Their efficiency may be offset by the required effort of iteration and matrix inversion. Furthermore, because the time step for flow calculation is usually smaller than that for sediment calculation, solving the nonlinear flow system might become a bottleneck and restrain the efficiency of the coupled models.

It is diffficult to give a quantitative criterion as to when the decoupled models are acceptable. Generally, in the lower flow regime with low sediment concentration, the decoupled models are applicable; otherwise, the coupled models should be used. Because the sediment concentration is usually low in most natural rivers, the decoupled models can still play an important role in river engineering analysis.

5.5　DATA REQUIREMENTS OF 1-D MODEL

The following data are commonly required by 1-D models. They are also required by 2-D and 3-D models, with higher spatial resolutions.

Study domain

The study domain usually covers the channel reach of interest and additional transition reaches in upstream and downstream. Its inlets and outlets should be located near gauge stations or control structures where measured flow and sediment data are available for determination of boundary conditions.

Computational grid

The study channel is represented by a suitable number of cross-sections. Each pair of consecutive cross-sections defines a reach between them. For a channel network, the cross-sections at channel confluences, splits, and hydraulic structures should be arranged according to the requirements of the used model.

Channel topography

Each cross-section is represented by finite points (stations), as shown in Figs. 5.3 and 5.4. The bed elevations and the distances to the left bank at all points should be measured. The reach lengths between cross-sections are also needed.

If hydraulic structures are involved, their geometries and hydraulic conditions should be provided.

Manning roughness coefficient

The Manning n is usually estimated using measured flow data. Empirical formulas may be used if no measurement data are available. The n values in streams with similar flow and sediment conditions may be used as reference.

Sediment particle properties

The specific gravity and shape factor of sediment particles should be measured. For most natural sands (quartz sands), the specific gravity is about 2.65, and the Corey shape factor is about 0.7.

The sediment size range should cover all sizes of bed load, suspended load, bed material, and bank material existing in the study domain. Wash load is sometimes also included. The entire size range is divided into a suitable number of size classes. The representative diameters and upper and lower bounds for all size classes should be determined.

Bed-material size and gradation

The initial bed-material gradation must be given for a realistic computation of stream behavior, particularly for determining scour and stability conditions. If only deposition is expected, such as sedimentation in reservoirs, the initial bed-material gradation is less important.

The bed-material porosity is also needed.

Bank-material properties

If bank erosion and mass failure are considered, bank-material properties, such as density and size composition, should be determined. For a cohesive bank, the cohesion and friction angle of the bank material, as well as the critical shear stress for bank toe erosion, are also needed.

Boundary conditions

Boundary conditions include inflow water discharge, water stage (at inlet or outlet depending on the flow regime), and inflow sediment discharge and size composition. The chosen time series of flow and sediment data should represent the average hydrological cycle in the study domain. Usually, the time series should be long enough and include high, intermediate, and low water years, with various recurrence frequencies.

Historical data

Historical measurement data of flow properties, sediment discharges, channel morphological changes, etc., should be collected and analyzed for better understanding of the study problem and calibration of the numerical model.

5.6 MODEL SENSITIVITY TO INPUT PARAMETERS

Out of all the model parameters, the adaptation length (coefficient) and mixing layer thickness are least understood and must be prescribed empirically in the sediment transport models described in Sections 5.3.2 and 5.3.3. Therefore, the concern here is to analyze the influence of these parameters on the model results. This analysis was performed in three typical cases by Wu and Vieira (2002), using the semi-coupled model described in Section 5.3.3.

Case 1. Channel degradation

The experiments performed by Ashida and Michiue (1971) for bed degradation and armoring processes due to clear water flow downstream of a dam were simulated. The experimental flume was 20 m long and 0.8 m wide. The flume bed was filled with non-uniform sediment with a median size of 1.5 mm and a standard deviation of 3.47. Clear water was pumped into the entrance of the flume at a constant discharge. In simulated experimental run 6, the flow discharge was 0.0314 m^3s^{-1}, and the initial bed slope was 0.01. The computational grid consisted of 40 elements with an equal spacing of 0.5 m, and the time step was 10 s. The experiments started from a flat bed. In order to account for the development of bed forms in the simulation, the bed form height was assumed to vary linearly with time. The Manning roughness coefficient for the fully developed bed was about 0.023. The bed-material porosity was calculated using the Komura (1963) formula. The sediment transport capacity was calculated using the Wu *et al.* (2000b) formula.

The sensitivity of the model results to the adaptation length was investigated using various functions $L_t = 7.3h$, $L_t = t$, and $L_t = 1 + 0.5t$ while keeping the mixing layer thickness constant as d_{50}, the median size of the parent sediment mixture. Here, h is the flow depth in meters, and t is the time in hours. Fig. 5.18 compares the measured and calculated bed scour depths at 7, 10, and 13 m upstream of the weir. The trends of intensive scour in the initial period and weak scour in the final equilibrium stage were reproduced well. The function $L_t = 7.3h$ provides the best results for the bed scouring process, especially regarding the time to reach the equilibrium state. The results for

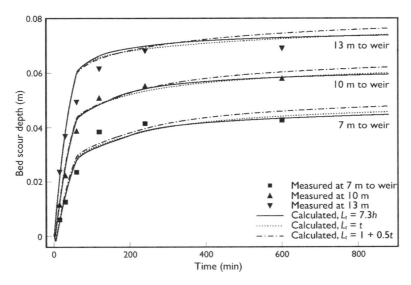

Figure 5.18 Bed scour depths using different adaptation lengths (Ashida and Michiue's Run 6).

$L_t = t$ and $L_t = 1 + 0.5t$ are also very close to the measured data, showing that the calculated scour depth is not very sensitive to L_t.

The influence of the mixing layer thickness on the calculated scour depth was examined by changing its value from d_{50} to $2d_{50}$, while keeping the adaptation length at $7.3h$. Fig. 5.19 shows that the thicker the mixing layer, the larger the equilibrium scour

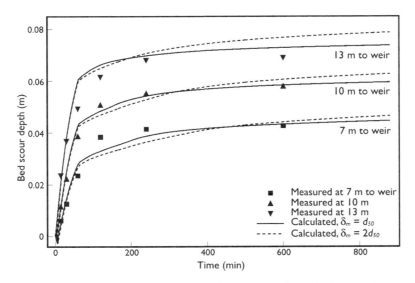

Figure 5.19 Bed scour depths using different mixing layer thicknesses δ_m (Ashida and Michiue's Run 6).

depth. The time to reach equilibrium increases as the mixing layer thickness increases. The choice of the mixing layer thickness is particularly important in the case of bed scour and armoring.

Case 2. Channel aggradation

The channel aggradation experiments performed by Seal *et al.* (1995) were calculated. The experimental setup is shown in Fig. 5.20. The flume was 45 m long and 0.305 m wide, with an initial bed slope of 0.002. The tailgate was kept at a constant height, so that an undular hydraulic jump was produced at the downstream end of the main gravel deposit. The input sediment was a weakly bimodal mixture comprising a wide range of sizes, from 0.125 to 64 mm. Due to sediment overloading, an aggradational wedge developed and its front gradually moved downstream while the upstream bed elevation continued to rise. In simulated experimental run 2, the water discharge was $0.049 \text{ m}^3\text{s}^{-1}$, the sediment feed rate was $5.65 \text{ kg} \cdot \text{min}^{-1}$, and the tailgate water stage was 0.45 m.

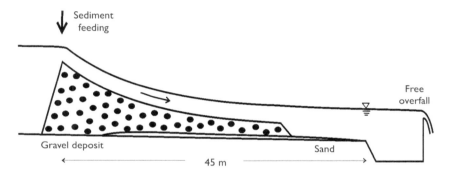

Figure 5.20 Sketch of channel aggradation experiments of Seal *et al.* (1995).

The model sensitivity to the adaptation length L_t was analyzed by specifying L_t as 0.5 m, 2 m, and $7.3h$ while setting the mixing layer thickness as half the dune height. Here, h is the average flow depth over the wedge from the inlet to the gravel deposit front, and $7.3h$ is approximately equal to 1 m. Fig. 5.21 compares the measured and predicted bed profiles at various times, and the water surface profiles at the final stage. The bed profiles were reproduced well, and the hydraulic jump downstream of the gravel deposit front was predicted qualitatively. It is shown that L_t has little influence on the height and celerity of the gravel deposit front, as well as the top slope of the wedge. The only significant impact is on the slope of the deposit front. The longer the adaptation length, the gentler the front slope.

Fig. 5.22 shows the calculated bed profiles when the mixing layer thickness was given as d_{50}, $6d_{50}$, and half the dune height and the adaptation length was kept at 0.5 m. The differences between the calculated bed profiles are very small. As the mixing layer thickness increases six times, the deposit front moves downstream by only 1.3%. The model is much less sensitive to the mixing layer thickness in the deposition case than in the previous erosion case.

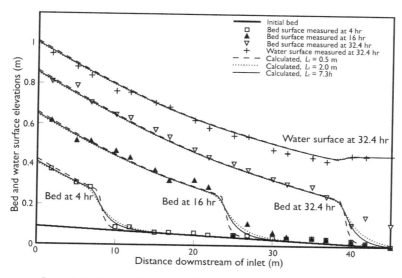

Figure 5.21 Bed profiles using different adaptation lengths (Run 2).

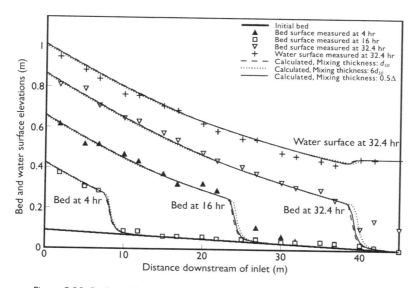

Figure 5.22 Bed profiles using different mixing layer thicknesses (Run 2).

Case 3. Sedimentation in the Danjiangkou Reservoir

The Danjiangkou Reservoir was constructed on the Hanjiang River, China, in 1968. Because the tributary Danjiang River joins the Hanjiang River just upstream of the dam, the Danjiangkou Reservoir has two branches with nearly equivalent storage capacities, as shown in Fig. 5.23. Although there are water and sediment exchanges between the two branches during flood seasons, the interactions are negligible for the

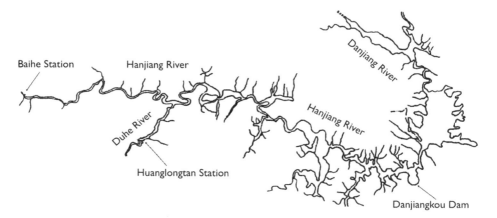

Figure 5.23 Sketch of Danjiangkou Reservoir.

study of the sedimentation process in the reservoir because the confluence is very close to the dam.

The sedimentation process in the Hanjiang River branch was simulated. The computational domain included a 188 km long reach in the main stream from the Baihe Hydrology Station to the dam, and a 12 km long reach in the tributary Duhe River, which joins the Hanjiang River 157 km upstream of the dam. Sixty-one cross-sections were distributed along the main stream and four cross-sections in the tributary. The simulation period encompassed 13 years, from 1968 to 1980. The Manning roughness coefficient and bed-material porosity were estimated using measurement data. The sediment transport capacity was determined using the Wu *et al.* (2000b) formula. The flocculation of fine sediments ($d < 0.01$ mm) was considered using the Migniot (1968) relation, which is described in Section 11.1.2.

Fig. 5.24 shows the calculated and measured annual sediment depositions in different years, while Fig. 5.25 shows the calculated and measured longitudinal distributions of sediment deposition accumulated from 1968 to 1979. In order for the influence of the adaptation coefficient α on the amounts of deposition to be investigated, it was given four constant values: 0.25, 0.5, 1.0, and 2.5, and calculated using the Armanini-di Silvio (1988) method. The values 0.5 and 1.0 and the Armanini-di Silvio method provide good predictions. In particular, the results obtained using the Armanini-di Silvio method and the constant value of 1.0 for α are very close.

The effect of the mixing layer thickness was also verified by specifying it as half the dune height and two constant values of 0.05 and 0.25 m. The calculated deposition amounts and longitudinal distributions are not sensitive to the mixing layer thickness, as found in Case 2. The simulation results are not shown here.

In summary, in the case of channel degradation, the computed equilibrium scour depth and bed-armoring process are not particularly sensitive to the adaptation length (coefficient), but are affected by the mixing layer thickness. In the case of channel

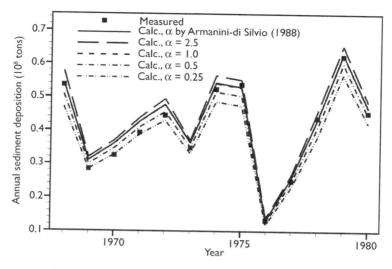

Figure 5.24 Annual sediment depositions using different α.

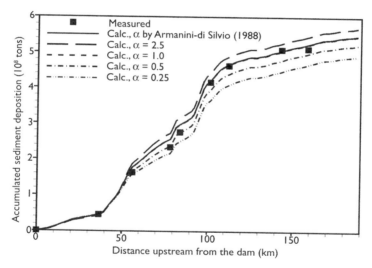

Figure 5.25 Longitudinal distributions of sediment deposition using different α.

aggradation, the simulated bed profiles are insensitive to both adaptation length (coefficient) and mixing layer thickness. However, this conclusion is based on the model of Wu and Vieira (2002), which adopts a coupled calculation procedure for sediment. Generally, the model sensitivity and reliability depend on the numerical schemes, calculation procedures, and empirical formulas used.

Chapter 6

2-D numerical models

If the vertical (or lateral) variations of flow and sediment quantities in a water body are sufficiently small or can be determined analytically, their variations in the horizontal plane (or longitudinal section) can be approximately described by a depth-averaged (or width-averaged) 2-D model. Presented in this chapter are the governing equations, boundary conditions, and numerical solutions of the depth-averaged and width-averaged 2-D models of flow and sediment transport in open channels, as well as the enhancement of the depth-averaged 2-D model to account for the effects of helical flow on fluvial processes in curved and meandering channels.

6.1 DEPTH-AVERAGED 2-D SIMULATION OF FLOW IN NEARLY STRAIGHT CHANNELS

6.1.1 Governing equations

For shallow water flows with low sediment concentration, the depth-averaged 2-D hydrodynamic equations are Eqs. (2.79), (2.82), and (2.83). In the case of nearly straight channels, the dispersion momentum transports due to the vertical non-uniformity of flow velocity are combined with the turbulent stresses, so these equations are rewritten as

$$\frac{\partial h}{\partial t} + \frac{\partial (hU_x)}{\partial x} + \frac{\partial (hU_y)}{\partial y} = 0 \tag{6.1}$$

$$\frac{\partial (hU_x)}{\partial t} + \frac{\partial (hU_x^2)}{\partial x} + \frac{\partial (hU_yU_x)}{\partial y} = -gh\frac{\partial z_s}{\partial x} + \frac{1}{\rho}\frac{\partial (hT_{xx})}{\partial x} + \frac{1}{\rho}\frac{\partial (hT_{xy})}{\partial y}$$
$$+ \frac{1}{\rho}(\tau_{sx} - \tau_{bx}) + f_c hU_y \tag{6.2}$$

$$\frac{\partial (hU_y)}{\partial t} + \frac{\partial (hU_xU_y)}{\partial x} + \frac{\partial (hU_y^2)}{\partial y} = -gh\frac{\partial z_s}{\partial y} + \frac{1}{\rho}\frac{\partial (hT_{yx})}{\partial x} + \frac{1}{\rho}\frac{\partial (hT_{yy})}{\partial y}$$
$$+ \frac{1}{\rho}(\tau_{sy} - \tau_{by}) - f_c hU_x \tag{6.3}$$

where x and y are the horizontal Cartesian coordinates (not necessarily along the longitudinal and transverse directions); and τ_{bx} and τ_{by} are the bed shear stresses, determined by

$$\tau_{bx} = \rho c_f m_b U_x \sqrt{U_x^2 + U_y^2}, \quad \tau_{by} = \rho c_f m_b U_y \sqrt{U_x^2 + U_y^2} \tag{6.4}$$

where $c_f = g n^2 / h^{1/3}$, with n being the Manning roughness coefficient of the channel bed; and m_b is the bed slope coefficient defined in Eq. (2.82). For a movable bed with sediment grains and bed forms, the Manning n can be evaluated using one of the empirical formulas introduced in Section 3.3.3; however, n is in general treated as a calibrated parameter because of its complexity, as discussed in Section 5.1.1.3. In addition, one may set the bed slope coefficient m_b as 1 and lump its effect into the Manning n.

Note that unlike the 1-D model, the depth-averaged 2-D model can simulate the effects of large-scale roughness structures, such as channel contraction, expansion, and curvature, on the flow field, using fine meshes. In addition, the depth-averaged 2-D model accounts for the effect of channel banks through boundary conditions and considers the effect of horizontal turbulent diffusion through the eddy viscosity. Therefore, the values of Manning n in the 1-D and 2-D models are not exactly the same.

In Eqs. (6.2) and (6.3), τ_{sx} and τ_{sy} represent the forces acting on the water surface, usually caused by wind driving:

$$\tau_{sx} = \rho_a c_{fa} U_{wind,x} \sqrt{U_{wind,x}^2 + U_{wind,y}^2}, \quad \tau_{sy} = \rho_a c_{fa} U_{wind,y} \sqrt{U_{wind,x}^2 + U_{wind,y}^2} \tag{6.5}$$

where $U_{wind,x}$ and $U_{wind,y}$ are the x- and y-components of wind velocity, ρ_a is the air density, and c_{fa} is the friction coefficient at the water surface.

The last terms in Eqs. (6.2) and (6.3) represent the Coriolis force due to the rotation of the earth. The Coriolis coefficient f_c is determined by

$$f_c = 2\varpi \sin \varphi \tag{6.6}$$

where ϖ is the rotation velocity of the earth in radians per second, and φ is the latitude degree of the water body of interest.

The Coriolis and wind driving forces are important in large water bodies, such as coastal waters, estuaries, and large lakes, but they are usually negligible in inland rivers.

The stresses T_{ij} ($i, j = x, y$), which include both viscous and turbulent effects, are determined using the Boussinesq assumption:

$$T_{xx} = 2\rho(\nu + \nu_t) \frac{\partial U_x}{\partial x} - \frac{2}{3}\rho k$$

$$T_{xy} = T_{yx} = \rho(\nu + \nu_t) \left(\frac{\partial U_x}{\partial y} + \frac{\partial U_y}{\partial x} \right)$$

$$T_{yy} = 2\rho(v + v_t)\frac{\partial U_y}{\partial y} - \frac{2}{3}\rho k \tag{6.7}$$

where v is the kinematic viscosity of water, and v_t is the eddy viscosity that needs to be determined using a turbulence model. Introduced below are the choices for determining v_t, including the depth-averaged parabolic model, modified mixing length model, and three depth-averaged linear k-ε turbulence models.

Averaging the parabolic eddy viscosity equation (2.49) over the flow depth yields

$$v_t = \alpha_1 U_* h \tag{6.8}$$

where α_1 is an empirical coefficient. Theoretically, α_1 should be equal to $\kappa/6$. However, it has been given various values in practice, because of the anisotropic structures of turbulence in horizontal and vertical directions and the effects of dispersion. According to experiments by Elder (1959), α_1 is about 0.23 for the longitudinal turbulent diffusion in laboratory channels. For transverse turbulent diffusion, Fischer et al. (1979) proposed that α_1 is about 0.15 in laboratory channels and 0.6 (0.3–1.0) in irregular waterways with weak meanders.

Eq. (6.8) is applicable in the region of main flow. Because the influence of horizontal shear is ignored, significant errors may arise when Eq. (6.8) is applied in regions close to rigid sidewalls. Improvement can be achieved through a combination of Eq. (6.8) and Prandtl's mixing length theory:

$$v_t = \sqrt{(\alpha_0 U_* h)^2 + (l_h^2 |\bar{S}|)^2} \tag{6.9}$$

where $|\bar{S}| = [2(\partial U_x/\partial x)^2 + 2(\partial U_y/\partial y)^2 + (\partial U_x/\partial y + \partial U_y/\partial x)^2]^{1/2}$; α_0 is an empirical coefficient similar to α_1 in Eq. (6.8) and has a value of about $\kappa/6$; and l_h is the horizontal mixing length, determined using $l_h = \kappa \min(y', c_m h)$, with y' being the distance to the nearest wall and c_m an empirical coefficient ranging between 0.4 and 1.2 (Wu et al., 2004b).

Rastogi and Rodi (1978) established a depth-averaged k-ε turbulence model through depth-integration of the 3-D standard k-ε model. The eddy viscosity v_t is still determined by Eq. (2.54), whereas the depth-averaged turbulent energy k and its dissipation rate ε are calculated using the following transport equations:

$$\frac{\partial k}{\partial t} + U_x\frac{\partial k}{\partial x} + U_y\frac{\partial k}{\partial y} = \frac{\partial}{\partial x}\left(\frac{v_t}{\sigma_k}\frac{\partial k}{\partial x}\right) + \frac{\partial}{\partial y}\left(\frac{v_t}{\sigma_k}\frac{\partial k}{\partial y}\right) + P_k + P_{kb} - \varepsilon \tag{6.10}$$

$$\frac{\partial \varepsilon}{\partial t} + U_x\frac{\partial \varepsilon}{\partial x} + U_y\frac{\partial \varepsilon}{\partial y} = \frac{\partial}{\partial x}\left(\frac{v_t}{\sigma_\varepsilon}\frac{\partial \varepsilon}{\partial x}\right) + \frac{\partial}{\partial y}\left(\frac{v_t}{\sigma_\varepsilon}\frac{\partial \varepsilon}{\partial y}\right) + c_{\varepsilon1}\frac{\varepsilon}{k}P_k + P_{\varepsilon b} - c_{\varepsilon2}\frac{\varepsilon^2}{k}$$

$$\tag{6.11}$$

where P_k is the production of turbulence due to the horizontal velocity gradients, defined as $P_k = v_t|\bar{S}|^2$; and P_{kb} and $P_{\varepsilon b}$ are the source terms, including all terms originating from non-uniformity of vertical profiles. The main contribution to P_{kb}

and $P_{\varepsilon b}$ stems from significant vertical velocity gradients near the bottom of the water body. These terms are related to the bed shear velocity by $P_{kb} = c_f^{-1/2} U_*^3/h$ and $P_{\varepsilon b} = c_{\varepsilon \Gamma} c_{\varepsilon 2} c_\mu^{1/2} c_f^{-3/4} U_*^4/h^2$. The standard values of coefficients c_μ, $c_{\varepsilon 1}$, $c_{\varepsilon 2}$, σ_k, and σ_ε are listed in Table 2.3, while the coefficient $c_{\varepsilon \Gamma}$ is given 3.6 for experimental cases and 1.8 for field cases (Rodi, 1993).

In analogy to Rastogi and Rodi's depth-averaged standard k-ε turbulence model, Wu et al. (2004b) adopted the concepts in the non-equilibrium k-ε turbulence model of Chen and Kim (1987) and the RNG k-ε turbulence model of Yakhot et al. (1992) in the depth-averaged 2-D simulation of shallow water flows. The k and ε equations are the same as Eqs. (6.10) and (6.11), with coefficients c_μ, $c_{\varepsilon 1}$, $c_{\varepsilon 2}$, σ_k, and σ_ε re-evaluated according to Table 2.3.

A comparison conducted by Wu et al. (2004b) shows that all five depth-averaged turbulence models described above can give reliable predictions for simple flows, but for complex flows, the three k-ε turbulence models generally provide more accurate results than the two zero-equation turbulence models. Among the three k-ε turbulence models, the non-equilibrium and RNG versions perform somewhat better than the standard version for recirculation flows.

6.1.2 Boundary conditions

Rigid wall boundary conditions

Near a rigid wall, which may be a bank or island as shown in Fig. 6.1, the flow is quite complex. A very thin viscous sublayer exists near a smooth wall, while roughness elements on a rough wall affect the flow significantly. Because the velocity gradient is quite high there, it is of high cost to resolve the flows in the viscous sublayer and around individual roughness elements. A wall-function approach is often used instead. The first grid point or cell center (denoted as P) adjacent to the wall is placed outside the viscous sublayer and above the roughness elements, and the resultant wall shear stress $\vec{\tau}_w$ is related to the flow velocity $\vec{U}_P$ at point P by

$$\vec{\tau}_w = -\lambda_w \vec{U}_P \tag{6.12}$$

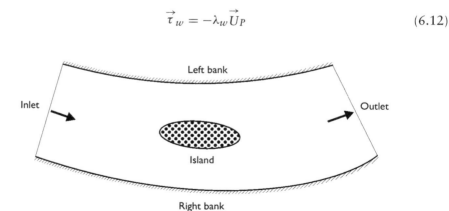

Figure 6.1 A typical horizontal 2-D computational domain.

where λ_w is a coefficient. In the k-ε turbulence models, λ_w is determined by $\lambda_w = \rho c_\mu^{1/4} k_P^{1/2} \kappa / \ln(E y_P^+)$ with $y_P^+ = c_\mu^{1/4} k_P^{1/2} y_P / \nu$. This relation of λ_w is derived using the log-law of velocity near the wall and the first relation in Eq. (6.15). Eq. (6.12) is applied in the region of $11.6 < y_P^+ <\sim 300$. In the zero-equation turbulence models, because the turbulent energy k is not solved, λ_w is determined by $\lambda_w = \rho u_* \kappa / \ln(E y_P^+)$ with $y_P^+ = u_* y_P / \nu$. Here, y_P is the distance from the wall to point P, u_* is the shear velocity on the wall defined as $u_* = \sqrt{\tau_w / \rho}$, and E is a roughness parameter. For a smooth wall, E is about 8.432. For a rough wall, E is related to the roughness Reynolds number $k_s^+ = u_* k_s / \nu$ by (Cebeci and Bradshaw, 1977)

$$E = \exp[\kappa (B_0 - \Delta B)] \tag{6.13}$$

where k_s is the equivalent roughness height on the wall, B_0 is an additive constant of 5.2, and ΔB is a function of k_s^+:

$$\Delta B = \begin{cases} 0 & k_s^+ < 2.25 \\ (B_0 - 8.5 + \frac{1}{\kappa} \ln k_s^+) \sin[0.4258(\ln k_s^+ - 0.811)] & 2.25 \leq k_s^+ < 90 \\ B_0 - 8.5 + \frac{1}{\kappa} \ln k_s^+ & k_s^+ \geq 90 \end{cases} \tag{6.14}$$

In the k-ε turbulence models, the turbulent energy and its dissipation rate at point P are specified as (Rodi, 1993)

$$k_P = \frac{u_*^2}{c_\mu^{1/2}}, \quad \varepsilon_P = \frac{u_*^3}{\kappa y_p} \tag{6.15}$$

which are derived by assuming that the local equilibrium of turbulence prevails near the wall.

However, the turbulent energy k_P may also be obtained by actually solving the k equation in the control volume near the wall, with the turbulence generation and dissipation rates specified as

$$P_{k,P} = \frac{\tau_w^2}{\kappa \mu y_P^+}, \quad \varepsilon_P = \frac{c_\mu^{3/4} k_P^{3/2}}{\kappa y_P} \tag{6.16}$$

The water level near a rigid wall is usually assumed to have a zero gradient in the direction normal to the boundary.

Inflow and outflow boundary conditions

As described in Section 5.1.1.2, for subcritical flow, a boundary condition is needed at each inlet and outlet in order to derive a well-imposed solution for Eqs. (6.1)–(6.3), while for supercritical flow, two boundary conditions should be specified at each inlet. For the sake of simplicity, only the subcritical flow case is considered below.

The inflow boundary condition is usually a time series of flow discharge. However, a lateral distribution of velocity at the inlet is required in the depth-averaged 2-D model. The streamwise (resultant) velocity U at each computational point of the inlet located in a nearly straight reach may be assumed to be proportional to the local flow depth, i.e., $U \propto h^r$. Here, r is an empirical exponent; $r \approx 2/3$ for uniform flow. A small r value means a fairly uniform distribution of velocity along the inlet cross-section. Therefore, for a given total inflow discharge $Q(= \int_0^B Uh dy')$, U is determined by

$$U = Qh^r \Big/ \int_0^B h^{1+r} dy' \tag{6.17}$$

where B is the channel width at the water surface, and y' is the transverse coordinate.

The inflow velocity direction must also be specified; it essentially determines the two components of velocity in the x- and y-directions at each point of the inlet.

The boundary condition at the outlet may be a time series of the measured water stage, a stage-discharge rating curve measured or generated using the uniform or critical flow condition, or a non-reflective wave condition, depending on the outlet configurations. For tidal flow, the tidal level may also be determined using the major astronomical constituents of tide in the study reach.

If a k-ε turbulence model is used, boundary conditions should be given for the turbulent energy and its dissipation rate at the inlet and outlet. At the inlet that is located in a nearly straight reach and far from hydraulic structures, the turbulence can be assumed to be at equilibrium; thus, Eqs. (6.10) and (6.11) are simplified to

$$P_{kv} - \varepsilon_{in} = 0 \tag{6.18}$$

$$P_{\varepsilon v} - c_{\varepsilon 2} \frac{\varepsilon_{in}^2}{k_{in}} = 0 \tag{6.19}$$

yielding

$$k_{in} = \frac{U_{*in}^2}{c_{\varepsilon\Gamma} c_\mu^{1/2} c_f^{1/4}}, \quad \varepsilon_{in} = \frac{U_{*in}^3}{c_f^{1/2} h_{in}} \tag{6.20}$$

where the subscript "in" denotes the quantities at the inlet.

At the outflow boundary, located in a reach with simple geometry and far from hydraulic structures, the gradients of flow velocity, turbulent energy, and dissipation rate can be given zero.

6.1.3 Numerical solutions

Unlike the Navier-Stokes equations described in Section 4.4, the shallow water equations (6.1)–(6.3) have a stronger linkage between velocity and pressure (water level), due to the appearance of flow depth in the depth-integrated continuity equation. It is apparently easier to solve Eqs. (6.1)–(6.3), as h, U_x, and U_y can be calculated by these three equations, respectively. However, special care is still needed in handling the convection and pressure gradient terms. Either a staggered grid approach or Rhie and Chow's (1983) momentum interpolation technique on the non-staggered grid is adopted. An upwind scheme is often used to discretize the convection terms. When the central difference scheme is used, artificial dissipations or TVD limiters are often used to suppress potential numerical oscillations. Some of these methods are used to simulate dam-break and overtopping flows, as discussed in Section 9.1. Described in this subsection are the SIMPLE(C) algorithm, the projection method, and the vorticity-based method, which are widely used in the simulation of common open-channel flows.

6.1.3.1 SIMPLE(C) algorithm

Discretization of governing equations

In the curvilinear coordinate system, Eqs. (6.1)–(6.3), (6.10), and (6.11) can be written in the following tensor notation form:

$$\frac{\partial}{\partial t}(\rho h J \phi) + \frac{\partial}{\partial \xi_m}\left(\rho h J \hat{u}_m \phi - h \Gamma_\phi J \alpha_j^m \alpha_j^m \frac{\partial \phi}{\partial \xi_m}\right) = h J S_\phi \qquad (6.21)$$

where ϕ stands for 1, U_x, U_y, k, and ε, depending on the equation considered; $\Gamma_\phi = \rho(v + v_t/\sigma_\phi)$ is the diffusivity of the quantity ϕ; S_ϕ is the source term in the equation of ϕ, including the cross-derivative diffusion terms; J is the Jacobian of the transformation between the Cartesian coordinate system x_i ($x_1 = x, x_2 = y$) and the curvilinear coordinate system ξ_m ($\xi_1 = \xi, \xi_2 = \eta$); $\hat{u}_m = \alpha_i^m U_i$; and $\alpha_i^m = \partial \xi_m/\partial x_i$.

As described in Section 4.4, the primary variables can be arranged in a staggered or non-staggered (collocated) pattern. The staggered grid approach for the depth-averaged 2-D model can be found in Lu and Zhang (1993) and Kim *et al.* (2003), whereas the non-staggered grid approach is applied here.

Eq. (6.21) is integrated over the control volume shown in Fig. 4.21. The convection terms can be discretized using one of the following schemes: hybrid, exponential, QUICK, HLPA or SOUCUP, presented in Section 4.3.1.1. The normal-derivative diffusion terms are usually discretized using the central difference scheme. The time-derivative term is discretized using the first-order backward scheme (4.23) or the three-level implicit scheme (4.43) and treated in analogy to Eq. (4.126). The discretized momentum equations give velocities $U_{i,P}^{n+1}$ ($i = 1, 2$) at cell center P as

$$U_{i,P}^{n+1} = \frac{1}{a_P^u}\left(\sum_{l=W,E,S,N} a_l^u U_{i,l}^{n+1} + S_{ui}\right) + D_i^1(p_w^{n+1} - p_e^{n+1}) + D_i^2(p_s^{n+1} - p_n^{n+1})$$

$$(6.22)$$

where $D_i^1 = h_P^{n+1}(J\alpha_i^1 \Delta \eta)_P/a_P^u$, $D_i^2 = h_P^{n+1}(J\alpha_i^2 \Delta \xi)_P/a_P^u$, and p is the pressure defined as $p = \rho g z_s$.

The relations of the velocity and pressure corrections in the depth-averaged 2-D model are similar to Eqs. (4.190), (4.191), (4.194), (4.195), and (4.202)–(4.204). Thus, they are not repeated here.

The depth-integrated continuity equation (6.1) is discretized as

$$p_P^{n+1} = p_P^n - g\frac{\Delta t}{\Delta A_P}(F_e - F_w + F_n - F_s) \tag{6.23}$$

where ΔA_P is the area of the cell centered by P; and F_e, F_w, F_n, and F_s are the convection fluxes across cell faces e, w, n, and s, defined as

$$F_w = (\rho h)_w^{n+1}(J\alpha_i^1\Delta\eta)_w U_{i,w}^{n+1} \tag{6.24}$$

$$F_s = (\rho h)_s^{n+1}(J\alpha_i^2\Delta\xi)_s U_{i,s}^{n+1} \tag{6.25}$$

It seems that the pressure (water level) can be calculated from the discretized continuity equation (6.23), but in fact node-to-node oscillations may exist on the non-staggered grid if the fluxes at the cell faces are linearly interpolated from the quantities stored at the cell centers, as explained in Section 4.4. To avoid this, Wenka (1992), Ye and McCorquodale (1997), and Minh Duc (1998) applied Rhie and Chow's (1983) momentum interpolation technique to evaluate the variable values at the cell faces from the quantities at the cell centers in the depth-averaged simulation of open-channel flows. In the formulations of Ye and McCorquodale (1997) and Minh Duc (1998), the pressure correction was defined as $p' = p^{n+1} - p^n$, which forms an explicit algorithm for pressure. To form a semi-implicit algorithm, which allows for longer time steps, the pressure correction was defined as $p' = p^{n+1} - p^*$ by Wu (2004). Wu's formulation is introduced below.

Using Rhie and Chow's (1983) momentum interpolation procedure as described in Section 4.4.4 yields the flux correction equations (4.196) and (4.197). For the depth-averaged 2-D SIMPLE algorithm, the coefficients a_W^p and a_S^p in these equations are derived as

$$a_W^p = \alpha_u(\rho h)_w^{n+1}(J\Delta\eta)_w(\alpha_{1,w}^1 Q_{1,w}^1 + \alpha_{2,w}^1 Q_{2,w}^1) \tag{6.26}$$

$$a_S^p = \alpha_u(\rho h)_s^{n+1}(J\Delta\xi)_s(\alpha_{1,s}^2 Q_{1,s}^2 + \alpha_{2,s}^2 Q_{2,s}^2) \tag{6.27}$$

where $Q_{i,w}^1 = [(1 - f_{x,P})/a_{PW}^u + f_{x,P}/a_P^u]h_w^{n+1}(J\alpha_i^1\Delta\eta)_w$; $Q_{i,s}^2 = [(1 - f_{y,P})/a_{PS}^u + f_{y,P}/a_P^u]h_s^{n+1}(J\alpha_i^2\Delta\xi)_s$; and a_{PW}^u and a_{PS}^u stand for a_P^u when Eq. (6.22) is applied in the control volumes centered by W and S, respectively.

For the depth-averaged 2-D SIMPLEC algorithm, the coefficients a_W^p and a_S^p are determined by Eqs. (6.26) and (6.27) with $Q_{i,w}^1$ and $Q_{i,s}^2$ replaced by $\tilde{Q}_{i,w}^1$ and $\tilde{Q}_{i,s}^2$ defined in Eqs. (4.203) and (4.204).

Inserting Eqs. (4.196) and (4.197), as well as two similar equations for F_e and F_n, into Eq. (6.23) yields the pressure correction equation:

$$a_P^p p_P' = a_W^p p_W' + a_E^p p_E' + a_S^p p_S' + a_N^p p_N' + S_p \tag{6.28}$$

where $a_P^p = \sum_{l=W,E,S,N} a_l^p + \Delta A_P/(g\Delta t)$, $S_p = -(p_P^* - p_P^n)\Delta A_P/(g\Delta t) - (F_e^* - F_w^* + F_n^* - F_s^*)$, and F_w^* and F_s^* are the fluxes determined using Eqs. (6.24) and (6.25) in terms of the approximate velocities $U_{i,w}^*$ abd $U_{i,s}^*$.

Implementation of boundary conditions

Near a rigid wall, the control volume is shown in Fig. 6.2. The velocity at point S, which is located on the wall, is non-slip and has a value of zero. When the x-momentum equation is integrated over this control volume, as demonstrated in Eq. (4.130), the convection flux should be zero and the shear stress τ_{xy} is determined using Eq. (6.12) at face s. This shear stress is moved into the source term, thus yielding a zero coefficient $a_S^{u_x}$ in Eq. (6.22).

When the y-momentum equation is integrated over the control volume in Fig. 6.2, the convection flux and the normal stress τ_{yy} at face s should be zero. Thus, the coefficient $a_S^{u_y}$ in Eq. (6.22) is zero as well.

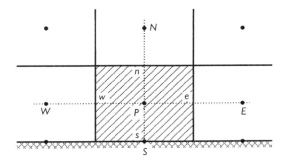

Figure 6.2 Control volume near rigid wall.

Because the flux F_s is zero, the pressure correction at face s is not needed and, naturally, a_S^p in Eq. (6.28) becomes zero. The pressure (water level) at the boundary point S can be extrapolated from the values at adjacent internal points.

As mentioned in Section 6.1.2, there are two approaches for handling k and ε at the wall boundary. One approach directly specifies the values of k and ε at center P in Fig. 6.2, according to Eq. (6.15). The other approach solves the k equation at the control volume near the wall. When the k equation is integrated over this control volume, the convection flux at face s and the coefficient a_S^k are set to zero, but the turbulence generation and dissipation rates at center P are given by Eq. (6.16).

At the inlet, the control volume is shown in Fig. 6.3(a), with face w being on the inflow side. For the specified total flow discharge Q, Eq. (6.17) cannot directly give a unique value for the inflow flux at each cell, due to the fact that the flow depth is also unknown. Iteration is usually needed. At first, a pressure is assumed at face w so that the inflow velocity and flux can be uniquely obtained using Eq. (6.17). Because the inflow flux is thereby obtained, the flux correction at face w is zero, and thus the pressure correction equation becomes

$$a_P^p p_P' = a_E^p p_E' + a_S^p p_S' + a_N^p p_N' + S_p \qquad (6.29)$$

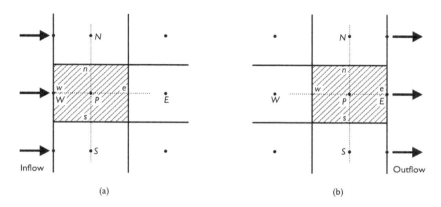

Figure 6.3 Control volumes near (a) inlet and (b) outlet.

where $a_P^p = a_E^p + a_S^p + a_N^p + \Delta A_P/(g\Delta t)$, and $S_p = -(p_P^* - p_P^n)\Delta A_P/(g\Delta t) - (F_e^* - F_w + F_n^* - F_s^*)$. The flow calculation can then be carried out over all internal points. After the internal pressure field has been obtained, the pressure at the w-face of each inlet cell can be extrapolated from the pressure values at adjacent internal points and a new inflow flux can then be obtained using Eq. (6.17). The above procedure is repeated until a convergent solution is obtained.

The turbulent energy and its dissipation rate can be directly specified at the center of the control volume near the inlet, according to Eq. (6.20). However, as an alternative, their fluxes may also be specified at the inflow face. When the relevant equation is integrated over the control volume near the inlet, the specified flux is moved into the source term and the coefficient at the inflow face is set to zero.

At the outlet, the pressure (water level) is specified for the subcritical flow. It may be specified at either the center or the outflow face of the control volume shown in Fig. 6.3(b). In the former case, the pressure correction should be zero at center P. In the latter case, an imaginary computational point (noted as E) without a control volume is set up at the outflow face, at which the pressure correction is zero. The pressure correction equation at point P is Eq. (6.28), but the coefficient a_E^p needs to be specially treated, as it cannot be determined in analogy to Eq. (6.26). The former approach is easier to implement.

The flow velocity, turbulent energy, and dissipation rate at the outlet can be extrapolated from the values at adjacent internal points. When the relevant differential equation is integrated over the control volume shown in Fig. 6.3(b), the diffusion flux at the outlet (face e) is zero due to the quantity's zero gradient; because the convection terms are usually discretized using an upwind scheme, the coefficient a_E^ϕ may actually be zero.

6.1.3.2 Projection method

Semi-implicit projection method

Casulli (1990) proposed a semi-implicit finite difference method for solving 2-D shallow water equations (6.1)–(6.3). The staggered grid shown in Fig. 4.24 is used.

The velocity divergence in the continuity equation as well as the bed friction and water level gradient terms in the momentum equations are discretized implicitly, while the other terms are discretized explicitly. The basic semi-implicit algorithm is formulated as

$$
z_{s,i,j}^{n+1} = z_{s,i,j}^n - \frac{\Delta t}{\Delta x}(h_{i+1/2,j}^n U_{x,i+1/2,j}^{n+1} - h_{i-1/2,j}^n U_{x,i-1/2,j}^{n+1})
$$

$$
- \frac{\Delta t}{\Delta y}(h_{i,j+1/2}^n U_{y,i,j+1/2}^{n+1} - h_{i,j-1/2}^n U_{y,i,j-1/2}^{n+1}) \tag{6.30}
$$

$$
U_{x,i+1/2,j}^{n+1} = F(U_{x,i+1/2,j}^n) - g\frac{\Delta t}{\Delta x}(z_{s,i+1,j}^{n+1} - z_{s,i,j}^{n+1}) - \Delta t \gamma_{i+1/2,j}^n U_{x,i+1/2,j}^{n+1} \tag{6.31}
$$

$$
U_{y,i,j+1/2}^{n+1} = F(U_{y,i,j+1/2}^n) - g\frac{\Delta t}{\Delta y}(z_{s,i,j+1}^{n+1} - z_{s,i,j}^{n+1}) - \Delta t \gamma_{i,j+1/2}^n U_{y,i,j+1/2}^{n+1} \tag{6.32}
$$

where $F(U_{x,i+1/2,j}^n)$ and $F(U_{y,i,j+1/2}^n)$ include all the remaining terms in the discretized momentum equations, and $\gamma = gn^2\sqrt{U_x^2 + U_y^2}/h^{4/3}$.

Substituting Eqs. (6.31) and (6.32) into Eq. (6.30) yields

$$
z_{s,i,j}^{n+1} - g\frac{\Delta t^2}{\Delta x^2}\left[\frac{h_{i+1/2,j}^n}{1+\gamma_{i+1/2,j}^n\Delta t}(z_{s,i+1,j}^{n+1} - z_{s,i,j}^{n+1}) - \frac{h_{i-1/2,j}^n}{1+\gamma_{i-1/2,j}^n\Delta t}(z_{s,i,j}^{n+1} - z_{s,i-1,j}^{n+1})\right]
$$

$$
- g\frac{\Delta t^2}{\Delta y^2}\left[\frac{h_{i,j+1/2}^n}{1+\gamma_{i,j+1/2}^n\Delta t}(z_{s,i,j+1}^{n+1} - z_{s,i,j}^{n+1}) - \frac{h_{i,j-1/2}^n}{1+\gamma_{i,j-1/2}^n\Delta t}(z_{s,i,j}^{n+1} - z_{s,i,j-1}^{n+1})\right]
$$

$$
= z_{s,i,j}^n - \frac{\Delta t}{\Delta x}\left[\frac{h_{i+1/2,j}^n}{1+\gamma_{i+1/2,j}^n\Delta t}F(U_{x,i+1/2,j}^n) - \frac{h_{i-1/2,j}^n}{1+\gamma_{i-1/2,j}^n\Delta t}F(U_{x,i-1/2,j}^n)\right]
$$

$$
- \frac{\Delta t}{\Delta y}\left[\frac{h_{i,j+1/2}^n}{1+\gamma_{i,j+1/2}^n\Delta t}F(U_{y,i,j+1/2}^n) - \frac{h_{i,j-1/2}^n}{1+\gamma_{i,j-1/2}^n\Delta t}F(U_{y,i,j-1/2}^n)\right] \tag{6.33}
$$

Eq. (6.33) constitutes a linear five-diagonal system of equations for water level. Because $h \geq 0$, this system is symmetric and positive definite and thus can be solved efficiently using many methods, such as the preconditioned conjugate gradient method (see Casulli, 1990).

The calculation procedure for this semi-implicit algorithm consists of the following steps: (i) calculate $F(U_{x,i+1/2,j}^n)$ and $F(U_{y,i,j+1/2}^n)$; (ii) solve Eq. (6.33) to obtain $z_{s,i,j}^{n+1}$; and (iii) calculate $U_{x,i+1/2,j}^{n+1}$ and $U_{y,i,j+1/2}^{n+1}$ using Eqs. (6.31) and (6.32). This algorithm can be considered an extension of Chorin's (1968) projection method described in Section 4.4.

A von Neumann analysis of this semi-implicit algorithm indicates that its stability depends only on the choice of difference operator F. Casulli (1990) suggested a Eulerian-Lagrangian approach. Generally speaking, many other schemes can also be used, but typically, an upwind scheme should be used for the convection terms.

Note that because the flow depth and velocity are evaluated at different time levels, the discretized continuity equation (6.30) is not strictly conservative. However, this semi-implicit algorithm has been shown to be computationally efficient.

Pressure-correction projection method

Jia *et al.* (2002) developed a depth-averaged 2-D model based on the projection method. The partially staggered grid shown in Fig. 6.4 is used. The pressure is defined at the cell centers, while both velocities U_x and U_y are at the cell corners. The governing equations are solved using the efficient element method (Wang and Hu, 1993). The convection terms in the momentum equations are discretized using the upwind interpolation scheme introduced in Section 4.2.4.2, while the other spatial derivative terms are discretized using the interpolation schemes (4.97) and (4.98). The time-derivative terms are discretized using the Euler scheme. The following pressure-correction method is used to achieve the coupling of velocity and pressure.

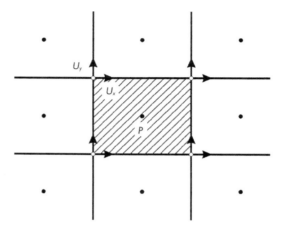

Figure 6.4 Partially staggered grid used by Jia *et al.* (2002).

The discretized momentum equations are arranged as

$$\vec{U}^{n+1} = \vec{U}^{n} + \Delta t\, \vec{G} - \frac{\Delta t}{\rho} \nabla(p^{n} + p') \tag{6.34}$$

where $\vec{G}$ includes all the remaining terms in Eqs. (6.2) and (6.3); ρ is the water density, which is assumed to be constant; p is the pressure, defined as $p = \rho g z_s$; and p' is the pressure correction, defined by

$$p^{n+1} = p^{n} + p' \tag{6.35}$$

The intermediate velocity is denoted as

$$\vec{U}^{*} = \vec{U}^{n} + \Delta t\, \vec{G} - \frac{\Delta t}{\rho} \nabla p^{n} \tag{6.36}$$

Subtracting Eq. (6.36) from Eq. (6.34) yields

$$\vec{U}^{n+1} = \vec{U}^* - \frac{\Delta t}{\rho}\nabla p' \tag{6.37}$$

Substituting Eq. (6.37) into Eq. (6.1) yields

$$\frac{\partial h}{\partial t} + \nabla \cdot (h\,\vec{U}^*) - \frac{\Delta t}{\rho}h\nabla^2 p' - \frac{\Delta t}{\rho}\nabla h \cdot \nabla p' = 0 \tag{6.38}$$

Neglecting the last term on the left-hand side of Eq. (6.38) and applying $\partial h/\partial t = (h^{n+1} - h^n)/\Delta t = p'/(\rho g\Delta t)$ yields the following Poisson equation for pressure correction:

$$(1 - gh\Delta t^2\nabla^2)p' = -\rho g\Delta t\nabla \cdot (h\vec{U}^*) \tag{6.39}$$

Eq. (6.39) is solved using the efficient element method for the Laplace operator on the left-hand side and the finite volume method for the term on the right-hand side. The resulting algebraic equations are solved using the SIP method.

The calculation procedure starts with calculating $\vec{U}^*$, using Eq. (6.36) with the known pressure p^n. Eq. (6.39) is then solved to obtain the pressure correction. Pressure and velocity at time level $n+1$ are obtained using the correction equations (6.35) and (6.37), respectively.

Because the pressure correction is defined by Eq. (6.35), the time step length is somewhat limited. This limitation can be relaxed by using $p^{n+1} = p^* + p'$, as demonstrated in Section 6.1.3.1. This modification requires p^n in Eq. (6.36) to be replaced by p^* and a term of $p^n - p^*$ to be added on the right-hand side of Eq. (6.39). Iteration is needed in the solution of the modified equations, but the time step can be longer and the mass balance is less affected by the omission of the last term on the left-hand side of Eq. (6.38), as $p' \to 0$ through the iteration.

6.1.3.3 Stream function and vorticity method

Stream function and vorticity equations

As described in Section 4.4.5, the stream function and vorticity method is highly convenient for solving the 2-D Navier-Stokes equations of incompressible flows. However, its application in the depth-averaged 2-D model is not straightforward, because the definition of stream function in Eq. (4.207) is not valid for the depth-integrated continuity equation (6.1). Wu *et al.* (1995) redefined the stream function and extended the stream function and vorticity method to the depth-averaged 2-D model for steady open-channel flows.

The governing equations for the depth-averaged 2-D steady shallow water flows are written as

$$\frac{\partial(hU_x)}{\partial x} + \frac{\partial(hU_y)}{\partial y} = 0 \tag{6.40}$$

$$U_x \frac{\partial U_x}{\partial x} + U_y \frac{\partial U_x}{\partial y} = -g \frac{\partial z_s}{\partial x} + \frac{1}{\rho} \frac{\partial T_{xx}}{\partial x} + \frac{1}{\rho} \frac{\partial T_{xy}}{\partial y} - \frac{\tau_{bx}}{\rho h} \qquad (6.41)$$

$$U_x \frac{\partial U_y}{\partial x} + U_y \frac{\partial U_y}{\partial y} = -g \frac{\partial z_s}{\partial y} + \frac{1}{\rho} \frac{\partial T_{yx}}{\partial x} + \frac{1}{\rho} \frac{\partial T_{yy}}{\partial y} - \frac{\tau_{by}}{\rho h} \qquad (6.42)$$

Corresponding to Eq. (6.40), the stream function ψ in the depth-averaged 2-D model is defined by

$$U_x = \frac{1}{h} \frac{\partial \psi}{\partial y}, \quad U_y = -\frac{1}{h} \frac{\partial \psi}{\partial x} \qquad (6.43)$$

The vorticity is still defined as

$$\Omega = \frac{\partial U_y}{\partial x} - \frac{\partial U_x}{\partial y} \qquad (6.44)$$

Therefore, the following stream-function equation is obtained by inserting Eq. (6.43) into Eq. (6.44):

$$\frac{\partial^2 \psi}{\partial x^2} + \frac{\partial^2 \psi}{\partial y^2} - \frac{1}{h} \frac{\partial h}{\partial x} \frac{\partial \psi}{\partial x} - \frac{1}{h} \frac{\partial h}{\partial y} \frac{\partial \psi}{\partial y} = -h\Omega \qquad (6.45)$$

Cross-differentiating Eqs. (6.41) and (6.42) with respect to y and x and subtracting them yields the vorticity equation:

$$\frac{\partial(U_x\Omega)}{\partial x} + \frac{\partial(U_y\Omega)}{\partial y} = \frac{\partial}{\partial x}\left[(\nu + \nu_t)\frac{\partial \Omega}{\partial x}\right] + \frac{\partial}{\partial y}\left[(\nu + \nu_t)\frac{\partial \Omega}{\partial y}\right] + S_\Omega \qquad (6.46)$$

where

$$S_\Omega = \left(\frac{\partial^2 \nu_t}{\partial x^2} - \frac{\partial^2 \nu_t}{\partial y^2}\right)\left(\frac{\partial U_x}{\partial y} + \frac{\partial U_y}{\partial x}\right) + 2\frac{\partial^2 \nu_t}{\partial x \partial y}\left(\frac{\partial U_y}{\partial y} - \frac{\partial U_x}{\partial x}\right)$$
$$+ \frac{\partial \nu_t}{\partial x}\left(\frac{\partial^2 U_y}{\partial x^2} + 2\frac{\partial^2 U_y}{\partial y^2} + \frac{\partial^2 U_x}{\partial x \partial y}\right) - \frac{\partial \nu_t}{\partial y}\left(2\frac{\partial^2 U_x}{\partial x^2} + \frac{\partial^2 U_x}{\partial y^2} + \frac{\partial^2 U_y}{\partial x \partial y}\right)$$
$$- \frac{\partial}{\partial x}\left(\frac{\tau_{by}}{\rho h}\right) + \frac{\partial}{\partial y}\left(\frac{\tau_{bx}}{\rho h}\right)$$

Similarly, differentiating Eq. (6.41) with respect to x and Eq. (6.42) with respect to y and adding them together leads to the following Poisson equation for the water level:

$$\frac{\partial^2 z_s}{\partial x^2} + \frac{\partial^2 z_s}{\partial y^2} = \frac{S_z}{g} \tag{6.47}$$

where

$$
\begin{aligned}
S_z = & -\left(\frac{\partial U_x}{\partial x}\right)^2 - 2\frac{\partial U_x}{\partial y}\frac{\partial U_y}{\partial x} - \left(\frac{\partial U_y}{\partial y}\right)^2 - U_x\left(\frac{\partial^2 U_x}{\partial x^2} + \frac{\partial^2 U_y}{\partial x \partial y}\right) \\
& - U_y\left(\frac{\partial^2 U_x}{\partial x \partial y} + \frac{\partial^2 U_y}{\partial y^2}\right) + \frac{1}{\rho}\left(\frac{\partial^2 T_{xx}}{\partial x^2} + 2\frac{\partial^2 T_{xy}}{\partial x \partial y} + \frac{\partial^2 T_{yy}}{\partial y^2}\right) \\
& - \frac{1}{\rho}\frac{\partial}{\partial x}\left(\frac{\tau_{bx}}{h}\right) - \frac{1}{\rho}\frac{\partial}{\partial y}\left(\frac{\tau_{by}}{h}\right)
\end{aligned}
$$

The stream function, vorticity, and water level equations (6.45)–(6.47) are apparently more complicated than Eqs. (6.40)–(6.42), but they are still typical partial differential equations and easy to solve. In addition, the mass is conserved automatically, and the difficulty in solving Eq. (6.40) is avoided.

Note that because the definition of stream function in Eq. (6.43) is not valid for Eq. (6.1), the aforementioned stream function and vorticity method cannot be used for the depth-averaged 2-D simulation of unsteady flows. However, it can be used in steady and quasi-steady cases. In particular, it can be used in the stepwise quasi-steady model for the long-term simulation of flow and sediment transport to reduce computational effort (Wu et al., 1995).

Boundary conditions for stream function and vorticity

For the river reach shown in Fig. 6.1, the stream function is set to zero on the left bank and Q on the right bank. Here, Q is the total flow discharge. The stream function along the island should have a constant value. The gradient of stream function along the flow direction at the outlet is set to zero. Corresponding to Eq. (6.17), the stream function at the inlet can be determined by

$$\Phi = \int_0^{y'} U h \, dy' = Q \int_0^{y'} h^{1+r} dy' \bigg/ \int_0^B h^{1+r} dy' \tag{6.48}$$

where the transverse coordinate y' starts from the left bank.

The water level is given at the outlet, as usual in the case of subcritical flow. The water level at the inlet should be extrapolated from the values at adjacent internal points. The water level gradient in the direction normal to rigid wall boundaries, such as banks and islands, can be set to zero.

The vorticity at the boundaries can be determined using Eq. (6.44), with the velocity derivatives calculated using the following one-sided schemes:

$$\frac{\partial U_x}{\partial y}\bigg|_1 = \frac{U_{x,2} - U_{x,1}}{\Delta y} \tag{6.49}$$

$$\frac{\partial U_x}{\partial y}\bigg|_1 = \frac{-3U_{x,1} + 4U_{x,2} - U_{x,3}}{2\Delta y} \tag{6.50}$$

or other higher-order approximations.

6.1.4 Wetting and drying techniques

In the calculation of flows in open channels with sloped banks, sand bars, and islands, the water edges change with time, and some part of the domain might be dry. A number of methods have been reported in the literature to handle this problem. They may be classified into two groups. One group tracks the moving water edges and adjusts the computational mesh to cover the wet domain. This group can use the boundary-fitted grid at each time (iteration) step and achieve better accuracy around the water edges. However, it results in complicated codes and perhaps requires more computational effort. The other group uses the fixed grid that covers the largest wet domain and treats dry nodes as part of the solution domain. The latter group includes the "small imaginary depth," "freezing," "porous medium," and "finite slot" methods.

The "small imaginary depth" method uses a threshold flow depth (a low value, such as 0.02 m in natural rivers and 0.001 m in experimental flumes) to judge drying and wetting at each time step. If the flow depth at a node is larger than the threshold value, this node is considered to be wet, and if the flow depth is lower than the threshold value, this node is dry. The dry nodes are assigned zero velocity. The water edges between the dry and wet areas can be treated as internal boundaries, at which the wall-function approach may be applied. The dry nodes can be excluded from the computation in an explicit algorithm, but must usually be included in an implicit algorithm. In the latter case, the "freezing" method is often adopted.

The "freezing" method also adopts a threshold flow depth to judge wetting and drying in the computational domain. At dry nodes, the Manning n or the coefficient a_P in Eq. (6.22) is given a very large value, such as 10^{30}; therefore, the calculated velocity is zero and the water level does not change (as it is frozen). The "freezing" method can include dry nodes in an implicit algorithm. However, it should be noted that the water level gradient may induce false flow motions at the dry nodes. To avoid this problem, a horizontal water level profile at the dry nodes may be assumed.

The "porous medium" method (Ghanem, 1995; Khan, 2000) assumes that the bed at the dry nodes is a porous medium and the flow can extend into the dry bed. Based on a specified minimum depth criterion, either the St. Venant or groundwater equations are applied at a particular computational point. The "finite slot" method proposed by Tao (1984) is similar to the "porous medium" method. In the "finite slot" method, a dry node is cut into two slots (with infinitesimal width and infinite depth) parallel to the x- and y-coordinates, respectively, in which the water is assumed to move. Thus, the water depth is kept positive artificially, even if the bed is dry. Different momentum equations are used at the dry nodes in the "porous medium" and "finite slot" methods, but the continuity equation at the dry nodes in both methods can be written as

$$f\frac{\partial h}{\partial t} + \nabla \cdot (h\vec{U}) = 0 \tag{6.51}$$

where f is the storativity in the "porous medium" method, or the slot width in the "finite slot" method. The slot width is given as

$$f = \begin{cases} \varepsilon_0 + (1 - \varepsilon_0)e^{a(z_s - z_b)} & z_s \leq z_b \\ 1 & z_s > z_b \end{cases} \tag{6.52}$$

where z_b is the bed elevation; ε_0 is the slot width, with a value between 0.02 and 0.05 when $z_s \ll z_b$; and a is a coefficient, which is usually larger than 2.0.

6.2 DEPTH-AVERAGED 2-D SIMULATION OF SEDIMENT TRANSPORT IN NEARLY STRAIGHT CHANNELS

6.2.1 Governing equations

As described in Section 5.1.2.1, sediment transport can be simulated by computing bed load and suspended load separately, or bed-material (total) load jointly. The depth-averaged 2-D sediment transport equations in both approaches are given below.

Bed-load and suspended-load transport model

The governing equations of the bed-load and suspended-load transport model in general situations were described in Section 2.7. For nearly straight channels, the dispersion terms in the suspended-load transport equation (2.157) are usually combined with the diffusion terms, thus yielding

$$\frac{\partial (hC_k/\beta_{sk})}{\partial t} + \frac{\partial (hU_x C_k)}{\partial x} + \frac{\partial (hU_y C_k)}{\partial y} = \frac{\partial}{\partial x}\left(E_{s,x}h\frac{\partial C_k}{\partial x}\right) + \frac{\partial}{\partial y}\left(E_{s,y}h\frac{\partial C_k}{\partial y}\right)$$
$$+ \alpha\omega_{sk}(C_{*k} - C_k) \quad (k = 1, 2, \ldots, N) \tag{6.53}$$

where $E_{s,x}$ and $E_{s,y}$ are the horizontal effective diffusion (mixing) coefficients of sediment in the x- and y-directions, respectively. If the dispersion effect is negligible, $E_{s,x}$ and $E_{s,y}$ are close to the turbulent diffusivity ε_s and can thus be related to the eddy viscosity ν_t. In general, the effective diffusivities depend on the flow, sediment, and channel conditions, and may have different values in the longitudinal and transverse directions. Their evaluation may refer to the methods for the horizontal effective diffusivities of heat and salinity introduced in Section 12.1.3.

The bed-load transport and bed change of size class k are governed by Eqs. (2.158) and (2.159), which are written below for convenience:

$$\frac{\partial(q_{bk}/u_{bk})}{\partial t} + \frac{\partial(\alpha_{bx}q_{bk})}{\partial x} + \frac{\partial(\alpha_{by}q_{bk})}{\partial y} = \frac{1}{L}(q_{b*k} - q_{bk}) \tag{6.54}$$

$$(1 - p'_m)\left(\frac{\partial z_b}{\partial t}\right)_k = \alpha\omega_{sk}(C_k - C_{*k}) + \frac{1}{L}(q_{bk} - q_{b*k}) \tag{6.55}$$

and the total rate of change in bed elevation is determined by Eq. (2.160). Note that the bed load is assumed to move along the direction of bed shear stress if the effect of bed slope in ignored; thus, the direction cosines of bed-load movement in a nearly straight channel are given by $\alpha_{bx} = U_x/U$ and $\alpha_{by} = U_y/U$ with $U = \sqrt{U_x^2 + U_y^2}$, according to Eq. (6.4). The effect of bed slope on bed-load transport is discussed in Section 6.3.4.

To close the set of equations (6.53)–(6.55), the equilibrium suspended-load concentration C_{*k} and the bed-load transport rate q_{b*k} need to be determined using empirical formulas, which can generally be written as

$$C_{*k} = p_{bk}C_k^*, \quad q_{b*k} = p_{bk}q_{bk}^* \tag{6.56}$$

where C_k^* is the potential equilibrium concentration of the kth size class of suspended load, q_{bk}^* is the potential equilibrium transport rate of the kth size class of bed load, and p_{bk} is the fraction of size class k in the mixing layer of the bed material.

The multiple-layer bed material sorting model introduced in Section 2.7.2 is applied here. For example, the bed-material gradation in the mixing layer is determined by

$$\frac{\partial(\delta_m p_{bk})}{\partial t} = \left(\frac{\partial z_b}{\partial t}\right)_k + p_{bk}^*\left(\frac{\partial \delta_m}{\partial t} - \frac{\partial z_b}{\partial t}\right) \tag{6.57}$$

Bed-material load transport model

The bed-material (total) load transport equation can be obtained by summing Eqs. (6.53) and (6.54) and using Eq. (2.149) for the sediment exchange at the bed. The resulting equation is written as

$$\frac{\partial}{\partial t}\left(\frac{hC_{tk}}{\beta_{tk}}\right) + \frac{\partial(hU_xC_{tk})}{\partial x} + \frac{\partial(hU_yC_{tk})}{\partial y} = \frac{\partial}{\partial x}\left[E_{s,x}h\frac{\partial(r_{sk}C_{tk})}{\partial x}\right]$$

$$+ \frac{\partial}{\partial y}\left[E_{s,y}h\frac{\partial(r_{sk}C_{tk})}{\partial y}\right] + \alpha_t\omega_{sk}(C_{t*k} - C_{tk}) \quad (k = 1, 2, \dots, N) \tag{6.58}$$

where C_{tk} and C_{t*k} are the actual and equilibrium (capacity) depth-averaged concentrations of the kth size class of bed-material load, respectively; β_{tk} is the correction factor determined using Eq. (2.92); α_t is the adaptation coefficient of bed-material load, defined as $\alpha_t = (Uh)/(L_t\omega_s)$ with L_t being the adaptation length; and r_{sk} is the ratio of the suspended load to the bed-material load, C_k/C_{tk}. To close Eq. (6.58), r_{sk} can be approximated by $r_{sk} = C_{*k}/C_{t*k}$. If the suspended load is dominant, r_{sk} is close to 1 and may be lumped into the diffusivity E_s.

The bed change is determined by

$$(1 - p'_m)\left(\frac{\partial z_b}{\partial t}\right)_k = \alpha_t \omega_{sk}(C_{tk} - C_{t*k}) \tag{6.59}$$

If the bed load is dominant, r_{sk} is close to 0 and the diffusion term in Eq. (6.58) can be ignored, thus yielding

$$\frac{\partial}{\partial t}\left(\frac{q_{tk}}{\beta_{tk}U}\right) + \frac{\partial(\alpha_{tx}q_{tk})}{\partial x} + \frac{\partial(\alpha_{ty}q_{tk})}{\partial y} = \frac{1}{L_t}(q_{t*k} - q_{tk}) \tag{6.60}$$

where q_{tk} and q_{t*k} are the actual and equilibrium (capacity) transport rates of the kth size class of bed-material load, respectively; and α_{tx} and α_{ty} are the direction cosines of bed-material load transport. Accordingly, the bed change is determined by

$$(1 - p'_m)\left(\frac{\partial z_b}{\partial t}\right)_k = \frac{1}{L_t}(q_{tk} - q_{t*k}) \tag{6.61}$$

The bed-material load transport capacity is determined using an equation similar to Eq. (6.56), and the bed material sorting is simulated using the previous multiple-layer model.

Because the bed-load and suspended-load model can cover the bed-material model in the numerical solution sense, only issues regarding the former model are introduced in the next subsections.

6.2.2 Boundary and initial conditions

Wall boundary conditions

At banks and islands, the bed-load transport rate and the suspended-load concentration gradient are set to zero:

$$q_{bk} = 0, \quad \frac{\partial C_k}{\partial n} = 0 \tag{6.62}$$

where n is the coordinate in the direction normal to the boundary.

Inflow boundary conditions

In the depth-averaged 2-D sediment transport simulation, the sediment discharge must be given at each point of the inflow boundary. In an unsteady case, a time series of the inflow sediment discharge is needed. For non-uniform sediment transport, the size distribution of the inflow sediment is also needed. Once the (fractional) bed-load and suspended-load discharges Q_{bk} and Q_{sk} have been given, they may be distributed laterally along the inlet located in a nearly straight reach by

$$q_{bk} = \frac{Q_{bk}qh^{r_b}}{\int_0^B qh^{r_b}dy'}, \quad q_{sk} = \frac{Q_{sk}qh^{r_s}}{\int_0^B qh^{r_s}dy'} \tag{6.63}$$

where q, q_{bk}, and q_{sk} are the flow, bed-load, and suspended-load discharges per unit channel width at each point; and r_b and r_s are empirical exponents. Note that Eq. (6.63) assumes $q_{bk} \propto qh^{r_b}$ and $q_{sk} \propto qh^{r_s}$.

Outflow boundary conditions

At the outflow boundary, calculating the bed load does not require any boundary condition, in principle. The suspended-load concentration gradient in the flow direction is set to zero:

$$\frac{\partial C_k}{\partial s} = 0 \tag{6.64}$$

where s is the coordinate in the flow direction.

Note that at a tidal boundary, the flow may go in and out alternately. The sediment transport rate needs to be provided during a flood tide (inflow), and the suspended-load concentration gradient in the flow direction is given zero during an ebb tide (outflow).

Initial conditions

The initial channel geometry, suspended-load concentration, and bed-load transport rate are required. The initial bed-material gradation in the entire solution domain must be given for the simulation of non-uniform sediment transport; it is particularly important for scour and channel stability analysis.

6.2.3 Numerical solutions

6.2.3.1 *Discretization of sediment transport equations*

Sediment transport equations can be discretized using the numerical methods introduced in Chapter 4. The finite volume method is chosen here as an example. The suspended-load transport equation (6.53) is written as Eq. (6.21) in the curvilinear coordinate system and discretized as

$$\frac{\Delta A_P}{\Delta t}\left(\frac{h_P^{n+1}C_{k,P}^{n+1}}{\beta_{sk,P}^{n+1}} - \frac{h_P^n C_{k,P}^n}{\beta_{sk,P}^n}\right) = a_W^C C_{k,W}^{n+1} + a_E^C C_{k,E}^{n+1} + a_S^C C_{k,S}^{n+1} + a_N^C C_{k,N}^{n+1}$$
$$- a_P^C C_{k,P}^{n+1} + \alpha\omega_{sk}\Delta A_P(C_{*k,P}^{n+1} - C_{k,P}^{n+1}) + S_{ck,P} \tag{6.65}$$

where a_E^C, a_W^C, a_N^C, a_S^C, and a_P^C are coefficients; and $S_{ck,P}$ includes the cross-derivative diffusion terms.

The bed-load transport equation (6.54) is integrated over the control volume shown in Fig. 4.21, with the convection terms discretized using the first-order upwind scheme or the QUICK scheme. The discretized bed-load transport equation is

$$\frac{\Delta A_P}{\Delta t}\left(\frac{q_{bk,P}^{n+1}}{u_{bk,P}^{n+1}} - \frac{q_{bk,P}^n}{u_{bk,P}^n}\right) = a_W^q q_{bk,W}^{n+1} + a_E^q q_{bk,E}^{n+1} + a_S^q q_{bk,S}^{n+1} + a_N^q q_{bk,N}^{n+1} - a_P^q q_{bk,P}^{n+1}$$
$$+ \frac{\Delta A_P}{L}(q_{b*k,P}^{n+1} - q_{bk,P}^{n+1}) \qquad (6.66)$$

Eqs. (6.65) and (6.66) can be iteratively solved using the Gauss-Seidel, ADI, or SIP method.

Note that the coefficient a_P^C in the discretized suspended-load transport equation (6.65) includes the term $F_e - F_w + F_n - F_s$, as shown in Eq. (4.135). This term can be treated using the discretized continuity equation (6.23) for better stability. However, the coefficient a_P^q in the discretized bed-load transport equation (6.66) cannot be treated thus. An alternative is to define a quantity $C_{bk} = q_{bk}/(Uh)$, substitute this relation into Eq. (6.54), and discretize the new bed-load transport equation in terms of C_{bk} as the dependent variable. The coefficient a_P in the resulting discretized equation has the term $F_e - F_w + F_n - F_s$, which can then be treated using Eq. (6.23).

To ensure mass conservation, the discretizations of the exchange terms in the bed change equation (6.55) and in the suspended-load and bed-load transport equations (6.53) and (6.54) should be consistent. Thus, Eq. (6.55) is discretized as

$$\Delta z_{bk,P} = \frac{\alpha \omega_{sk} \Delta t}{1 - p_m'}(C_{k,P}^{n+1} - C_{*k,P}^{n+1}) + \frac{\Delta t}{(1 - p_m')L}(q_{bk,P}^{n+1} - q_{b*k,P}^{n+1}) \qquad (6.67)$$

where Δz_{bk} is the change in bed elevation due to the kth size class of sediment at time step Δt.

After the fractional change in bed elevation has been calculated, the total change is obtained as

$$\Delta z_{b,P} = \sum_{k=1}^N \Delta z_{bk,P} \qquad (6.68)$$

and the bed elevation is then updated by

$$z_{b,P}^{n+1} = z_{b,P}^n + \Delta z_{b,P} \qquad (6.69)$$

The bed material sorting equation (6.57) is discretized as

$$p_{bk,P}^{n+1} = \frac{\Delta z_{bk,P} + \delta_{m,P}^n p_{bk,P}^n + p_{bk,P}^{*n}(\delta_{m,P}^{n+1} - \delta_{m,P}^n - \Delta z_{b,P})}{\delta_{m,P}^{n+1}} \qquad (6.70)$$

where $p_{bk,P}^{*n}$ is defined as $p_{bk,P}^{n}$, the fraction of size class k in the mixing layer of bed material if $\Delta z_{b,P} + \delta_{m,P}^{n} \geq \delta_{m,P}^{n+1}$, and as the fraction of size class k in the second layer if $\Delta z_{b,P} + \delta_{m,P}^{n} < \delta_{m,P}^{n+1}$.

6.2.3.2 Solution of discretized sediment transport equations

Fully decoupled model

Like the 1-D sediment transport model in Section 5.3, the depth-averaged 2-D sediment transport model can be solved in a decoupled or coupled form. In the decoupled model, the bed-material gradation in Eq. (6.56) is treated explicitly:

$$C_{*k,P}^{n+1} = p_{bk,P}^{n} C_{k,P}^{*n+1}, \quad q_{b*k,P}^{n+1} = p_{bk,P}^{n} q_{bk,P}^{*n+1} \tag{6.71}$$

The decoupled sediment transport model is usually decoupled from the flow model, thus yielding a fully decoupled procedure for flow and sediment calculations. The calculations in the fully decoupled model are executed as follows:

(1) Calculate the flow field;
(2) Calculate C_{*k}^{n+1} and q_{b*k}^{n+1} using Eq. (6.71);
(3) Calculate C_{k}^{n+1} using Eq. (6.65);
(4) Calculate q_{bk}^{n+1} using Eq. (6.66);
(5) Determine Δz_{bk} and Δz_{b} using Eqs. (6.67) and (6.68);
(6) Calculate p_{bk}^{n+1} using Eq. (6.70);
(7) Update the bed topography using Eq. (6.69) and the bed-material gradations in the subsurface layers;
(8) Return to step (1) for the next time step until a specified time is reached.

Semi-coupled model

In the semi-coupled model, the flow and sediment calculations are decoupled, but the three components of the sediment model – sediment transport, bed change, and bed material sorting – are coupled. To couple the sediment calculations, the bed-material gradation in Eq. (6.56) is treated implicitly:

$$C_{*k,P}^{n+1} = p_{bk,P}^{n+1} C_{k,P}^{*n+1}, \quad q_{b*k,P}^{n+1} = p_{bk,P}^{n+1} q_{bk,P}^{*n+1} \tag{6.72}$$

The discretized suspended-load transport equation (6.65), bed-load transport equation (6.66), bed change equations (6.67) and (6.68), bed material sorting equation (6.70), and sediment transport capacity equation (6.72) need to be solved simultaneously through iteration. One iteration procedure is to set the bed-material gradation $p_{bk,P}^{n}$ as the initial estimate for $p_{bk,P}^{n+1}$, and solve the above discretized equations, in the sequence Eqs. (6.72), (6.65), (6.66), (6.67), (6.68), and (6.70), to obtain a new estimate for $p_{bk,P}^{n+1}$. This is repeated until a convergent solution is reached. This iteration procedure is simple, but the level of coupling among sediment transport, bed change, and bed material sorting is relatively low. Therefore, Wu (2004) suggested another iteration procedure, which is described below.

Substituting Eqs. (6.70) and (6.72) into Eq. (6.67) yields

$$
\begin{aligned}
\Delta z_{bk,P} = {} & \frac{\alpha \omega_{sk} \Delta t \delta_{m,P}^{n+1} C_{k,P}^{n+1} + \Delta t \delta_{m,P}^{n+1} q_{bk,P}^{n+1}/L}{(1-p_m')\delta_{m,P}^{n+1} + \alpha \omega_{sk} \Delta t C_{k,P}^{*n+1} + \Delta t q_{bk,P}^{*n+1}/L} \\[2mm]
& - \frac{[\alpha \omega_{sk} \Delta t C_{k,P}^{*n+1} + \Delta t q_{bk,P}^{*n+1}/L][\delta_{m,P}^{n} p_{bk,P}^{n} + (\delta_{m,P}^{n+1} - \delta_{m,P}^{n}) p_{bk,P}^{*n}]}{(1-p_m')\delta_{m,P}^{n+1} + \alpha \omega_{sk} \Delta t C_{k,P}^{*n+1} + \Delta t q_{bk,P}^{*n+1}/L} \\[2mm]
& + \frac{[\alpha \omega_{sk} \Delta t C_{k,P}^{*n+1} + \Delta t q_{bk,P}^{*n+1}/L] p_{bk,P}^{*n}}{(1-p_m')\delta_{m,P}^{n+1} + \alpha \omega_{sk} \Delta t C_{k,P}^{*n+1} + \Delta t q_{bk,P}^{*n+1}/L} \Delta z_{b,P} \qquad (6.73)
\end{aligned}
$$

Summing Eq. (6.73) over all size classes and using Eq. (6.68) yields the following equation for the total change in bed elevation:

$$
\begin{aligned}
\Delta z_{b,P} = {} & \left\{ \sum_{k=1}^{N} \frac{\alpha \omega_{sk} \Delta t \delta_{m,P}^{n+1} C_{k,P}^{n+1} + \Delta t \delta_{m,P}^{n+1} q_{bk,P}^{n+1}/L}{(1-p_m')\delta_{m,P}^{n+1} + \alpha \omega_{sk} \Delta t C_{k,P}^{*n+1} + \Delta t q_{bk,P}^{*n+1}/L} \right. \\[2mm]
& \left. - \sum_{k=1}^{N} \frac{[\alpha \omega_{sk} \Delta t C_{k,P}^{*n+1} + \Delta t q_{bk,P}^{*n+1}/L][\delta_{m,P}^{n} p_{bk,P}^{n} + (\delta_{m,P}^{n+1} - \delta_{m,P}^{n}) p_{bk,P}^{*n}]}{(1-p_m')\delta_{m,P}^{n+1} + \alpha \omega_{sk} \Delta t C_{k,P}^{*n+1} + \Delta t q_{bk,P}^{*n+1}/L} \right\} \Bigg/ \\[2mm]
& \left\{ 1 - \sum_{k=1}^{N} \frac{[\alpha \omega_{sk} \Delta t C_{k,P}^{*n+1} + \Delta t q_{bk,P}^{*n+1}/L] p_{bk,P}^{*n}}{(1-p_m')\delta_{m,P}^{n+1} + \alpha \omega_{sk} \Delta t C_{k,P}^{*n+1} + \Delta t q_{bk,P}^{*n+1}/L} \right\} \qquad (6.74)
\end{aligned}
$$

Thus, the semi-coupled flow and sediment calculations using the suggested iteration procedure are executed as follows:

(1) Calculate the flow field;
(2) Determine $\Delta z_{b,P}$ using Eq. (6.74) with estimated C_k^{n+1} and q_{bk}^{n+1};
(3) Compute $\Delta z_{bk,P}$ using Eq. (6.73);
(4) Calculate p_{bk}^{n+1} using Eq. (6.70);
(5) Determine C_{*k}^{n+1} and q_{b*k}^{n+1} using Eq. (6.72);
(6) Calculate C_k^{n+1} using Eq. (6.65);
(7) Calculate q_{bk}^{n+1} using Eq. (6.66);
(8) Use the calculated C_k^{n+1} and q_{bk}^{n+1} as new estimates and repeat steps (2)–(7) until the convergent solution is obtained;
(9) Update the bed topography using Eq. (6.69) and the bed-material gradations in the subsurface layers;
(10) Return to step (1) for the next time step until a specified time is reached.

Eq. (6.74) constitutes a tighter correlation among the unknown sediment variables.

Therefore, this iteration procedure forms a higher level of coupling for the sediment model.

Fully coupled model

The depth-averaged 2-D equations of flow, sediment transport, bed change, and bed material sorting can be solved in an iteratively or fully coupled form, as described in Section 5.4 on 1-D models. For non-uniform sediment transport, an iteratively coupled model can be implemented more conveniently than a fully coupled model. The flow model in Section 6.1 and the sediment model in this section can be designed as flow and sediment loops, and then a simultaneous solution of flow and sediment transport can be obtained using the iteration procedure introduced in Section 5.4.3. Such iteratively coupled 2-D models have been reported by Spasojevic and Holly (1990) and Kassem and Chaudhry (1998).

More generally, a fully coupled flow and sediment transport model should consider the effects of sediment transport and bed change on the flow field, using Eqs. (2.119)–(2.121) as the governing equations of flow. However, this type of fully coupled 2-D model has rarely been applied in practice.

6.2.3.3 Implementation of sediment boundary conditions

The finite difference method approximates boundary conditions using difference operators. Generally, for Dirichlet boundary conditions, the known values of the suspended-load concentration and bed-load transport rate are specified at the boundary points and nothing further is necessary. For Neumann or mixed-type boundary conditions, the gradients must be evaluated using difference schemes, such as the one-sided schemes similar to (6.49) and (6.50) or higher-order approximations. The algebraic equations resulting from the discretization of boundary conditions are solved together with the discretized governing equations at the internal points.

As described in Section 6.1.3.1, the finite volume method usually plugs the sediment boundary conditions into the transport equations integrated over the control volumes near boundaries in order to insure mass balance. The details are given below.

Near a rigid sidewall, when the suspended-load and bed-load transport equations are integrated over the control volume shown in Fig. 6.2, as demonstrated in Eq. (4.130), the convection and diffusion fluxes at the wall face should be zero, thus yielding a zero coefficient a_S^ϕ in the discretized equations.

At the inlet, the suspended-load and bed-load fluxes can be specified at the inflow cell face, according to Eq. (6.63). When the relevant governing equation is integrated over the control volume near the inlet shown in Fig. 6.3(a), the specified flux is arranged in the source term and the coefficient a_W^ϕ is set to zero. Note that the specified suspended-load flux is equal to the sum of the convection and diffusion fluxes, and the specified bed-load flux is equal to the convection flux at the inflow face (w).

At the outlet, the suspended-load concentration and bed-load transport rate can be extrapolated or copied from the values at adjacent internal points. When the relevant governing equation is integrated over the control volume shown in Fig. 6.3(b), the convection terms are usually discretized using an upwind scheme, and the suspended-load diffusion flux is zero. Thus, the coefficient a_E^ϕ may actually be zero.

The bed changes at the inlet and outlet can be calculated using Eq. (6.67). The bed change along a vertical rigid wall may be extrapolated from the values at adjacent internal points. The bed change at the water edge near a sloped bank or island may be set to zero, because the flow depth is almost zero there; however, the bank or island may deform (collapse) due to the effect of gravity.

Bed-material gradations at boundary points can be calculated using Eq. (6.70) in analogy to those at internal points.

6.2.4 Examples

Two examples are cited here to demonstrate the verification and application of the depth-averaged 2-D model.

Case 1. Erosion in a basin due to clear water inflow

The erosion process in a rectangular basin due to clear water inflow from a narrow channel experimentally investigated by Thuc (1991) was numerically simulated by Minh Duc *et al.* (2004) and Wu (2004). The basin was 5 m long and 4 m wide, connected with a 0.2 m wide and 2 m long channel in the upstream and a 1.2 m wide and 1.0 m long channel in the downstream. The basin bed was covered with a 0.16 m thick layer of fine sand. The sand had a settling velocity of $0.013 \, \text{m} \cdot \text{s}^{-1}$. The inflow velocity in the upstream channel was $0.6 \, \text{m} \cdot \text{s}^{-1}$, and the water depth at the outlet was 0.15 m during the experiment. The two simulations used similar flow models based on the finite volume method, with slight difference in pressure correction as described in Section 6.1.3.1. Both applied the non-equilibrium transport model of bed load and suspended load described in Section 6.2.1; however, Minh Duc *et al.* used the sediment exchange model (2.128), the saltation step length for bed-load adaptation length, and the van Rijn formulas for sediment transport capacity, while Wu used the sediment exchange model (2.132), the sand dune length for bed-load adaptation length, and the Wu *et al.* (2000b) formulas for sediment transport capacity. The computational meshes in the basin consisted of 62×62 nodes, and the grid spacing around the basin centerline was refined in each mesh. The time step for sediment calculation was 5 sec

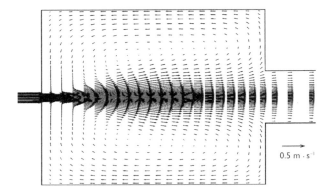

0.5 m · s⁻¹

Figure 6.5 Simulated flow field at 4 hr (Wu, 2004).

and 1 min in the simulations of Minh Duc *et al.* and Wu, respectively. The Manning roughness coefficient in the basin was given a value of 0.022 in both simulations.

Fig. 6.5 shows the flow pattern in the basin after 4 hr simulated by Wu. One can see that two symmetric recirculation eddies appeared. Fig. 6.6 shows the bed elevation change patterns in the inflow region after 4 hr calculated by two models, and Fig. 6.7 shows the measured and simulated bed elevation changes along the longitudinal centerline. Erosion occurred due to the inflow of clear water, and the eroded sediment moved downstream and deposited, forming a mound. Wu's simulation predicted faster erosion and wider deposition than that of Minh Duc *et al.* Both simulated maximum erosion depths are in fairly good agreement with the measured data.

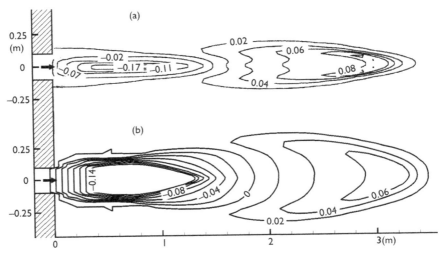

Figure 6.6 Contours of bed elevation change (m) at 4 hr: (a) simulated by Minh Duc *et al.* (2004) and (b) simulated by Wu (2004).

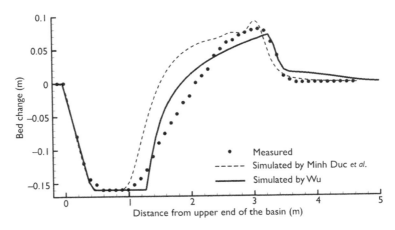

Figure 6.7 Measured and simulated bed elevation changes at 4 hr along basin centerline.

Case 2. Sediment transport in the lower Yellow River

Wu *et al.* (2006) simulated the flow and sediment transport in the lower Yellow River during the 1982 flood using the flow model in Section 6.1.3.1 and the semi-coupled sediment model in Section 6.2.3. The computational domain was the 103 km long reach between the Huayuankou and Jiahetan gauge stations. The Huayuankou station, located 259 km downstream of the Sanmenxia Hydroproject, was set as the inlet. The computational mesh is shown in Fig. 6.8, and consists of 201 and 21 points in the longitudinal and transverse directions, respectively. The measured time series of flow discharge and sediment concentration at Huayuankou, shown in Fig. 6.9(a), were used as inflow boundary conditions, while the measured time series of water stage at Jiahetan was used as the outlet boundary condition. The peak flow discharge of this flood at Huayuankou was $15{,}300\,\mathrm{m}^3\mathrm{s}^{-1}$, while the peak sediment concentration was $66.6\,\mathrm{kg}\cdot\mathrm{m}^{-3}$. The sediment was non-uniform, with sizes ranging from 0.002 to 0.18 mm. Five size classes were used to represent the non-uniform sediment mixture. The Manning roughness coefficient was between 0.009 and 0.015, with bigger values for the rising stage and smaller values for the falling stage of the flood. The computational period was from July 30 to August 11, 1982. The time step was 15 minutes. The adaptation coefficient α was 0.25. The effect of sediment concentration was considered by modifying the settling velocity of sediment particles according to the Richardson-Zaki formula (3.19).

Fig. 6.9(b) shows the measured and simulated flow discharges and sediment concentrations at Jiahetan (outlet). The simulated results generally agree well with the

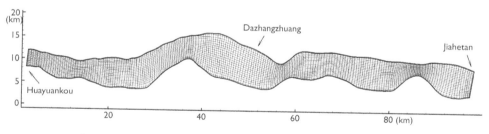

Figure 6.8 Computational mesh between Huayuankou and Jiahetan.

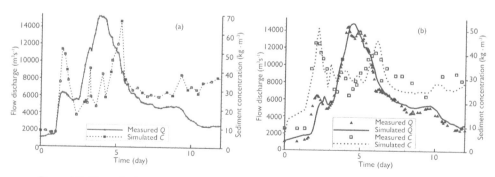

Figure 6.9 Flow discharges and sediment concentrations at (a) Huayuankou (inlet) and (b) Jiahetan (outlet) (Wu *et al.*, 2006).

measured data. Because of deposition, the simulated peak sediment concentration at Jiahetan corresponding to the peak flow decreased to $45.57 \, \text{kg} \cdot \text{m}^{-3}$, compared to the measured value of $40.7 \, \text{kg} \cdot \text{m}^{-3}$. The time delay between peak flow and sediment concentration was 34.5 and 37.5 hr at the inlet and outlet, respectively, and exhibited a trend of increasing downstream. Fig. 6.10 shows the simulated flow field corresponding to a flow discharge of 4,000 m^3s^{-1} at Huayuankou. The vectors represent the flow direction and magnitude, while the contours denote the flow depth. It can be seen that the main flow meanders in the river and interacts with the flow in floodplains. Fig. 6.11 compares the measured and simulated maximum water levels in various stations during this flood, showing a generally good agreement.

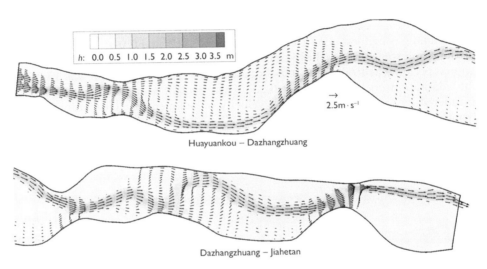

Figure 6.10 Simulated flow field at flow discharge of 4,000 m^3s^{-1} at Huayuankou (Wu et al., 2006).

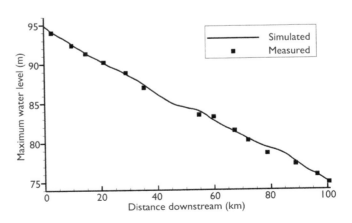

Figure 6.11 Measured and simulated maximum water levels.

6.3 DEPTH-AVERAGED 2-D SIMULATION OF FLOW AND SEDIMENT TRANSPORT IN CURVED AND MEANDERING CHANNELS

As described in Section 2.4.1, in the derivation of the depth-integrated momentum and suspended-load transport equations, dispersion terms arise from the split of three-dimensional quantities into their depth-averaged values and the remainders. For nearly straight channels, the dispersion terms can be combined with the turbulent stress/flux terms. However, for curved channels, the dispersion terms are of great significance, due to the presence of helical flow (Flokstra, 1977). Thus, among the depth-averaged flow and sediment transport equations introduced in Sections 6.1 and 6.2, the depth-integrated momentum equations (6.2) and (6.3) and suspended-load transport equation (6.53) are replaced by

$$\frac{\partial(hU_x)}{\partial t} + \frac{\partial(hU_x^2)}{\partial x} + \frac{\partial(hU_yU_x)}{\partial y} = -gh\frac{\partial z_s}{\partial x} + \frac{1}{\rho}\frac{\partial[h(T_{xx}+D_{xx})]}{\partial x}$$
$$+ \frac{1}{\rho}\frac{\partial[h(T_{xy}+D_{xy})]}{\partial y} - \frac{\tau_{bx}}{\rho} \qquad (6.75)$$

$$\frac{\partial(hU_y)}{\partial t} + \frac{\partial(hU_xU_y)}{\partial x} + \frac{\partial(hU_y^2)}{\partial y} = -gh\frac{\partial z_s}{\partial y} + \frac{1}{\rho}\frac{\partial[h(T_{yx}+D_{yx})]}{\partial x}$$
$$+ \frac{1}{\rho}\frac{\partial[h(T_{yy}+D_{yy})]}{\partial y} - \frac{\tau_{by}}{\rho} \qquad (6.76)$$

$$\frac{\partial(hC_k/\beta_{sk})}{\partial t} + \frac{\partial(hU_xC_k)}{\partial x} + \frac{\partial(hU_yC_k)}{\partial y} = \frac{\partial}{\partial x}\left[h\left(\varepsilon_s\frac{\partial C_k}{\partial x} + D_{sxk}\right)\right]$$
$$+ \frac{\partial}{\partial y}\left[h\left(\varepsilon_s\frac{\partial C_k}{\partial y} + D_{syk}\right)\right]$$
$$+ \alpha\omega_{sk}(C_{*k} - C_k) \qquad (6.77)$$

where D_{ij} and $D_{sik}(i,j=x,y)$ are the dispersion momentum transports and sediment fluxes defined in Eqs. (2.82), (2.83), and (2.86). Note that the wind driving and Coriolis forces are omitted in Eqs. (6.75) and (6.76), for simplicity.

The helical flow also affects the bed-load transport. Introduced in this section are the methods used to evaluate the dispersion fluxes in Eqs. (6.75)–(6.77) and account for the effect of helical flow on the bed load.

6.3.1 Flow properties in curved channels

As shown in Fig. 6.12, the major secondary flow observed in the cross-section of a channel bend is the helical flow, which exists due to the difference between the centrifugal forces in the upper and lower flow layers, and points to the outer bank in the upper layer and to the inner bank in the lower layer. Other secondary flow cells may also exist. For example, one often appears in the upper corner along the outer bank due to anisotropic turbulence, and more may exist in trapezoidal and

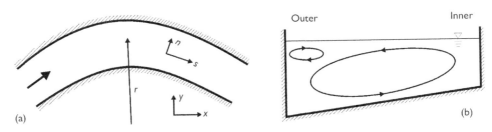

Figure 6.12 (a) Coordinate systems and (b) secondary flows in a curved channel.

compound cross-sections; however, they are usually much smaller in size and less important than the helical flow in a channel bend. Therefore, only the helical flow is considered here.

Though the main flow is affected by the aforementioned secondary flows, the vertical distribution of the streamwise flow velocity in curved channels can still be assumed to follow the logarithmic law (Rozovskii, 1957) or the power law (Zimmermann and Kennedy, 1978). A function for the vertical distribution of the helical flow velocity was derived by Rozovskii (1957), but its original formulation is complex and inconvenient to use. If the Chezy coefficient is larger than 50, the Rozovskii distribution can be simplified to a linear distribution, which is used here to evaluate the transverse velocity (Odgaard, 1986):

$$u_n = U_n + b_s I \left(2\frac{z}{h} - 1 \right) \tag{6.78}$$

where U_n is the depth-averaged cross-stream velocity, u_n is the local cross-stream velocity at height z, b_s is a coefficient with a value of about 6.0, and I is the intensity of helical flow. Note that for the sake of simplicity, z is defined here as the vertical coordinate above the channel bed rather than an arbitrary datum.

Theoretically $I = U_s h / r$ at the channel centerline (Rozovskii, 1957; de Vriend, 1977), in which r is the local radius of curvature. De Vriend (1981a) proposed an equation for approximately determining the helical flow intensity I in the entire bend:

$$\frac{\partial (hI)}{\partial t} + \frac{\partial (hU_x I)}{\partial x} + \frac{\partial (hU_y I)}{\partial y} = \frac{\partial}{\partial x}\left(D_I h \frac{\partial I}{\partial x}\right) + \frac{\partial}{\partial y}\left(D_I h \frac{\partial I}{\partial y}\right)$$
$$- \frac{h}{T_a}\left(I - \frac{\beta_I h U_s}{r}\right) \tag{6.79}$$

where D_I is a coefficient representing the diffusion and dispersion of I, T_a is the adaptation time of I, and β_I is a coefficient that is 1.0 in de Vriend's original equation but usually ranges from 1.0 to 2.0.

Eq. (6.79) needs to be solved numerically. This means that one partial differential equation is added to the set of shallow water equations, and the computational effort is increased. To avoid this, by neglecting the time-derivative term, convection terms, longitudinal diffusion term, and other high-order terms, Wu and Wang (2004a) simplified Eq. (6.79) to an ordinary differential equation for the helical flow intensity in the developed regions:

$$\frac{D_I}{B^2}\frac{d^2 I}{d\eta^2} = \frac{1}{T_a}\left(I - \frac{\beta_I h U_s}{r}\right) \tag{6.80}$$

where η is the dimensionless transverse coordinate (y'/B), with $\eta = 0$ at the inner bank and 1 at the outer bank; and B is the channel width at the water surface.

By assuming constant D_I, T_a, and source term in Eq. (6.80) and applying boundary conditions $I = 0$ at $\eta = 0$ and $\eta = 1$, a solution for Eq. (6.80) was then derived as

$$\frac{rI}{\beta_I h U_s} = 1 - \frac{1 - e^{-B/\sqrt{T_a D_I}}}{e^{B/\sqrt{T_a D_I}} - e^{-B/\sqrt{T_a D_I}}}e^{B\eta/\sqrt{T_a D_I}} - \frac{e^{B/\sqrt{T_a D_I}} - 1}{e^{B/\sqrt{T_a D_I}} - e^{-B/\sqrt{T_a D_I}}}e^{-B\eta/\sqrt{T_a D_I}}$$

$$\tag{6.81}$$

Eq. (6.81) shows that the cross-stream profile of I is determined by parameters B, T_a, D_I, and β_I. Among these parameters, the channel width B is predefined. D_I can be determined using $D_I = \alpha_d U_* h$, in which α_d is an empirical coefficient. If the radius of curvature at the channel centerline r_c and the average flow velocity U are used to represent the adaptation length and velocity scales of I, respectively, it can be assumed that the adaptation time scale $T_a = \alpha_a r_c/U$. Here, α_a is a dimensionless coefficient. Therefore, the product of T_a and D_I can be determined by

$$T_a D_I = \lambda_t r_c n \sqrt{g} h^{5/6} \tag{6.82}$$

where λ_t is the product of α_a and α_d.

Therefore, the helical flow intensity profile along the cross-section is determined by two parameters: β_I and λ_t. Usually, β_I determines the magnitude of I, while λ_t determines its lateral distribution. According to calibrations using many laboratory and field measurements, β_I is in the range of 1.0–2.0, and λ_t has a value of about 3.0.

Fig. 6.13 compares the secondary flow intensity calculated using Eq. (6.81) and that measured by de Vriend (1981b) in an 180° bend with a rectangular cross-section

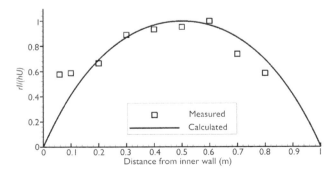

Figure 6.13 Transverse distribution of secondary flow intensity in de Vriend's bend (Wu and Wang, 2004a).

and a rough bed. The channel width was 1.7 m, and the radius of curvature at the centerline was 4.25 m. The flow discharge was $0.19 \, \mathrm{m^3 s^{-1}}$, and the overall average flow depth was about 0.2 m. The Manning roughness coefficient was 0.024. The measured secondary flow intensity was determined by least-square fitting of Eq. (6.78) with the secondary flow velocities measured at cross-section 180°. The helical flow intensity profile calculated using Eq. (6.81) with $\lambda_t = 3.0$ and $\beta_I = 1.5$ matches the general trend of the measurement data. Errors exist in regions close to the sidewalls, due to the influence of the walls and the possible appearance of the other secondary flow at the top corner close to the outer wall.

6.3.2 Dispersion of flow momentum

Jin and Steffler (1993) derived differential equations for dispersion momentum transports from the 3-D moment-of-momentum equations. However, their equations are complex and require additional computational effort. The simpler approach is to use algebraic expressions (e.g., Flokstra, 1977). Wu and Wang (2004a) derived a general algebraic formulation for the dispersion transports, which is introduced below.

By using the power law (3.27) for the streamwise flow velocity and the linear model (6.78) for the helical flow velocity, the x- and y-components of local velocity in a curved channel are evaluated as

$$u_x = \alpha_{11} u_s + \alpha_{12} u_n = \alpha_{11} \frac{m+1}{m} U_s \left(\frac{z}{h}\right)^{1/m} + \alpha_{12} \left[U_n + b_s I \left(2\frac{z}{h} - 1\right)\right]$$

(6.83)

$$u_y = \alpha_{21} u_s + \alpha_{22} u_n = \alpha_{21} \frac{m+1}{m} U_s \left(\frac{z}{h}\right)^{1/m} + \alpha_{22} \left[U_n + b_s I \left(2\frac{z}{h} - 1\right)\right]$$

(6.84)

where α_{ii} are the coefficients of transformation between the (x, y) and (s, n) coordinate systems shown in Fig. 6.12(a); U_s is the depth-averaged velocity in the streamwise direction; u_s is the local streamwise velocity at height z; and m is usually about 7.

Substituting Eqs. (6.83) and (6.84) into the definition expressions of dispersion transports in Eqs. (2.82) and (2.83) leads to

$$D_{xx} = -\rho \left[\frac{1}{m(m+2)}\alpha_{11}\alpha_{11} U_s^2 + \frac{2b_s}{2m+1}\alpha_{11}\alpha_{12} I U_s + \frac{b_s^2}{3}\alpha_{12}\alpha_{12} I^2\right]$$

(6.85)

$$D_{xy} = -\rho \left[\frac{1}{m(m+2)}\alpha_{11}\alpha_{21} U_s^2 + \frac{b_s}{2m+1}(\alpha_{11}\alpha_{22} + \alpha_{12}\alpha_{21}) I U_s + \frac{b_s^2}{3}\alpha_{12}\alpha_{22} I^2\right]$$

(6.86)

$$D_{yy} = -\rho \left[\frac{1}{m(m+2)}\alpha_{21}\alpha_{21}U_s^2 + \frac{2b_s}{2m+1}\alpha_{21}\alpha_{22}IU_s + \frac{b_s^2}{3}\alpha_{22}\alpha_{22}I^2 \right] \quad (6.87)$$

Note that the dispersion transports can also be derived using the logarithmic law and another helical flow model. The dispersion terms can be treated as additional source terms in the momentum equations (6.75) and (6.76).

Physically, the helical flow transfers the upper layer of water, which has stronger momentum, toward the outer bank and the lower layer of water, which has weaker momentum, toward the inner bank, and thus shifts the main flow to the outer bank in the bend. If this effect is ignored, the main flow in a curved channel may not be simulated correctly. Fig. 6.14 shows the simulated flow patterns in a 270° bend, and Fig. 6.15 compares the measured and simulated velocities at six cross-sections. The flow was measured by Steffler (1984) in a flume with a flat bed and a rectangular cross-section. The flume width was 1.07 m, the radius of curvature at the centerline was 3.66 m, and the bed slope was 0.00083. The Manning n was 0.0125. The inflow discharge was 0.0235 m^{3}s^{-1}, and the water depth at the outlet was 0.061 m. The simulations were conducted with and without consideration for the dispersion effects in the depth-averaged momentum equations (Wu and Wang, 2004a). The parameters β_I and λ_t were evaluated as 2.0 and 3.0, respectively, using the measurement data of secondary flow velocity. It can be seen that when the helical flow effect is not taken into account, the calculated main flow is along the inner wall over the entire bend; and when the helical flow effect is considered, the main flow shifts from the inner wall to the outer wall and the accuracy of the simulated flow velocities is much improved.

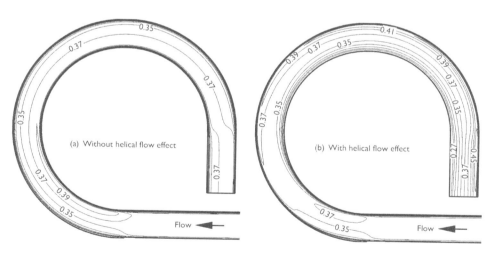

Figure 6.14 Calculated velocity contours without and with helical flow effect in Steffler's 270° bend (Wu and Wang, 2004a).

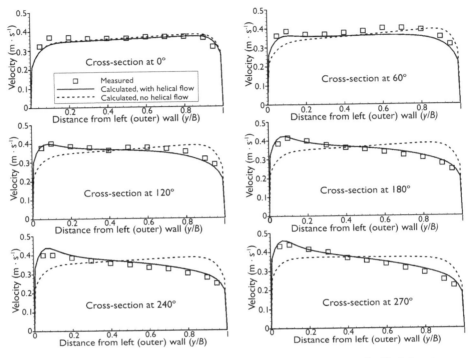

Figure 6.15 Measured and calculated velocities at cross-sections in Steffler's bend (Wu and Wang, 2004a).

6.3.3 Dispersion of suspended load

The distribution of suspended-load concentration along the flow depth can be written as

$$c = Cf(z) \qquad (6.88)$$

where c is the local suspended-load concentration, C is the depth-averaged suspended-load concentration, and $f(z)$ is a distribution function. By using $U_s hC = \int_\delta^h u_s c dz$ with the Lane-Kalinske distribution for the suspended-load concentration c and the power law for the flow velocity u_s, one can derive $f(z) = \frac{m}{m+1}\Pi^{-1}\exp\left(-\frac{15\omega_s}{U_*}\frac{z}{h}\right)$ with $\Pi = \int_{\delta/h}^1 \zeta^{1/m}e^{-15\omega_s\zeta/U_*}d\zeta$. Here, δ is the thickness of the bed-load layer.

Integrating the x-direction suspended-load convection flux along the flow depth yields

$$\int_\delta^h u_x c dz = \alpha_{11}\frac{m+1}{m}U_s C\int_\delta^h \left(\frac{z}{h}\right)^{1/m}f(z)dz + \alpha_{12}U_n C\int_\delta^h f(z)dz$$

$$+ \alpha_{12}b_s IC\int_\delta^h \left(2\frac{z}{h}-1\right)f(z)dz \qquad (6.89)$$

Using relations $\frac{m+1}{m}\int_\delta^h \left(\frac{z}{h}\right)^{1/m}f(z)dz = h$ and $\int_\delta^h f(z)dz \approx h$ in Eq. (6.89) leads to the dispersion flux in the x-direction:

$$D_{sx} = U_x C - \frac{1}{h} \int_\delta^h u_x c\, dz \approx -\alpha_{12} b_s IC \frac{1}{h} \int_\delta^h \left(2\frac{z}{h} - 1\right) f(z)\, dz \qquad (6.90)$$

Similarly, the dispersion flux in the y-direction can be derived as

$$D_{sy} = U_y C - \frac{1}{h} \int_\delta^h u_y c\, dz \approx -\alpha_{22} b_s IC \frac{1}{h} \int_\delta^h \left(2\frac{z}{h} - 1\right) f(z)\, dz \qquad (6.91)$$

Using the Lane-Kalinske distribution, the integral in Eqs. (6.90) and (6.91) is evaluated as

$$\int_\delta^h \left(2\frac{z}{h} - 1\right) f(z)\, dz = -\frac{m}{m+1} \frac{\Pi^{-1} U_* h}{15\omega_s} \left[1 + e^{-15\omega_s/U_*} - \frac{2U_*}{15\omega_s}(1 - e^{-15\omega_s/U_*})\right]$$

$$(6.92)$$

As alternatives, one may also use the Rouse distribution for the suspended load and the logarithmic law for the streamwise flow velocity in the above derivation. This is left to the interested reader.

6.3.4   Bed-load transport in curved channels

The bed-load movement direction deviates from the main flow direction due to the helical flow effect. Engelund (1974), Zimmermann and Kennedy (1978), and Odgaard (1986) proposed empirical formulas for evaluating this deviation. The Engelund formula is

$$\tan\delta_b = 7\frac{h}{r} \qquad (6.93)$$

where δ_b is the angle between the bed-load movement and main flow directions.
In Odgaard's method, the direction of bed-load movement is calculated by

$$\tan\delta_{bs} = \frac{u_{by}}{u_{bx}} \qquad (6.94)$$

which is equivalent to

$$\alpha_{bx} = u_{bx} \Big/ \sqrt{u_{bx}^2 + u_{by}^2}, \quad \alpha_{by} = u_{by} \Big/ \sqrt{u_{bx}^2 + u_{by}^2} \qquad (6.95)$$

where u_{bx} and u_{by} are the x- and y-components of bed-load velocity or the flow velocity near the bed, which can be converted from u_{bs} and u_{bn} determined using Eqs. (6.83) and (6.84).
The effect of gravity on bed-load transport in a sloped bend is also important. Kikkawa *et al.* (1976) derived analytically the lateral bed-load transport affected by

both helical flow and gravity. Parker (1984) simplified the result of Kikkawa *et al.* as

$$\frac{q_{bn}}{q_{bs}} = \tan \delta_b + \frac{1 + \alpha_p \mu_c}{\lambda_s \mu_c} \sqrt{\frac{\Theta_c}{\Theta}} \tan \varphi \tag{6.96}$$

where q_{bs} and q_{bn} are the bed-load transport rates along the longitudinal and transverse directions, respectively; φ is the lateral inclination of the bed; Θ is the Shields number; Θ_c is the critical Shields number $= 0.04$; μ_c is the dynamic coefficient of Coulomb friction; α_p is the ratio of lift coefficient to drag coefficient; and λ_s is a sheltering coefficient. Kikkawa *et al.* (1976) evaluated μ_c, α_p, and λ_s as 0.43, 0.85, and 0.59, respectively.

Struiksma *et al.* (1985) and Sekine and Parker (1992) also proposed a similar relation as

$$\frac{q_{bn}}{q_{bs}} = \tan \delta_b - \beta_b \frac{\partial z_b}{\partial n} \tag{6.97}$$

where β_b is a coefficient. Struiksma *et al.* (1985) suggested $\beta_b = 1/[\chi_0 (U'_*)^2]$ with χ_0 varying between 1 and 2, while Sekine and Parker (1992) gave $\beta_b = 0.75(\Theta_c/\Theta)^{1/4}$.

By applying Eq. (3.132) to the x and y directions, Wu (2004) derived a method that replaces the bed-load transport direction cosines α_{bx} and α_{by} by $\alpha_{bx,e}$ and $\alpha_{by,e}$:

$$\frac{\alpha_{bx,e}}{\alpha_{by,e}} = \frac{\tau'_b \alpha_{bx} + \lambda_0 \tau_c \sin \varphi_x / \sin \phi_r}{\tau'_b \alpha_{by} + \lambda_0 \tau_c \sin \varphi_y / \sin \phi_r} \tag{6.98}$$

where φ_x and φ_y are the bed angles with the horizontal along x- and y-directions (with positive values denoting downslope beds), respectively. Eq. (6.98) can also be written as Eq. (6.97) with $\beta_b = \lambda_0(\tau_c/\tau'_b)/\sin \phi_r$. Note that τ'_b in Eq. (6.98) may be replaced by the total bed shear stress τ_b if the bed-load transport capacity is determined by formulas that consider τ_b as the tractive force for bed-load transport.

The dispersion fluxes in the suspended-load transport equation and the adjustment of the bed-load transport angle tend to move sediment from the outer bank toward the inner bank. Therefore, with such enhancements, the depth-averaged 2-D model can reasonably predict erosion along the outer bank and deposition along the inner bank. This is demonstrated in the following example.

Wu and Wang (2004a) simulated the sediment transport and morphological change in an 180° bend under unsteady flow conditions, which were experimentally investigated by Yen and Lee (1995). The width of the flume was 1 m, the radius of curvature at the centerline was 4 m, and the initial bed slope was 0.002. The flow hydrograph was triangular. The base flow discharge was 0.02 m³s⁻¹, and the base flow depth, h_o, was 0.0544 m. In the simulated case (Run 4), the peak flow discharge was 0.053 m³s⁻¹, and the duration was 240 min. The peak of the hydrograph was set at the first third of its duration. The sediment was non-uniform and had a median diameter of 1.0 mm and a standard deviation of 2.5. The Manning roughness coefficient was given as $d_{50}^{1/6}/20$ in the simulation, with d_{50} being the median size of the bed material in the mixing layer. The two parameters in the helical flow intensity model (6.81) were set as $\lambda_t = 3.0$ and $\beta_s = 1.0$. The computational mesh in the bend reach consisted of 91 and 31 points in the longitudinal and transverse directions, respectively. The time step was 1 min. Fig. 6.16 compares the measured and simulated bed change contours in the bend, and Fig. 6.17 shows the lateral profiles of the bed changes at four cross-sections. The general patterns of the deeper channel along the outer wall and the point bar along the inner wall are reproduced well by the model. The calculated bed changes are in agreement with the measured data. Without considering the helical flow effect, one cannot obtain such reasonable results.

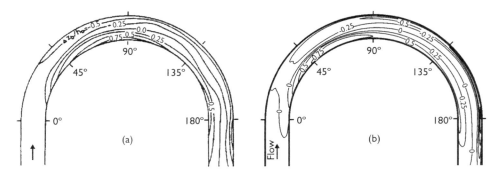

Figure 6.16 (a) Measured and (b) calculated bed change contours ($\Delta z_b/h_0$) (Wu and Wang, 2004a).

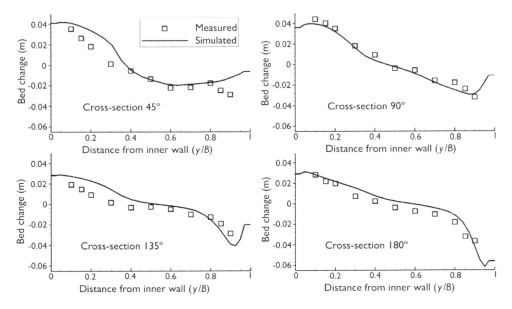

Figure 6.17 Bed changes at cross-sections in Yen and Lee's (1995) bend (Wu and Wang, 2004a).

6.3.5 Channel meandering process

To simulate the channel meandering process, a depth-averaged 2-D model must be capable of: (1) considering the helical flow effect in meandering channels; (2) simulating bank erosion; and (3) handling the moving boundary problem. Because the channel meandering process is usually much slower than the sediment transport and bed change processes, the aforementioned strategies of considering the helical flow effect in curved channels with fixed banks can be applied in the simulation of flow and sediment transport in meandering channels.

Erosion mechanisms differ for banks with cohesive and non-cohesive materials. For a cohesive bank, the bank material fails in blocks. The bank erosion models described in Section 5.3.6.1 for the 1-D simulation can be extended to the depth-averaged 2-D simulation. Bank-toe erosion can be computed using the method of Arulanandan *et al.* (1980), and mass failures can be calculated using the method of Osman and Thorne (1988) or Simon *et al.* (2000). These are not repeated here.

For a non-cohesive bank, the bank material fails in particles. Once the bank slope exceeds the repose angle, the bank particles will slide to the bank toe and form a new slope with the repose angle. In the simulation of this sliding process, mass conservation should be satisfied. The following algorithm is recommended for handling the non-cohesive bank sliding process. It is also applicable to the sliding of a non-cohesive bed with a steep slope.

Consider a cluster comprising of the cell centered by point P and eight adjacent cells, shown in Fig. 6.18(a). If the slope between point P and one of the eight adjacent points exceeds the repose angle, the bank particles will slide and form a new bank slope with the repose angle between the two points. Fig. 6.18(b) shows the sliding process between point P and an adjacent point denoted as i. This process can be described mathematically by

$$\frac{(z_{bi} + \Delta z_{bi}) - (z_{bP} + \Delta z_{bP})}{\Delta l_{Pi}} = \pm \tan \phi_r \qquad (6.99)$$

where z_{bP} and z_{bi} are the bed elevations at points P and i; Δz_{bP} and Δz_{bi} are the changes in bed elevations due to sliding; Δl_{Pi} is the distance between points P and

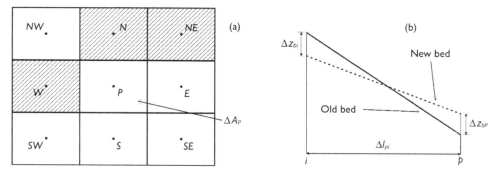

Figure 6.18 Non-cohesive bank sliding model: (a) plan view; (b) side view.

i; and ϕ_r is the repose angle. The positive sign on the right-hand side of Eq. (6.99) is used in the case of downslope from point i to P, i.e., $z_{bi} > z_{bP} + \Delta l_{Pi} \tan \phi_r$, while the negative sign is used in the case of upslope from point i to P, i.e., $z_{bi} < z_{bP} - \Delta l_{Pi} \tan \phi_r$.

Suppose that I points out of the eight adjacent points exceed the sliding criterion. Eq. (6.99) should be applied to each of these I points. These I points may be divided into two groups according to whether they form downslopes or upslopes to point P. Denote I_{up} as the number of points that form upslopes to point P. Mass balance between cell P and these I_{up} cells reads

$$\Delta z_{bP} \Delta A_P + \sum_{i=1}^{I_{up}} \Delta z_{bi} \Delta A_i = 0 \qquad (6.100)$$

where ΔA_P and ΔA_i are the areas of the cells centered at points P and i, respectively.

The equation set consisting of Eq. (6.99) for all I_{up} points and the mass balance equation (6.100) provides a unique solution for Δz_{bi} ($i = 1, 2, \ldots, I_{up}$) and Δz_{bP}, which can be derived as follows.

Eq. (6.99) is rearranged as

$$\Delta z_{bi} = \Delta z_{bP} + z_{bP} - z_{bi} - \Delta l_{Pi} \tan \phi_r \qquad (6.101)$$

Inserting Eq. (6.101) into Eq. (6.100) yields

$$\Delta z_{bP} = \left[\sum_{i=1}^{I_{up}} \Delta A_i (z_{bi} - z_{bP} + \Delta l_{Pi} \tan \phi_r) \right] \Big/ \left(\Delta A_P + \sum_{i=1}^{I_{up}} \Delta A_i \right) \qquad (6.102)$$

Once Δz_{bP} has been determined using Eq. (6.102), Δz_{bi} ($i = 1, 2, \ldots, I_{up}$) can then be obtained from Eq. (6.101).

In the same way, an equation similar to (6.100) can be obtained for mass balance among the group of points that form downslopes to point P, and the solution for Δz_{bi} and Δz_{bP} for this group can then be found.

The above algorithm should be performed by sweeping over the entire domain or region of interest. Because one sweep may not get the solution for the entire domain, it must be repeated until all slopes are gentler than the repose angle. Moreover, the sweeping sequence should be alternated between the positive and negative x and y directions.

Because of bank erosion, the wetted area varies with time during the channel meandering process. The methods of handling drying and wetting processes described in Section 6.1.4 can be used here. Either a fixed grid covering the entire area in which the channel may migrate, or a moving grid that conforms to the migrating channel can be used. Nagata et al. (2000) and Duan et al. (2001) adopted moving grids. In their approaches, flow, sediment transport, bed change, and bank erosion are computed on the old mesh at each time step. After the bank lines have been moved by erosion and deposition, a new mesh conforming to the new bank lines is created, and the flow field and bed topography are interpolated from the old mesh to the new mesh. Then the computations of flow, sediment transport, bed change, and bank erosion are continued on the new mesh at the next time step.

Fig. 6.19 shows the channel bank migration process simulated by Duan *et al.* (2001) using a depth-averaged 2-D model, which generally agrees with the experimental data measured by Nagata *et al.* (2000). Fig. 6.20 shows a simulation example performed by Duan *et al.*, in which a slightly curved channel develops into a strongly meandering channel. The results are plausible.

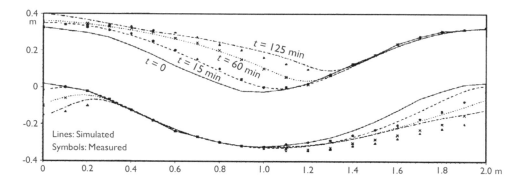

Figure 6.19 Simulated and measured channel migration processes (Duan *et al.*, 2001).

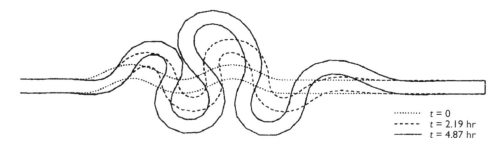

Figure 6.20 Simulated channel meandering process (Duan *et al.*, 2001).

6.4   WIDTH-AVERAGED 2-D MODEL OF FLOW AND SEDIMENT TRANSPORT

6.4.1   Width-averaged 2-D hydrodynamic model

6.4.1.1   Governing equations

In the width-averaged 2-D model, the open-channel flows with low sediment concentration are governed by Eqs. (2.95)–(2.97), which can be written as follows when the dispersion momentum transports are combined with the turbulent stresses and "$\sim$", representing the width-averaged quantities, is omitted:

$$\frac{\partial(bU_x)}{\partial x} + \frac{\partial(bU_z)}{\partial z} = 0 \tag{6.103}$$

$$\frac{\partial(bU_x)}{\partial t} + \frac{\partial(bU_x^2)}{\partial x} + \frac{\partial(bU_zU_x)}{\partial z} = -b\frac{1}{\rho}\frac{\partial p}{\partial x} + \frac{1}{\rho}\frac{\partial(bT_{xx})}{\partial x} + \frac{1}{\rho}\frac{\partial(bT_{xz})}{\partial z}$$
$$- \frac{1}{\rho}(m_1\tau_{x1} + m_2\tau_{x2}) \tag{6.104}$$

$$\frac{\partial(bU_z)}{\partial t} + \frac{\partial(bU_xU_z)}{\partial x} + \frac{\partial(bU_z^2)}{\partial z} = -bg - b\frac{1}{\rho}\frac{\partial p}{\partial z} + \frac{1}{\rho}\frac{\partial(bT_{zx})}{\partial x} + \frac{1}{\rho}\frac{\partial(bT_{zz})}{\partial z}$$
$$- \frac{1}{\rho}(m_1\tau_{z1} + m_2\tau_{z2}) \tag{6.105}$$

where x is the longitudinal coordinate; z is the vertical coordinate above a datum; U_x and U_z are the width-averaged flow velocities in the x- and z-directions, respectively; p is the width-averaged pressure; and T_{ij} $(i,j = x, z)$ are the width-averaged stresses:

$$T_{xx} = 2\rho(\nu + \nu_t)\frac{\partial U_x}{\partial x} - \frac{2}{3}\rho k$$

$$T_{xz} = T_{zx} = \rho(\nu + \nu_t)\left(\frac{\partial U_x}{\partial z} + \frac{\partial U_z}{\partial x}\right) \tag{6.106}$$

$$T_{zz} = 2\rho(\nu + \nu_t)\frac{\partial U_z}{\partial z} - \frac{2}{3}\rho k$$

where the eddy viscosity ν_t can be determined using Prandtl's mixing length model (2.48), the parabolic model (2.49), or Eq. (2.54) in the linear k-ε turbulence models. In the width-averaged 2-D k-ε turbulence models, the turbulent energy k and its dissipation rate ε are determined by

$$\frac{\partial k}{\partial t} + U_x\frac{\partial k}{\partial x} + U_z\frac{\partial k}{\partial z} = \frac{\partial}{\partial x}\left(\frac{\nu_t}{\sigma_k}\frac{\partial k}{\partial x}\right) + \frac{\partial}{\partial z}\left(\frac{\nu_t}{\sigma_k}\frac{\partial k}{\partial z}\right) + P_k + P_{kl} - \varepsilon \tag{6.107}$$

$$\frac{\partial \varepsilon}{\partial t} + U_x\frac{\partial \varepsilon}{\partial x} + U_z\frac{\partial \varepsilon}{\partial z} = \frac{\partial}{\partial x}\left(\frac{\nu_t}{\sigma_\varepsilon}\frac{\partial \varepsilon}{\partial x}\right) + \frac{\partial}{\partial z}\left(\frac{\nu_t}{\sigma_\varepsilon}\frac{\partial \varepsilon}{\partial z}\right) + c_{\varepsilon1}\frac{\varepsilon}{k}(P_k + c_{\varepsilon l}P_{kl}) - c_{\varepsilon2}\frac{\varepsilon^2}{k}$$

$$\tag{6.108}$$

where P_k is the production of turbulence due to shear, defined as $P_k = \nu_t[2(\partial U_x/\partial x)^2 + 2(\partial U_z/\partial z)^2 + (\partial U_x/\partial z + \partial U_z/\partial x)^2]$; P_{kl} accounts for the generation of turbulence due to bank shear, modeled by $P_{kl} = c_{kl}[(m_1\tau_{x1} + m_2\tau_{x2})U_x + (m_1\tau_{z1} + m_2\tau_{z2})U_z]/\rho$; and $c_{\varepsilon1}, c_{\varepsilon2}, c_{kl}, c_{\varepsilon l}, \sigma_k$, and σ_ε are coefficients. The values of $c_{\varepsilon1}, c_{\varepsilon2}, \sigma_k$, and σ_ε are listed in Table 2.3, while c_{kl} and $c_{\varepsilon l}$ are about 1.0 and 1.33, respectively.

If the flow depth is very small in comparison with the flow width, the bank shear stresses τ_{xi} and τ_{zi} are negligible; otherwise, they should be approximated by friction formulas. In general, the bank shear stresses are determined by

$$\tau_{xl} = \rho c_{fw} U_x \sqrt{U_x^2 + U_z^2}, \quad \tau_{zl} = \rho c_{fw} U_z \sqrt{U_x^2 + U_z^2} \quad (l = 1, 2) \qquad (6.109)$$

where c_{fw} is the friction coefficient.

If $U_z \ll U_x$, which is valid for gradually varied flows, the components of bank shear stresses in the vertical direction are usually ignored, while the longitudinal components can be related to the total friction force by using Einstein's division of hydraulic radius (Wu, 1992):

$$\tau_x = \rho g \frac{\widehat{U} |\widehat{U}|}{C_b^2} \chi \left/ \left[\left(\frac{n_b}{n_w}\right)^{3/2} \chi_b + \chi_w \right] \right. \qquad (6.110)$$

where $\widehat{U}$ is the velocity averaged over the cross-section; χ_b, χ_w, and χ are the wetted perimeters of the bed, banks, and entire cross-section, respectively; and n_b and n_w are the Manning roughness coefficients for the bed and banks, respectively.

In the case of $B = 1$ and without bank shear stresses, Eqs. (6.103)–(6.105) reduce to the idealized vertical 2-D model equations. Because the width-averaged 2-D model considers the variations of channel width in the longitudinal and vertical directions, it is more often used in practice than the idealized vertical 2-D model. On the other hand, the width-averaged 2-D model is an extension of the idealized vertical 2-D model; therefore, many numerical techniques developed for the idealized vertical 2-D model can be applied here.

Note that if the lateral expansion or contraction of channel width is too large (larger than about 7°), the flow may detach from the two side boundaries and the width-averaged 2-D model may not be applicable. However, if the side separation zones are excluded, the width-averaged flow model can still be approximately applied in the main flow regions.

6.4.1.2 Boundary conditions

At the water surface, the kinematic condition is applied:

$$\frac{\partial z_s}{\partial t} + U_{hx} \frac{\partial z_s}{\partial x} = U_{hz} \qquad (6.111)$$

where U_{hx} and U_{hz} are the x- and z-components of velocity at the water surface.

In the presence of wind, the wind shear force results in a gradient of flow velocity at the water surface:

$$\left. \frac{\partial U_s}{\partial n} \right|_{z=z_s} = \frac{\tau_s}{\rho \nu_t} \qquad (6.112)$$

where U_s is the velocity in the tangential direction of water surface, n is the coordinate along the direction normal to the water surface, and τ_s is the streamwise component of wind shear stress.

At the channel bottom, the wall-function approach is applied. Its implementation in the finite volume method has been described in Section 6.1.2. In the finite difference model, Wu (1992) used the logarithmic law at the bottom point P:

$$U_{s,P} = \frac{U_*}{\kappa} \ln(Ez_P^+) \tag{6.113}$$

where E is the roughness parameter, defined in Eq. (6.13); U_* is the bed shear velocity, related to the cross-section-averaged velocity $\widehat{U}$ by $U_* = \sqrt{g}\,\widehat{U}/C_h$; and $z_P^+ = U_* z_P'/\nu$, with z_P' being the height above the bed.

At the outlet, a time series of water stage, a stage-discharge rating curve, or a non-reflective wave condition is applied. The U_x-velocity at the outlet points is usually extrapolated or copied from adjacent internal points. Vasiliev (2002) suggested that the vertical velocity component changes linearly from $U_z = U_x \partial z_b / \partial x$ at the bottom to $U_z = \partial z_s / \partial t$ at the surface:

$$U_z = \frac{\partial z_s}{\partial t}\frac{z - z_b}{h} + U_x\frac{\partial z_b}{\partial x}\frac{z_s - z}{h} \tag{6.114}$$

6.4.1.3 Numerical solutions

Hydrostatic pressure model

For gradually varied open-channel flows, the hydrostatic pressure assumption is often adopted (Blumberg, 1977; Wu, 1992; Edinger *et al.*, 1994; Li *et al.*, 1994; Vasiliev, 2002). Under this assumption, the z-momentum equation (6.105) is simplified to Eq. (2.64), and the x-momentum equation (6.104) takes the following form:

$$\frac{\partial(bU_x)}{\partial t} + \frac{\partial(bU_x^2)}{\partial x} + \frac{\partial(bU_zU_x)}{\partial z} = -gb\frac{\partial z_s}{\partial x} + \frac{\partial}{\partial x}\left(b\nu_t\frac{\partial U_x}{\partial x}\right) + \frac{\partial}{\partial z}\left(b\nu_t\frac{\partial U_x}{\partial z}\right)$$
$$- \frac{1}{\rho}(m_1\tau_{x1} + m_2\tau_{x2}) \tag{6.115}$$

To overcome the difficulty of handling the free surface, the stretching coordinate transformation (4.91) is applied, under which Eqs. (6.103) and (6.115) are converted to (Wu, 1992)

$$\left(b + \zeta\frac{\partial b}{\partial \zeta}\right)\frac{\partial h}{\partial \tau} + \frac{\partial(bh\widehat{U}_\xi)}{\partial \xi} + h\frac{\partial(b\widehat{U}_\zeta)}{\partial \zeta} = 0 \tag{6.116}$$

$$\frac{\partial(bU_x)}{\partial \tau} + \widehat{U}_\xi\frac{\partial(bU_x)}{\partial \xi} + \widehat{U}_\zeta\frac{\partial(bU_x)}{\partial \zeta} = -gb\frac{\partial z_s}{\partial \xi} + \frac{\partial}{\partial \xi}\left(b\nu_t\frac{\partial U_x}{\partial \xi}\right)$$
$$+ \frac{H^2}{h^2}\frac{\partial}{\partial \zeta}\left(b\nu_t\frac{\partial U_x}{\partial \zeta}\right) - \frac{1}{\rho}(m_1\tau_{x1} + m_2\tau_{x2}) \tag{6.117}$$

where $\widehat{U}_\xi = U_x$, and $\widehat{U}_\zeta = \frac{H}{b}U_z - \frac{\zeta}{b}\left(\frac{\partial b}{\partial t} + U_x\frac{\partial b}{\partial x}\right) - \frac{H}{b}U_x\frac{\partial z_b}{\partial x}$. Because only gradually varied flows are considered here, several high-order terms related to $\partial b/\partial x$ and $\partial z_b/\partial x$ are neglected in Eq. (6.117) in order to simplify the problem.

Under the coordinate transformation (4.91), the computational domain is turned to a fixed rectangle with a constant height of H, and the kinematic condition (6.111) at the free surface is converted to $\widehat{U}_\zeta = 0$. Therefore, many numerical methods may be used to solve the transformed governing equations (6.116) and (6.117) in the (ξ, ζ) coordinate system. These methods may be based on a staggered grid (Blumberg, 1977) or non-staggered grid (Wu, 1992). The staggered grid is used here as an example.

As shown in Fig. 6.21, scalar quantities are defined at the cell centers, while velocities are arranged in a staggered pattern so that the vertical velocity is defined on the top and bottom faces of each grid cell and the horizontal velocity is on the two sides. The water level, which varies only longitudinally, is defined at the cell center for each vertical line. On this staggered grid, the continuity equation (6.116) is discretized as

$$\left(b_{i,j}^{n+1} + \frac{m_{i,j}^{n+1}b_i^{n+1}\zeta_j}{H}\right)\frac{b_i^{n+1} - b_i^n}{\Delta\tau} + \frac{\left(bb\widehat{U}_\xi\right)_{i+1/2,j}^{n+1} - (bb\widehat{U}_\xi)_{i-1/2,j}^{n+1}}{\Delta\xi}$$

$$+ b_i^{n+1}\frac{(b\widehat{U}_\zeta)_{i,j+1/2}^{n+1} - (b\widehat{U}_\zeta)_{i,j-1/2}^{n+1}}{\Delta\zeta} = 0 \qquad (6.118)$$

where $m = \partial b/\partial z$.

Eq. (6.118) is actually used to determine the vertical velocity. To close the problem, the flow depth needs to be calculated using the free-surface kinematic condition (6.111) or the 1-D continuity equation. Both methods were compared by Wu (1992), and the 1-D continuity equation was found to be more robust.

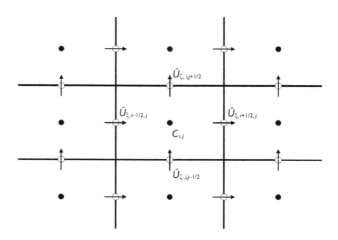

Figure 6.21 Staggered grid in vertical 2-D model.

Non-hydrostatic pressure model

The governing equations in the non-hydrostatic pressure model are Eqs. (6.103)–(6.105). Some approaches used for solving the 2-D Navier-Stokes equations in Section 4.4 can be applied here. For example, Karpik and Raithby (1990) proposed the SIMPLED algorithm, which is a modification of the SIMPLE algorithm on the staggered grid, to solve this set of equations. In analogy to the depth-averaged 2-D model, the width-averaged 2-D model can also be solved using the SIMPLE(C) algorithms on the non-staggered grid. The details are not presented here, because the formulations in both 2-D models are similar.

The stream function and vorticity method can be extended to the width-averaged 2-D model by defining the stream function ψ corresponding to the continuity equation (6.103) as

$$U_x = \frac{1}{b}\frac{\partial \psi}{\partial z}, \quad U_z = -\frac{1}{b}\frac{\partial \psi}{\partial x} \tag{6.119}$$

and the vorticity as

$$\Omega = \frac{\partial U_z}{\partial x} - \frac{\partial U_x}{\partial z} \tag{6.120}$$

Therefore, the following equation for stream function is obtained by inserting Eq. (6.119) into Eq. (6.120):

$$\frac{\partial^2 \psi}{\partial x^2} + \frac{\partial^2 \psi}{\partial z^2} - \frac{1}{b}\frac{\partial b}{\partial x}\frac{\partial \psi}{\partial x} - \frac{1}{b}\frac{\partial b}{\partial z}\frac{\partial \psi}{\partial z} = -b\Omega \tag{6.121}$$

Cross-differentiating Eqs. (6.104) and (6.105) with respect to z and x and subtracting them yields the vorticity equation:

$$\frac{\partial \Omega}{\partial t} + \frac{\partial (U_x \Omega)}{\partial x} + \frac{\partial (U_z \Omega)}{\partial z} = \frac{\partial}{\partial x}\left[(v + v_t)\frac{\partial \Omega}{\partial x}\right] + \frac{\partial}{\partial z}\left[(v + v_t)\frac{\partial \Omega}{\partial z}\right] + S_\Omega \tag{6.122}$$

where S_Ω includes all the remaining terms.

Eqs. (6.121) and (6.122) constitute the governing equations of the width-averaged 2-D stream function and vorticity model. Unlike the depth-averaged 2-D model, the width-averaged 2-D stream function and vorticity model is applicable to both steady and unsteady flows.

A special task in the width-averaged 2-D model is handling the free surface. The techniques introduced in Section 7.1 for the 3-D model can be used here.

6.4.2 Width-averaged 2-D sediment transport model

In the case of low sediment concentration, applying the width-averaged 2-D transport equation (2.99) to the non-uniform suspended load yields

$$\frac{\partial(bC_k)}{\partial t} + \frac{\partial(bU_x C_k)}{\partial x} + \frac{\partial(bU_z C_k)}{\partial z} - \frac{\partial(b\omega_{sk}C_k)}{\partial z}$$
$$= \frac{\partial}{\partial x}\left(bE'_{s,x}\frac{\partial C_k}{\partial x}\right) + \frac{\partial}{\partial z}\left(bE'_{s,z}\frac{\partial C_k}{\partial z}\right) + S_{ck} \quad (k=1,2,\ldots,N) \quad (6.123)$$

where C_k is the width-averaged concentration of the kth size class of suspended load; and $E'_{s,x}$ and $E'_{s,y}$ are the effective diffusion (mixing) coefficients of sediment in the longitudinal and vertical directions.

At the water surface, the net vertical sediment flux is zero. At the interface between the bed-load and suspended-load layers, the deposition rate is $D_{bk} = \omega_{sk}C_{bk}$ and the entrainment rate E_{bk} is

$$E_{bk} = -E'_{s,z}\frac{\partial C_k}{\partial z}\bigg|_{z=z_b+\delta} = \omega_{sk}C_{b*k} \quad (6.124)$$

where C_{b*k} is the equilibrium concentration at the interface (reference level) determined using one of the existing formulas introduced in Section 3.5.2.

In the width-averaged 2-D model, the bed-load transport is in fact a one-dimensional process, which is determined by Eq. (5.28).

Extending Eq. (2.152) to the width-averaged 2-D model yields the bed change equation:

$$(1-p'_m)\left(\frac{\partial A_b}{\partial t}\right)_k = B_b(D_{bk} - E_{bk}) + \frac{1}{L}(Q_{bk} - Q_{b*k}) \quad (6.125)$$

where B_b is the channel width at the interface between the bed-load and suspended-load zones.

Bed material sorting is simulated using the same multiple-layer model used in the 1-D model. For example, the bed-material gradation of the mixing layer is determined by Eq. (5.32).

For a well-posed solution to the above set of equations, the vertical distribution of the suspended-load concentration and the bed-load transport rate for each size class have to be specified at the inflow boundary, and the suspended-load concentration gradient in the flow direction is given zero at the outflow boundary.

The width-averaged suspended-load transport equation (6.123) is a typical convection-diffusion equation, which can easily be solved by applying the numerical methods introduced in Sections 4.2 and 4.3. The near-bed sediment exchange in the width-averaged 2-D model is similar to that in the 3-D model. The method for handling it is discussed in Section 7.3.

Note that the width-averaged 2-D flow and sediment transport models described above have been applied by Blumberg (1977), Karpik and Raithby (1990), Wu (1992),

Edinger *et al.* (1994), Li et al. (1994), Vasiliev (2002), and others in numerous studies. Though the relevance of such 2-D models to engineering practice becomes less and less important as 3-D models are more and more popular, knowing how to develop and use them would help modelers and engineers with more choices. Interested readers may find application examples in the relevant references.

Chapter 7

3-D numerical models

Flows in curved and braided channels and near in-stream structures usually exhibit complex three-dimensional features that significantly affect sediment transport and morphological evolution processes. Realistically simulating these complex near-field phenomena must rely on 3-D models rather than 1-D or 2-D models. Introduced in this chapter are 3-D modeling approaches for open-channel flows, sediment transport in general situations, local scour near in-stream structures, and headcut migration.

7.1 FULL 3-D HYDRODYNAMIC MODEL

7.1.1 Governing equations

As described in Section 2.2.3, the influence of sediment transport on the flow field is assumed to be negligible in the case of low sediment concentration. However, the bed sediment affects the flow by forming bed roughness elements, such as particles, ripples, and dunes. This will be accounted for through bed boundary conditions in the 3-D model. Therefore, the flow field is determined by the Reynolds-averaged continuity and Navier-Stokes equations (2.42) and (2.43), which are written in the Cartesian coordinate system shown in Fig. 2.6 as follows:

$$\frac{\partial u_i}{\partial x_i} = 0 \tag{7.1}$$

$$\frac{\partial u_i}{\partial t} + \frac{\partial (u_i u_j)}{\partial x_j} = F_i - \frac{1}{\rho}\frac{\partial p}{\partial x_i} + \frac{1}{\rho}\frac{\partial \tau_{ij}}{\partial x_j} \tag{7.2}$$

where u_i ($i = 1, 2, 3$) are the components of mean flow velocity; F_i are the components of external forces, such as gravity and Coriolis force, per unit volume; p is the mean pressure; and τ_{ij} are the stresses, including both viscous and turbulent effects. If Boussinesq's eddy viscosity concept is adopted, the stresses are determined by

$$\tau_{ij} = \rho(\nu + \nu_t)\left(\frac{\partial u_i}{\partial x_j} + \frac{\partial u_j}{\partial x_i}\right) - \frac{2}{3}\rho k \delta_{ij} \tag{7.3}$$

The eddy viscosity ν_t can be determined by the parabolic model, mixing length model, or linear (standard, non-equilibrium, and RNG) k-ε turbulence models.

The parabolic model uses Eq. (2.49). The 3-D mixing length model is

$$v_t = l_m^2 |S| \tag{7.4}$$

where $|\bar{S}| = [(\partial u_i/\partial x_j + \partial u_j/\partial x_i)\partial u_i/\partial x_j]^{1/2}$, and l_m is the mixing length described in Section 2.3.2.

In the linear k-ε turbulence models, the eddy viscosity is calculated by Eq. (2.54), and the turbulent energy k and its dissipation rate ε are determined by Eqs. (2.55) and (2.56). The turbulent stresses can also be determined using the nonlinear k-ε turbulence model, algebraic Reynolds stress model, Reynolds stress model, etc. The details and relevant references can be found in Section 2.3.

7.1.2 Boundary conditions

Water surface

In early developed full 3-D hydrodynamic models, the water surface is treated as a rigid lid; thus, the computational domain is fixed and the problem is simplified. On the rigid lid, the normal velocity must be set to zero, and the pressure is no longer atmospheric. This rigid lid approach encounters difficulties in the case of long river reach under unsteady flow conditions where the water surface varies in time and space. Therefore, in several recently developed full 3-D models (e.g., Wu *et al.*, 2000a; Jia *et al.*, 2001), the variation of water surface is simulated as part of the solution. At the water surface, the pressure is given the atmospheric value, and the free-surface kinematic condition (2.71) is applied.

The free surface approach is physically more reasonable than the rigid lid approach. However, the free surface approach requires more computational effort because the user must solve a movable boundary problem.

When wind shear appears, the wind driving force is determined using Eq. (6.5). Note that the wind driving force is added as a source term in the depth-averaged 2-D model, whereas it is treated as a boundary condition in the 3-D model. In analogy to Eq. (6.112), a flow velocity gradient that forms a shear stress equating to the wind driving force is applied near the water surface.

In the absence of wind shear, the net normal fluxes of horizontal momentum and turbulent kinetic energy at the water surface are set to zero, and the dissipation rate ε can be calculated using the relation given by Rodi (1993):

$$\varepsilon = \frac{k^{3/2}}{0.43h} \tag{7.5}$$

River bed and banks

On the river bed, banks, and other solid boundaries, the wall-function approach described in Section 6.1.2 is applied. In particular, the movable bed roughness is quantified by the equivalent roughness height k_s. For a stationary flat bed, k_s is usually set to the median diameter d_{50} of bed material, but in practice, higher values are also adopted, e.g., $3d_{90}$ by van Rijn (1984c). For a sand-wave bed, k_s should be related to the height of bed forms and can be determined using an empirical formula, such as Eq. (3.58) proposed by van Rijn. However, the effects of large-scale bed forms, such as point bars and even sand dunes, can be simulated using fine grids with a 3-D model. This implies that k_s should consider only the roughness elements that are somewhat smaller than the grid spacing. In addition, the near-wall values of turbulent energy k and dissipation rate ε are given by Eq. (6.15).

Inflow and outflow boundary conditions

Flow conditions at the inlet can be either the flow discharge or a detailed 3-D distribution of flow velocity. For a given discharge at the inlet, the cross-stream distribution of the depth-averaged velocity can be determined using Eq. (6.17), and then, the vertical distribution of local flow velocity can be specified according to the logarithmic or power law.

The inflow direction should also be specified, which essentially determines three velocity components at each point of the inlet.

If the inlet is located in a nearly straight reach with simple geometry and far from hydraulic structures, the turbulent energy and its dissipation rate can be determined using the relations of Nezu and Nakagawa (1993):

$$k_{in} = 4.78 U_*^2 e^{-2z'/h}, \quad \varepsilon_{in} = E_1 \frac{U_*^3}{h} \left(\frac{z'}{h}\right)^{-1/2} e^{-3z'/h} \tag{7.6}$$

where z' is the vertical coordinate above the bed, and E_1 is a coefficient related to the Reynolds number. At moderate Reynolds numbers of 10^4 to 10^5, E_1 is approximately equal to 9.8.

The specification of outflow boundary conditions in the 3-D model is similar to that in 1-D and 2-D models. If the flow is subcritical, the water level is required at the outlet. The gradients of flow velocity, turbulent energy, and dissipation rate in the streamwise direction can be set to zero at an outlet located in a reach with simple geometry and far from hydraulic structures.

7.1.3 Numerical solutions

The 2-D MAC, projection, and SIMPLE algorithms described in Section 4.4 can be easily extended to solve the full 3-D Navier-Stokes equations (7.1) and (7.2). However, for open-channel flows, special care has to be taken in handling the free surface. A number of techniques have been used to solve this moving boundary problem. They may be grouped under two main categories: surface tracking and volume tracking (Shyy *et al.*, 1996). A surface tracking method usually adopts a moving (adaptive) grid in which at least one grid line is along the free surface so that the surface shape is exactly simulated. Examples of the surface tracking method are given in Sections 7.1.3.2 and 7.1.3.3. A volume tracking method usually uses a fixed grid and defines the shape and location of the free surface through the volume of fluid at each grid cell. Examples include the MAC, volume-of-fluid (VOF), and level set methods. The details on the level set method can be found in Osher and Sethian (1988), while the MAC and VOF methods are introduced below.

7.1.3.1   MAC and VOF methods

The general 2-D MAC method of Harlow and Welch (1965) introduced in Section 4.4.1 can be easily extended to the 3-D case. The numerical discretization and calculation procedure are not repeated here. The technique of handling the free surface, which is not included in Section 4.4.1, is described below.

The MAC method adopts a fixed, Eulerian grid, which covers the fluid (water) area and surrounding void (air) area. The location of fluid within the grid is determined by a set of marker particles that move with the fluid, as shown in Fig. 7.1. Grid cells containing markers are considered occupied by fluid, while those without markers are void. A free surface is defined to exist in any grid cell that contains markers and has at least one neighboring grid cell that is void. Evolution of the free surface is calculated by moving the markers with locally interpolated flow velocities. At the free surface, the air pressure is assigned to all surface cells, and velocity components are assigned on or immediately outside the surface to satisfy the conditions of incompressibility and zero shear stress.

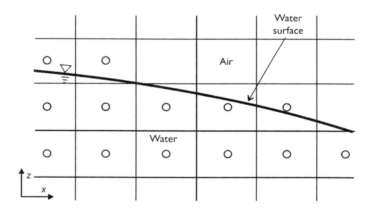

Figure 7.1 Computational grid in MAC method.

The MAC method can successfully handle the breakup and coalescence of fluid masses. The reason is that the markers track fluid volumes rather than surfaces directly. Surfaces are simply the boundaries of the volumes so that surfaces may appear, merge, or disappear as volumes break apart or coalesce. However, the MAC method has been used primarily for 2-D models because it requires considerable memory and CPU time to accommodate the necessary number of marker particles. Typically, an average of 16 markers in each grid cell is needed to insure an accurate tracking of surfaces undergoing large deformations. In addition, this method is inefficient in regions involving converging and diverging flows with stagnation points.

To avoid the disadvantages of the MAC method, Hirt and Nichols (1981) proposed the VOF method, which also employs volume tracking. The VOF method adopts the

MAC numerical discretization and solution procedure, but uses a continuous function, the fluid volume fraction f, instead of discrete marker particles to identify the domain of fluid (Fig. 7.2). The values of f are set to 1 and 0 for the cells with water and air, respectively. The interfacial cells are then identified as those with fractional values of f. The volume fraction f is advected with the local flow velocity and governed by the following kinematic equation:

$$\frac{\partial f}{\partial t} + u_x \frac{\partial f}{\partial x} + u_y \frac{\partial f}{\partial y} + u_z \frac{\partial f}{\partial z} = 0 \tag{7.7}$$

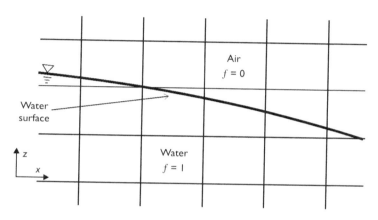

Figure 7.2 Computational grid in VOF method.

The VOF method has the advantages of the MAC method, but uses less memory and CPU time. It can also handle very complicated interfacial phenomena, such as droplet and wave breaking. Furthermore, because this method uses a continuous function, it does not suffer from the lack of divisibility that discrete markers exhibit.

However, the VOF method has a disadvantage in that it cannot precisely treat the water surface at which the distribution of volume fraction f has sharp, step-function-like features. A straightforward numerical approximation cannot be used to solve Eq. (7.7) because numerical diffusion and dispersion errors may destroy this sharp distribution. Special care needs to be taken to recover the surface shape. Grid refinements are sometimes needed along the water surface.

7.1.3.2 SIMPLE algorithm

Numerical discretization

Since the location of water surface is part of the solution and can in general also change with time, a moving, curvilinear grid that adjusts to the changing free surface is used. Eqs. (7.1), (7.2), (2.55), and (2.56) can be written as Eq. (4.152) in the moving, curvilinear coordinate system, with ϕ standing for 1, u_i, k, and ε depending on the equation considered and ξ_m ($m = 1, 2, 3$) being the curvilinear coordinates.

As discussed in Section 4.4.4, the non-staggered grid is more convenient than the staggered grid in the 3-D model for flows in channels with complex geometry. The SIMPLE algorithm on the non-staggered grid used by Peric (1985) and Majumdar *et al.* (1992) for general flows is herein applied to acquire the pressure-velocity coupling in the solution of open-channel flows (Wu *et al.*, 2000a). The 2-D version of this SIMPLE algorithm on fixed grid has been described in Section 4.4.4, and its 3-D extension on moving grid is described below.

The discretized equations are obtained by integrating (4.152) over the control volume shown in Fig. 4.22. The resulting algebraic equations for velocities $u_{i,P}^{n+1}$ ($i = 1$, 2, 3) are

$$
u_{i,P}^{n+1} = \frac{1}{a_P^u} \left(\sum_{l=W,E,S,N,B,T} a_l^u u_{i,l}^{n+1} + S_{ui} \right) + D_i^1(p_w^{n+1} - p_e^{n+1})
$$
$$
+ D_i^2(p_s^{n+1} - p_n^{n+1}) + D_i^3(p_b^{n+1} - p_t^{n+1}) \tag{7.8}
$$

where $D_i^1 = (J\alpha_i^1 \Delta\eta\Delta\zeta)_P/a_P^u$, $D_i^2 = (J\alpha_i^2\Delta\xi\Delta\zeta)_P/a_P^u$, and $D_i^3 = (J\alpha_i^3\Delta\xi\Delta\eta)_P/a_P^u$.

The pressure correction is defined in Eq. (4.191). In analogy to Eq. (4.190), the velocity correction at cell center P in the 3-D case is:

$$
u_{i,P}^{n+1} = u_{i,P}^* + \alpha_u[D_i^1(p_w' - p_e') + D_i^2(p_s' - p_n') + D_i^3(p_b' - p_t')] \tag{7.9}
$$

Using Rhie and Chow's (1983) momentum interpolation technique and the flux definition in Eq. (4.144) yields the following relations for flux corrections at cell faces:

$$
F_w = F_w^* + a_W^p(p_W' - p_P') \tag{7.10}
$$
$$
F_s = F_s^* + a_S^p(p_S' - p_P') \tag{7.11}
$$
$$
F_b = F_b^* + a_B^p(p_B' - p_P') \tag{7.12}
$$

where F_w^*, F_s^*, and F_b^* are the fluxes in terms of the approximate velocities $u_{i,w}^*$, $u_{i,s}^*$, and $u_{i,b}^*$; $a_W^p = \alpha_u\rho_w^{n+1}(J\alpha_i^1\Delta\eta\Delta\zeta)_w Q_{i,w}^1$, $a_S^p = \alpha_u\rho_s^{n+1}(J\alpha_i^2\Delta\xi\Delta\zeta)_s Q_{i,s}^2$, and $a_B^p = \alpha_u \rho_b^{n+1}(J\alpha_i^3\Delta\xi\Delta\eta)_b Q_{i,b}^3$ with $Q_{i,w}^1 = [(1-f_{x,P})/a_{PW}^u + f_{x,P}/a_P^u](J\alpha_i^1\Delta\eta\Delta\zeta)_w$, $Q_{i,s}^2 = [(1-f_{y,P})/a_{PS}^u + f_{y,P}/a_P^u](J\alpha_i^2\Delta\xi\Delta\zeta)_s$, and $Q_{i,b}^3 = [(1 - f_{z,P})/a_{PB}^u + f_{z,P}/a_P^u](J\alpha_i^3\Delta\xi\Delta\eta)_b$.

Integrating the 3-D continuity equation over the control volume and discretizing the time-derivative term by the backward difference scheme yields

$$
\frac{\rho_P^{n+1}\Delta V_P^{n+1} - \rho_P^n\Delta V_P^n}{\Delta\tau} + F_e - F_w + F_n - F_s + F_t - F_b = 0 \tag{7.13}
$$

where $\Delta\tau$ is the time step in the moving, curvilinear grid system.

Note that the variation of flow density is considered in Eq. (7.13) for more general applications where its effect needs to be considered (see Section 12.1).

Substituting Eqs. (7.10)–(7.12) and three similar equations for F_e, F_n, and F_t into the discretized continuity equation (7.13) results in the following equation for pressure correction:

$$a^p_P p'_P = a^p_W p'_W + a^p_E p'_E + a^p_S p'_S + a^p_N p'_N + a^p_B p'_B + a^p_T p'_T + S_p \qquad (7.14)$$

where $a^p_P = \sum_{l=W,E,S,N,B,T} a^p_l$, and $S_p = -(\rho^{n+1}_P \Delta V^{n+1}_P - \rho^n_P \Delta V^n_P)/\Delta\tau - (F^*_e - F^*_w + F^*_n - F^*_s + F^*_t - F^*_b)$.

Water level calculation

The water level can be calculated using the free-surface kinematic condition (2.71) or the 2-D depth-integrated continuity equation (6.1). It can also be determined using the 2-D Poisson equation of water level proposed by Wu et al. (2000a), which is expressed as Eq. (6.47) with the term $-\partial(\partial U_x/\partial x + \partial U_y/\partial y)/\partial t$ added into the source term S_z in the case of unsteady flows. However, all the depth-averaged velocities and stresses appearing on the right-hand side of Eq. (6.47) are determined by depth-integrating the local quantities computed from the present 3-D model.

To the author's knowledge, the Poisson equation (6.47) is more stable than the free-surface kinematic condition (2.71) and the 2-D depth-integrated continuity equation (6.1) in the calculation of water level. However, Eq. (6.47) is valid only for gradually varied open-channel flows because it is derived under the hydrostatic pressure assumption. Nevertheless, many tests have shown that it can approximately be applied to rapidly varied flows where no obvious hydraulic jump occurs.

Grid adjustment

In general, to conform to the water surface profile, the computational grid should be regenerated once the water level changes. If a boundary-fitted grid is used, the Poisson equations (4.74) need to be solved at each time step. This is relatively time-consuming. The local coordinate transformation (4.89) on moving grids can be extended to the 3-D case, which may be more efficient. For a channel with simple geometry, the grid needs to be adjusted in only one or two directions or in part of the domain, and thus, the grid can be regenerated using simple algebraic methods, such as the σ-coordinate (4.91).

Once the grid is adjusted, the parameters related to geometry should be updated.

7.1.3.3   Projection method

In analogy to the depth-averaged 2-D model in Section 6.1.3.2, Jia et al. (2001) developed a 3-D model based on the projection method. The partially staggered grid shown in Fig. 6.4 is extended to the 3-D case with the pressure stored on the base grid (cell centers) and velocities u_x, u_y, and u_z on the staggered grid (cell corners). The governing equations are discretized and solved using the same methods mentioned in Section 6.1.3.2. The pressure-correction algorithm is slightly modified below to achieve the coupling of velocity and pressure in the 3-D model.

The velocity correction is written as follows:

$$\vec{u}^{n+1} = \vec{u}^* - \frac{\Delta t}{\rho}\nabla p' \tag{7.15}$$

where p' is defined in Eq. (6.35). The intermediate velocity $\vec{u}^*$ is determined by

$$\vec{u}^* = \vec{u}^n + \Delta t\,\vec{G} - \frac{\Delta t}{\rho}\nabla p^n \tag{7.16}$$

with $\vec{G}$ representing all the remaining terms in the discretized momentum equations. Substituting Eq. (7.15) into Eq. (7.1) yields

$$\frac{\Delta t}{\rho}\nabla^2 p' = \nabla \cdot \vec{u}^* \tag{7.17}$$

The water level is determined using the kinematic condition (2.71), and the grid is adjusted in the vertical direction to track the temporal variation of water surface.

7.2 3-D FLOW MODEL WITH HYDROSTATIC PRESSURE ASSUMPTION

Based on the hydrostatic pressure assumption introduced in Section 2.4, the inertia and diffusion effects in the vertical momentum equation of gradually varied (shallow water) flows can be ignored. The resulting 3-D governing equations are written as

$$\frac{\partial u_x}{\partial x} + \frac{\partial u_y}{\partial y} + \frac{\partial u_z}{\partial z} = 0 \tag{7.18}$$

$$\frac{\partial u_x}{\partial t} + \frac{\partial (u_x^2)}{\partial x} + \frac{\partial (u_y u_x)}{\partial y} + \frac{\partial (u_z u_x)}{\partial z} = -g\frac{\partial z_s}{\partial x} + \frac{1}{\rho}\frac{\partial \tau_{xx}}{\partial x} + \frac{1}{\rho}\frac{\partial \tau_{xy}}{\partial y} + \frac{1}{\rho}\frac{\partial \tau_{xz}}{\partial z} + f_c u_y \tag{7.19}$$

$$\frac{\partial u_y}{\partial t} + \frac{\partial (u_x u_y)}{\partial x} + \frac{\partial (u_y^2)}{\partial y} + \frac{\partial (u_z u_y)}{\partial z} = -g\frac{\partial z_s}{\partial y} + \frac{1}{\rho}\frac{\partial \tau_{yx}}{\partial x} + \frac{1}{\rho}\frac{\partial \tau_{yy}}{\partial y} + \frac{1}{\rho}\frac{\partial \tau_{yz}}{\partial z} - f_c u_x \tag{7.20}$$

where f_c is the Coriolis coefficient, determined by Eq. (6.6).

The hydrostatic pressure assumption simplifies the full three-dimensional hydrodynamic equations (7.1) and (7.2) significantly. However, this assumption is valid only for gradually varied flows (bottom slope less than about 5%), and a full 3-D model with dynamic (non-hydrostatic) pressure should be used for rapidly varied flows around instream structures, such as bridge piers, spur-dikes, and bendway weirs. Nevertheless, gradually varied flows exist widely in rivers, lakes, estuaries, and coastal waters; thus, the hydrostatic pressure assumption is often adopted. The 3-D models developed by Sheng (1983), Wang and Adeff (1986), Blumberg and Mellor (1987), Casulli and

Cheng (1992), Jankowski *et al.* (1994), Lin and Falconer (1996), and Shanhar *et al.* (2001) are based on this assumption.

Note that in the set of equations (7.18)–(7.20), the water level replaces the pressure in the momentum equations, but it does not appear in the continuity equation. This weak linkage between water level and velocity may produce node-to-node (checkerboard) numerical oscillations if not handled carefully. Approaches used to solve this set of equations are described below.

7.2.1   Layer-integrated model

The layer-integrated model divides the water depth into a number of layers, as shown in Fig. 7.3. The layer interfaces are usually horizontal planes with constant altitudes, and the z-coordinate directs vertically. The thicknesses of the top and bottom layers are variable to track the temporal and spatial changes in water surface and channel bed. The grid is staggered with the vertical velocity stored at layer interfaces. Integrating Eqs. (7.18)–(7.20) over each layer yields (Shanhar *et al.*, 2001)

$$u_{z,l+1/2} = u_{z,l-1/2} - \frac{\partial}{\partial x}(h_l u_x) - \frac{\partial}{\partial y}(h_l u_y) \tag{7.21}$$

$$\frac{\partial}{\partial t}(h_l u_x) + \frac{\partial}{\partial x}(h_l u_x^2) + \frac{\partial}{\partial y}(h_l u_y u_x) + (u_z u_x)_{l+1/2} - (u_z u_x)_{l-1/2} = -g h_l \frac{\partial z_s}{\partial x}$$

$$+ \frac{1}{\rho}\frac{\partial}{\partial x}(h_l \tau_{xx}) + \frac{1}{\rho}\frac{\partial}{\partial y}(h_l \tau_{xy}) + \left(\frac{\tau_{xz}}{\rho}\right)_{l+1/2} - \left(\frac{\tau_{xz}}{\rho}\right)_{l-1/2} + f_c h_l u_y \tag{7.22}$$

$$\frac{\partial}{\partial t}(h_l u_y) + \frac{\partial}{\partial x}(h_l u_x u_y) + \frac{\partial}{\partial y}(h_l u_y^2) + (u_z u_y)_{l+1/2} - (u_z u_y)_{l-1/2} = -g h_l \frac{\partial z_s}{\partial y}$$

$$+ \frac{1}{\rho}\frac{\partial}{\partial x}(h_l \tau_{yx}) + \frac{1}{\rho}\frac{\partial}{\partial y}(h_l \tau_{yy}) + \left(\frac{\tau_{yz}}{\rho}\right)_{l+1/2} - \left(\frac{\tau_{yz}}{\rho}\right)_{l-1/2} - f_c h_l u_x \tag{7.23}$$

where h_l is the thickness of the lth layer.

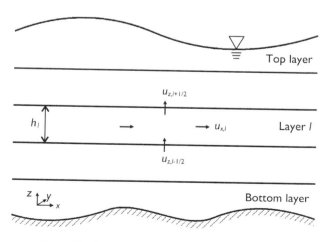

Figure 7.3 Configuration of layer-integrated model.

Reynolds stresses τ_{xx}, τ_{xy}, τ_{yx}, and τ_{yy} in Eqs. (7.22) and (7.23) are determined using the Boussinesq assumption with the eddy viscosity determined by a turbulence model. The shear stresses at layer interfaces are determined by

$$\tau_{xz} = \rho v_t \frac{\partial u_x}{\partial z}, \quad \tau_{yz} = \rho v_t \frac{\partial u_y}{\partial z} \tag{7.24}$$

The wind driving force determined by Eq. (6.5) is applied at the water surface, and the bed shear stress at the channel bottom is determined by

$$\tau_{bx} = \rho c_{fb} u_{bx} \sqrt{u_{bx}^2 + u_{by}^2}, \quad \tau_{by} = \rho c_{fb} u_{by} \sqrt{u_{bx}^2 + u_{by}^2} \tag{7.25}$$

where u_{bx} and u_{by} are the horizontal components of velocity at the bottom layer, and c_{fb} is the bottom friction coefficient.

The water level can be determined using the kinematic condition (2.71), the depth-integrated continuity equation (6.1), or the Poisson equation (6.47).

Eqs. (7.21)–(7.23) can be discretized using the finite difference method, finite element method, or finite volume method. Normally, the convection terms should be discretized using upwind schemes, and the other spatial derivatives can be discretized using the central difference schemes or other similar schemes. The time derivatives can be discretized explicitly or implicitly. Examples of this layer-integrated model can be found in Lin and Falconer (1996) and Shanhar et al. (2001).

7.2.2 Splitting of internal and external modes

Sheng (1983) and Blumberg and Mellor (1987) developed 3-D models for estuarine and coastal systems based on the hydrostatic pressure assumption. The significant feature of these models is the splitting of internal and external modes. The internal mode handles the slower vertical baroclinic flow structures, while the external mode computes the depth-integrated quantities that are governed by the fast barotropic dynamics.

In Sheng's model, the momentum equations (7.19) and (7.20) are rewritten as

$$\frac{\partial u_x}{\partial t} = -g \frac{\partial z_s}{\partial x} + \frac{\partial}{\partial z} \left(v_t \frac{\partial u_x}{\partial z} \right) + B_x \tag{7.26}$$

$$\frac{\partial u_y}{\partial t} = -g \frac{\partial z_s}{\partial y} + \frac{\partial}{\partial z} \left(v_t \frac{\partial u_y}{\partial z} \right) + B_y \tag{7.27}$$

where B_x and B_y include all the remaining terms.

As described in Section 2.4.1, vertically integrating Eqs. (7.18)–(7.20) leads to Eqs. (2.79), (2.82), and (2.83), which are the governing equations of the external mode. They are rewritten as

$$\frac{\partial h}{\partial t} + \frac{\partial q_x}{\partial x} + \frac{\partial q_y}{\partial y} = 0 \tag{7.28}$$

$$\frac{\partial q_x}{\partial t} = -gh\frac{\partial z_s}{\partial x} + D_x \tag{7.29}$$

$$\frac{\partial q_y}{\partial t} = -gh\frac{\partial z_s}{\partial y} + D_y \tag{7.30}$$

where q_x and q_y are the depth-integrated specific discharges in the x- and y-directions, and D_x and D_y include all the remaining terms in the momentum equations.

The perturbation (not necessarily infinitesimal) velocities are defined as $\hat{u}_x = u_x - q_x/h$ and $\hat{u}_y = u_y - q_y/h$. Subtracting the vertically-integrated momentum equations (7.29) and (7.30) from the 3-D momentum equations (7.26) and (7.27) yields

$$\frac{1}{h}\frac{\partial(h\hat{u}_x)}{\partial t} = B_x - \frac{D_x}{h} + \frac{\partial}{\partial z}\left[v_t\frac{\partial}{\partial z}\left(\frac{h\hat{u}_x + q_x}{h}\right)\right] \tag{7.31}$$

$$\frac{1}{h}\frac{\partial(h\hat{u}_y)}{\partial t} = B_y - \frac{D_y}{h} + \frac{\partial}{\partial z}\left[v_t\frac{\partial}{\partial z}\left(\frac{h\hat{u}_y + q_y}{h}\right)\right] \tag{7.32}$$

Eqs. (7.31) and (7.32) are the governing equations of the internal mode. They do not contain the surface-slope terms, and thus, a large time step (much larger than the limit imposed by the gravity wave propagation) may be used in the computation of the internal mode.

The vertical σ-coordinate transformation (4.91) and a horizontal stretching coordinate transformation are adopted to handle the variation of free surface and the complexity of horizontal geometry. The water level and velocity are stored on a staggered grid. In the external mode, all terms in the continuity equation (7.28) and the time-derivative and surface-slope terms in the momentum equations (7.29) and (7.30) are treated implicitly. The resulting discretized equation system is factorized in the x- and y-directions and solved consecutively by inversion of tridiagonal matrices. In the internal mode, Eqs. (7.31) and (7.32) are discretized by a two-level scheme with the vertical diffusion terms treated implicitly. The vertical implicit scheme is essential, as it increases the time step significantly. The bottom friction terms are also treated implicitly for unconditional numerical stability in shallow water. After the depth-averaged and perturbation velocities are solved in the external and internal modes, the local velocities u_x and u_y are obtained using $u_x = \hat{u}_x + q_x/h$ and $u_y = \hat{u}_y + q_y/h$. The vertical local velocity u_z is then determined using the discretized 3-D continuity equation.

In Blumberg and Mellor's model, the external mode also solves Eqs. (7.28)–(7.30) to obtain the depth-averaged quantities of tidal motions, but the internal mode directly solves Eqs. (7.26) and (7.27) to compute the local velocity rather than the perturbation velocity. Explicit schemes are used in both internal and external modes except that the vertical diffusion terms in Eq. (7.26) and (7.27) are treated implicitly. Because numerical stability is controlled by surface gravity waves in the external mode but by advection and diffusion in the internal mode, the time step in the external mode is much shorter than that in the internal mode. To enhance the computational efficiency, a special time-marching strategy is adopted. The external mode solutions are first obtained with the terms D_x and D_y in Eqs. (7.29) and (7.30) held fixed in time, and after a large number of time steps, of the order of 100, an internal mode calculation is performed. The external mode provides $\partial z_s/\partial x$ and $\partial z_s/\partial y$ for insertion into the internal mode equations, which are then solved with a much longer time step. Once the vertical structures have been determined, the terms D_x and D_y in Eqs. (7.29) and (7.30) are updated and another external mode solution begins.

In both models described above, there may be a slow tendency that the vertical integral of the internal mode velocity differs from the external mode velocity. This arises because of different truncation errors in each mode. To prevent accumulated mismatch, the vertical mean of the internal velocity is replaced at every time step by the external mode velocity.

7.2.3 Projection method

Casulli and Cheng (1992) proposed an algorithm that uses an implicit scheme in the vertical direction and a semi-implicit scheme in the horizontal directions. The grid is staggered. The momentum equations (7.19) and (7.20) are rewritten as

$$\frac{\partial u_x}{\partial t} + \text{ADVU} + g\frac{\partial z_s}{\partial x} = 0 \tag{7.33}$$

$$\frac{\partial u_y}{\partial t} + \text{ADVV} + g\frac{\partial z_s}{\partial y} = 0 \tag{7.34}$$

where ADVU and ADVV include all the remaining terms.

Time differencing of Eqs. (7.33) and (7.34) leads to

$$u_x^{n+1} = F(u_x) - g\Delta t\frac{\partial z_s^{n+1}}{\partial x} \tag{7.35}$$

$$u_y^{n+1} = F(u_y) - g\Delta t\frac{\partial z_s^{n+1}}{\partial y} \tag{7.36}$$

In the above equations, $F(u_x)$ and $F(u_y)$ represent the solutions to Eqs. (7.33) and (7.34) at time level $n+1$ where the contribution from the water surface gradient terms has been deferred.

The depth-integrated continuity equation is written as

$$\frac{\partial z_s}{\partial t} + \frac{\partial}{\partial x}\int_{z_b}^{z_s} u_x^{n+1}dz + \frac{\partial}{\partial y}\int_{z_b}^{z_s} u_y^{n+1}dz = 0 \tag{7.37}$$

Substituting the momentum equations (7.35) and (7.36) into Eq. (7.37) and ignoring the term $-g\Delta t^2(\partial h/\partial x \cdot \partial z_s^{n+1}/\partial x + \partial h/\partial y \cdot \partial z_s^{n+1}/\partial y)$ leads to the Poisson equation for the water level at time level $n+1$:

$$(1 - gh\Delta t^2\nabla^2)z_s^{n+1} = z_s^n - \Delta t\left[\frac{\partial}{\partial x}\int_{z_b}^{z_s} F(u_x)\,dz + \frac{\partial}{\partial y}\int_{z_b}^{z_s} F(u_y)dz\right] \tag{7.38}$$

Eq. (7.38) is symmetric and positive definite and can be solved by many methods.

7.2.4 SIMPLE algorithm

Eqs. (7.18)–(7.20) can be written as Eq. (4.152) and discretized using the finite volume method presented in Section 7.1.3.2. Discretizing the momentum equations (7.19) and (7.20) leads to the following equation for horizontal velocities $u_{i,P}^{n+1}$ $(i = 1, 2)$:

$$u_{i,P}^{n+1} = \frac{1}{a_P^u} \left(\sum_{l=W,E,S,N,B,T} a_l^u u_{i,l}^{n+1} + S_{ui} \right) + D_i^1(p_w^{n+1} - p_e^{n+1}) + D_i^2(p_s^{n+1} - p_n^{n+1})$$

(7.39)

where $D_i^1 = (J\alpha_i^1 \Delta\eta\Delta\zeta)_P/a_P^u$, $D_i^2 = (J\alpha_i^2 \Delta\xi\Delta\zeta)_P/a_P^u$, and $p = \rho g z_s$.

In analogy to Eq. (7.9), the velocity correction at cell center P is as follows:

$$u_{i,P}^{n+1} = u_{i,P}^* + \alpha_u[D_i^1(p_w' - p_e') + D_i^2(p_s' - p_n')]$$

(7.40)

Suppose that the ζ–grid lines are along the vertical direction and cell faces w and s are on vertical planes. Using Rhie and Chow's momentum interpolation technique and the flux definition in Eq. (4.144) yields the horizontal flux corrections at cell faces w and s, as described in Eqs. (7.10) and (7.11).

Integrating the 3-D continuity equation over each control volume leads to Eq. (7.13). Summing it over all control volumes along a vertical grid line and using the boundary conditions at the water surface and channel bed yields

$$\frac{p_P^{n+1} - p_P^n}{g\Delta\tau} \Delta A_P + \sum_{l=1}^{K} F_{e,l} - \sum_{l=1}^{K} F_{w,l} + \sum_{l=1}^{K} F_{n,l} - \sum_{l=1}^{K} F_{s,l} = 0$$

(7.41)

where K is the total number of control volumes at the vertical grid line, the subscript l is the control volume index in the vertical direction, and ΔA_P is the area of the 2-D control volume that is obtained by projecting the 3-D control volume onto the horizontal plane.

Substituting Eqs. (7.10) and (7.11) into Eq. (7.41) leads to the following equation for pressure correction:

$$b_P^p p_P' = b_W^p p_W' + b_E^p p_E' + b_S^p p_S' + b_N^p p_N' + S_p$$

(7.42)

where $b_W^p = \sum_{l=1}^{K} a_{W,l}^p$, $b_E^p = \sum_{l=1}^{K} a_{E,l}^p$, $b_S^p = \sum_{l=1}^{K} a_{S,l}^p$, $b_N^p = \sum_{l=1}^{K} a_{N,l}^p$, $b_P^p = b_W^p + b_E^p + b_S^p + b_E^p + \Delta A_P/(g\Delta\tau)$, and

$$S_p = -(p_P^* - p_P^n)\Delta A_P/(g\Delta\tau) - \left(\sum_{l=1}^{K} F_{e,l}^* - \sum_{l=1}^{K} F_{w,l}^* + \sum_{l=1}^{K} F_{n,l}^* - \sum_{l=1}^{K} F_{s,l}^* \right).$$

Once the pressure correction is calculated using Eq. (7.42), the horizontal fluxes at cell faces w and s are corrected using Eqs. (7.10) and (7.11). Because the vertical flux at the bottom face of the first cell closest to the bed is zero, the vertical fluxes at the top faces of all cells along each vertical grid line are determined using Eq. (7.13) by sweeping from 1 to K. At cell center P, the horizontal velocities are computed using Eq. (7.39), and the vertical velocity can be calculated from the known vertical fluxes at faces b and t.

7.3 3-D SEDIMENT TRANSPORT MODEL

7.3.1 Governing equations and boundary conditions

Lin and Falconer (1996), Olsen and Kjellesvig (1998), and Fang and Wang (2000) developed 3-D suspended-load transport models, whereas Wang and Adeff (1986), van Rijn (1987), Spasojevic and Holly (1993), Wu *et al.* (2000a), and Olsen (2003) established 3-D total-load transport models. In general, a total-load transport model divides the moving sediment into suspended load and bed load, hence the solution domain into the bed-load layer with a thickness of δ and the suspended-load layer with a thickness of $h - \delta$, as shown in Fig. 2.6. Applying Eq. (2.72) to non-uniform suspended-load transport yields

$$\frac{\partial c_k}{\partial t} + \frac{\partial[(u_j - \omega_{sk}\delta_{j3})c_k]}{\partial x_j} = \frac{\partial}{\partial x_j}\left(\varepsilon_s \frac{\partial c_k}{\partial x_j}\right) \quad (k = 1, 2, \dots, N) \tag{7.43}$$

where c_k is the local concentration of the kth size class of suspended load, and ε_s is the turbulent diffusivity of sediment.

At the water surface, condition (2.73) is written as

$$\left(\varepsilon_s \frac{\partial c_k}{\partial z} + \omega_{sk}c_k\right)_{z=z_s} = 0 \tag{7.44}$$

A number of researchers (Wang and Adeff, 1986; Olsen and Kjellesvig, 1998; Olsen, 2003) have used the "concentration" boundary condition (2.74) at the lower limit of the suspended-load layer for solving Eq. (7.43). This implies that the near-bed concentration is typically defined as the equilibrium concentration and evaluated using one of the empirical formulas introduced in Section 3.5.2. However, Celik and Rodi (1988) suggested that this treatment is inadequate under non-equilibrium conditions. In general, the "gradient" boundary condition (2.75) should be used at the interface between the bed-load and suspended-load layers. Van Rijn (1987), Lin and Falconer (1996), and Wu *et al.* (2000a) determined the deposition rate as $D_{bk} = \omega_{sk}c_{bk}$ and the entrainment rate as

$$E_{bk} = -\left.\varepsilon_s \frac{\partial c_k}{\partial z}\right|_{z=z_b+\delta} = \omega_{sk}c_{b*k} \tag{7.45}$$

where c_{b*k} is the equilibrium suspended-load concentration at the interface. Therefore, the net sediment flux $D_{bk} - E_{bk} = \omega_{sk}(c_{bk} - c_{b*k})$ is prescribed at the interface.

The bed load is simulated using the equilibrium transport model (Wang and Adeff, 1986; van Rijn, 1987; Spasojevic and Holly, 1993; Olsen, 2003) or the non-equilibrium transport model (Wu *et al.*, 2000a). As described in Section 2.6, the non-equilibrium transport model is more adequate. Because the bed-load layer is very thin, the bed-load transport equation in the 3-D model has the same formulation as the horizontal 2-D model equation (2.158):

$$\frac{\partial(q_{bk}/u_{bk})}{\partial t} + \frac{\partial(\alpha_{bx}q_{bk})}{\partial x} + \frac{\partial(\alpha_{by}q_{bk})}{\partial y} = \frac{1}{L}(q_{b*k} - q_{bk}) \tag{7.46}$$

with slight difference in determining the direction cosines α_{bx} and α_{by} of bed-load transport. Because secondary flows, such as the helical flow in curved channels, can be simulated somewhat, their effects on sediment transport are automatically taken into account in the 3-D model when α_{bx} and α_{by} are set as the direction cosines of the calculated bed shear stress. However, if the bed slope is steep, the effect of gravity should be considered by adjusting α_{bx} and α_{by} using the methods introduced in Section 6.3.4.

The bed change is determined by

$$(1 - p'_m)\left(\frac{\partial z_b}{\partial t}\right)_k = D_{bk} - E_{bk} + \frac{1}{L}(q_{bk} - q_{b*k}) \tag{7.47}$$

or by the overall sediment balance equation:

$$(1 - p'_m)\left(\frac{\partial z_b}{\partial t}\right)_k + \frac{\partial}{\partial t}\left(\frac{q_{bk}}{u_{bk}} + \int_{z_b+\delta}^{z_s} c_k dz\right) + \frac{\partial q_{tkx}}{\partial x} + \frac{\partial q_{tky}}{\partial y} = 0 \tag{7.48}$$

where q_{tkx} and q_{tky} are the specific fluxes of total load in the x- and y-directions:

$$q_{tkx} = \alpha_{bx}q_{bk} + \int_{z_b+\delta}^{z_s}\left(u_x c_k - \varepsilon_s \frac{\partial c_k}{\partial x}\right)dz$$

$$q_{tky} = \alpha_{by}q_{bk} + \int_{z_b+\delta}^{z_s}\left(u_y c_k - \varepsilon_s \frac{\partial c_k}{\partial y}\right)dz \tag{7.49}$$

As compared with Eq. (7.47), Eq. (7.48) more easily ensures mass conservation but is more complex.

The equilibrium near-bed suspended-load concentration and bed-load transport rate need to be determined using the empirical relations introduced in Sections 3.4 and 3.5. In general, these formulas can be written as

$$c_{b*k} = p_{bk}c^*_{bk}, \quad q_{b*k} = p_{bk}q^*_{bk} \tag{7.50}$$

where p_{bk} is the fraction of size class k in the mixing layer of bed material, c^*_{bk} is the potential equilibrium concentration of the kth size class of suspended load at the interface, and q_{bk}^* is the potential equilibrium transport rate of the kth size class of bed load.

The bed material sorting in the 3-D model is handled with the same multiple-layer approach used in the depth-averaged 2-D model. For example, the bed-material gradation of the mixing layer is determined by Eq. (6.57).

For a well-posed solution, the bed-load transport rates and suspended-load concentrations of all size classes have to be specified at the inflow boundary. If the total discharges of bed load and suspended load are specified at the inlet, their specific discharges at each vertical grid line of the inlet can be determined using Eq. (6.63), and then, the vertical distribution of local suspended-load concentration can be determined according to the Rouse or Lane-Kalinske distribution.

The sediment boundary conditions at solid and outflow boundaries and initial conditions in the 3-D model are similar to those in the depth-averaged 2-D model.

7.3.2 Discretization of sediment transport equations

To solve the suspended-load transport equation (7.43), the sediment settling term $\partial(\omega_{sk} c_k)/\partial z$ can be treated as a source term or combined with the vertical convection term. After considerable testing, Wu $et\ al.$ (2000a) suggested the former approach might be better. This term can be evaluated using the central or forward difference scheme in the vertical direction. The central difference scheme has better accuracy, but the forward difference scheme has better stability.

Eq. (7.43) can be discretized using the numerical methods introduced in Sections 4.2 and 4.3. The finite volume method is chosen here as an example. The discretized suspended-load transport equation is

$$\frac{\Delta V_P^{n+1} c_{k,P}^{n+1} - \Delta V_P^n c_{k,P}^n}{\Delta t} = a_W c_{k,W}^{n+1} + a_E c_{k,E}^{n+1} + a_S c_{k,S}^{n+1} + a_N c_{k,N}^{n+1}$$
$$+ a_B c_{k,B}^{n+1} + a_T c_{k,T}^{n+1} - a_P c_{k,P}^{n+1} + S_{k,P} \qquad (7.51)$$

As presented by Wu $et\ al.$ (2000a), boundary conditions (7.44) and (7.45) are implemented by prescribing fluxes at the water and bed surfaces, respectively, which are depicted in Fig. 7.4. An important choice has to be made as to the reference level at which the equilibrium concentration c_{*b} and hence the entrainment rate E_b are determined. In general, the reference level is set at the top of the bed-load layer.

To determine the deposition rate D_{bk}, it is necessary to calculate the concentration c_{bk} at $z = z_b + \delta$ from c_k values at neighboring grid points. Wu $et\ al.$ (2000a) assumed that the concentration distribution between $z = z_b + \delta$ and the first grid (point 2 in Fig. 7.4) is governed by the following equation, which is simplified from Eq. (7.43) by ignoring the storage, convection, and horizontal diffusion effects:

$$\frac{\partial}{\partial z}\left(\varepsilon_s \frac{\partial c_k}{\partial z} + \omega_{sk} c_k\right) = 0 \qquad (7.52)$$

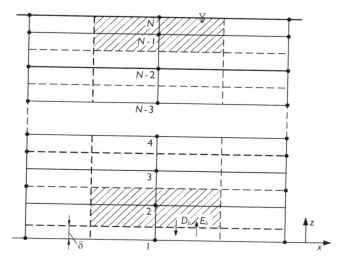

Figure 7.4 Control volumes near bed and water surfaces.

which has the following analytical solution:

$$c_k = a_1 + a_2 e^{-z\omega_{sk}/\varepsilon_s} \tag{7.53}$$

where a_1 and a_2 are two constants determined by applying conditions: $c_k = c_{2k}$ at $z = z_2$ (point 2) and $c_k = c_{bk}$ at $z = z_b + \delta$ (interface). Inserting Eq. (7.53) with the obtained a_1 and a_2 into Eq. (7.45) yields the following relation for the near-bed concentration c_{bk}:

$$c_{bk} = c_{2k} + c_{b*k}[1 - e^{-(z_2 - z_b - \delta)\omega_{sk}/\varepsilon_s}] \tag{7.54}$$

If $z_2 - z_b - \delta$ is small, Eq. (7.54) may be approximated with the following linear relation:

$$c_{bk} = c_{2k} + c_{b*k}(z_2 - z_b - \delta)\frac{\omega_{sk}}{\varepsilon_s} \tag{7.55}$$

The bed-load transport equation (7.46) is a 2-D partial differential equation. It is discretized by integrating over the horizontal 2-D control volume shown in Fig. 4.21 with the values of q_b at cell faces given by a first-order or higher-order upwind scheme. Note that the 2-D control volume is obtained by projecting the 3-D control volume onto the horizontal plane, as described in Section 7.2.4. The discretized bed-load transport equation is

$$\frac{\Delta A_P}{\Delta t}\left(\frac{q_{bk,P}^{n+1}}{u_{bk,P}^{n+1}} - \frac{q_{bk,P}^{n}}{u_{bk,P}^{n}}\right) = b_W q_{bk,W}^{n+1} + b_E q_{bk,E}^{n+1} + b_S q_{bk,S}^{n+1} + b_N q_{bk,N}^{n+1}$$
$$- b_P q_{bk,P}^{n+1} + S_{bk,P} \tag{7.56}$$

where ΔA_P is the area of the horizontal 2-D control volume.

The discretized equations (7.51) and (7.56) can be solved using the ADI or SIP method described in Section 4.5.

It is important to note that as shown in Fig. 7.4, the first control volume near the bed extends only to $z = z_b + \delta$ for solving the suspended-load transport equation (7.43), while it extends to the bed $(z = z_b)$ for solving the hydrodynamic equations. It is cumbersome that the near-bed control volumes for flow and suspended-load calculations are not identical. A simpler practice is to set the top face of the near-bed control volume at the lower limit of the suspended-load zone. This means that the bed-load layer occupies the entire first control volume near the bed, and the computational domain of suspended load starts from the second control volume. Moreover, the bed-load transport rate q_{bk} in Eq. (7.46) can be replaced by the sediment concentration in the bed-load layer: $c_{2k} = q_{bk}/(\delta u_2)$. Here, the subscript "2" denotes the center of the first control volume in Fig. 7.4, and u_2 is the resultant flow velocity at this cell center. This arrangement allows the bed-load equation (7.56) to be solved together with the suspended-load equation (7.51) using a 3-D iteration solver.

The bed change equation (7.47) is discretized as

$$\Delta z_{bk,P} = \frac{\Delta t}{1 - p'_m}\left[D_{bk,P}^{n+1} - E_{bk,P}^{n+1} + \frac{1}{L}(q_{bk,P}^{n+1} - q_{b*k,P}^{n+1}) \right] \tag{7.57}$$

If the overall sediment balance equation (7.48) is used to calculate the bed change, it is integrated in the horizontal 2-D control volume, and the resulting discretized equation is

$$\Delta z_{bk,P} = \frac{1}{1 - p'_m}\left(\frac{q_{bk}^n}{u_{bk}^n} - \frac{q_{bk}^{n+1}}{u_{bk}^{n+1}} + \int_{z_b+\delta}^{z_s} (c_k^n - c_k^{n+1})dz \right)$$
$$+ \frac{\Delta t}{(1 - p'_m)\Delta A_P}(\tilde{q}_{tk,w}^{n+1} - \tilde{q}_{tk,e}^{n+1} + \tilde{q}_{tk,s}^{n+1} - \tilde{q}_{tk,n}^{n+1}) \tag{7.58}$$

where $\tilde{q}_{tk,w}^{n+1}$, $\tilde{q}_{tk,e}^{n+1}$, $\tilde{q}_{tk,s}^{n+1}$, and $\tilde{q}_{tk,n}^{n+1}$ are the total-load fluxes at faces w, e, s, and n.

The total change in bed elevation is determined by

$$\Delta z_{b,P} = \sum_{k=1}^{N} \Delta z_{bk,P} \tag{7.59}$$

After the bed change is calculated, the bed elevation is updated by

$$z_{b,P}^{n+1} = z_{b,P}^n + \Delta z_{b,P} \tag{7.60}$$

or by the following Lax-type scheme (van Rijn, 1987):

$$z_{b,i,j}^{n+1} = (1 - \psi_x - \psi_y)z_{b,i,j}^n + \frac{\psi_x}{2}(z_{b,i+1,j}^n + z_{b,i-1,j}^n)$$
$$+ \frac{\psi_y}{2}(z_{b,i,j+1}^n + z_{b,i,j-1}^n) + \Delta z_{b,i,j} \tag{7.61}$$

where ψ_x and ψ_y are weighting coefficients, which are positive and satisfy $0 \leq \psi_x + \psi_y \leq 1$.

As compared with Eq. (7.60), Eq. (7.61) can enhance the stability of the sediment model, but it may encounter numerical diffusion and mass imbalance. Wu *et al.* (2000a) suggested giving the transverse coefficient ψ_y zero and the longitudinal coefficient ψ_x a small value. However, the author's later work shows that if the adaptation length L is given properly as recommended in Section 2.6.3, the model stability can be enhanced and the use of Eq. (7.61) can be avoided.

As described in the depth-averaged 2-D model, the dicretized equation for the bed material sorting in the mixing layer is Eq. (6.70).

7.3.3 Solution of discretized sediment transport equations

As demonstrated in 1-D and depth-averaged 2-D models, the bed-material gradation in Eq. (7.50) can be treated explicitly, and then, a decoupled procedure can be established to solve the discretized 3-D equations of sediment transport, bed change, and bed material sorting. The sediment model is often decoupled with the flow model. If the SIMPLE algorithm is used in the flow model, the fully decoupled calculations are executed in the following sequence:

1) Start from the initial channel bed and flow field;
2) Solve the momentum equations with the estimated pressure p^* and then the pressure and velocity correction equations to obtain p^{n+1} and u_i^{n+1};
3) Solve the k- and ε-equations and update the eddy viscosity v_t;
4) Calculate the water surface profile and then adjust the grid if any change in water surface occurs;
5) Treat the obtained pressure p^{n+1} as a new estimate, return to step (2), and repeat the above flow calculation until a converged solution is obtained;
6) Calculate c_{b*k}^{n+1} and q_{b*k}^{n+1} using Eq. (7.50) with the known p_{bk}^n;
7) Calculate c_{bk}^{n+1} and q_{bk}^{n+1} using Eqs. (7.51) and (7.56);
8) Calculate Δz_{bk} and Δz_b using Eqs. (7.57) and (7.59);
9) Calculate p_{bk}^{n+1} using Eq. (6.70);
10) Update the bed topography using Eq. (7.60) and then adjust the grid if any change in bed elevation occurs;
11) Return to (2) and repeat the above calculations for the next time step until a specified time is reached.

If the bed-material gradation in Eq. (7.50) is treated implicitly, from Eqs. (7.50), (7.57), (7.59), and (6.70), one can derive equations similar to (6.73) and (6.74) to determine the fractional and total bed changes and establish a coupled procedure for the 3-D calculation of sediment transport, bed change, and bed material sorting. This coupled sediment model can be decoupled from the flow model to constitute a semi-coupled model or coupled with the flow model to form a fully coupled model. The calculation sequences in the 3-D semi-coupled and fully coupled flow and sediment transport models are similar to those in the depth-averaged 2-D models described in Section 6.2.3.2. The details are not repeated here.

7.3.4 Examples

Case 1. Sediment transport in an 180° channel bend

The flow, sediment transport, and bed change processes in an 180° channel bend investigated experimentally by Odgaard and Bergs (1988) was simulated by Wu and Wenka (1998), Wu *et al.* (2000a), and Zeng *et al.* (2005) using 3-D models. The channel bend was 80 m long and 2.44 m wide, connected with 20 m long straight sections upstream and downstream. The cross-section was trapezoidal with vertical sidewalls, and the channel bed was filled with a 30 cm thick layer of sand with an initially flat surface. The sand had a median diameter of 0.3 mm and a geometric standard deviation of 1.45. The experiment was carried out at a discharge of $0.153\,\mathrm{m^3 s^{-1}}$ with an average water depth of 0.15 m and average velocity of $0.45\,\mathrm{m \cdot s^{-1}}$. The sediment moved through the channel mainly as bed load at a rate of $3.7\,\mathrm{g \cdot cm^{-1} min^{-1}}$, as measured by a bed-load sampler.

Wu and Wenka (1998) and Wu *et al.* (2000a) used the 3-D flow model introduced in Section 7.1.3.2 with the standard k-ε turbulence closure, whereas Zeng *et al.* (2005) used a 3-D flow model with the k-ω turbulence closure. Wu *et al.* and Zeng *et al.*

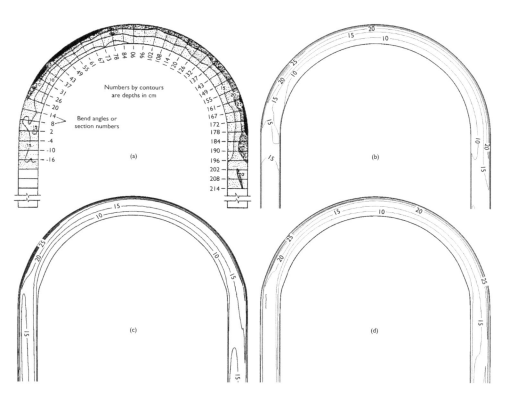

Figure 7.5 Flow depth contours in an 180° bend: (a) measured by Odgaard and Bergs (1988), (b) calculated by Zeng *et al.* (2005), (c) calculated by Wu and Wenka (1998), and (d) calculated by Wu *et al.* (2000a).

used two total-load transport models similar to the decoupled model introduced in Section 7.3, whereas Wu and Wenka used only the bed-load model component. The computational meshes used by Wu and Wenka (1998) and Wu *et al.* (2000a) were the same and had 121, 22, and 15 grid points in the streamwise, lateral, and vertical directions, respectively. Fig. 7.5 shows the measured and calculated contours of water depth at the end of the experiment. The morphological developments in the bend obtained by the three simulations are in fairly good agreement with the measurement. The maximum depths calculated by Wu and Wenka (1998) and Wu *et al.* (2000a) were 25.59 and 25.68 cm, respectively, at around 45°, as compared with the measured value of 27 cm at the section around 55°. Both total-load models predicted that the bed load was about 80% of the total load, so that the results from the bed-load simulation of Wu and Wenka are not significantly different from those of the two total-load simulations.

Fig. 7.6 displays the secondary flow velocity vectors and streamwise velocity contours at various cross-sections through the bend calculated by Wu *et al.* (2000a).

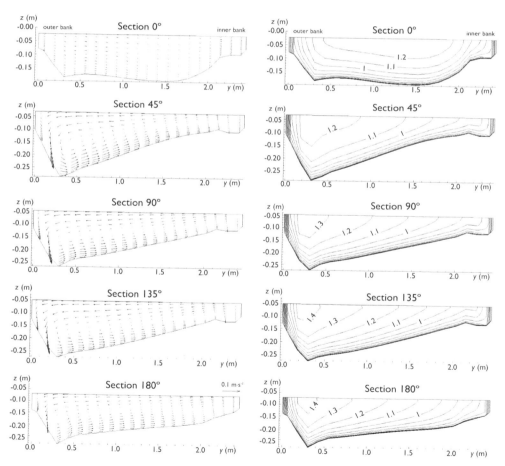

Figure 7.6 Calculated secondary flow velocity vectors and normalized streamwise velocity U/U_{in} contours at cross-sections in Odgaard and Bergs' bend (Wu *et al.*, 2000a).

This figure also shows clearly how the channel bed developed in the asymptotic state from a near-trapezoidal shape at the bend entrance to a closely triangular shape with a scour hole near the outer bank and a point bar near the inner bank in the bend.

Case 2. Sedimentation in the upstream neighborhood of TGP dam

Fang and Rodi (2000) and Fang and Wang (2000) calculated the flow and sediment transport processes in the neighborhood of Three Gorges dam on the Yangtze River in a period of 76 years after the dam started operating. This problem was also investigated in a laboratory experiment at Tsinghua University (1996). Fang and Rodi (2000) applied the 3-D model described in Sections 7.1.3.2 and 7.3 presented by Wu *et al.* (2000a), while Fang and Wang (2000) used a different 3-D model based on the finite analytic method. Since both simulation results are similar, only those of Fang and Rodi (2000) are introduced here. As the sediment transport was almost entirely due to suspended load, only the suspended-load model was used in the calculation. The computational domain was a 16.7 km long reach upstream of the dam, as shown in Fig. 7.7. This domain was represented by a numerical grid that had 234, 42, and 22 points in the streamwise, lateral, and vertical directions, respectively. The initial

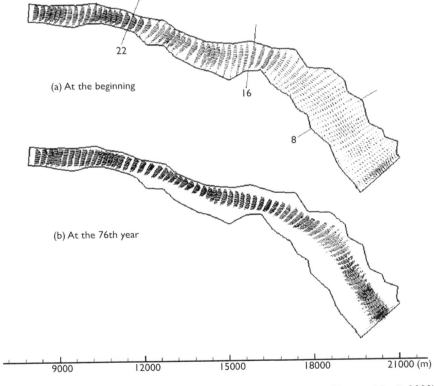

Figure 7.7 Calculated surface velocity vectors upstream of TGP dam (Fang and Rodi, 2000).

bed geometry was the natural river. The time series of flow and sediment input at the inflow section was basically taken the same as in the experiment, but it was somewhat smoothed by averaging over certain periods. In addition, the experimental water depth at the downstream boundary (dam) was prescribed.

Fig. 7.7 shows the calculated surface velocities at the beginning and after 76 years of dam operation. At the beginning, when the natural bed still existed, the flow occupied the entire calculation domain and obviously was faster in the narrower upstream reach, and had a lower velocity and a more complex pattern in the vicinity of the dam where the cross-section was considerably wider. After 76 years, much sediment has deposited on the sides, and the river flows only in a fairly narrow channel similar to its behavior before the dam was erected. The changes in the bed and hence the flow in several selected cross-sections can be seen from Fig. 7.8 where the bed profiles are given for the year zero and for the 54th year, showing clearly the rise of bed elevation due to sediment deposition over the years. The figure compares the calculated bed profiles and

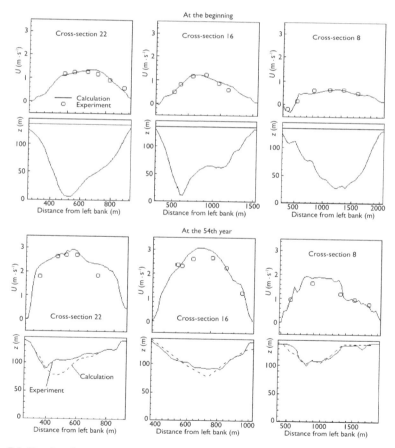

Figure 7.8 Simulated and measured surface velocity and bed profiles upstream of TGP dam (Fang and Rodi, 2000).

surface velocities with those measured in the laboratory experiment, and the agreement can be seen to be quite good.

7.4   3-D SIMULATION OF LOCAL SCOUR AROUND IN-STREAM STRUCTURES

In-stream structures, such as bridge piers, abutments, spur-dikes, sluice gates, spillways, weirs, and headcuts, have considerable potential to cause significant local scour in their vicinities. The scour not only endangers the channel bed stability, but also might have devastating effects on the structures. The 3-D numerical models used to predict the local scour around in-stream structures are introduced here.

7.4.1   Complexity of local scour processes around in-stream structures

Flows around in-stream structures are truly three-dimensional. Fig. 7.9 shows the flow pattern around a bridge pier. The boundary-layer flow approaches from upstream to the pier, and a stagnation pressure establishes in front of the pier. Due to this stagnation pressure, the water surface increases, forming a bow wave and inducing a downward flow there. The strong pressure gradient around the pier diverts the downstream flow laterally. The downward flow and pressure gradient are responsible for the initiation and development of local scour. If the flow strength increases to a certain level, the three-dimensional boundary layer at the pier undergoes a separation. A horseshoe-vortex system forms at the base of the pier and stretches into the downstream direction; it removes bed material from around the pier and intensifies the local scour. A trailing wake-vortex system forms behind the pier over the entire flow depth; it increases the turbulence intensity and consequently enhances sediment transport. However, the horseshoe and wake vortices diminish their strengths rapidly, and thus, the sediment transported from around the pier deposits immediately downstream of the scour hole.

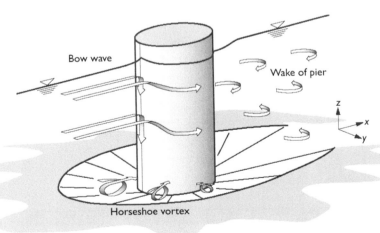

Figure 7.9 3-D flow pattern around a bridge pier (Graf and Altinakar, 1998).

The flow pattern around a bridge abutment (or spur-dike) also involves downward flow, localized pressure gradient, and vortices, which induce local scour around the base of the abutment. The obstruction of the flow forms a horizontal vortex starting at the upstream end and running along the toe of the abutment and a vertical trailing wake vortex at the downstream end. The horizontal and wake vortices at the abutment are similar to the horseshoe and wake vortices at the pier.

Fig. 7.10 shows the flow features often existing downstream of weirs, spillways, sluice gates, and headcuts. Overflow appears in form of a jet and plunges into the downstream pool, while underflow appears as a submerged or free horizontal jet. Flow may separate laterally due to the sudden expansion of channel width. Considerable scour occurs in the downstream pool. Vortices exist and affect the magnitude of scour. The sediment eroded from the scour hole may deposit as a mound downstream, forming an "S" shaped bed profile.

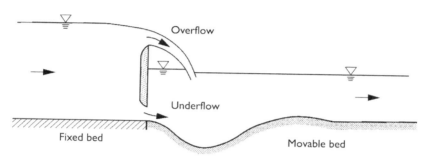

Figure 7.10 Flow over/under a structure.

The total scour around in-stream structures includes three components: (a) general degradation or aggradation due to natural or man-induced causes in upstream channels and watershed; (b) contraction scour due to reduction of the flow area by the structures and approach embankments; and (c) local scour caused by flow acceleration, jet impingement, and vortices induced by the structures. The contraction scour and local scour may be clear-water or live-bed (see Fig. 7.11). The clear-water scour indicates that no sediment comes from the approach reach, or the sediments transported are so fine that they wash through the reach. In the clear-water scour process, the scour depth increases gradually and approaches an asymptotic value when the capacity of transport out of the scour hole is zero. In the live-bed scour, the scour depth increases rapidly and attains an equilibrium value when the capacity of sediment transport out of the scour hole is equal to the one into the scour hole. The live-bed scour can be cyclic in nature, typically scouring during the rising stage of a flood event and refilling during the falling stage.

In addition, the channel migration in floodplains may affect the scour around bridge piers and erode abutments, spurs, and embankments by changing the course of main flow and the flow angle of attack. Debris accumulation and vegetation growth around structures may also affect the scour significantly.

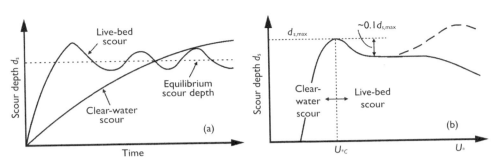

Figure 7.11 Clear-water and live-bed scours at bridge piers (Raudkivi and Ettema, 1983).

Traditionally, the maximum scour depths at hydraulic structures are predicted using empirical formulas calibrated with laboratory experiments and limited field observations. For example, HEC-18 (Federal Highway Administration, [FHWA], 1995) adopts the modified CSU equation for the scour at bridge piers and the Froehlich formula for the scour at bridge abutments. More formulas for the local scour at bridge piers and abutments and downstream of sluice gates and spillways are summarized in Simons and Senturk (1992), Graf and Altinakar (1998), and others. However, these formulas usually predict only the maximum scour depth. For detailed information on the scour processes, numerical modeling should be used. This is discussed in the following subsections.

7.4.2 Simulation of sediment transport and local scour near in-stream structures

Because of the complexity of rapidly varied flows and local scour processes around in-stream structures, only the full 3-D model is suitable for a realistic simulation of such phenomena. In general, the hydrodynamic equations include the continuity equation (7.1) and momentum equations (7.2) with a k-ε or more advanced turbulence closure to simulate the complex turbulent flows near the structures. The governing equations for sediment transport around in-stream structures are the same as those in general situations, including the suspended-load transport equation (7.43), bed-load transport equation (7.46), bed change equation (7.47), and bed material sorting equation (6.57). The numerical solution methods introduced in Sections 7.1 and 7.3 can be used here to solve these equations. Body-fitted grids are usually required to obtain accurate solutions.

However, some model parameters, such as sediment transport capacity and adaptation length, may be different for gradually and rapidly varied flows. In particular, the sediment transport capacity of rapidly varied flows is affected by complex flow features, such as downward flow, horseshoe vortex, localized pressure gradient, and turbulence intensity. The empirical formulas developed for the sediment transport capacity of gradually varied flows are not applicable to the local scour case. They must be modified to consider the complex features of rapidly varied flows.

Olsen and Kjellesvig (1998) and Nagata *et al.* (2005) demonstrated the feasibility of applying 3-D models in the local scour simulation, but they did not investigate the sediment transport capacity under rapidly varied flow conditions. After having analyzed the mechanism of local scour at bridge piers, Dou (1997) related the sediment transport capacity for local scour to the mean flow, downward flow, vorticity, and turbulence intensity as follows:

$$C_* = a_1 \frac{U^3}{ghw_s} + a_2 \frac{|W| - |W|_{app}}{U} + a_3 \frac{(\Omega - \Omega_{app})b}{U_*} + a_4 \frac{i - i_{app}}{w_s} \qquad (7.62)$$

where C_* is the sediment transport capacity for local scour, U is the depth-averaged velocity, U_* is the bed shear velocity, $|W|$ is the magnitude of downward flow, Ω is the vorticity, b is the pier width, i is the turbulence intensity, a_i ($i = 1, 2, 3, 4$) are empirical coefficients, and the subscript "app" denotes the approaching flow quantities.

In Eq. (7.62), the non-dimensional parameter $U^3/(ghw_s)$ represents the contribution of mean flow, which is used in the Zhang formula for general sediment transport capacity, as introduced in Section 3.5.3. The term $(|W| - |W|_{app})/U$ represents the influence of downward flow, $(\Omega - \Omega_{app})b/U_*$ accounts for the effect of vorticity, and $(i - i_{app})/w_s$ takes into account the effect of turbulence intensity. Eq. (7.62) considers the significant factors that affect the local scour process. However, in reality, it is difficult to use the linear formulation in Eq. (7.62) to reflect all factors reasonably, and the empirical coefficients a_i need to be calibrated extensively.

Based on the analysis of the forces acting on sediment particles near the bed exerted by rapidly varied flows, Wu and Wang (2005) modified the van Rijn (1984a & b) formulas (3.70) and (3.95) of equilibrium bed-load transport rate and near-bed suspended-load concentration for the simulation of local scour by determineing the transport stage number T with $T = \tau_{be}/\tau_{cr} - 1$. Here, τ_{be} is the effective tractive force and τ_{cr} is the critical shear stress for sediment incipient motion in rapidly varied flows. τ_{cr} is determined by

$$\tau_{cr} = K_p K_d K_g \tau_c \qquad (7.63)$$

where τ_c is the critical shear stress for sediment incipient motion, determined using the Shields curve; and K_p, K_d, and K_g are the correction factors for the effects of vertical dynamic pressure gradient, downward flow, and bed slope, respectively.

Fig. 7.12 depicts the localized dynamic pressure field due to the impingement of a jet onto a channel bed. This localized dynamic pressure also exists in front of bridge piers, abutments, and spur-dikes. The gradient of fluid pressure causes a pressure-difference force on sediment particles, which is also called the general buoyancy. This pressure-difference force on a single particle is determined by (Liu, 1993)

$$\vec{f}_p = -\frac{1}{6}\pi d^3 \nabla p \qquad (7.64)$$

where d is the particle diameter.

The vertical component of the pressure-difference force changes the effective weight of sediment in water and, in turn, the critical shear stress required for sediment incipient

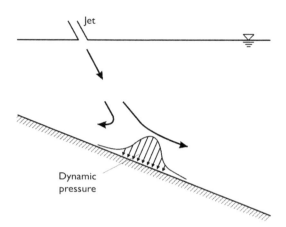

Figure 7.12 Localized dynamic pressure due to jet impingement.

motion. Using Eqs. (7.64) and (3.1) yields the following relation for K_p:

$$K_p = 1 + \frac{1}{(\rho_s - \rho)g} \frac{\partial p_d}{\partial z} \tag{7.65}$$

where p_d is the dynamic (non-hydrostatic) pressure. Because the effect of the hydro-static pressure has already been considered in general sediment transport formulas, only the effect of the dynamic pressure needs to be considered in K_p.

The downward flow increases the acting area of tractive force on sediment particles and then decreases the critical shear stress for sediment incipient motion. Therefore, the correction factor K_d can be determined by

$$K_d = \frac{1}{1 + \sin \beta} \tag{7.66}$$

where β is the impact angle of flow to the bed, defined as the angle between the near-bed resultant flow and the bed.

As discussed in Section 3.7, gravity affects the incipient motion of sediment on a steep slope. The correction factor K_g is defined as

$$K_g = \frac{\sin(\phi_r - \varphi)}{\sin \phi_r} \tag{7.67}$$

where φ is the streamwise bed slope angle with the horizontal (positive values denoting downslope beds), and ϕ_r is the repose angle of submerged bed material. It should be noted that when the bed slope angle is close to the repose angle, K_g is close to zero. This must be limited by imposing a small lower bound, such as 0.1, to K_g.

Eq. (7.67) considers only the effect of bed slope along the flow direction; it was applied in the simulation of the vertical 2-D headcut migration by Wu and Wang (2005). For general applications, the effect of bed slope in the direction normal to the flow can also be added, as shown in Eq. (3.47). However, the side slope affects sediment transport within the scour hole in two counteractive ways: reducing the critical shear stress for sediment incipient motion, but tending to move sediment toward the scour hole center.

The effective tractive force τ_{be} includes the bed shear stress and the horizontal component of the dynamic-pressure-difference force:

$$\tau_{be} = \tau_b' - \frac{a\pi}{6} d \frac{\partial p_d}{\partial s} \tag{7.68}$$

where τ_b' is the bed shear stress due to grain roughness, $\partial p_d/\partial s$ is the streamwise gradient of dynamic pressure near the bed, and a is a coefficient assumed as $4/\pi$. Eq. (7.68) is derived by assuming that the shear stress on a sediment particle is $\tau_b' d^2/a$ and determing the dynamic-pressure-difference-force on this particle with Eq. (7.64).

Because the method for computing τ_b' used by van Rijn (1984a & b) for uniform flow is not appropriate for rapidly varied flows, τ_b' is directly set to the bed shear stress calculated by the flow model. However, to be consistent with the original van Rijn formulas, the equivalent bed roughness height k_s used in the wall-boundary approach in the flow model is set to the grain roughness $3d_{90}$. Because only the grain roughness is considered, this approach is applicable to situations without bed forms. Usually, most clear-water scour cases belong to such situations. For more general applications, all roughness elements may be considered in the wall-boundary approach, and then, the grain shear stress is separated from the computed total bed shear stress using the approaches introduced in Section 3.3.2.

In addition, the effect of gravity on sediment transport over a steep slope may be considered by adding the streamwise component of gravity to the effective tractive force τ_{be} rather than applying the correction factor K_g to the critical shear stress, as shown in Eq. (3.132). Thus, Eqs. (7.63) and (7.68) can be modified as

$$\tau_{cr} = K_p K_d \tau_c \tag{7.69}$$

$$\tau_{be,i} = \tau_{b,i}' + \lambda_0 \tau_c \sin \varphi_i / \sin \phi_r - \frac{a\pi}{6} d \frac{\partial p_d}{\partial x_i} \quad (i = x, y) \tag{7.70}$$

In analogy to Eq. (6.98), an equation can be derived from Eq. (7.70) for the bed-load transport direction cosines $\alpha_{bx,e}$ and $\alpha_{by,e}$.

If the slope angle in the scour hole is larger than the repose angle of sediment, a loose bed will collapse due to gravity. This physical phenomenon can be calculated by adjusting the steeper bed slope to the repose angle according to mass conservation. The non-cohesive bank or bed sliding algorithm introduced in Section 6.3.5 can be adopted here.

The approaches presented in Eqs. (7.63) and (7.68)–(7.70) were tested in the simulation of local scour process at bridge piers by this author using the 3-D flow model introduced in Section 7.1.3.2 and the sediment transport model introduced in

Section 7.3. The sediment adaptation length was determined by $L = \min(0.4t, 7.3h_0)$, in which t is the elapsed time in hours and h_0 is the approach flow depth in meters. The computational mesh consisted of 43 and 25 points in the transverse and vertical directions, respectively, and a suitable number of points in the longitudinal direction depending on the flume length. The plan view of the mesh around a cylindrical pier is shown in Fig. 7.13. The vertical grid spacing was refined near the bed.

Fig. 7.14 compares the simulated and measured scour holes at a cylindrical bridge pier for Yanmaz and Altinbilek's Run 3 with a pier diameter (D) of 6.7 cm, a sediment

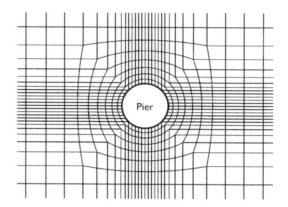

Figure 7.13 Computational mesh near a cylindrical pier (plan view).

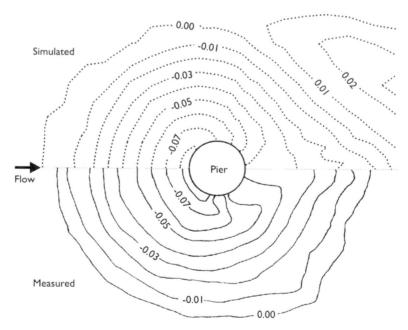

Figure 7.14 Measured and simulated scour depth contours (m) at a cylindrical pier for Yanmaz and Altinbilek's Run 3 at 100 min.

size of 1.07 mm, a flow discharge of $30 \, l \cdot s^{-1}$, and an approach flow depth of 0.135 m. The simulated scour depth contours in the hole agree well with those measured. The simulation predicted deposition downstream of the scour hole, but the measurement lacked this information. Fig. 7.15 compares the simulated and measured deepest scour depths varying with time for Yanmaz and Altinbilek's (1991) Run 3, Ettema's (1980) experiment with $D = 0.24$ m and $d_{50} = 1.9$ mm, and the Run 7 ($D = 0.91$ m, $d_{50} = 2.9$ mm) of Sheppard *et al.* Durations (t_e) of these three runs were 5, 14.5, and 188 hr, respectively. Erosion was very intensive at first and then reduced gradually. The erosion processes were reproduced well by the numerical model.

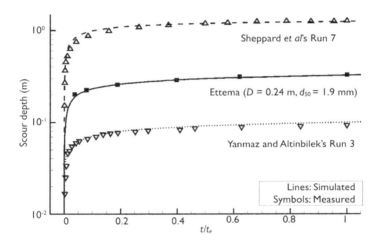

Figure 7.15 Temporal variation of the deepest scour depth at cylindrical piers.

Tests have also shown that the approach using Eqs. (7.63) and (7.68) for τ_{cr} and τ_{be} and that using Eqs. (7.69) and (7.70) do not have significant difference in terms of the predicted maximum scour depth at bridge piers. This is due to the fact that the local scour is induced by other factors, such as downward flow and localized dynamic pressure gradient. In contrast, the bed slope effect may reduce the predicted local scour depth because of the inverse bed slope in the downstream of the scour hole. The results shown in Figs. 7.14 and 7.15 were obtained with the approach using Eqs. (7.63) and (7.68).

7.4.3   Headcut migration model

A headcut is a vertical or near-vertical drop or discontinuity on the channel bed of a stream, rill, or gully at which a free overfall flow often occurs, as shown in Fig. 7.16. The headcut is usually eroded by the action of hydraulic shear, basal sapping, weathering, or a combination of these processes. Headcut erosion can accelerate soil loss, increase sediment yields in streams, damage earthen spillways, and disturb bank stability.

De Ploey (1989) and Temple (1992) established empirical formulas by relating the headcut migration rate to the energy head change at the headcut. The coefficients

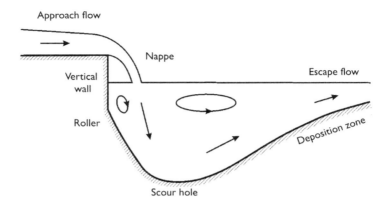

Figure 7.16 Fluvial and geomorphic features of headcut (side view).

in these formulas depend on soil properties, such as void ratio, moisture content, saturation degree, bulk density, and clay content (Hanson *et al.*, 1997; Wu and Wang, 2005). Kitamura *et al.* (1999) and Wu and Wang (2005) developed numerical models to simulate the detailed erosion and migration processes of headcut. The numerical model of Wu and Wang is introduced below.

The flow model introduced in Section 7.1.3.2 with the standard k-ε turbulence closure is used to compute the flow in the plunge pool. The jet impingement velocity and angle are estimated using the empirical formulas proposed by Robinson (1996).

There are three modes of erosion occurring at the headcut. The first mode is the surficial erosion along the vertical headwall due to the hydraulic shear of flow. The second mode is the toe erosion due to the development of scour hole in the plunge pool. After the surficial erosion and toe erosion develop to a certain extent, the headwall will exceed the criterion of stability, and a mass failure will occur, which is the third mode of headcut erosion. In reality, the mass failure occurs periodically, and each event finishes in a very short period. Its occurrence may be predicted using the stability models suggested by Barfield *et al.* (1991) and Robinson and Hanson (1994), which are similar to that for bank failure shown in Fig. 5.17. However, simulating such a discontinuous phenomenon using a hydrodynamic model is difficult, as the collapsed block of soil will strongly disturb the flow field in the plunge pool. For simplicity, a time-averaged headcut migration model is adopted, as shown in Fig. 7.17.

The erosion rate on the headwall surface due to hydraulic shear, dl_s/dt (m $\cdot$ s^{-1}), is determined by (Wu and Wang, 2005)

$$\frac{dl_s}{dt} = \begin{cases} 0.0000625\dfrac{\tau_{vm}}{M}, & \dfrac{\tau_{vm}}{M} < 8 \\ 0.00977 - 0.00238\dfrac{\tau_{vm}}{M} + 0.000153\left(\dfrac{\tau_{vm}}{M}\right)^2, & \dfrac{\tau_{vm}}{M} \geq 8 \end{cases} \quad (7.71)$$

where τ_{vm} is the maximum shear stress on the headwall surface (pa); M is a material-dependent parameter, related to the soil bulk density ρ'_b and clay content p_c by

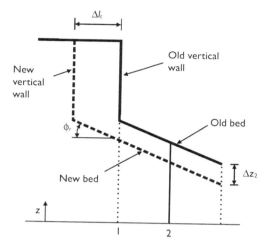

Figure 7.17 Headcut migration due to toe scour.

$M = p_c^{0.4}/(1 - \rho_b'/\rho_s)^2$; and ρ_s is the soil density $(\mathrm{kg \cdot m^{-3}})$. At each time step Δt, Eq. (7.71) gives the headwall retreat length Δl_s due to hydraulic shear.

The sediment transport in the plunge pool, which is highly affected by the rapidly-varied jet impinging flow, is simulated using the 3-D model introduced in Section 7.3 with the sediment transport capacity determined using Eqs. (3.70) and (3.95) with the corrections described in Eqs. (7.63) and (7.68). From this calculation, the bed change Δz_2 at the center of the first control volume near the toe of headwall surface at the time step Δt can be obtained, and then, the headcut retreat length due to this toe scour is given as $\Delta l_t = \Delta z_2/\tan\phi_r$, as shown in Fig. 7.17. Here, ϕ_r is the repose angle of sediment. The actual headcut migration length is the larger of Δl_s and Δl_t.

The time-averaged model described above simulates the headcut migration caused by the headwall surficial erosion and toe erosion. These two erosion modes are the main factors of headcut migration, as they induce the mass failure and wash out the wasted sediment debris. Therefore, this time-averaged model can acquire the main features of headcut migration.

The headcut boundary propagates upstream and expands sideward. A moving grid technique is implemented to capture this moving boundary in the horizontal directions, combined with the original one to track the water surface change and bed deformation in the vertical direction. At each time step or iteration step, the computational grid is adjusted after the calculations of water level, bed deformation, and headcut migration.

The established headcut migration model was tested by Wu and Wang (2005) using the experiments of Bennett *et al.* (1997). The experiments were carried out in a flume 5.5 m long. The test part was a cavity 2 m long and 0.165 m wide, filled with 0.25 m deep soil. The bed slope of the flume was 1%. The used soil was the parent material of the Ruston silt loam after being crushed and air-dried and consisted of 20.0% clay, 2.9% silt, and 77.1% sand. A preformed 0.025 m high headcut was constructed at the downstream end of the cavity. Application of simulated rain produced a surface seal layer to negate any detachment of the soil material by the subsequent overland

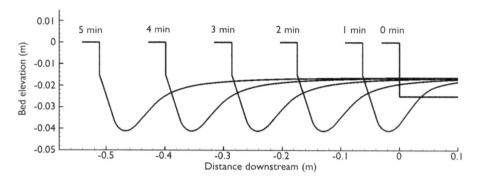

Figure 7.18 Simulated headcut migration process (Run 2) (Wu and Wang, 2005).

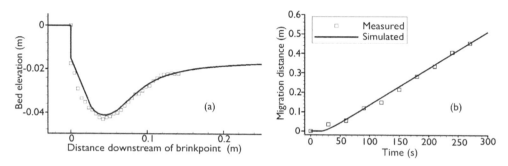

Figure 7.19 (a) Steady-state scour hole and (b) brinkpoint migration distance (Run 2)
(Wu and Wang, 2005).

flow. A scour hole developed due to jet impingement, and the headcut migrated upstream. Fig. 7.18 shows the calculated headcut migration process for experimental run 2. The scour hole maintained a quasi-steady profile as the headcut migrated upstream. Figs. 7.19(a) and (b) show the measured and calculated scour hole morphologies and brinkpoint migration distances. The scour hole profile and headcut migration rate were predicted well.

Chapter 8

Domain decomposition and model integration

Many problems in river engineering exhibit geometric irregularities and multiple length and time scales in flow velocity and mass concentration. To handle these characteristics effectively, the computational domain is often decomposed into subdomains that may be represented by meshes with different grid densities and topologies, and treated by models with different dimensions and complexities. The multiblock algorithm, coupling of 1-D, 2-D and 3-D models, and integration of channel and watershed models are introduced in this chapter.

8.1 MULTIBLOCK METHOD

8.1.1 General considerations

The multiblock method divides the entire computational domain into blocks, as shown in Fig. 8.1, and generates the mesh for each individual block independently. The governing equations are first solved in each block with information exchanged at block boundaries, and the results in all blocks are then assembled to obtain the global solution. The multiblock method allows much greater grid flexibility and local refinement than the single-block method does.

To facilitate the exchange of information, an interface between two adjacent blocks needs to be constructed. The grids on the blocks can be patched or overlapped around the interface, as shown in Fig. 8.2, depending on the numerical methods used. On the

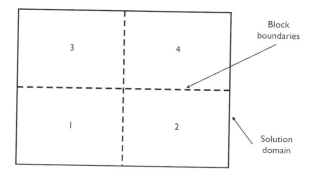

Figure 8.1 Sketch of domain decomposition.

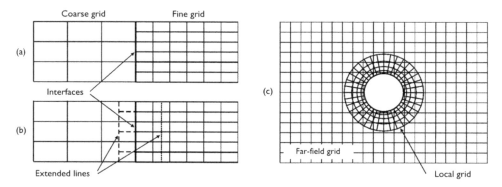

Figure 8.2 Interface arrangements for multiblock grid: (a) patched, (b) simple overlapping, and (c) complex overlapping.

patched grids, two neighboring blocks are connected at a common grid line without overlapping, while on the overlapping grids, the blocks can be superimposed arbitrarily on each other to cover the domain of interest. Compared to the patched grids, the overlapping grids have more flexibility in grid generation but may be less accurate due to interpolation errors as information is exchanged between blocks.

The discretization of the governing equations in each block is the same as in a single-block domain. However, for a complete solution, internal boundary conditions should be applied at the interfaces. To avoid errors and the generation of spurious numerical oscillations, the conservation laws should be satisfied at the interfaces. During the solution process, the information updated at each time or iteration step needs to be transferred between the blocks. Therefore, the key issues that affect the performance of the multiblock method are interface treatment and information exchange between blocks.

8.1.2 Multiblock method for 1-D problems

Let us consider the 1-D steady convection-diffusion problem (4.15) as an example of the multiblock method (Shyy *et al.*, 1997). For the sake of simplicity, only two grid blocks are used. The grids in these two blocks are shown in Fig. 8.3. Let $\phi_{l,i}$ denote the discrete approximation to ϕ at the point $i\,(= 1, 2, \ldots, N_k)$ of the component grid block $l\,(=1, 2)$. Eq. (4.15) can be discretized using the generic numerical scheme introduced in Chapter 4 as

$$a_{Pl,i}\phi_{l,i} = a_{Wl,i}\phi_{l,i-1} + a_{El,i}\phi_{l,i+1} \tag{8.1}$$

where $a_{Pl,i}$, $a_{El,i}$, and $a_{Wl,i}$ are coefficients.

The discretized equations on each block can be solved with a direct or iterative method. However, boundary conditions have to be provided at the interface to connect the solutions on the two blocks so that a global solution can be obtained. In the solution

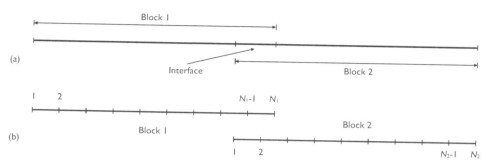

Figure 8.3 (a) Domain decomposition and (b) grid arrangement in a 1-D model.

of block 1, the boundary value of ϕ_{1,N_1} is unknown, and in the solution of block 2, the boundary value of $\phi_{2,1}$ is unknown. They can be interpolated using the values of ϕ on the other block by the following interpolation schemes:

$$\phi_{1,N_1} = \sum_{j=L_1}^{L_2} \gamma_j \phi_{2,j}, \quad 1 \leq L_1 \leq L_2 \leq N_2 \tag{8.2a}$$

$$\phi_{2,1} = \sum_{j=M_1}^{M_2} \beta_j \phi_{1,j}, \quad 1 \leq M_1 \leq M_2 \leq N_1 \tag{8.2b}$$

where γ_j and β_j are interpolation coefficients; and L_1, L_2, M_1, and M_2 denote the bounds of the grid points used in the interpolation. The number of points involved depends on the order of the chosen interpolation formula.

The evaluation of the coefficient a_W for the point near the interface on block 2 will involve some quantities interpolated from block 1. Similarly, the evaluation of a_E for the point near the interface on block 1 will involve some quantities interpolated on block 2.

With the above interface treatment, an iteration process between the two blocks is conducted to solve the equations over the entire domain.

8.1.3 Multiblock method for multidimensional problems

Interpolation and conservative correction at the interface

Shyy *et al.* (1997) introduced a multiblock algorithm to solve the Navier-Stokes equations using the finite volume method on the staggered grid. It is straightforward to extend their method to the depth-averaged 2-D flow and sediment transport model. Fig. 8.4 shows a typical configuration of the interface between two blocks. For simplicity, the non-staggered grid is used here instead. The interface is set at the common face of the neighboring coarse and fine control volumes. The coarse and

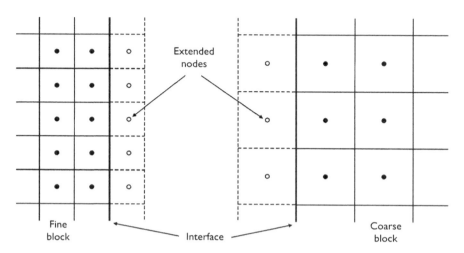

Figure 8.4 Interface between blocks for non-staggered FVM grid.

fine grids are extended one layer (or more layers if needed for methods of high accuracy) to the adjacent block for the convenience of information transfer and numerical discretization at the interface.

To solve the flow and sediment transport equations in each block, the boundary conditions of velocity, pressure (or pressure correction), suspended-load concentration, and bed-load transport rate are required at the interface. Other variables that have to be transferred are flow depth, bed level, bed shear stress, sediment transport capacity, bed change, and bed-material gradation. All these variables should be interpolated from the corresponding points on the adjacent block. The linear or quadratic interpolation method may be used. However, to obtain continuous flow and mass fields, the interpolation should satisfy the conservation laws at the interface.

Fig. 8.5 shows a portion of the interface corresponding to the width of a single control volume on the coarse grid, and to the width of several control volumes, indexed from $i = 1$ to $i_{\max}$, on the fine grid. The conservation law for flow flux reads

$$U_c h_c l_c = \sum_{i=1}^{i_{\max}} U_{fi} h_{fi} l_{fi} \tag{8.3}$$

where l_c is the length of the interface on the coarse grid, h_c and U_c are the flow depth and velocity at l_c, l_{fi} is the length of the interface of cell i on the fine grid, and h_{fi} and U_{fi} are the flow depth and velocity at l_{fi}. Note that the velocities U_c and U_{fi} are normal to the corresponding cell faces.

As information is exchanged from the fine grid to the coarse grid (Fig. 8.5), the flow flux and, in turn, the flow velocity can be uniquely obtained with Eq. (8.3), which satisfies the conservation law. However, it is not straightforward to obtain the flow velocity from the coarse grid to the fine grid. The conservation law (8.3) does

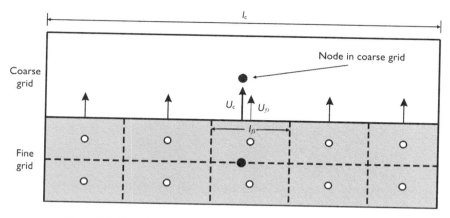

Figure 8.5 Flow fluxes at interface between coarse and fine grids.

not provide an unique value for U_{fi}. A distribution has to be assumed to determine U_{fi}. As an approximation, U_{fi} may be first obtained using the linear (or quadratic) interpolation of the velocity on the coarse grid. The interpolated quantity is denoted as $\tilde{U}_{fi}$, which may not satisfy the mass conservation law. Then the flow velocity on the fine grid is scaled so that the obtained total flux is $U_c h_c l_c$. Therefore, U_{fi} is calculated by

$$U_{fi} = |\tilde{U}_{fi}|(U_c h_c l_c) \Big/ \left(\sum_{i=1}^{i_{\max}} |\tilde{U}_{fi}| h_{fi} l_{fi} \right) \tag{8.4}$$

The flow velocity on the fine grid determined using Eq. (8.4) satisfies the mass conservation law, and its distribution is close to the interpolated one.

At the interface between two blocks, the conservation law for suspended load is

$$U_c h_c l_c C_c = \sum_{i=1}^{i_{\max}} U_{fi} h_{fi} l_{fi} C_{fi} \tag{8.5}$$

and that for bed load is

$$q_{bc} l_c = \sum_{i=1}^{i_{\max}} q_{bfi} l_{fi} \tag{8.6}$$

where C_c and C_{fi} are the suspended-load concentrations on the coarse and fine grids, respectively; and q_{bc} and q_{bfi} are the bed-load transport rates per unit width correspondingly.

Like the flow flux, the suspended-load concentration and bed-load transport rate can be uniquely, conservatively obtained from the fine grid to the coarse grid, but the conservation laws are not satisfied when interpolating these quantities from the coarse grid to the fine grid. Conservative corrections should be applied. If the interpolated suspended-load concentration is denoted as $\widetilde{C}_{fi}$ and the interpolated bed-load transport rate is $\widetilde{q}_{bfi}$, the relevant corrections are

$$C_{fi} = \widetilde{C}_{fi}(U_c h_c l_c C_c) \bigg/ \left(\sum_{i=1}^{i_{max}} U_{fi} h_{fi} l_{fi} \widetilde{C}_{fi} \right) \tag{8.7}$$

$$q_{bfi} = \widetilde{q}_{bfi}(q_{bc} l_c) \bigg/ \left(\sum_{i=1}^{i_{max}} \widetilde{q}_{bfi} l_{fi} \right) \tag{8.8}$$

If a cell face on the coarse grid does not exactly match to an integer number of cell faces on the fine grid at the interface, the cells on the fine grid can be split into smaller subcells. The above conservative correction can still be applied in the conversion from the coarse cell to the subcells on the finer grid.

Interface treatment for governing equations

If the hybrid upwind/center scheme or the exponential scheme is used for discretizing the convection terms and the center difference scheme for the diffusion terms in the momentum and scale transport equations on the non-staggered grid, an overlapping interface with one extended layer on each side of the interface is enough, as shown in Fig. 8.4. However, if higher-order schemes, such as QUICK and HLPA, are used, more extended layers on each side may be needed; otherwise, a lower-order scheme has to be used to substitute the higher-order schemes near the interface. The convection flux F and diffusion parameter D at the interface are determined using the interpolated variables, if needed.

As Shyy et al. (1997) described, when solving the pressure-correction equation on each block, either the flow flux or the pressure correction interpolated from the adjacent blocks can be used as the boundary condition. If the interpolated flow flux is used, as shown in Fig. 8.6(a), the solution of the pressure-correction equation is a Neumann-type problem, and then the pressure correction is governed by Eq. (6.29) in the depth-averaged 2-D model. If the interpolated pressure correction is used as the boundary condition, as shown in Fig. 8.6(b), the problem is of the Dirichlet-type, and the pressure-correction equation is the same as Eq. (6.28).

For the Dirichlet-type boundary, after Eq. (6.28) is solved, the flow flux F_w^* at the interface in Fig. 8.6(b) is to be corrected using Eq. (4.196), while for the Neumann-type boundary, the flow flux F_w at the interface does not need correction.

If the Neumann-type boundary is used on two sides of the interface, the pressure fields in the neighboring blocks are independent and thus may be discontinuous. On the other hand, if the Dirichlet-type boundary is used on both sides, the flow flux at the interface is corrected twice, which may lead to inconsistency and discontinuity in flow flux. Therefore, special care should be taken in these two cases to ensure continuous (smooth) pressure and flux fields on two adjacent blocks. One remedy is to use the Dirichlet-type boundary on one side of the interface and the Neumann-type boundary on the other side.

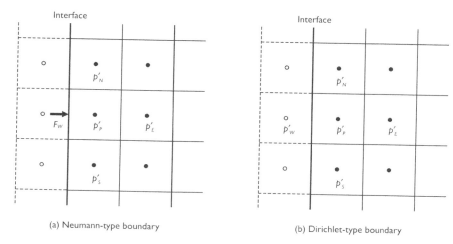

(a) Neumann-type boundary   (b) Dirichlet-type boundary

Figure 8.6 Treatment of pressure-correction at the interface.

Solution procedure

After the governing equations are discretized in a multiblock domain, the algebraic equations on each block can be solved in the same way as for a single block problem. However, an additional level of iteration among grid blocks must be introduced. Two different iteration strategies can be established. The first strategy is to form an outer loop for the iteration among blocks and an inner loop for the iteration among equations. In the outer loop, information on all relevant variables is transferred. Denoting the total number of blocks as nb, the procedure of the multiblock SIMPLE algorithm at each time step is:

for blocks 1 to n_b

(1) Guess the pressure field p^*;
(2) Solve the momentum equations to obtain U_x^* and U_y^*;
(3) Solve the pressure-correction equation to obtain p';
(4) Calculate p^{n+1}, U_x^{n+1}, and U_y^{n+1};
(5) Treat the corrected pressure p^{n+1} as a new guessed pressure p^*, and repeat the procedure from step 2 to step 6 until a converged solution is obtained;
(6) Conduct the calculation of sediment transport and bed change, if needed.

end for

The second strategy forms an outer loop for the iteration among equations and an inner loop among blocks. In the inner loop, information on all relevant variables is transferred. The corresponding procedure for the multiblock SIMPLE algorithm at each time step is:

(1) Guess the pressure field p^*, for blocks 1 to nb;
(2) Solve the momentum equations to obtain U_x^* and U_y^*, for blocks 1 to nb;

(3) Solve the pressure-correction equation to obtain p', for blocks 1 to nb;
(4) Calculate p^{n+1}, U_x^{n+1}, and U_y^{n+1}, for blocks 1 to nb;
(5) Treat the corrected pressure p^{n+1} as a new guessed pressure p^*, and repeat the procedure from step 2 to step 6 until a converged solution is obtained;
(6) Conduct the calculation of sediment transport and bed change, if needed, for blocks 1 to nb.

In principle, both strategies can be successful. Which one performs better is problem-dependent. Usually, the first strategy needs less memory than the second strategy, because the second strategy needs to store the coefficients for all blocks while the first strategy needs to store the coefficients for only one block.

8.1.4 Efficiency of multiblock method

For a complex problem, the multiblock method requires less computer memory and has more flexibility for grid generation than the single-block method. The task of computations in each block can be assigned to a processor and, thus, the multiblock method can be run on a parallel computer. However, the computations on all blocks should be synchronized, and the information transferred between processors may not be the "latest"; thus, some multiblock algorithms designed for serial (single-processor) computers may not run efficiently on parallel computers. The efficiency depends on the problem, the method of solving it, and the computer used.

The explicit schemes for an unsteady problem can be easily extended from single block to multiblock algorithms. Because all operations are performed on data from the previous time step, all that is needed is to transfer the data at the interface between neighboring blocks after the completion of each time step. The sequences of operation are identical when using single block and multiblock methods on either serial or parallel computers; so are the results. For solving the algebraic equations that result from a steady problem or from implicit schemes for an unsteady problem, the Jacobi iteration method at each iteration step only needs the "old" values of the previous iteration step, which is very similar to the manipulation of explicit schemes in the sense of time step. Therefore, the explicit schemes and the Jacobi iteration method are very easy to parallelize, with very efficient speed-up.

On serial computers, the multiblock Gauss-Seidel iteration method has the same performance as the single-block version, if the sweeping order among blocks is in such a sequence that the "latest" values of the variable in one block can be used for the calculation in the adjacent block. For example, if the sweeping order for the point-to-point iteration in each block is from the lower left point to the lower right point and thence to the upper left point, the sweeping order among blocks must be also from the lower left block to the lower right block and thence to the upper left block. However, the multiblock Gauss-Seidel iteration method cannot be extended straightforwardly from serial computers to parallel computers, because the "latest" values at the interface between blocks are not available during the synchronized computations on all the assigned processors. To solve this problem, a specially designed "red-black" ordering method is often used (see Golub and Ortega, 1993). On a structured grid, the points are imaged to be "colored" in the same way as a checkerboard, as shown in Fig. 8.7.

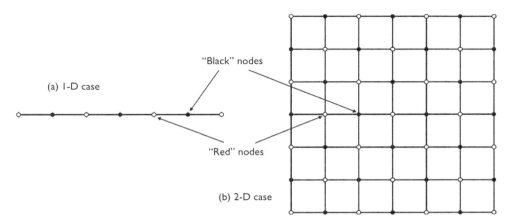

(a) I-D case

"Black" nodes

"Red" nodes

(b) 2-D case

Figure 8.7 "Red-black" ordering method.

The method consists of two Jacobi steps: black points are updated first, and then the red points. When the values at black points are updated, only the "old" values at red points are used, see Eq. (4.219); in the next step, red values are calculated using the updated black points. This alternate application of the Jacobi method to the two sets of points produces a method with the same convergence properties as the Gauss-Seidel method. When applying this "red-black" ordering method on parallel computers, only the results of the previous step are needed, and thus the computations of new values on either set of points can be performed in parallel. Communication between processors working on neighboring blocks takes place twice per iteration step — after each set of points is updated.

It is not straightforward to extend the TDMA (TriDiagonal-Matrix Algorithm) from single block to multiple blocks, because it is recursive. This can be demonstrated by using the 1-D problem shown in Fig. 8.3. If a single block is used for the entire domain, the boundary-condition information from the ends of the domain is transmitted at once to the interior through the double sweeps in each iteration step. This results in fast convergence of iteration. However, if multiple blocks are used for the domain and a separate TDMA is used in each block, the boundary-condition information at the ends of the domain is transmitted block by block to the interior; thus, the convergence speed decreases. To avoid this, one may decompose the computations at each sweep into multiblocks, and keep the double sweeps over the entire domain.

For a multidimensional problem, the ADI method, which uses alternately the TDMA in different directions, requires special partition of the solution domain for extension from single block to multiple blocks. One of the most efficient partitions is shown in Fig. 8.8. The calculation consists of two steps. In the first step when the implicit scheme is applied on i-lines and the explicit scheme is on j-lines, the entire domain is divided into horizontal strips (blocks) that extend along i-lines and end at west and east boundaries. The TDMA can be used at the strips along i-lines. Similarly, in the second step when the implicit scheme is applied on j-lines and the explicit scheme is on i-lines, the entire domain is divided into vertical strips that extend along j-lines and end at south and north boundaries. The TDMA can be used at the strips along j-lines. The sweeping order among the strips in each step is arranged in such a sequence that the "latest" values at the interface can be transferred from one strip to the next strip. This multiblock ADI method for a 2-D problem has almost the same efficiency as the single-block version. However, the domain is divided differently in two steps, which requires the transpose of the matrices.

The multiblock ADI method shown in Fig. 8.8 can be run on parallel computers. Its efficiency can be enhanced by applying the "red-black" ordering method among blocks. The blocks are "colored" as red and black alternately, and the calculation sequence among blocks follows the two steps similar to those used in the "red-black" ordered Gauss-Seidel method.

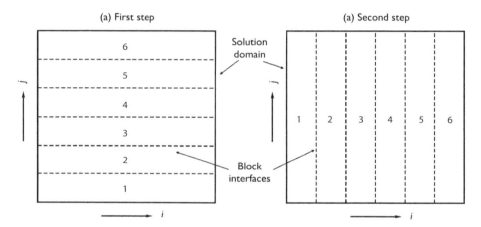

Figure 8.8 Strategy of domain decomposition for ADI method.

The SIP method is recursive, making its extension from single to multiple blocks less straightforward. One may divide the entire domain into several, such as four sub-domains shown in Fig. 8.1. The global coefficient matrix is correspondingly split into a system of diagonal blocks A_{ii}, which contain the elements connecting the points that belong to the ith subdomain, and off-diagonal blocks A_{ij} $(i \neq j)$, which represent the interaction of subdomains i and j. Then the SIP method is applied in each subdomain, while the terms related to the blocks A_{ij} are put in the source term. The global iteration matrix is selected so that the blocks are decoupled, i.e., $M_{ij} = 0$ for $i \neq j$. Each diagonal block matrix M_{ii} is decomposed into L and U matrices in the normal way. The multiblock SIP method can be parallelized. However, the information on boundary conditions is transmitted more slowly to the interior in a multiblock domain than in a single-block domain; thus, the convergence speed may deteriorate with the increase in the number of blocks. The details can be found in Ferziger and Peric (1995).

In addition, the speed-up of parallel computations also depends on the information transfer among processors, which is machine-dependent. A detailed analysis of this problem can be found in Golub and Ortega (1993) and Shyy et al. (1997).

8.2 COUPLING OF 1-D, 2-D, AND 3-D MODELS

8.2.1 General considerations

As described in Chapters 6 and 7, the flow and sediment transport in rivers with complex geometries and hydraulic structures should be simulated using 2-D or 3-D models rather than 1-D models. However, it may not be feasible to use 2-D and 3-D models in the simulation of the fluvial processes in a long river reach during a long period, because they require much more computation time than 1-D models. Therefore, it is cost-effective to couple 1-D, 2-D, and/or 3-D models (McAnally *et al.*, 1986; Wu and Li, 1992; Vieira, 1995; Zhang, 1999). The basic idea is to divide the entire study domain into subdomains (reaches), and apply a 1-D model in less important subdomains with simple geometries and a 2-D or 3-D model in more important subdomains with complex geometries. For convenience, the subdomains (reaches) handled by 1-D, 2-D, and 3-D models are herein called 1-D, 2-D, and 3-D subdomains (reaches), respectively.

The concept of coupling 1-D, 2-D, and 3-D models in a generic river system is illustrated in Fig. 8.9. The upstream portion with dams can be simulated using a 1-D model, thus simplifying the boundary conditions. The broad floodplains can be calculated adequately using a depth-averaged 2-D model. The portion with bridge crossing should be simulated using a 3-D model. In the estuary, the stratified flow and salinity intrusion can be simulated using a width-averaged 2-D or 3-D model. In the estuary entrance, the interaction between river flow and tidal current is very complex and should be calculated using a 3-D model. Such a coupled modeling can provide a feasible solution for the entire river system.

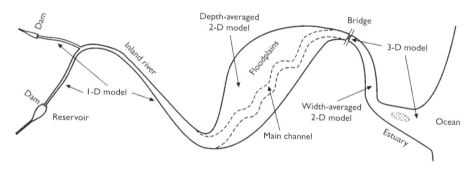

Figure 8.9 Concept of coupling 1-D, 2-D, and 3-D models.

Two approaches have been employed in the literature for combining 1-D, 2-D, and 3-D models. One approach is to use a simpler (1-D or 2-D) model in the entire solution domain and a more complex (2-D or 3-D) model in confined subdomains. The simpler model provides boundary conditions for the more complex model, but they are not coupled. This approach is called the hybrid modeling (McAnally *et al.*, 1986). The other approach is to fully couple 1-D, 2-D, and/or 3-D models (Wu and Li, 1992; Zhang, 1999) by simultenously solving all component models. Because the fully coupling approach is more general, it is the main concern here. The discretization and solution schemes in each subdomain usually are the same as those used in a single domain, so that the key issues are handling of the interfaces between subdomains and arrangement of the overall calculations. These are discussed in the next subsections.

8.2.2   Connection conditions

The solution domain can be partitioned into 1-D, 2-D, and 3-D subdomains in several ways, depending on its shape. For a relatively narrow channel, the domain is often divided into reaches as subdomains, as shown in Figs. 8.10(a) and (b). The 3-D model is applied in the near-field reaches where the flow is strongly three-dimensional, the 1-D model is used in the far-field reaches, and the 2-D model is used in the reaches between them. For a wide water body, the partition method shown in Fig. 8.10(c) is often used instead. The 3-D model is applied only in the regions around local structures, while the 2-D (or 1-D) model is used in the other regions or over the entire domain providing boundary conditions for the 3-D simulation around local structures. Like the multiblock method in Section 8.1, the 1-D, 2-D, and 3-D subdomains can be patched or overlapping at interfaces. For simplicity, only the patched interface is considered here.

Flow and sediment transport should satisfy continuity conditions at the interfaces between 1-D, 2-D, and 3-D subdomains, as discussed below (Wu, 1991; Wu and Li, 1992).

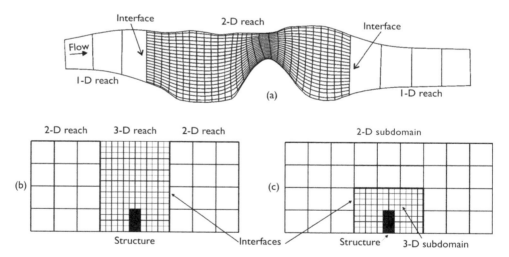

Figure 8.10 Connections among 1-D, 2-D, and 3-D subdomains.

Water level

The water levels at the interfaces should satify

$$Bz_{s,1d} = \int_0^B z_{s,2d}dy = \int_0^B z_{s,3d}dy \qquad (8.9)$$

where $z_{s,1d}$, $z_{s,2d}$, and $z_{s,3d}$ are the water levels calculated in the 1-D, 2-D, and 3-D subdomains, respectively; B is the width of the interface; and y is the coordinate along the interface.

Eq. (8.9) contains the conditions for connections between 1-D and depth-averaged 2-D models, between depth-averaged 2-D and 3-D models, and between 1-D and 3-D models. The connections with the width-averaged model are left to the interested reader.

In the 3-D model that solves the full Navier-Stokes equations, the dynamic pressure needs to be provided at the interfaces, whereas the 1-D and depth-averaged 2-D models assume a hydrostatic pressure distribution there. To overcome this problem, the interfaces should be located in the regions where the flow varies gradually and the hydrostatic pressure assumption is valid.

Flow discharge

The flow discharges at the interfaces should satisfy the continuity condition:

$$Q_{1d} = \int_0^B U_{2d} h_{2d} dy = \oiint_\Omega u_{3d} dy dz \tag{8.10}$$

where Q_{1d} is the flow discharge calculated in the 1-D subdomain, h_{2d} and U_{2d} are the flow depth and depth-averaged velocity in the 2-D subdomain, u_{3d} is the local velocity in the 3-D subdomain, Ω denotes the interface in the 3-D subdomain, and z is the vertical coordinate.

Flow resistance

The bed shear stresses at the interfaces satisfy

$$\chi \tau_{b,1d} = \int_0^\chi \tau_{b,2d} d\chi = \int_0^\chi \tau_{b,3d} d\chi \tag{8.11}$$

where $\tau_{b,1d}$, $\tau_{b,2d}$, and $\tau_{b,3d}$ are the bed shear stresses calculated in the 1-D, 2-D, and 3-D subdomains, respectively; and χ is the wetted perimeter at the interfaces.

In 1-D and 2-D models, the bed shear stress is determined using either the Manning equation or the log-law approach with the equivalent roughness height. Inserting the Manning equations for 1-D and 2-D uniform flows into condition (8.11) yields

$$n_{1d}^2 = \frac{R_{1d}^{4/3}}{A_{1d} U_{1d}^2} \int_0^\chi \frac{n_{2d}^2 u_{2d}^2}{h_{2d}^{1/3}} d\chi \tag{8.12}$$

where R_{1d} and U_{1d} are the hydraulic radius and the section-averaged flow velocity calculated in the 1-D reach, respectively; and n_{1d} and n_{2d} are the Manning coefficients in the 1-D and 2-D models, respectively.

It can be seen from Eq. (8.12) that in order to satisfy condition (8.11) at the interface between 1-D and 2-D reaches, n_{1d} and n_{2d} may have to be given different values.

Moreover, the bed shear stress is usually determined using the wall-function approach in the 3-D model. Thus, condition (8.11) may not be satisfied at the interface between 2-D (1-D) and 3-D reaches. Correction is often needed to satisfy this condition exactly.

Sediment discharge

The suspended-load and bed-load discharges at the interfaces should satisfy the mass balance conditions:

$$Q_{b,1d} = \int_0^B q_{b,2d} dy = \int_0^B q_{b,3d} dy \tag{8.13}$$

$$Q_{1d} C_{1d} = \int_0^B U_{2d} h_{2d} C_{2d} dy = \oiint_\Omega u_{3d} c_{3d} dy dz \tag{8.14}$$

where $Q_{b,1d}$, $q_{b,2d}$, and $q_{b,3d}$ are the total and unit bed-load transport rates; and C_{1d}, C_{2d}, and c_{3d} are the section-averaged, depth-averaged, and local suspended-load concentrations in the 1-D, 2-D, and 3-D subdomains, respectively.

Sediment transport capacity

Analogously to Eqs. (8.13) and (8.14), the bed-load and suspended-load transport capacities at the interfaces have relations:

$$Q_{b*,1d} = \int_0^B q_{b*,2d} dy = \int_0^B q_{b*,3d} dy \tag{8.15}$$

$$Q_{1d} C_{*,1d} = \int_0^B U_{2d} h_{2d} C_{*,2d} dy = \int_0^B \frac{1}{\alpha_{c*}} U_{3d} h_{3d} c_{b*,3d} dy \tag{8.16}$$

where $Q_{b*,1d}$, $q_{b*,2d}$, and $q_{b*,3d}$ are the total and unit equilibrium bed-load transport rates; $C_{*,1d}$, $C_{*,2d}$, and $c_{b*,3d}$ are the section-averaged, depth-averaged, and local equilibrium suspended-load concentrations in the 1-D, 2-D, and 3-D subdomains; h_{3d} and U_{3d} are the flow depth and depth-averaged velocity calculated in the 3-D subdomain; and α_{c*} is the adaptation coefficient introduced in Section 2.5.

At the interface between 1-D and 2-D subdomains, assuming $C_{*,1d} = K_{1d} [U_{1d}^3/(gR_{1d}\omega_s)]^m$ and $C_{*,2d} = K_{2d}[U_{2d}^3/(gh_{2d}\omega_s)]^m$ and substituing these relations into Eq. (8.16) yields

$$\frac{K_{1d}}{K_{2d}} = \frac{\int_0^B U_{2d} h_{2d} (U_{2d}^3/h_{2d})^m dy}{Q_{1d}(U_{1d}^3/R_{1d})^m} \tag{8.17}$$

Eq. (8.17) implies that if the same formula is used to determine the suspended-load transport capacities in 1-D and 2-D models, its coefficients need to be adjusted to satisfy condition (8.16). At the interface between 2-D and 3-D subdomains, the situation is even more complicated, because the 3-D model involves the equilibrium near-bed suspended-load concentration that is related to local flow features. This is also true for the bed-load transport capacities at the interfaces between 1-D, 2-D, and 3-D subdomains. Therefore, conservative correction is often made to satisfy conditions (8.15) and (8.16).

Bed change

The bed changes at the interfaces satisfy

$$\frac{\partial A_{b,1d}}{\partial t} = \int_0^B \frac{\partial z_{b,2d}}{\partial t} dy = \int_0^B \frac{\partial z_{b,3d}}{\partial t} dy \qquad (8.18)$$

where $\partial A_{b,1d}/\partial t$ is the rate of change in bed area calculated in the 1-D subdomain; and $\partial z_{b,2d}/\partial t$ and $\partial z_{b,3d}/\partial t$ are the rates of change in bed elevation calculated in the 2-D and 3-D subdomains, respectively.

At the interface between the 1-D and 2-D subdomains, bed change equations (5.36) and (6.61) show that if the actual discharges and transport capacities of sediment satisfy conditions (8.13)–(8.16) and if the adaptation length has the same value in the 1-D and 2-D domains, the bed changes automatically satisfy condition (8.18) at the interface. However, to maintain the same cross-sectional geometry at the interface, the redistribution of the 1-D bed area change along the channel width should result in the same bed elevation change as that calculated in the 2-D subdomain.

However, because the adapatation length L and coefficient α are often treated as calibrated parameters (as discussed in Sections 2.5 and 2.6) and different methods are used to calculate the 2-D and 3-D bed-load and suspended-load transport capacities, difficulties exist in satisfying condition (8.18) at the interface between 2-D and 3-D subdomains. The simplest way to solve this problem is to correct one of the bed changes calculated in the adjoining two subdomains to make sure both have the same value at the interface.

Bed-material gradation

The bed-material gradations at the interfaces satisfy

$$B p_{bk,1d} = \int_0^B p_{bk,2d} dy = \int_0^B p_{bk,3d} dy \qquad (8.19)$$

where $p_{bk,1d}$, $p_{bk,2d}$, and $p_{bk,3d}$ represent the bed-material gradations calculated in the 1-D, 2-D, and 3-D subdomains, respectively.

It is found from Eqs. (5.32) and (6.57) that if the mixing layer thickness has the same value in two neighboring subdomains, satisfaction of condition (8.18) for all size classes will guarantee satisfaction of condition (8.19).

Among all the above connection conditions, conditions (8.9), (8.10), (8.13), and (8.14) are more essential because they are internal boundary conditions and guarantee the continuity of flow and sediment transport between subdomains, while the other conditions affect only locally and can be satisfied by correction. However, there is a problem in the use of conditions (8.9), (8.10), (8.13), and (8.14). It is straightforward to convert the calculated water level, flow velocity, bed-load transport rate, and suspended-load concentration from 3-D to 2-D and then to 1-D subdomains using these four conditions, but the reverse conversion is not unique. To solve this problem, a common practice is to locate the interfaces at nearly straight prismatic channels or regions far away from hydraulic structures so that the following approximations are acceptable: the water level can be assumed to have a uniform lateral distribution; the depth-averaged flow velocity can be determined by Eq. (6.17); the unit transport rates of bed load and suspended load can be given by Eq. (6.63) at the interface from 1-D to 2-D subdomains; and the local flow velocity can be assumed to follow the log or power law and the local suspended-load concentration can be determined using the Rouse or Lane-Kalinske distribution along the depth at the interface from 2-D to 3-D subdomains. In addition, if the 2-D and 3-D grid points do not match at the interface, interpolation and conservative correction are needed in the conversion of 2-D and 3-D quantities.

8.2.3 Calculation procedures

For steady flow in a channel shown in Figs. 8.10(a) and (b), the flow discharges in all reaches are readily known. For simplicity, only subcritical flow is considered here. Thus, the calculation may be conducted reach by reach from downstream to upstream, with the water level at the outlet of each reach provided by the adjacent downstream reach and the lateral distribution of depth-averaged flow velocity at the inlets of 2-D and 3-D reaches determined by Eq. (6.17). However, for unsteady flow in such a channel, the hydrodynamic equations in all reaches are related and should usually be solved in a coupled manner. Conditions (8.9)–(8.11) should be satisfied at the interfaces.

For sediment transport in a channel shown in Figs. 8.10(a) and (b) under both steady and unsteady conditions, the calculation is conducted reach by reach from upstream to downstream, with the upstream reach providing boundary conditions for the downstream reach. Conditions (8.13)–(8.16), (8.18), and (8.19) should be satisfied at the interfaces.

For flow and sediment transport in a wide water body shown in Fig. 8.10(c), the governing equations in all subdomains should usually be solved together with conditions (8.9), (8.10), (8.13), and (8.14) at the interfaces. In addition to the solution procedures in individual component models, an iteration loop among subdomains should be introduced.

The computational time steps allowed by numerical stability in 1-D, 2-D, and 3-D models are usually different. Therefore, the overall time step should be carefully selected. One choice is to use the shortest time step allowed by all component models. The other choice is to use different time steps in different component models and convert the quantities at interfaces in different times by interpolation. The former choice is less efficient, whereas the latter choice is cumbersome for fully coupling unsteady 1-D, 2-D, and 3-D models.

8.2.4 Examples

Coupled 1-D/2-D/3-D models have been applied in numerous case studies. For example, McAnally *et al.* (1986) used a mixed 2-D/3-D model to study the salinity intrusion in New York Harbor. Because most of the harbor is a well-mixed estuary, a depth-averaged 2-D model was first used to verify the tidal propagation and overall current velocities, and then a 3-D model was applied using the coefficients from the 2-D verification as a starting point. This approach worked well since the 2-D model was used in a large portion of the domain and the 3-D model was only used in a small region where the flow was highly three-dimensional.

Wu and Li (1992) applied the coupled 1-D and depth-averaged 2-D quasi-steady model in the study of sedimentation problems in the fluctuating backwater region of China's Three Gorges Project (TGP) 30 years after the dam's construction. The study domain included a 176 km long reach in the main stream of the Yangtze River and a 13 km long reach in the Jialing River tributary. The entire study domain was divided into four 1-D reaches and four 2-D reaches. The simulation results were qualitatively consistent with the physical model results.

Zhang (1999) applied a 1-D unsteady model to simulate the flow and sediment transport in the channel and a depth-averaged 2-D unsteady model in the offshore area near the Yellow River mouth, as shown in Fig. 8.11. The coupled simulation gave plausible results for sediment concentration and bed deformation under the effects of runoff, tide, and wind-driven currents and waves.

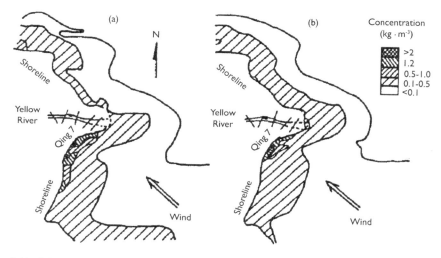

Figure 8.11 Calculated contours of sediment concentration caused by wind surges near the Yellow River mouth: (a) flood tide and (b) ebb tide (Zhang, 1999).

8.3 INTEGRATION OF CHANNEL AND WATERSHED MODELS

Human activities, such as urbanization, agricultural practices, vegetative clearing, dam construction, and river restoration, affect the equilibrium of watershed-channel systems. Prediction of these impacts at a watershed scale is very important. Traditionally, this prediction has been split into two parts: watershed and channel simulations, which compute the rainfall-runoff and soil erosion in upland fields and the flow routing and sediment transport in channels, respectively. This splitting has practical merits. However, because channel and watershed are deeply interrelated, watershed and channel simulations should be integrated for a successful watershed-scale study. The two parts in the integrated modeling system can complement each other to produce a better prediction. Watershed simulation provides a good description of upland runoff and soil erosion, especially for those ungauged watersheds, while channel simulation enhances the watershed-scale study by providing more accurate flow routing, sediment transport, and morphological evolution in channels.

8.3.1 Modeling components

A watershed model can be integrated with a 1-D, 2-D, or 3-D channel model in various ways. Here, the integration between the channel network model CCHE1D and the watershed model AGNPS or SWAT, as shown in Fig. 8.12, is used as an example to illustrate the concepts and methodologies of integration, which can be readily extended to other models. This integrated watershed-channel modeling system includes three model components: landscape analysis, watershed simulation, and channel simulation (Vieira and Wu, 2002), as described below.

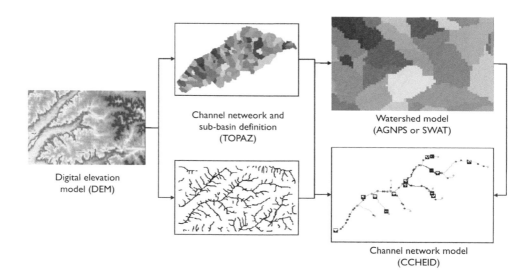

Figure 8.12 Integration of channel and watershed models.

Landscape analysis

The landscape analysis tool used here is the Topographic PArameteriZation (TOPAZ) program (Garbrecht and Martz, 1995), which comprehensively analyzes the raster Digital Elevation Models (DEM) to segment watersheds, define drainage divides, identify drainage networks, and parameterize subwatersheds.

The Deterministic Eight-neighbor (D8) method is used in TOPAZ to determine the landscape properties for each individual raster cell. The relative slope to each adjacent raster cell is calculated, and the flow is assigned in the direction of the steepest slope. After the flow directions are established, upstream drainage areas are determined for each cell. Channels are then defined as those cells with upstream drainage areas greater than a critical source area (CSA). The CSA represents the drainage area necessary to produce enough runoff to form a permanent channel. However, if only the CSA is considered, it is possible to generate a few very short exterior (source) channels. TOPAZ thus provides an option to remove these spurious channels by applying a minimum source channel length (MSCL). The defined channels are further processed to determine the Strahler orders and assign identification numbers in the channel network. The subwatersheds draining into source nodes and into the left and right banks of each channel are also identified. The connectivity of the derived channels and subwatersheds is established and a variety of properties and parameters of them are deduced, which will be used in the watershed and channel simulations.

Watershed simulation

The digital representation of the delineated subwatersheds can be used as computational elements by one of the existing watershed models, such as AGNPS (Agricultural Non-Point Source model, Bingner and Theurer, 2001) and SWAT (Soil and Water Assessment Tool, Arnold *et al.*, 1993). The flow, sediment and pollutant loads in each subwatershed are computed continuously for a series of storm events.

The basic model components in AGNPS and SWAT are hydrology, erosion, sediment transport, and chemical transport. In the hydrology component, runoff volume is calculated by the SCS (U.S. Soil Conservation Service) curve number procedure. Peak flow rate is estimated using empirical equations, which take into account the effects of drainage area, channel slope, runoff volume, and watershed length-width ratio. AGNPS uses the Revised Universal Soil Loss Equation (RUSLE) to calculate sediment yield, while SWAT uses the Modified Universal Soil Loss Equation (MUSLE), in which the rainfall factor is replaced with a runoff factor to predict sediment yield on an event basis.

The outputs of the watershed model, which consist of daily water runoff and sediment loads for all subwatersheds, are transferred to the CCHE1D channel network model as inflow boundary conditions.

Channel simulation

The flow and sediment transport in the channel network are simulated using the 1-D model CCHE1D (Wu and Vieira, 2002), which is described in Sections 5.2.2 and 5.3.3. The flow and sediment transport equations are discretized using the Preissmann implicit scheme. The discretized flow equations are solved using the Thomas algorithm, while the discretized sediment transport equations are solved using a direct method. The runoff and sediment loads from upland fields calculated by the watershed model are used as inflow discharges at the source nodes or as side discharges at the internal nodes in the channel network. The effects of in-stream hydraulic structures are taken into account by solving the corresponding stage-discharge equations in the flow model. Bank toe erosion and mass failure are simulated using Osman and Thorne's (1988) method. The eroded bank materials are treated as side discharge in the sediment transport equations. The outputs of the channel model are the runoff and sediment yields from both uplands and channels, channel morphological changes, etc.

8.3.2 Integration approaches

The three modeling components (landscape analysis, watershed simulation, and channel simulation) can be integrated in several ways. The first approach is to combine them into a single modeling system. Such an integrated system is very convenient for data transfer and manipulation but very expensive to develop. The second approach is to develop a separate subsystem for each modeling component, with common data file formats being set up for data transfer between the modeling components. The third approach is to develop two subsystems: one combines the landscape analysis and channel simulation, and the other combines the landscape analysis and watershed simulation. Data transfer is still needed between the two subsystems. In order to make the two subsystems compatible, the landscape analysis tools used by them should be identical or very similar. The integration of CCHE1D and AGNPS (or SWAT) adopts the third approach (Vieira and Wu, 2002).

8.3.3 Scale issues

Time scale

The watershed model uses a daily time step, while the channel model employs a time step in the order of minutes. It is not efficient to use a common time step for both components in the integrated modeling system. The more widely adopted approach uses different time steps. However, it is needed to convert the daily runoff and sediment loads calculated by the watershed model to continuous hydrographs for channel simulation. For runoff, a triangular or gamma-function hydrograph shown in Fig.8.13 is often used. To define the triangular or gamma-function hydrograph, the base flow discharge (Q_0), peak flow discharge (Q_p), starting time (t_0), and time to peak (t_p) should be provided. The duration (t_d) is additionally required for the triangular hydrograph, and a shape factor for the gamma-function hydrogragh. These parameters depend

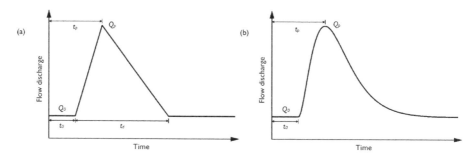

Figure 8.13 Typical hydrographs: (a) triangular and (b) gamma-function.

on precipitation, drainage area, land use, land cover, and soil properties. They are determined by the watershed model.

The daily sediment loads from upland fields can also be converted to the triangular or gamma-function sedigraphs, by assuming the sediment discharge to be proportional to the flow discharge. In addition, the watershed model often gives only fractional loads for clay, silt, and sand. These three size classes perhaps are too few for channel simulation and need to be subdivided into more size classes.

Space scale

Because the watershed and channel models have different computational domains, a space scale problem exists in the integrated simulation. Fig. 8.14(a) shows the configuration commonly used in the watershed model, in which a channel usually starts from a source node or a junction node and ends at the watershed outlet or another junction node. The subwatersheds are segmented corresponding to the source nodes and to the left and right sides of channels. The runoff and sediment loads at each subwatershed are simulated. However, the channels used in the watershed model are usually too long to be used as computational elements in the channel model and thus should be subdivided into shorter reaches by adding more computational points, as shown in Fig. 8.14(b). This requires subdivision of the runoff and sediment loads from the subwatersheds to the reaches.

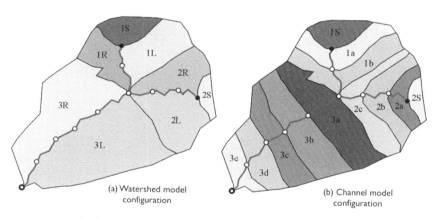

(a) Watershed model configuration

(b) Channel model configuration

Figure 8.14 Computational domains in watershed and channel models.

There are three ways to subdivide the runoff and sediment loads at subwatersheds. One is to put all loads on the source node or the junction node, i.e., the first node of each channel. However, this approach may not be accurate enough. The second way is to distribute the loads uniformly along the channel length. The third way is to define the subcatchment for each reach and then distribute the loads according to the subcatchment area, as shown in Fig. 8.14(b). The third way is the most reasonable.

8.3.4 Application to Goodwin Creek Watershed

The integrated watershed-channel simulation was applied in the study of sediment transport in the Goodwin Creek Experimental Watershed established in 1977 as a prototype of the much larger Demonstrate Erosion Control (DEC) watersheds in the US. The drainage area above the watershed outlet was 21.3 km². Most of the channels were ephemeral, with perennial flows occurring only in the lower reaches of the watershed. The sediment transported in the channels ranged from silt (<0.062 mm) to sand to gravel (<65 mm). TOPAZ was used to delineate a network

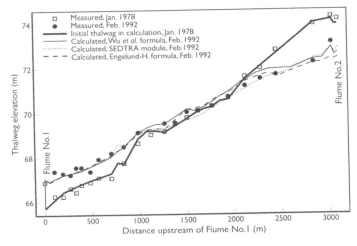

Figure 8.15 Thalweg changes in the lower reach of Goodwin Creek.

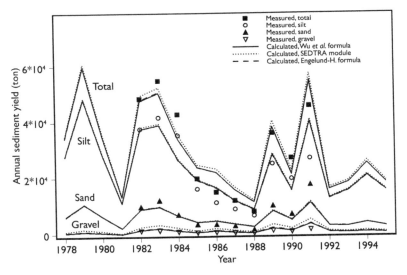

Figure 8.16 Sediment yields at the outlet of Goodwin Creek Watershed.

of 19 channels and 138 subbasins from a 30-meter resolution DEM, as illustrated in Fig. 8.12.

Bingner *et al.* (1997) used SWAT to simulate the runoff and sediment loads in the watershed for 1192 storm events from 1978 to 1995. Wu and Vieira (2002) computed the flow routing, sediment transport, and morphological evolution in the channels using CCHE1D, with the watershed model results given as inputs. The channel simulation considered the effects of ten in-stream measuring flumes and four box culverts. Fig. 8.15 shows the calculated and measured thalweg changes in the lower 3-km reach upstream of the watershed outlet (flume No. 1) from 1978 to 1992. The model reproduced well the patterns of erosion and deposition. Fig. 8.16 shows the calculated and measured sediment yields at the watershed outlet. The level of observed agreement is good.

Chapter 9

Simulation of dam-break fluvial processes

Flood due to dam failure can cause catastrophic damage of properties and loss of life. Prediction of dam-break flow and the associated sediment transport is very important for dam operation, flood control, disaster mitigation, water infrastructure safety assessment, etc. This chapter covers the numerical algorithms developed for simulation of dam-break and overtopping flows over fixed and movable beds.

9.1 SIMULATION OF DAM-BREAK FLOW OVER FIXED BEDS

Although the St. Venant (or shallow water) equations were derived based on the hydrostatic pressure assumption, many studies have revealed that these equations are approximately applicable to dam-break flow over fixed beds. The 1-D St. Venant equations (5.1) and (5.2) are written in conservative form as follows:

$$\frac{\partial \mathbf{\Phi}}{\partial t} + \frac{\partial \mathbf{F}(\mathbf{\Phi})}{\partial x} = \mathbf{S}(\mathbf{\Phi}) \tag{9.1}$$

where $\mathbf{\Phi}, \mathbf{F}(\mathbf{\Phi})$, and $\mathbf{S}(\mathbf{\Phi})$ represent the vectors of unknown variables, fluxes, and source terms:

$$\mathbf{\Phi} = \begin{bmatrix} h \\ q \end{bmatrix}, \quad \mathbf{F}(\mathbf{\Phi}) = \begin{bmatrix} q \\ q^2/h + gh^2/2 \end{bmatrix}, \quad \mathbf{S}(\mathbf{\Phi}) = \begin{bmatrix} 0 \\ gh(S_0 - S_f) \end{bmatrix} \tag{9.2}$$

where q is the flow discharge per unit channel width; S_0 is the bed slope; and S_f is the friction slope: $S_f = n^2 q|q|/h^{10/3}$.

The convection term in Eq. (9.1) can be rewritten as

$$\frac{\partial \mathbf{F}(\mathbf{\Phi})}{\partial x} = \frac{\partial \mathbf{F}(\mathbf{\Phi})}{\partial \mathbf{\Phi}} \frac{\partial \mathbf{\Phi}}{\partial x} = \mathbf{M} \frac{\partial \mathbf{\Phi}}{\partial x} \tag{9.3}$$

where $\mathbf{M}$ is the Jacobian matrix:

$$\mathbf{M} = \begin{pmatrix} 0 & 1 \\ -q^2/h^2 + gh & 2q/h \end{pmatrix} \tag{9.4}$$

Note that the use of primary variables h and q in Eq. (9.2) is restricted in rectangular channels. However, this arrangement provides convenience for extension of a 1-D algorithm to the 2-D model.

The 2-D shallow water equations (6.1)–(6.3) are written in the following conservative form:

$$\frac{\partial \boldsymbol{\Phi}}{\partial t} + \frac{\partial \mathbf{F}(\boldsymbol{\Phi})}{\partial x} + \frac{\partial \mathbf{G}(\boldsymbol{\Phi})}{\partial y} = \mathbf{S}(\boldsymbol{\Phi}) \tag{9.5}$$

where $\boldsymbol{\Phi}, \mathbf{F}(\boldsymbol{\Phi}), \mathbf{G}(\boldsymbol{\Phi})$, and $\mathbf{S}(\boldsymbol{\Phi})$ represent the vectors of unknown variables, fluxes, and source terms:

$$\boldsymbol{\Phi} = \begin{bmatrix} h \\ hU_x \\ hU_y \end{bmatrix}, \quad \mathbf{F}(\boldsymbol{\Phi}) = \begin{bmatrix} hU_x \\ hU_x^2 + gh^2/2 \\ hU_xU_y \end{bmatrix},$$

$$\mathbf{G}(\boldsymbol{\Phi}) = \begin{bmatrix} hU_y \\ hU_xU_y \\ hU_y^2 + gh^2/2 \end{bmatrix}, \quad \mathbf{S}(\boldsymbol{\Phi}) = \begin{bmatrix} 0 \\ gh(S_{0x} - S_{fx}) \\ gh(S_{0y} - S_{fy}) \end{bmatrix} \tag{9.6}$$

The stress effects are usually omitted in Eq. (9.5) for dam-break flow. The convection terms can be rewritten as

$$\frac{\partial \mathbf{F}(\boldsymbol{\Phi})}{\partial x} = \mathbf{A}\frac{\partial \boldsymbol{\Phi}}{\partial x}, \quad \frac{\partial \mathbf{G}(\boldsymbol{\Phi})}{\partial y} = \mathbf{B}\frac{\partial \boldsymbol{\Phi}}{\partial y} \tag{9.7}$$

where $\mathbf{A}$ and $\mathbf{B}$ are the Jacobian matrices:

$$\mathbf{A} = \begin{bmatrix} 0 & 1 & 0 \\ gh - U_x^2 & 2U_x & 0 \\ -U_xU_y & U_y & U_x \end{bmatrix}, \quad \mathbf{B} = \begin{bmatrix} 0 & 0 & 1 \\ -U_xU_y & U_y & U_x \\ gh - U_y^2 & 0 & 2U_y \end{bmatrix} \tag{9.8}$$

Many traditional numerical schemes, such as the Preissmann (1961) scheme, designed for solving the 1-D shallow water equations in common flow situations, are insufficient for dam-break flow simulation, producing non-physical oscillations. Numerous numerical schemes based on finite volume, finite difference, and finite element methods have been developed recently for simulation of dam-break flow. As an example, a finite volume approach is introduced here, which uses the non-staggered grid shown in Fig. 9.1 for the 1-D problem. The computational domain is divided into I segments. Each segment is a control volume (cell) embraced by two faces. The primary variables h and q are defined at cell centers and represent the average values over each cell, while the fluxes are defined at cell faces.

Integrating Eq. (9.1) over the ith control volume yields

$$\int_{x_{i-1/2}}^{x_{i+1/2}} \frac{\partial \boldsymbol{\Phi}}{\partial t} dx + \int_{x_{i-1/2}}^{x_{i+1/2}} \frac{\partial \mathbf{F}(\boldsymbol{\Phi})}{\partial x} dx = \int_{x_{i-1/2}}^{x_{i+1/2}} \mathbf{S}(\boldsymbol{\Phi}) dx \tag{9.9}$$

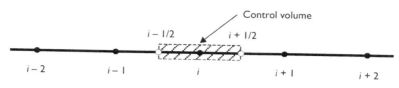

Figure 9.1 1-D finite volume mesh.

Applying the Green theorem to Eq. (9.9) and using the Euler scheme for the time derivative results in the following discretized equation:

$$\Phi_i^{n+1} = \Phi_i^n - \frac{\Delta t}{\Delta x_i}(F_{i+1/2}^n - F_{i-1/2}^n) + \Delta t S_i^n \qquad (9.10)$$

where $F_{i+1/2}^n$ is the intercell flux at face $i + 1/2$, Δx_i is the length of the ith control volume, Δt is the time step, and the superscript n is the time step index.

A rectangular (quadrilateral) or triangular mesh may be used in the numerical solution of the 2-D shallow water equations. For simplicity, the rectangular mesh shown in Fig. 9.2 is used here. Integrating Eq. (9.5) over the 2-D control volume numbered as (i, j) and using the Euler scheme for the time derivative yields the following discretized equation:

$$\Phi_{i,j}^{n+1} = \Phi_{i,j}^n - \frac{\Delta t}{\Delta x_{i,j}}(F_{i+1/2,j}^n - F_{i-1/2,j}^n) - \frac{\Delta t}{\Delta y_{i,j}}(G_{i,j+1/2}^n - G_{i,j-1/2}^n) + \Delta t S_{i,j}^n \qquad (9.11)$$

where $\Delta x_{i,j}$ and $\Delta y_{i,j}$ are the lengths of the control volume in the x- and y-directions, respectively.

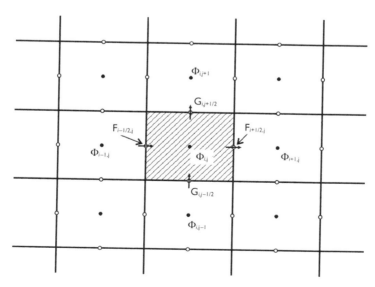

Figure 9.2 2-D finite volume mesh.

The Euler scheme for the time-derivative terms results in simple, explicit time-marching algorithms in Eqs. (9.10) and (9.11), even though it is only first-order accurate in time. The accuracy can be enhanced by applying the MacCormack scheme (Fennema and Chaudhry, 1990), Range-Kutta method, etc.

To complete the discretization in Eqs. (9.10) and (9.11), numerical schemes are needed to evaluate the intercell fluxes. The often used schemes are the central difference scheme and MacCormack scheme with artificial diffusion fluxes (e.g., Fennema and Chaudhry, 1990; Molls and Chaudhry, 1995), approximate Riemann solvers (Godunov, 1959; Roe, 1981; Osher and Solomon, 1982; Harten et al., 1983; Toro et al., 1994), TVD (Total Variation Diminishing) schemes (Harten, 1983; Yee, 1987; Garcia-Navarro et al., 1992; Wang et al., 2000), and upwind flux schemes (Ying et al., 2004). Some of them are introduced briefly below.

9.1.1 Central difference scheme with artificial diffusion flux

If the intercell fluxes are evaluated using the central difference scheme, spurious oscillations may appear in the solution near regions with sharp gradients and discontinuities (see Molls and Chaudhry, 1995). To eliminate these oscillations, an artificial diffusion (and/or dissipation) flux is usually added, thus yielding

$$\mathbf{F}_{i+1/2}^n = \frac{1}{2}(\mathbf{F}_i^n + \mathbf{F}_{i+1}^n) + \mathbf{D}_{i+1/2}^n \tag{9.12}$$

where $\mathbf{D}_{i+1/2}^n$ is the artificial diffusion flux. Various formulations for $\mathbf{D}_{i+1/2}^n$ can be found in the literature. The one suggested by Martinelli and Jameson (1988) consists of second-order and fourth-order terms:

$$\mathbf{D}_{i+1/2}^n = \rho(\mathbf{M})_{i+1/2}[-\varepsilon_{i+1/2}^{(2)}(\mathbf{\Phi}_{i+1}^n - \mathbf{\Phi}_i^n) + \varepsilon_{i+1/2}^{(4)}(\mathbf{\Phi}_{i+2}^n - 3\mathbf{\Phi}_{i+1}^n + 3\mathbf{\Phi}_i^n - \mathbf{\Phi}_{i-1}^n)] \tag{9.13}$$

where $\rho(\mathbf{M})$ is the spectral radius of the Jacobian matrix $\mathbf{M}$, defined as $\rho(\mathbf{M}) = |q|/h + \sqrt{gh}$; and $\varepsilon_{i+1/2}^{(2)}$ and $\varepsilon_{i+1/2}^{(4)}$ are non-linear functions:

$$\varepsilon_{i+1/2}^{(2)} = \kappa^{(2)}\max(\chi_i, \chi_{i+1}), \quad \varepsilon_{i+1/2}^{(4)} = \max[0, (\kappa^{(4)} - \varepsilon_{i+1/2}^{(2)})] \tag{9.14}$$

with $\kappa^{(2)}$ and $\kappa^{(4)}$ being coefficients, and $\chi_i = |h_{i-1} - 2h_i + h_{i+1}|/(h_{i-1} + 2h_i + h_{i+1})$.

In the 2-D case, in analogy to Eq. (9.12), the intercell fluxes are evaluated as

$$\mathbf{F}_{i+1/2,j}^n = \frac{1}{2}(\mathbf{F}_{i,j}^n + \mathbf{F}_{i+1,j}^n) + \rho(\mathbf{A})_{i+1/2,j}[-\varepsilon_{i+1/2,j}^{(2)}(\mathbf{\Phi}_{i+1,j}^n - \mathbf{\Phi}_{i,j}^n)$$
$$+ \varepsilon_{i+1/2,j}^{(4)}(\mathbf{\Phi}_{i+2,j}^n - 3\mathbf{\Phi}_{i+1,j}^n + 3\mathbf{\Phi}_{i,j}^n - \mathbf{\Phi}_{i-1,j}^n)] \tag{9.15a}$$

$$G_{i+1/2,j}^n = \frac{1}{2}(G_{i,j}^n + G_{i,j+1}^n) + \rho(\mathbf{B})_{i,j+1/2}[-\varepsilon_{i,j+1/2}^{(2)}(\Phi_{i,j+1}^n - \Phi_{i,j}^n)$$

$$+ \varepsilon_{i,j+1/2}^{(4)}(\Phi_{i,j+2}^n - 3\Phi_{i,j+1}^n + 3\Phi_{i,j}^n - \Phi_{i,j-1}^n)] \qquad (9.15\text{b})$$

where $\rho(\mathbf{A})$ and $\rho(\mathbf{B})$ are the spectral radii of matrices $\mathbf{A}$ and $\mathbf{B}$, respectively.

9.1.2 Approximate Riemann solvers

For the 1-D problem, Godunov (1959) suggested determining the intercell fluxes in Eq. (9.10) as

$$\mathbf{F}_{i+1/2} = \mathbf{F}(\Phi_{i+1/2}(0)) \qquad (9.16)$$

where $\Phi_{i+1/2}(0)$ is the exact similarity solution $\Phi_{i+1/2}(x/t)$ of the Riemann problem

$$\begin{cases} \dfrac{\partial \Phi}{\partial t} + \dfrac{\partial \mathbf{F}(\Phi)}{\partial x} = 0 \\[2mm] \Phi(x,0) = \begin{cases} \Phi_L, & \text{if } x < 0 \\ \Phi_R, & \text{if } x > 0 \end{cases} \end{cases} \qquad (9.17)$$

evaluated at $x/t = 0$.

Solving the exact Riemann problem (9.17) is rather complicated. The frequently used methods are approximate Riemann solvers. Many of them were developed in computational aerodynamics (Godunov, 1959; Roe, 1981; van Leer, 1982; Harten et al., 1983; Osher and Solomon, 1982; etc.) and later adopted in free-surface flow simulation (Glaister, 1988; Alcrudo and Garcia-Navarro, 1993; Jha et al., 2000; etc.). Here, the HLL and HLLC approximate Riemann solvers are introduced as examples. More Riemann solvers were summarized by Toro (2001).

Harten, Lax, and van Leer (HLL, 1983) suggested an approximate Riemann solver by directly finding an approximation to the intercell flux (9.16). The HLL approach assumes estimates S_L and S_R for the smallest and largest signal velocities in the solution of the Riemann problem (9.17) with data $\Phi_L = \Phi_i^n$, $\Phi_R = \Phi_{i+1}^n$ and corresponding fluxes $\mathbf{F}_L = \mathbf{F}(\Phi_L)$, $\mathbf{F}_R = \mathbf{F}(\Phi_R)$, as shown in Fig. 9.3. The following HLL numerical flux is derived by applying the integral form of the conservation laws in appropriate control volumes:

$$\mathbf{F}_{i+1/2} = \begin{cases} \mathbf{F}_L & \text{if } S_L \geq 0 \\[2mm] \mathbf{F}^{hll} \equiv \dfrac{S_R \mathbf{F}_L - S_L \mathbf{F}_R + S_R S_L(\Phi_R - \Phi_L)}{S_R - S_L} & \text{if } S_L \leq 0 \leq S_R \\[2mm] \mathbf{F}_R & \text{if } S_R \leq 0 \end{cases} \qquad (9.18)$$

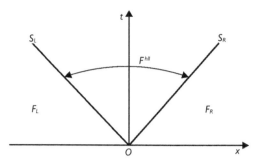

Figure 9.3 Wave structure assumed in HLL approach.

There are several ways to determine the wave speed estimates S_L and S_R. One choice was suggested by Toro (2001) as follows:

$$S_L = u_L - a_L \lambda_L, \quad S_R = u_R + a_R \lambda_R \qquad (9.19)$$

where $u_K (K = L, R)$ is the velocity, a_K is the celerity, and λ_K is given as

$$\lambda_K = \begin{cases} \sqrt{\dfrac{1}{2} \dfrac{(h_* + h_K) h_*}{h_K^2}} & \text{if } h_* > h_K \\[3mm] 1 & \text{if } h_* \le h_K \end{cases} \qquad (9.20)$$

where h_* is an estimate for the exact solution of h in the star region and can be evaluated as

$$h_* = \frac{1}{2}(h_L + h_R) - \frac{1}{4}(u_R - u_L)(h_L + h_R)/(a_L + a_R) \qquad (9.21)$$

The HLL approximate Riemann solver ignores intermediate waves, such as shear waves and contact discontinuities, which arise when scalar transport equations are added to the basic shallow water equations. Toro *et al.* (1994) proposed a modification of the HLL scheme to account for the influence of intermediate waves. This new approach is called the HLLC Riemann solver. Fig. 9.4 illustrates the assumed wave structure in the HLLC scheme. S_* denotes the estimate of the speed of the middle wave. In the exact Riemann solver, $S_* = u_*$. Unlike the case in Fig. 9.3, there are two distinct fluxes in the star region in Fig. 9.4. The HLLC numerical flux is thus determined by

$$\mathbf{F}_{i+1/2} = \begin{cases} \mathbf{F}_L & \text{if } S_L \ge 0 \\ \mathbf{F}_{*L} \equiv \mathbf{F}_L + S_L(\boldsymbol{\Phi}_{*L} - \boldsymbol{\Phi}_L) & \text{if } S_L \le 0 \le S_* \\ \mathbf{F}_{*R} \equiv \mathbf{F}_R + S_R(\boldsymbol{\Phi}_{*R} - \boldsymbol{\Phi}_R) & \text{if } S_* \le 0 \le S_R \\ \mathbf{F}_R & \text{if } S_R \le 0 \end{cases} \qquad (9.22)$$

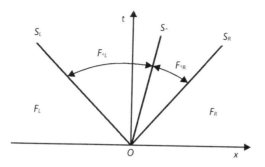

Figure 9.4 HLLC Riemann solver for *x*-split 2-D shallow water equations.

with the states $\mathbf{\Phi}_{*L}$ and $\mathbf{\Phi}_{*R}$ given by

$$\mathbf{\Phi}_{*K} = h_K \left(\frac{S_K - u_K}{S_K - S_*} \right) \begin{pmatrix} 1 \\ S_* \\ \phi_K \end{pmatrix} \qquad (9.23)$$

where ϕ is the variable representing the scalar quantity.

The wave speed estimates S_L and S_R are given by Eq. (9.19). The estimate S_* for the middle wave speed can be provided as the particle speed u_* that is estimated below:

$$u_* = \frac{1}{2}(u_L + u_R) - (h_R - h_L)(a_L + a_R)/(h_L + h_R) \qquad (9.24)$$

Riemann solvers for the 2-D problem can also be established, but they are usually very complicated. The often used method is to split the 2-D shallow water equations (9.5) into two augmented 1-D equations along the *x*- and *y*-directions as

$$\begin{cases} \dfrac{\partial \mathbf{\Phi}}{\partial t} + \dfrac{\partial \mathbf{F}(\mathbf{\Phi})}{\partial x} = \mathbf{S}_x(\mathbf{\Phi}) \\[2mm] \dfrac{\partial \mathbf{\Phi}}{\partial t} + \dfrac{\partial \mathbf{G}(\mathbf{\Phi})}{\partial y} = \mathbf{S}_y(\mathbf{\Phi}) \end{cases} \qquad (9.25)$$

where $\mathbf{S}_x$ and $\mathbf{S}_y$ are the source terms split from $\mathbf{S}$.

Applying the finite volume discretization scheme (9.10) for Eq. (9.25) yields

$$\mathbf{\Phi}_{ij}^{n+1/2} = \mathbf{\Phi}_{ij}^{n} - \frac{\Delta t}{\Delta x_{i,j}}(\mathbf{F}_{i+1/2,j}^{n} - \mathbf{F}_{i-1/2,j}^{n}) + \Delta t \mathbf{S}_{xi}^{n} \qquad (9.26a)$$

$$\mathbf{\Phi}_{ij}^{n+1} = \mathbf{\Phi}_{ij}^{n+1/2} - \frac{\Delta t}{\Delta y_{i,j}}(\mathbf{G}_{i,j+1/2}^{n+1/2} - \mathbf{G}_{i,j-1/2}^{n+1/2}) + \Delta t \mathbf{S}_{yi}^{n+1/2} \qquad (9.26b)$$

The computation procedure consists of two sweeps. In the first sweep, Eq. (9.26a) is solved along the i-direction to obtain $\Phi_{ij}^{n+1/2}$ from Φ_{ij}^{n}; in the second sweep, Eq. (9.26b) is solved along the j-direction to obtain Φ_{ij}^{n+1} from $\Phi_{ij}^{n+1/2}$. Both sweeps have a common time step, Δt. In each sweep, Eq. (9.22) can be applied to evaluate the intercell fluxes $\mathbf{F}$ and $\mathbf{G}$.

The aforementioned splitting scheme is first-order accurate in time. More first- and second-order accurate splitting schemes can be found in Toro (2001) and other references.

9.1.3   TVD schemes

9.1.3.1   TVD schemes for scalar problems

Consider a one-dimensional function ϕ in the domain shown in Fig. 9.1. The total variation at time level t^n is defined as

$$TV(\phi^n) = \sum_{i=1}^{I-1} |\phi_{i+1}^n - \phi_i^n| \tag{9.27}$$

A numerical scheme $\phi_i^{n+1} = H(\phi_{i-k}^n, \ldots, \phi_i^n, \ldots, \phi_{i+l}^n)$ is said to be Total Variation Diminishing (TVD) if

$$TV(\phi^{n+1}) \leq TV(\phi^n) \quad \forall n \tag{9.28}$$

A TVD scheme can preserve the monotonicity of the solution. This means that if the data $\{\phi_i^n\}$ is monotone, the solution $\{\phi_i^{n+1}\}$ by a TVD scheme is monotone in the same sense (Harten, 1983).

Suppose that numerical schemes written as

$$\phi_i^{n+1} = \phi_i^n - C_{i-1/2}(\phi_i^n - \phi_{i-1}^n) + D_{i+1/2}(\phi_{i+1}^n - \phi_i^n) \tag{9.29}$$

are used to solve the following equation for the scalar hyperbolic conservation law:

$$\frac{\partial \phi}{\partial t} + \frac{\partial f(\phi)}{\partial x} = 0 \tag{9.30}$$

where $f(\phi)$ is the flux. According to Harten (1983), the sufficient conditions for any scheme written as Eq. (9.29) to be TVD are

$$C_{i-1/2} \geq 0, \quad D_{i+1/2} \geq 0, \quad \text{and} \quad 0 \leq C_{i-1/2} + D_{i+1/2} \leq 1 \tag{9.31}$$

Harten (1983), Sweby (1984), and Yee (1987) developed a variety of TVD schemes that respect conditions (9.31). Most of the TVD schemes are essentially constructed by combining high-order and low-order fluxes with a limiter imposed to constrain the gradients of flux functions and to prevent the formation of local extrema. One typical approach is given below.

Integrating Eq. (9.30) over the ith control volume in Fig. 9.1 yields the following general conservative scheme:

$$\phi_i^{n+1} = \phi_i^n - \frac{\Delta t}{\Delta x_i}(f_{i+1/2}^n - f_{i-1/2}^n) \tag{9.32}$$

The intercell flux $f_{i+1/2}^n$ is constructed as (see Wang $et\ al.$, 2000)

$$f_{i+1/2}^n = \frac{1}{2}\left\{f_i^n + f_{i+1}^n - \left[\frac{\Delta t}{\Delta x_i}(a_{i+1/2})^2\varphi(r) + (1 - \varphi(r))\psi(a_{i+1/2})\right]\Delta\phi_{i+1/2}\right\} \tag{9.33}$$

where $\Delta\phi_{i+1/2} = \phi_{i+1}^n - \phi_i^n$; $a_{i+1/2}$ is the characteristic speed, defined as $a_{i+1/2} = \Delta f_{i+1/2}/\Delta\phi_{i+1/2}$ when $\Delta\phi_{i+1/2} \neq 0$ and $a_{i+1/2} = a(\phi_i)$ when $\Delta\phi_{i+1/2} = 0$; and $\psi(a_{i+1/2})$ is the dissipative function, defined by Harten (1983) as

$$\psi(a_{i+1/2}) = \begin{cases} |a_{i+1/2}|, & |a_{i+1/2}| \geq \varepsilon \\ [(a_{i+1/2})^2 + \varepsilon^2]/2\varepsilon \text{ or } \varepsilon, & |a_{i+1/2}| < \varepsilon \end{cases} \tag{9.34}$$

where ε is a small positive number.

The function $\varphi(r)$ in Eq. (9.33) is a limiter, with r defined as

$$r = \frac{[|a_{i+1/2-\sigma}| - \Delta t(a_{i+1/2-\sigma})^2/\Delta x_{i-\sigma}]\Delta\varphi_{i+1/2-\sigma}}{[|a_{i+1/2}| - \Delta t(a_{i+1/2})^2/\Delta x_i]\Delta\varphi_{i+1/2}}, \quad \sigma = \text{sign}(a_{i+1/2}) \tag{9.35}$$

The limiter function $\varphi(r)$ is essential to obtain monotonic solutions. Many limiters have been published in the literature. For example, $\varphi(r) = 0$ yields the upwind scheme that is a first-order TVD scheme. The commonly used limiters in second-order TVD schemes include Roe's minmod limiter

$$\varphi(r) = \min \text{mod}(1, r) \tag{9.36}$$

van Leer's monotonic limiter

$$\varphi(r) = \frac{r + |r|}{1 + |r|} \tag{9.37}$$

van Leer's MUSCL limiter

$$\varphi(r) = \max[0, \min(2r, 2), 0.5(1 + r)] \tag{9.38}$$

and Roe's superbee limiter

$$\varphi(r) = \max[0, \min(2r, 1), \min(r, 2)] \tag{9.39}$$

9.1.3.2 TVD schemes for St. Venant equations

Many investigators, e.g., Garcia-Navapro *et al.* (1992) and Wang *et al.* (2000), have extended TVD schemes to solve 1-D and 2-D shallow water equations. Here, the method of Wang *et al.* (2000) is presented.

For the 1-D problem governed by Eq. (9.1), the eigenvalues of $\mathbf{M}$ in Eq. (9.4) are

$$a^1 = q/h - \sqrt{gh}, \quad a^2 = q/h + \sqrt{gh} \tag{9.40}$$

and the corresponding right eigenvectors are

$$\mathbf{R}^1 = \begin{pmatrix} 1 \\ a^1 \end{pmatrix}, \quad \mathbf{R}^2 = \begin{pmatrix} 1 \\ a^2 \end{pmatrix} \tag{9.41}$$

Define $\Delta \boldsymbol{\Phi}_{i+1/2} = \boldsymbol{\Phi}_{i+1}^n - \boldsymbol{\Phi}_i^n$, which can be written as

$$\Delta \boldsymbol{\Phi}_{i+1/2} = \sum_{l=1}^{2} \alpha_{i+1/2}^l \mathbf{R}_{i+1/2}^l \tag{9.42}$$

where $\alpha_{i+1/2}^l$ represents the component of $\Delta \boldsymbol{\Phi}_{i+1/2}$ in the coordinate system $\mathbf{R}_{i+1/2}^l$.

In analogy to Eq. (9.33), the following intercell flux used in the scheme (9.10) for solving Eq. (9.1) can be constructed by using Eq. (9.42):

$$\mathbf{F}_{i+1/2}^n = \frac{1}{2} \left\{ \mathbf{F}_i^n + \mathbf{F}_{i+1}^n \right.$$

$$\left. - \sum_{l=1}^{2} \left[\frac{\Delta t}{\Delta x_i} (a_{i+1/2}^l)^2 \varphi(r) + (1 - \varphi(r)) \psi(a_{i+1/2}^l) \right] \alpha_{i+1/2}^l \mathbf{R}_{i+1/2}^l \right\} \tag{9.43}$$

For the 2-D problem, Wang *et al.* (2000) split Eq. (9.5) into two augmented 1-D systems similar to Eq. (9.25) along the x- and y-directions. These two 1-D systems are discretized using a scheme similar to Eq. (9.43) except that the summation is increased from two to three terms corresponding to the right eigenvectors of matrices $\mathbf{A}$ and $\mathbf{B}$ in Eq. (9.8).

9.1.4 WAF schemes

The weighted average flux (WAF) method was first proposed for the Euler equations by Toro (1989) and then applied to the shallow water equations by Toro (1992) and others. It is a second-order extension of the Godunov upwind method and may also be interpreted as a Riemann-problem-based extension of the Lax-Wendroff method.

For a wave structure like Fig. 9.3 or 9.4, the basic WAF method gives the intercell flux as

$$\mathbf{F}_{i+1/2} = \sum_{k=1}^{N+1} w_k \mathbf{F}_{i+1/2}^{(k)} \tag{9.44}$$

where w_k is the weight given by

$$w_k = \frac{1}{2}(c_k - c_{k-1}) \tag{9.45}$$

with c_k as the Courant number for wave k, $c_k = S_k \Delta t / \Delta x$, $c_0 = -1$, and $c_{N+1} = 1$. Here, S_k is the speed of wave k, and N is the number of conservation laws or the number of waves in the solution of the Riemann problem. $\mathbf{F}_{i+1/2}^{(k)}$ is the value of the flux vector $\mathbf{F}(\mathbf{\Phi}_{i+1/2}^{(k)})$ in the interval k of length w_k shown in Fig. 9.5, and can be determined using Eq. (9.18) or (9.22). Inserting Eq. (9.45) into Eq. (9.44) yields an alternative expression:

$$\mathbf{F}_{i+1/2} = \frac{1}{2}(\mathbf{F}_i + \mathbf{F}_{i+1}) - \frac{1}{2}\sum_{k=1}^{N} c_k \Delta \mathbf{F}_{i+1/2}^{(k)} \tag{9.46}$$

where $\Delta \mathbf{F}_{i+1/2}^{(k)} = \mathbf{F}_{i+1/2}^{(k+1)} - \mathbf{F}_{i+1/2}^{(k)}$ is the flux jump across wave k.

For a linear convection equation, the WAF scheme (9.46) reproduces identically the Lax-Wendroff method, which is second-order accurate in space and time. Spurious oscillations in the vicinity of high gradients are expected. Such non-physical oscillations can be avoided by enforcing the TVD constraint on the scheme. The TVD version of the WAF scheme is

$$\mathbf{F}_{i+1/2} = \frac{1}{2}(\mathbf{F}_i + \mathbf{F}_{i+1}) - \frac{1}{2}\sum_{k=1}^{N} \text{sign}(c_k) A_k \Delta \mathbf{F}_{i+1/2}^{(k)} \tag{9.47}$$

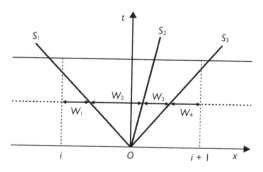

Figure 9.5 Weights in the WAF scheme.

where A_k is a WAF limiter function. There are various choices for A_k, such as

$$A_k(r, |c_k|) = \begin{cases} 1 & \text{if } r \leq 0 \\ 1 - 2(1 - |c_k|)r & \text{if } 0 < r \leq 1/2 \\ |c_k| & \text{if } 1/2 < r \leq 1 \\ 1 - (1 - |c_k|)r & \text{if } 1 < r \leq 2 \\ 2|c_k| - 1 & \text{if } r > 2 \end{cases} \quad (9.48)$$

$$A_k(r, |c_k|) = \begin{cases} 1 & \text{if } r \leq 0 \\ 1 - 2(1 - |c_k|)r/(1 + r) & \text{if } r > 0 \end{cases} \quad (9.49)$$

$$A_k(r, |c_k|) = \begin{cases} 1 & \text{if } r \leq 0 \\ 1 - (1 - |c_k|)r(1 + r)/(1 + r^2) & \text{if } r > 0 \end{cases} \quad (9.50)$$

$$A_k(r, |c_k|) = \begin{cases} 1 & \text{if } r \leq 0 \\ 1 - (1 - |c_k|)r & \text{if } 0 < r \leq 1 \\ |c_k| & \text{if } r > 1 \end{cases} \quad (9.51)$$

where r is the ratio of the upwind change to the local change in a scalar quantity f:

$$r = \begin{cases} \Delta f_{i-1/2}^{(k)}/\Delta f_{i+1/2}^{(k)} & \text{if } c_k > 0 \\ \Delta f_{i+3/2}^{(k)}/\Delta f_{i+1/2}^{(k)} & \text{if } c_k < 0 \end{cases} \quad (9.52)$$

with $\Delta f_{i+1/2}^{(k)} = f_{i+1}^{(k)} - f_i^{(k)}$. For the x-split two-dimensional shallow water equations, $f = h$ for the non-linear waves, and $f = v$, the tangential velocity component, for the shear wave. For other passive scalars, f is set as the corresponding state quantities.

The WAF limiter functions (9.48)–(9.51) are entirely equivalent to the conventional superbee limiter, van Leer's limiter, van Albada's limiter, and minbee limiter, respectively (Toro, 2001).

The numerical schemes introduced above have been tested extensively in the literature. For example, the performances of the central difference scheme (9.12), the HLL scheme (9.18), the TVD scheme (9.43) with van Leer's monotonic limiter, and the TVD WAF scheme (9.47) with van Albada's limiter are demonstrated in the following simulation of dam-break flow in a straight rectangular channel with a horizontal bed. The channel is 1200 m long, and a dam is located at 500 m from the upstream end. The water in the reservoir is 10 m deep, while the initial downstream water depth is given as 1 and 0.001 m to test the schemes in cases of wet and dry beds. The dam is assumed to be instantaneously, completely removed. The channel is set to be sufficiently wide so that the flow is uniform along the transverse direction and an analytical solution in the frictionless case can be derived (see Graf and Altinakar, 1998). The Manning n is set as 0. The longitudinal grid length is 10 m. For the central difference scheme with artificial diffusion flux, a time step of 0.1 s is used in both wet- and dry-bed cases. For the HLL, TVD, and TVD WAF schemes, the time step is 0.6 s for the wet-bed case and 0.3 s for the dry-bed case. Figs. 9.6 and 9.7 compare the calculated water surface profiles with the analytical solutions at 30 s after dam failure. It can be seen that the central difference scheme (9.12) with artificial diffusion flux has errors near the vicinity

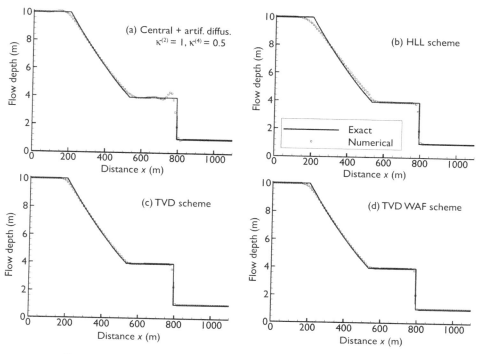

Figure 9.6 Dam-break waves over a wet bed at an elapsed time of 30 s.

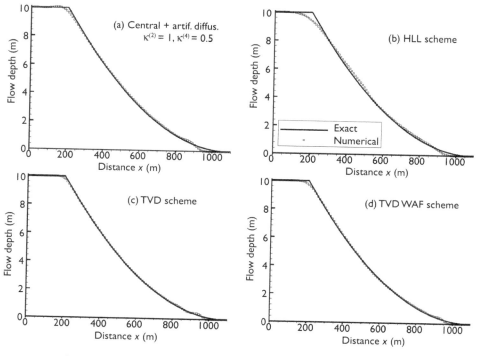

Figure 9.7 Dam-break waves over a dry bed at an elapsed time of 30 s.

of sharp gradients, which are essentially affected by the coefficients $\kappa^{(2)}$ and $\kappa^{(4)}$. The HLL scheme (9.18) provides a smooth solution with first-order accuracy. The TVD scheme (9.43) and the TVD WAF scheme (9.47), which are second-order accurate, provide better results than the central difference scheme and the HLL scheme.

9.1.5 Upwind flux schemes

Ying et al. (2004) proposed an upwind flux scheme to solve the St. Venant equations. To handle the 1-D dam-break flow in natural rivers, the flow area A and discharge Q are used as primary variables. The depth gradient and bed slope are combined as the water surface gradient, which is arranged into the source term. This treatment can avoid the use of different discretization schemes for the depth gradient and bed slope terms. The governing equations are still written as Eq. (9.1) with the unknown variables, fluxes, and source terms:

$$\Phi = \left[\begin{array}{c} A \\ Q \end{array} \right], \quad F(\Phi) = \left[\begin{array}{c} Q \\ Q^2/A \end{array} \right], \quad S(\Phi) = \left[\begin{array}{c} 0 \\ -gA\dfrac{\partial z_s}{\partial x} - g\dfrac{n^2 Q|Q|}{AR^{4/3}} \end{array} \right]$$

$$(9.53)$$

The mesh system is the same as that shown in Fig. 9.1. Integrating Eq. (9.1) over the ith control volume yields the discretized equation (9.10). The intercell flux $F^n_{i+1/2}$ is evaluated using the first-order upwind method:

$$F^n_{i+1/2} = \left[\begin{array}{c} Q^n_{i+k} \\ (Q^n_{i+k})^2/A^n_{i+k} \end{array} \right]$$

$$(9.54)$$

where $k = 0$, if $Q_i > 0$ and $Q_{i+1} > 0$; $k = 1$, if $Q_i < 0$ and $Q_{i+1} < 0$; and $k = 1/2$ for others. Here, the subscript $i + 1/2$ represents the average of values at grid points i and $i + 1$.

The source term in the momentum equation is evaluated as

$$S_i(Q) = -gA^{n+1}_i \left[w_1 \frac{z^{n+1}_{s,i+1-k} - z^{n+1}_{s,i-k}}{x_{i+1-k} - x_{i-k}} + w_2 \frac{z^{n+1}_{s,i+k} - z^{n+1}_{s,i-1+k}}{x_{i+k} - x_{i-1+k}} \right] - g\frac{n^2_i Q^n_i |Q^n_i|}{A^n_i (R^n_i)^{4/3}} \quad (9.55)$$

where w_1 and w_2 are weighting factors. If $w_1 = 0.5$ and $w_2 = 0.5$, Eq. (9.55) is equivalent to the central difference scheme for water surface gradient. This may result in non-physical oscillations. In order to improve numerical accuracy, these weighting factors are related to the Courant number. One of the relations suggested by Ying et al. is

$$w_1 = 1 - \frac{\Delta t}{2}(|U_{i+1-k}| + |U_{i-k}|)/(x_{i+1-k} - x_{i-k})$$

$$w_2 = \frac{\Delta t}{2}(|U_{i+k}| + |U_{i-1+k}|)/(x_{i+k} - x_{i-1+k}) \quad (9.56)$$

where U is the flow velocity.

The performance of the method of Ying *et al.* is demonstrated in the simulation of the previous dam-break flow cases shown in Figs. 9.6 and 9.7. The computational time step is 0.6 and 0.3 s for the wet- and dry-bed cases, respectively. The Ying *et al.* scheme and the central difference scheme for water surface gradient are used, i.e., either the weighting factors w_1 and w_2 are determined with Eq. (9.56) or both are set as 0.5. Fig. 9.8 compares the numerical and analytical solutions at 30 s after dam failure. The Ying *et al.* scheme for water surface gradient provides better results than the central difference scheme.

The method of Ying *et al.* is only first-order accurate, but it is simple and robust. Extension of it to the solution of the 2-D problem (9.5) can be found in Ying and Wang (2004).

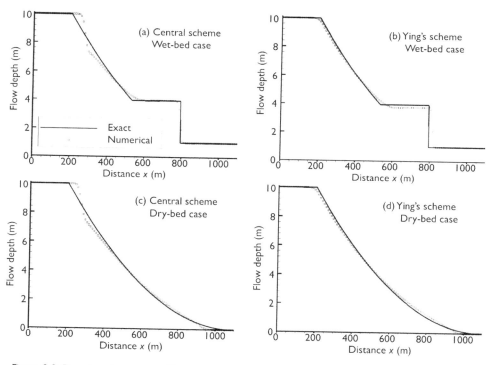

Figure 9.8 Dam-break waves at 30 s simulated using first-order upwind flux scheme with various schemes for water surface gradient.

9.1.6 Stability and accuracy of explicit and implicit schemes

The aforementioned central difference scheme with artificial diffusion fluxes, approximate Riemann solvers, TVD schemes, and upwind flux schemes are built in explicit algorithms, and thus, the time step should be limited to satisfy the Courant-Friedrichs-Lewy (CFL) stability condition. In general, the CFL condition for the 1-D problem is

$$\frac{|q/h| + \sqrt{gh}}{\Delta x} \Delta t < 1 \qquad (9.57)$$

and for the 2-D problem is

$$\left(\frac{|U_x| + \sqrt{gh}}{\Delta x} + \frac{|U_y| + \sqrt{gh}}{\Delta y} \right) \Delta t < 1 \qquad (9.58)$$

However, the stability of each numerical scheme also may rely on its own formulation, and the longest time step allowed for a stable solution varies among different schemes.

To relax the restriction on time step imposed by the CFL conditions, Molls and Chaudhry (1995) adopted the ADI method and Delis *et al.* (2000) established implicit TVD schemes for 1-D St. Venant equations. It has been found that longer time steps can be used in implicit algorithms, and the computational cost may be reduced. However, implicit schemes may induce numerical diffusion. This is demonstrated in the following example using the SIMPLEC algorithm

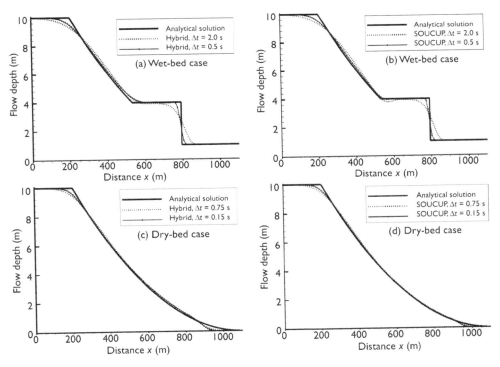

Figure 9.9 Dam-break waves at 30 s simulated using SIMPLEC.

The SIMPLEC algorithm described in Section 6.1.3.1 is developed primarily to simulate common open-channel flows, but its superior numerical stability makes it feasible for simulation of dam-break flow. This is tested by simulating the previous wet- and dry-bed cases shown in Figs. 9.6 and 9.7. To investigate the effect of time step on the simulation results, the time step is set as 2.0 and 0.5 s for the wet-bed case and 0.75 and 0.15 s for the dry-bed case. The hybrid (Spalding, 1972) and SOUCUP (Zhu and Rodi, 1991) schemes are used for the convection terms. The results are shown in Fig. 9.9. The model reproduces reasonably well the upstream negative wave and downstream positive wave. The hybrid scheme produces slightly larger errors in water level than the SOUCUP scheme. One can see that a smaller time step gives a better prediction. As the time step increases, the wave front becomes less sharp. This is due to numerical diffusion, which is particularly significant near regions with sharp gradients. This was also observed in the implicit method of Delis *et al.* (2000). There is a trade-off between the time step length (i.e., computational efficiency) and numerical accuracy when implicit algorithms are used in the simulation of dam-break flow where discontinuities and sharp gradients exist.

9.2 SIMULATION OF DAM-BREAK FLOW OVER MOVABLE BEDS

Simulation of dam-break flow over movable beds is much more challenging than that over fixed beds. One of the problems encountered in the movable-bed case is that the sediment concentration is so high and the bed varies so rapidly that the effects of sediment transport and bed change on the flow cannot be ignored. Another problem is that the sediment transport in the higher flow regime, such as dam-break flow, is little understood, and the existing sediment transport formulas may not be applicable.

Ferreira and Leal (1998), Fraccarollo and Armanini (1998), and Yang and Greimann (1999) established movable-bed dam-break flow models. However, some of these models ignore the effects of sediment transport and bed change on the flow, and some use the assumption of local equilibrium sediment transport that is no longer valid in the case of dam-break flow. Fraccarollo and Capart (2002) proposed a two-layer model of movable-bed dam-break flow in which the clear water in the upper layer and the mixture of sediment and water in the lower layer are simulated separately. The applicability of this two-layer model is limited because a constant sediment concentration is assumed in the lower layer. Capart and Young (1998), Cao *et al.* (2004), and Wu and Wang (2007) developed more advanced models for dam-break flow over movable beds, considering the non-equilibrium sediment transport and the effects of sediment concentration and bed change on the flow. However, the sediment exchange model proposed by Capart and Young does not consider the effect of sediment size, and thus, its applicability is restricted. The model of Cao *et al.* (2004) simulates only the suspended-load transport and needs to be tested quantitatively. Wu and Wang's model simulates the total-load transport and has been verified using available experimental data. It is introduced below.

9.2.1 Governing equations

Hydrodynamic equations

To account for the interactions among flow, sediment transport, and bed change, the generalized 1-D shallow water equations (5.177) and (5.178) should be used, which are written as follows by setting $\beta = 1$ and neglecting side flows:

$$\frac{\partial(\rho A)}{\partial t} + \frac{\partial(\rho Q)}{\partial x} + \rho_b \frac{\partial A_b}{\partial t} = 0 \tag{9.59}$$

$$\frac{\partial}{\partial t}(\rho Q) + \frac{\partial}{\partial x}\left(\frac{\rho Q^2}{A}\right) + \rho g A \frac{\partial z_s}{\partial x} + \frac{1}{2} g A h_p \frac{\partial \rho}{\partial x} + \rho g \frac{n^2 Q|Q|}{A R^{4/3}} = 0 \tag{9.60}$$

The effect of alluvial bed roughness can be accounted for through the dependence of Manning n on flow and sediment conditions, but constant Manning n values are adopted here for simplicity. The sensitivity analysis performed by Wu and Wang (2007) shows that the Manning n affects very little on the model results, and constant Manning n values in a dam-break event can be justified.

Sediment transport equations

As described in Section 2.1.2.3, the total-load sediment in natural rivers may be divided into bed load and suspended load as per sediment transport mode or into bed-material load and wash load as per sediment source. The former approach is adopted here and, thus, the total-load transport rate, Q_t, is computed by

$$Q_t = QC_t = QC + Q_b \tag{9.61}$$

where C is the suspended-load volumetric concentration averaged over the cross-section, and Q_b is the bed-load transport rate. The 1-D suspended-load transport equation (2.108) is written as

$$\frac{\partial}{\partial t}(AC) + \frac{\partial}{\partial x}(QC) = B(E_b - D_b) \tag{9.62}$$

where D_b and E_b are the sediment deposition and entrainment rates at the interface between the bed-load and suspended-load layers, defined as $D_b = \omega_{sm} c_b$ and $E_b = \omega_{sm} c_{b*}$ with ω_{sm} being the settling velocity of sediment particles in turbid water. To consider the effect of sediment concentration, ω_{sm} is determined using the Richardson-Zaki (1954) formula (3.19), which is written as

$$\omega_{sm} = (1 - C_t)^n \omega_s \tag{9.63}$$

where ω_s is the settling velocity of single particles in clear water, determined using the Zhang formula (3.12) if no measurement data is provided; and n is an empirical exponent of about 4.0.

The near-bed concentration c_b is related to the average concentration C by $c_b = \alpha_c C$, as described in Section 2.5.1. Following Cao et al. (2004), α_c is determined by

$$\alpha_c = \min\left[\alpha_0, (1 - p'_m)/C\right] \tag{9.64}$$

where α_0 is given a value of 2.0. Note that Eq. (9.64) limits c_b up to a maximum value equal to $1 - p'_m$.

The 1-D bed-load transport equation is

$$\frac{\partial}{\partial t}\left(\frac{Q_b}{U_b}\right) + \frac{\partial Q_b}{\partial x} = \frac{1}{L}(Q_{b*} - Q_b) \tag{9.65}$$

where Q_{b*} is the equilibrium (capacity) bed-load transport rate; U_b is the bed-load velocity, which can be evaluated using the van Rijn (1984a) formula (3.136) or Fig. (3.27) but is set to be the flow velocity here for simplicity; and L is the adaptation length of sediment, determined using Eq. (2.155).

The bed change can be determined by

$$(1 - p'_m)\frac{\partial A_b}{\partial t} = B(D_b - E_b) + \frac{1}{L}(Q_b - Q_{b*}) \tag{9.66}$$

Empirical formulas

To close the aforementioned sediment transport model, additional empirical formulas are required to determine the equilibrium bed-load transport rate Q_{b*} and suspended-load near-bed concentration c_{b*}. The van Rijn (1984a & b) formulas (3.70) and (3.95) herein are used. However, these formulas, which were calibrated in the lower flow regime, should be modified for extension to the situation of dam-break flow.

According to the observation of Fraccarollo and Capart (2002), the sediment concentration in the lower layer near the bed under dam-break flow conditions is very high (nearly the same as that of bed material). Therefore, Wu and Wang (2007) introduced a correction factor for the transport stage number in the van Rijn (1984a & b) formulas by replacing the water density with the mixture density near the bed:

$$\frac{\tau_b}{\tau_c} = \frac{U_{*b}^2}{\theta_c(\rho_s/\rho_{mb} - 1)gd} = k_t\frac{U_{*b}^2}{\theta_c(\rho_s/\rho_f - 1)gd} \tag{9.67}$$

where θ_c is the critical Shields number for sediment incipient motion, determined using the Shields diagram; ρ_{mb} is the density of the water and sediment mixture near the bed; and k_t is the correction factor, expressed as $k_t = 1 + c_a\rho_s/[(1 - c_a)\rho_f]$, with c_a being a concentration of sediment near the bed. In principle, c_a can be related to the average sediment concentration in the bed-load layer or the depth-averaged suspended-load concentration, but this usually requires iteration in the sediment module. On the other hand, the dynamic pressure in the dam-break wave front might affect sediment entrainment and transport significantly, but it is very difficult to consider this effect in the framework of a 1-D model. As an approximation, this effect is lumped into the correction factor k_t in Eq. (9.67). By trial and error, it is found that $k_t \approx 1 + 1.5\rho_s/\rho_f$ is adequate for the two test cases presented in Section 9.2.3. This value of k_t is equivalent to a value of 0.6 for c_a, which qualitatively agrees with the observation of Fraccarollo and Capart (2002). Because the parameter c_a is not included explicitly in this relation of k_t, the solution procedure in the sediment module is simplified significantly.

Note that the reference level for near-bed suspended-load concentration in the original van Rijn formula is defined at the equivalent roughness height or half the dune height, which is not investigated well in the case of dam-break flow. Because the correction factor k_t is a lumped parameter in the modified sediment transport capacity formulas, this reference level should also be interpreted as an empirical parameter. For simplicity, the reference level herein is set at $\max(2d, 0.005h)$, in which d is the sediment size. In addition, in the range of high shear stress, the near-bed equilibrium concentration of suspended load determined using the van Rijn (1984b) formula may be larger than $1 - p'_m$; this is not physically reasonable and is eliminated in the simulation by imposing an upper bound of $1 - p'_m$ to c_{b*}.

9.2.2　Numerical methods

Eqs. (9.59), (9.60), (9.62), (9.65), and (9.66) constitute a hyperbolic system, which can be solved numerically using the shock-capturing schemes introduced in Section 9.1. As an example, the upwind flux scheme is used here.

To establish an explicit algorithm, Eqs. (9.59) and (9.60) are reformulated by eliminating the flow density on their left-hand sides using the relation $\rho = \rho_f(1 - C_t) + \rho_s C_t$ and Eqs. (9.62), (9.65), and (9.66). The derived continuity and momentum equations are

$$\frac{\partial A}{\partial t} + \frac{\partial Q}{\partial x} = \frac{1}{1 - p'_m}\left[B(E_b - D_b) + \frac{1}{L}(Q_{b*} - Q_b)\right] \tag{9.68}$$

$$\frac{\partial Q}{\partial t} + \frac{\partial}{\partial x}\left(\frac{Q^2}{A}\right) = -gA\frac{\partial z_s}{\partial x} - \frac{1}{2}gAh_p\frac{1}{\rho}\frac{\partial \rho}{\partial x} - g\frac{n^2 Q|Q|}{AR^{4/3}}$$

$$- \frac{\rho_s - \rho_f}{\rho}U\left(1 - \frac{C_t}{1 - p'_m}\right)\left[B(E_b - D_b) + \frac{1}{L}(Q_{b*} - Q_b)\right] \tag{9.69}$$

Eqs. (9.68), (9.69), (9.62), and (9.65) are written in conservative form as Eq. (9.1), in which $\mathbf{\Phi}$ and $\mathbf{F}(\mathbf{\Phi})$ represent the vectors of unknown variables and fluxes:

$$\mathbf{\Phi} = \begin{bmatrix} A \\ Q \\ AC \\ Q_b/U_b \end{bmatrix}, \quad \mathbf{F}(\mathbf{\Phi}) = \begin{bmatrix} Q \\ Q^2/A \\ QC \\ Q_b \end{bmatrix} \tag{9.70}$$

and $\mathbf{S}(\mathbf{\Phi})$ includes the remaining terms in each equation.

Using the finite volume method in Section 9.1, one can derive the discretized equation (9.10), with the intercell fluxes determined using the first-order upwind scheme as

$$
F_{i+1/2}^n = \begin{bmatrix} Q_i^n \\ (Q_i^n)^2/A_i^n \\ Q_i^n C_i^n \\ Q_{b,i}^n \end{bmatrix}, \quad \text{if } Q \geq 0; \qquad F_{i+1/2}^n = \begin{bmatrix} Q_{i+1}^n \\ (Q_{i+1}^n)^2/A_{i+1}^n \\ Q_{i+1}^n C_{i+1}^n \\ Q_{b,i+1}^n \end{bmatrix}, \quad \text{if } Q < 0
$$

$$(9.71)$$

The friction term and sediment exchange terms on the right-hand sides of Eqs. (9.68) and (9.69) are evaluated in a pointwise manner using the values at cell center i, and the flow density gradient term in Eq. (9.69) is discretized using the central difference scheme at time level n. The water surface gradient term in Eq. (9.69) is discretized using the scheme in Eq. (9.55) proposed by Ying et al. (2004).

The bed change equation (9.66) is discretized in time as

$$
\Delta A_{b,i} = \frac{\Delta t}{1 - p_m'} \left[B_i^n (D_{b,i}^n - E_{b,i}^n) + \frac{1}{L} (Q_{b,i}^n - Q_{b*,i}^n) \right]
$$

$$(9.72)$$

The numerical solution is obtained through the following steps: (a) Solve the continuity equation (9.68) to obtain the flow area and, in turn, the water level; (b) Solve the momentum equation (9.69) to obtain the flow discharge and, in turn, the velocity; (c) Calculate the equilibrium suspended-load near-bed concentration and bed-load transport rate; (d) Solve the sediment transport equations (9.62) and (9.65) to obtain the actual suspended-load concentration and bed-load transport rate; (e) Calculate the bed change using Eq. (9.72); and (f) Continue steps (a)–(e) for the next time step until the entire time period is finished.

Because the aforementioned solution procedure is explicit, the computational time step should be limited by the Courant-Friedrichs-Lewy (CFL) condition for flow computation and additional numerical stability conditions for sediment transport and bed change computations.

9.2.3   Examples

The movable-bed dam-break flow model described above was verified by Wu and Wang (2007) using two sets of laboratory experiments performed in Taipei (University of Taiwan) and Louvain (Université Catholique de Louvain) (Capart and Young, 1998; Fraccarollo and Capart, 2002). Both experiments concerned small-scale dam-break waves over movable beds in prismatic channels with rectangular cross-sections. They differed primarily in the used sediment materials. In the Taipei test, the sediment particles were artificial spherical pearls covered with a shiny white coating, having a diameter of 6.1 mm, a density of $1048 \, \text{kg} \cdot \text{m}^{-3}$, and a settling velocity of about $7.6 \, \text{cm} \cdot \text{s}^{-1}$. In the Louvain test, the sediment particles were cylindrical PVC pellets having a diameter of 3.2 mm, a height of 2.8 mm (hence an equivalent spherical diameter of 3.5 mm), a density of $1540 \, kg \cdot m^{-3}$, and a settling velocity of about $18 \, cm \cdot s^{-1}$. The test reach in the Taipei experiment was 1.2 m long and 0.2 m wide, and in the Louvain experiment was 2.5 m long and 0.1 m wide. In both experiments, the upstream initial water depth, h_0, was 0.1 m.

In the numerical simulations for both experiments, a uniform mesh with a grid spacing of 0.005 m was used, and the time step was 0.001 t_0. Here, $t_0 = (h_0/g)^{1/2}$ is the hydrodynamic time scale and has a value of about 0.101 s. To handle the dry-bed problem, an initial flow depth of 0.0005 m was set in the downstream of the dam. The sediment porosity was estimated as 0.28 and 0.3 for the Taipei and Louvain tests, respectively, using Eq. (2.20). The bed-load adaptation length L_b was given as 0.25 m. The Manning roughness coefficient was set as 0.01 for sidewalls and 0.025 for the flume bed in both test cases.

According to initial trials using the original van Rijn (1984a & b) sediment transport capacity formulas, the numerical model significantly under-predicts bed erosion in both cases. With the modification described in Eqs. (9.63) and (9.67), it performs

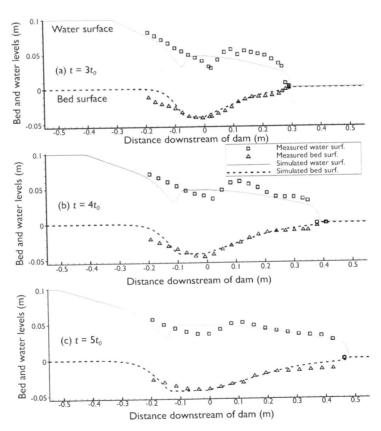

Figure 9.10 Bed and water surface profiles in Taipei case (Wu and Wang, 2007).

much better. The water and bed surface profiles at various times in both the Louvain and Taipei cases simulated using the modified van Rijn formulas are compared with the measured data in Figs. 9.10 and 9.11. The agreement between simulations and measurements is fairly good. The erosion magnitudes and wave front locations in both test cases are predicted well by the numerical model. A hydraulic jump in the water surface forms around the initial dam site in both test cases. Its location is predicted reasonably well in the Louvain case, where both simulation and measurement show that the hydraulic jump propagates upstream. However, the location of the hydraulic jump in the Taipei case is predicted less accurately. The hydraulic jump moves upstream in the simulation, but this movement was not clearly observed in the Taipei experiment.

Sensitivities of the simulation results to model parameters, such as the suspended-load adaptation coefficient α_0, bed-load adaptation length L_b, Manning n, and correction factor k_t, were analyzed (Wu and Wang, 2007). When each parameter was considered, only it was adjusted, and all other parameters were given the same values as used in the model testing just described above. Fig. 9.12 shows how the simulation results respond to adjustment of each parameter for the Taipei case. As α_0

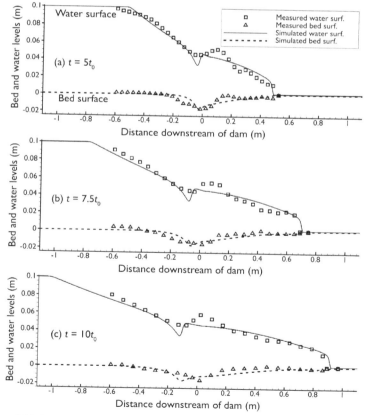

Figure 9.11 Bed and water surface profiles in Louvain case (Wu and Wang, 2007).

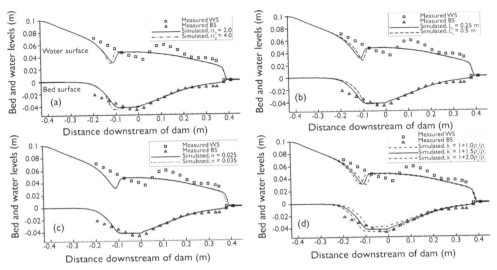

Figure 9.12 Sensitivities of model results in Taipei case to (a) adaptation coefficient α_0, (b) adaptation length L_b, (c) Manning n, and (d) coefficient k_t (Wu and Wang, 2007).

increases from 2.0 to 4.0 and L_b increases from 0.25 to 0.5 m, the simulated water and bed surface profiles change locally, especially upstream of the dam site. The influences of coefficients α_0 and L_b on the simulated water and bed surfaces are thus at limited levels. As the Manning n increases from 0.025 to 0.035, the simulated water and bed surfaces vary very little, and the wave front slightly slows down. The model results are not particularly sensitive to variation in the Manning n values. As the correction factor k_t increases from $1 + 1.0\rho_s/\rho_f$ to $1 + 2.0\rho_s/\rho_f$, the maximum erosion depth increases by 21%. The evaluation of k_t is important to the erosion magnitude. The test shows that $k_t = 1 + 1.5\rho_s/\rho_f$ provides reasonable results.

9.3 SIMULATION OF DAM SURFACE EROSION DUE TO OVERTOPPING FLOW

Flow overtopping earth dams and levees can cause serious erosion and even wash out the structures. Compared to the dam-break flow case, the sediment transport and morphological change due to overtopping flow may be less intensive. However, the overtopping flow is usually in mixed regimes, and the interactions among flow, sediment transport, and bed change are still appreciable. These should be considered in the simulation of overtopping flow and the associated erosion process.

Tingsanchali and Chinnarasri (1999) simulated the dam surface erosion process due to overtopping flow, but their model ignores the effects of sediment transport and bed change on the flow and uses the assumption of local equilibrium sediment transport. This erosion process herein is simulated using a more advanced model, which computes the flow using Eqs. (9.59) and (9.60) and the bed-material load transport using Eqs. (5.34) and (5.36). The sediment transport capacity is determined using the Wu *et al.* (2000b) formulas (3.80) and (3.102) without the correction described in Eq. (9.67). The numerical algorithm presented in Section 9.2 is used to solve these equations. In particular, the hydrodynamic equations are solved using the method of Ying *et al.* (2004) with slight modification to accommodate the differences between two sediment transport models. The details are not repeated here.

The model is tested against the experiments of Chinnarasri *et al.* (2003) on the dam (dike) surface erosion process due to overtopping flow. The experimental setup is sketched in Fig. 9.13. The experimental flume was 35 m long and 1 m wide. A dam across the flume width was located at 17.5 m downstream of the inlet. The dam was 0.8 m high, and its crest was 0.3 m wide. The upstream and downstream slopes of the dam were 1V:3H and 1V:2.5H for the simulated experimental run 2. The dam was made of medium sand with a diameter of 0.86 mm. With a constant inflow discharge of $1.23 \text{l} \cdot \text{s}^{-1}$, the upstream water level was increased initially to a desired height (0.03 m above the dam crest) and held by a vertical plate at the crest. The vertical plate was then instantaneously lifted up to allow overtopping flow to start. The overtopping flow discharge and erosion rate were small, due to a low water height above the crest at the beginning, which increased rapidly after a certain period and then decreased. The dam surface became wavelike in shape as anti-dunes appeared and moved upstream. The bed-material porosity is estimated as 0.35, the Manning n is 0.018, and the sediment adaptation length L_t is 0.05 m in the simulation.

A uniform mesh covering the entire flume is used. According to trials using various grid spacings, the occurrence of anti-dune waves in the simulation depends on the ratio of grid spacing and adaptation length, $\Delta x / L_t$. If this ratio is less than about 0.6, anti-dune waves appear. This implies that the computational mesh should be fine enough to simulate the anti-dunes. A 1-D model might not capture the anti-dunes accurately, and the measurement did not quantify them; thus, only the mean flow and sediment transport patterns are simulated by using a coarse mesh with a grid spacing of 0.05 m (i.e., $\Delta x / L_t = 1$).

Fig. 9.14 shows the longitudinal profiles of the simulated water level, Froude number, and sediment concentration at elapsed times of 30 and 150 s. The flow is subcritical in the upstream reservoir; it changes to supercritical on the downstream slope and then to subcritical in the downstream tailwater. A hydraulic jump appears at the end of the downstream slope. The sediment concentration on the downstream

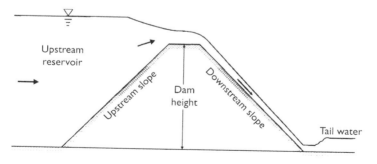

Figure 9.13 Sketch of overtopping flow over a dam (side view).

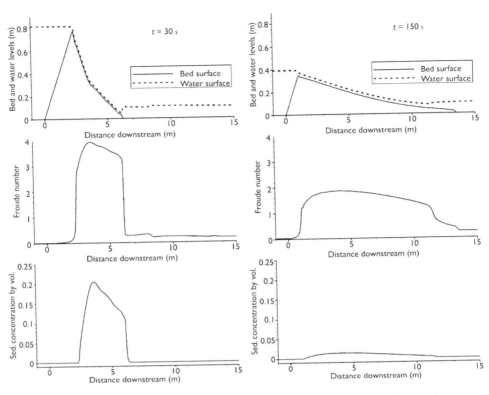

Figure 9.14 Water level, Froude number, and sediment concentration on dam surface.

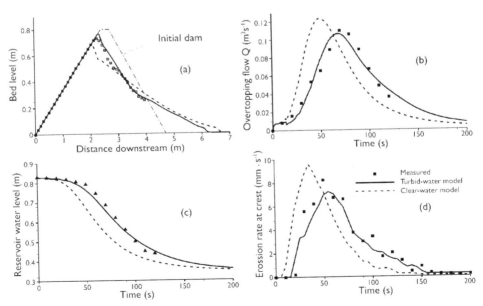

Figure 9.15 Results (Run 2) of turbid- and clear-water models: (a) bed profiles at 30 s, (b) overtopping flow discharges, (c) reservoir water levels, and (d) erosion rates at dam crest.

slope is higher at the early stage and becomes lower with time. Fig. 9.15(a) compares the simulated (solid lines) and measured bed profiles at 30 s, which are in fairly good agreement. Figs. 9.15(b)–(d) show the simulated and measured overtopping flow discharges, reservoir water levels, and erosion rates at the dam crest during a time period of 200 s. The erosion process is predicted generally well.

In addition, the "turbid-water" model presented above is compared with the traditional "clear-water" movable-bed model, in which the effects of sediment concentration and bed change on the flow are neglected by setting ρ as the pure water density and removing the third term in Eq. (9.59) and the fourth term in Eq. (9.60). Fig. 9.15 shows the simulation results of these two models for Run 2. It can be seen that the "clear-water" model significantly over-predicts the erosion process. This proves that the effects of sediment transport and bed change on the flow should be considered in cases of strong sediment transport and rapid bed change.

Chapter 10

Simulation of flow and sediment transport in vegetated channels

Vegetation can markedly affect the fluvial processes in streams. To understand and analyze this, fundamental research and numerical modeling of flow and sediment transport in vegetated channels have been conducted in recent years. The concepts, governing equations, and numerical methods of the vegetation effect models are introduced in this chapter.

10.1 EFFECTS OF VEGETATION ON FLOW AND SEDIMENT TRANSPORT

10.1.1 Geometric characteristics of vegetation

Vegetation height and diameter

Vegetation is classified to be either flexible or rigid, based on whether it is subject to deformation by the flow. Herbaceous species, such as grasses, usually are flexible, whereas woody species, such as trees, usually are rigid. However, it is recognized that different portions of the plant and the same plant in different stages of its life cycle can behave in significantly different ways. For example, the stems of trees and shrubs are often rigid, while their branches, twigs, and leaves are flexible. Saplings usually act as flexible stems until they mature sufficiently to be able to withstand deformation by the flow.

Because the shape of vegetation is highly irregular, it is challenging to represent a vegetation element with simple geometry. As an approximation, a vegetation stem (such as tree trunk) is often conceptualized as a cylinder with a height, h_v, and a representative diameter, D, as shown in Fig. 10.1. The stem height h_v is defined as the one without any deformation by the flow. Thus, the height of a rigid stem can be readily measured. However, a flexible stem bends under the shear of flow, as shown in Fig. 10.2, and its actual height, h'_v, is related to stem properties and flow conditions (Kouwen and Li, 1980).

The representative diameter D can be related to the stem volume, V, by $D = \sqrt{4V/\pi h_v}$. In practice, one may directly measure the stem diameter. Due to the fact that the diameter of many species (such as shrubs) may change along the height, the convention of foresters and ecologists suggests that the stem diameter of trees should be measured at the breast height. As an approximation, the stem diameter may be measured at $h_v/4$ (Freeman *et al.*, 2000).

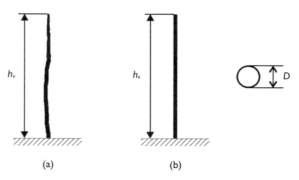

Figure 10.1 (a) Natural vegetation stem and (b) conceptualized cylinder.

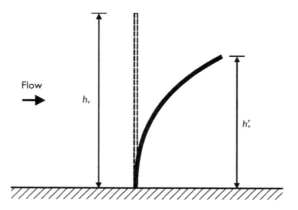

Figure 10.2 Flexible stem under flow shear.

Vegetation density

Vegetation density, denoted as N_a, is often defined as the number of vegetation elements per unit area in the horizontal plane. It can be determined by directly counting the number of vegetation elements or measuring their average spacings, as shown in Fig. 10.3. N_a is related to the longitudinal and transverse average spacings, l_s and l_n, as

$$N_a = \frac{1}{l_s l_n} \tag{10.1}$$

Projected area of vegetation

The projected area is defined as the frontal area of a vegetation element projected to the plane normal to the streamwise flow direction. Because vegetation may be emergent or submerged, as shown in Fig. 10.4, the projected area of the wetted portion is more

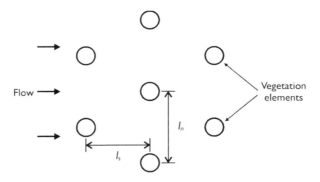

Figure 10.3 Notation of vegetation spacings (plan view).

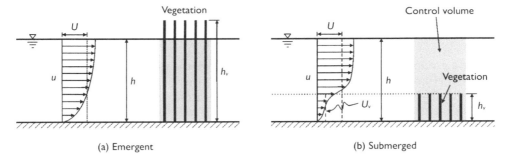

(a) Emergent   (b) Submerged

Figure 10.4 Vegetation in open channels (side view).

often used. For a rigid stem, the wetted projected area is

$$A_v = \alpha_v D \min(h_v, h) \tag{10.2}$$

where h is the flow depth, and α_v is the shape factor of vegetation and equals 1 for a rigid cylinder. When the stem is partially submerged, D represents the diameter of the wetted portion.

For a flexible stem, the wetted projected area is

$$A_v = \alpha_v D \min(h'_v, h) \tag{10.3}$$

One may also use Eq. (10.2) for the projected area of a flexible stem by lumping the factor h'_v/h_v into α_v.

For vegetation with limbs and leaves shown in Fig. 10.5, the conceptual model of single cylinder is no longer realistic. In this case, the projected area is often defined as the blockage area of the limbs and leaves. However, the limbs and leaves deform under flow shear, and thus the blockage area changes with flow conditions.

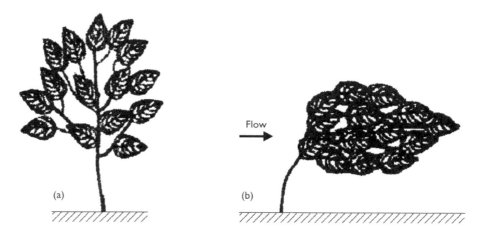

Figure 10.5 Leafy vegetation: (a) without and (b) with flow shear.

For a group of homogeneous vegetation elements, the total projected area per unit area is defined as the product of the vegetation density and the projected area of an individual stem, $N_a A_v$. However, in natural cases, vegetation elements may be heterogeneous, and thus the sum of the projected areas of all elements, $\sum_i A_{vi}$, is often considered. Here, A_{vi} is the project area of the individual vegetation element numbered as i.

Vegetation concentration

Consider a control volume extending from the stream bed to the water surface and holding the mixture of the water and a group of vegetation stems, as shown in Fig. 10.4. The volumetric concentration of vegetation, c_v, is defined as the ratio of the volume of vegetation to the total volume of the mixture. For homogeneous, rigid vegetation, c_v can be expressed as

$$c_v = N_a \frac{\pi D^2 \min(h_v, h)}{4h} \tag{10.4}$$

Note that c_v is the vegetation concentration averaged over the entire flow depth. Naturally, the local concentration of vegetation in the vegetation layer is defined as

$$c_{v0} = N_a \frac{\pi D^2}{4} \tag{10.5}$$

For the emergent vegetation shown in Fig. 4(a), $c_v = c_{v0}$, and for the submerged vegetation shown in Fig. 4(b), $c_v = c_{v0} h_v / h$.

Comparing Eqs. (10.2) and (10.4) leads to a relation between $N_a A_v$ and c_v:

$$N_a A_v = \frac{4\alpha_v c_v h}{\pi D} \tag{10.6}$$

$N_a A_v$ and c_v have also been used to represent the vegetation density by Shimizu and Tsujimoto (1994) and Wu and Wang (2004b). To avoid confusion, only N_a is called the vegetation density in this book.

10.1.2 Flow resistance due to vegetation

Drag force on vegetation

The drag force exerted on a vegetation element is expressed as

$$\vec{F}_d = \frac{1}{2} C_d \rho A_v \, |U_v| \, \vec{U}_v \tag{10.7}$$

where C_d is the drag coefficient of vegetation element, $\vec{U}_v$ is the vector of flow velocity acting on the vegetation element, and $|U_v|$ (or simply written as U_v) is the magnitude of $\vec{U}_v$. For emergent vegetation, $\vec{U}_v$ is the depth-averaged flow velocity $\vec{U}$, while for submerged vegetation, $\vec{U}_v$ should be the velocity averaged only over the vegetation layer, as shown in Fig. 10.4.

Note that the drag force defined in Eq. (10.7) acts on the entire wetted vegetation height, which may be used in the 1-D and depth-averaged 2-D models. For the analysis of local flow pattern in a 3-D model, the drag force per unit vegetation height is defined as

$$\vec{f}_d = \frac{1}{2} C_d \rho a_v \, |u| \, \vec{u} \tag{10.8}$$

where $\vec{u}$ is the vector of the local flow velocity acting on the vegetation element; and a_v is the projected area of the vegetation element per unit height, related to the stem diameter by $a_v = \alpha_v D$.

The drag coefficient is the key parameter in Eqs. (10.7) and (10.8). White (1991) summarized the experimental data of Wieselberger and Tritton, and obtained the drag coefficient of a single cylinder in an ideal two-dimensional flow as a function of the Reynolds number. This relation is shown in Fig. 10.6 and approximated by

$$C_{d0} = 1 + \frac{10}{R_e^{2/3}} \tag{10.9}$$

where C_{d0} is the drag coefficient for a single cylinder, and $R_e = U_v D / \nu$. Eq. (10.9) is in fair agreement with the measurement data up to $R_e \approx 2.5 \times 10^5$. More sophisticated curve-fit formulas for C_{d0} can also be found in White (1991). Typically, C_{d0} has values from 1.0 to 1.2 for the Reynolds numbers between $10^2 - 2.5 \times 10^5$.

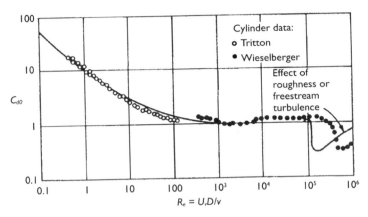

Figure 10.6 Drag coefficient for single cylinder (White, 1991).

Li and Shen (1973) investigated the drag coefficient for a group of cylinders with various setups. They identified four factors that need to be considered to determine the drag coefficient: (1) flow turbulence, (2) non-uniform velocity profile, (3) free surface, and (4) blockage. Lindner (1982) suggested that in densely vegetated channels, the first two of these factors are of minor importance and can be neglected. Lindner extended the work of Li and Shen, and established the following formula to compute the drag coefficient C_d for a single cylinder in a group:

$$C_d = \left(1 + 1.9\frac{D}{l_n}C_{d0}\right)\left[0.2025\left(\frac{l_s}{D}\right)^{0.46}C_{d0}\right] + \left(\frac{2l_n}{l_n - D} - 2\right) \quad (10.10)$$

The two terms on the right-hand side of Eq. (10.10) represent the blockage and free surface effects, respectively. The experiments were conducted with PVC cylinders of 10 mm in diameter and 150 mm in height.

Based on Lindner's approach and further experiments, Pasche and Rouve (1985) presented a semi-empirical iterative process to determine C_d. Many other investigators, e.g., Klaassen and Zwaard (1974) and Jarvela (2002), suggested the drag coefficient C_d would have values close to 1.5 for most practical cases.

The drag coefficient C_d in Eq. (10.7) is based on the apparent velocity U_v. According to Stone and Shen (2002), the drag coefficient C_{dm} based on the constricted cross-sectional velocity U_{vm} shown in Fig. 10.7 is more appropriate than C_d. This is because C_{dm} is closer to the drag coefficient of single cylinder and has less variation for a wide range of values for vegetation density, stem size, and cylinder Reynolds number in comparison with C_d. The relation between C_d and C_{dm} is

$$C_d = C_{dm}\frac{U_{vm}^2}{U_v^2} \quad (10.11)$$

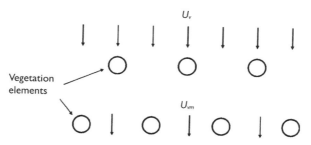

Figure 10.7 Definitions of U_v and U_{vm} in a matrix of vegetation elements.

If the vegetation stems with a diameter of D are distributed uniformly in the lateral direction with a spacing of l_n, then $U_v = U_{vm}(1 - D/l_n)$ and Eq. (10.11) can be written as

$$C_d = C_{dm}/(1 - D/l_n)^2 \qquad (10.12)$$

Furthermore, if the vegetation stems are arranged in a staggered pattern with equal spacing in both the longitudinal and transverse directions shown in Fig. 10.7, $U_v = U_{vm}(1 - D\sqrt{N_a})$. Using Eq. (10.5), one can write Eq. (10.11) as

$$C_d = C_{dm}/\left(1 - \sqrt{4c_{v0}/\pi}\right)^2 \qquad (10.13)$$

Note that the drag force is expected to increase with the velocity squared in Eq. (10.7). This is valid for rigid vegetation. Eq. (10.7) may still be used to compute the drag force on flexible vegetation, but the projected area should be computed using the deformed height as expressed in Eq. (10.3) (Tsujimoto and Kitamura, 1998), or the drag coefficient has to be related to flow conditions. More methods for flexible vegetation roughness are discussed later in this section.

Roughness of emergent rigid vegetation

Consider a steady, uniform flow in a channel with a plane bed covered with uniformly distributed emergent, rigid vegetation. For the control volume over a unit bed area extending from the stream bed to the water surface shown in Fig. 10.4, the total resistance, τ, consists of the bed shear stress, τ_b, and the drag force of vegetation, $N_a F_d$:

$$(1 - c_v)\tau = (1 - c_v)\tau_b + N_a F_d \qquad (10.14)$$

Note that the factor $1 - c_v$ appears in Eq. (10.14) to account for only the water column and bed area occupied by the flow. If the vegetation is relatively sparse, $1 - c_v$ is close to 1 and can be eliminated from Eq. (10.14).

Define the total resistance and the bed shear stress as

$$\tau = \rho c_f U^2 = \frac{\rho g n^2 U^2}{R_s^{1/3}} \tag{10.15}$$

$$\tau_b = \rho c_{fb} U^2 = \frac{\rho g n_b^2 U^2}{R_s^{1/3}} \tag{10.16}$$

where c_f and n are the friction factor and Manning coefficient corresponding to the total roughness, c_{fb} and n_b are the friction factor and Manning coefficient corresponding to the bed roughness, and R_s is the hydraulic radius of the bed with vegetation. The hydraulic radius R_s has been defined differently in the literature. Many models simply set R_s as the flow depth h, while Barfield *et al.* (1979) considered the effect of vegetation on the flow "eddy size" and suggested the following relation:

$$R_s = \frac{h l_n}{2h + l_n} \tag{10.17}$$

where l_n is the lateral spacing of vegetation elements.

Substituting Eqs. (10.7), (10.15), and (10.16) into Eq. (10.14) yields

$$n^2 = n_b^2 + \frac{1}{2g(1 - c_v)} C_d N_a A_v R_s^{1/3} \tag{10.18}$$

For the channel with densely distributed vegetation, the drag of vegetation becomes the major contributor to the total resistance, and thus the term of n_b in Eq. (10.18) can be eliminated.

Eq. (10.18) does not include the effect of side banks, so it is used only for a wide channel or in a depth-averaged 2-D model. When the effect of side banks needs to be considered, R_s in Eq. (10.18) is set as A/P_w, or the following equation derived by Petryk and Bosmajian (1975) may be used:

$$n^2 = n_b^2 + \frac{1}{2g} C_d \frac{\sum A_{vi}}{A} \left(\frac{A}{\chi}\right)^{4/3} \tag{10.19}$$

where A is the cross-sectional area of flow, χ is the wetted perimeter, and $\sum A_{vi}$ is the total frontal area of vegetation elements blocking the flow per unit channel length. Note that $1 - c_v \approx 1$ and C_d represents an average drag coefficient of all vegetation elements in Eq. (10.19). If the heterogeneity of vegetation elements is considered, a variable C_{di} should be used for each element.

Eq. (10.19) is applicable to emergent, rigid vegetation distributed relatively uniformly in the lateral direction.

Roughness of submerged rigid vegetation

For submerged, rigid vegetation, the "acting" flow velocity (which should be used for computing drag) is the average velocity in the vegetation layer, as shown in Fig. 10.4(b). This velocity can be determined using Stone and Shen's (2002) method:

$$U_v = \eta_v U \left(\frac{h_v}{h}\right)^{1/2}$$
(10.20)

where η_v is a coefficient of about 1.0.

Besides the bed shear stress, a shear stress exists at the top of vegetation elements. The total bottom shear stress can be approximated by

$$\tau_b = \rho c_{fv} U_v^2 + \rho c_{fu} (U_u - U_v)^2$$
(10.21)

where U_u is the average resultant velocity in the water column above the vegetation elements, which can be derived by applying the mass continuity equation $U_u(h - h_v) + U_v h_v = Uh$; and c_{fv} and c_{fu} are the friction factors on the channel bed and at the top of vegetation elements, respectively. However, evaluation of the two coefficients c_{fv} and c_{fu} is relatively complicated. For simplicity, one may lump them into one parameter, and determine τ_b using Eq. (10.16), with R_s given as h or by

$$R_s = \frac{h_v l_n}{2h_v + l_n} + h - h_v$$
(10.22)

Eq. (10.22) reduces to Eq. (10.17) for emergent vegetation ($h_v = h$), to $R_s = h$ when there is no vegetation ($h_v = 0$), and to $R_s \approx h$ when the vegetation is sparse, i.e., $l_n \gg 2h_v$. For densely distributed, submerged vegetation, Eq. (10.16) may not be as physically realistic as Eq. (10.21). However, in such a case the bed shear becomes negligible in comparison with the vegetation drag force, and thus the use of Eq. (10.16) does not induce significant errors in the flow calculation.

Therefore, substituting Eqs. (10.7), (10.15), (10.16), and (10.20) into Eq. (10.14) yields

$$n^2 = n_b^2 + \frac{1}{2g(1 - c_{v0})} C_d N_a A_v \eta_v^2 \frac{h_v}{h} R_s^{1/3}$$
(10.23)

Similarly, Petryk and Bosmajian's equation (10.19) can be modified for submerged, rigid vegetation as

$$n^2 = n_b^2 + \frac{1}{2g} C_d \eta_v^2 \frac{\bar{h}_v}{h} \frac{\sum A_{vi}}{A} \left(\frac{A}{\chi}\right)^{4/3}$$
(10.24)

where $\bar{h}_v$ represents the averaged height of vegetation elements blocking the flow.

Roughness of flexible vegetation

According to Kouwen and Li (1980), the resistance to flow by flexible vegetation in channels can be determined using a relative roughness approach similar to the widely accepted resistance relationships developed for rigid roughness in pipes and channels. Because flexible vegetation bends when subjected to shear, its roughness height is a function of vegetation properties and flow parameters. The significant vegetation properties are the stem density M and the flexural rigidity in bending, given by $J = EI$. Here, E is the stem's modulus of elasticity and I is the second moment of inertia of the stem area. The stem density M is defined as the ratio of the stem count to a reference number of stems per unit area. The reference number is arbitrary, but for convenience it is taken to be 1 stem per square meter, thus yielding $M = N_a$. Note that M is dimensionless.

Based on laboratory experiments on flow over flexible plastic strips, Kouwen and Li (1980) showed that the roughness height k_s would vary as a function of the amount of drag exerted by the flow and the parameter MEI:

$$k_s = 0.14 h_v \left[\frac{(MEI/\tau_b)^{0.25}}{h_v} \right]^{1.59} \tag{10.25}$$

where h_v is in meters, τ_b is in $N \cdot m^{-2}$, and MEI is in $N \cdot m^2$. Tsujimoto and Kitamura (1998) conducted numerical analysis using a hydrodynamic model coupled with a cantilever beam model describing bending of vegetation, and obtained a similar formula for the flexible vegetation height under the shear of flow.

Kouwen and Li (1980) suggested the use of the semi-logarithmic resistance equation to determine the Darcy-Weisbach friction factor λ:

$$\frac{1}{\sqrt{\lambda}} = a + b \log \left(\frac{R}{k_s} \right) \tag{10.26}$$

where R is the hydraulic radius of the channel; and a and b are two fitted parameters, depending on the relative magnitude of the shear velocity U_* and a critical value U_{*crit}. In a numerical model test, Darby (1999) used the form of Hey's (1979) equation:

$$\frac{1}{\sqrt{\lambda}} = 2.03 \log \left(\frac{a_s R}{k_s} \right) \tag{10.27}$$

where a_s is a dimensionless shape correction factor, given by $a_s = 11.1(R/h_{max})^{-0.314}$, with h_{max} being the maximum flow depth in the cross-section.

The key aspect of successful application of Eq. (10.25) appears to lie in the measurement of MEI. This parameter can be measured directly for different species using the "board drop" test. In this test, a standard wooden board is dropped onto a vegetated surface to impart a frictional force that deflects the stems in a manner similar to flowing water. The distance between the ground and the bottom edge of the fallen board, which reflects the ability of the vegetation to resist bending under flow conditions, is measured and then used to calculate the parameter *MEI* from the standard calibration curves (Kouwen, 1988).

Based on laboratory experiments, Kouwen (1988, also see Temple, 1987) related the parameter *MEI* with the vegetation height for growing and dormant grass species, respectively, as

$$MEI = 319h_v^{3.3}, \quad MEI = 25.4h_v^{2.26} \tag{10.28}$$

Application of Eq. (10.28) should be restricted to those grasses that have been tested, including alfalfa, Bermuda grass, buffalo grass, blue grass, weeping love grass, Kentucky grass, Serica lespezeda, Sudan grass, and Rhodes grass. Experimental data encompass stem heights in range of 0.04–1.0 m, *MEI* values in range of 0.007–212 $N \cdot m^2$, and stem densities (M) in range of 140–11,600.

Apparently Eq. (10.25) is only applicable to submerged flexible vegetation, such as grasses. Kouwen and Fathi-Maghadam (2000) conducted flume experiments using coniferous tree sapling and air experiments using large coniferous trees to investigate the relation between the friction factor and mean flow velocity, and proposed a method to estimate the friction factor for emergent woody vegetation:

$$\lambda = 4.06 \left(\frac{U}{\sqrt{\xi E/\rho}} \right)^{-0.46} \frac{h}{h_v} \tag{10.29}$$

where ξ accounts for all aspects of deformation of the plant as a result of an increasing flow velocity. The parameter ξE is called the "vegetation index", which is obtained from the resonant frequency, mass, and height of a tree using a mathematical model based on the works of Niklas and Moon (1988) and Fathi-Maghadam (1996):

$$\xi E = Nf_1^2 \frac{m_s}{h_v} \tag{10.30}$$

where m_s is the total mass, and Nf_1 is the natural frequency of the tree. Nf_1 is measured by rigidly mounting the tree, followed by flexing the top of the tree sideways prior to releasing the tree to swing freely. An accelerometer is attached part way along the stem to record the frequency of the swings. Table 10.1 lists the average vegetation indices for four species of coniferous trees measured by Fathi-Maghadam (1996).

Table 10.1 Vegetation indices for coniferous trees (Fathi-Maghadam, 1996)

Species	$\xi E \ (N \cdot m^{-2})$
Cedar	2.07
Spruce	3.36
White pine	2.99
Austrian pine	4.54

For flexible vegetation, such as trees and bushes, the foliage, whether broad or needle-like, is the major contributor to the total drag. For example, in Jarvela's (2002) experiments on willows, the drag coefficient for leafy willows was three to seven times that of the leafless willows, depending primarily on flow velocity. Considering this fact, Jarvela (2004) suggested the use of the leaf area index (i_{la}) in determining the friction factor. The leaf area index is conventionally defined as the ratio of the upper-side projected area of the leaves in canopy to the area of the surface under the canopy, i.e., the one-sided area of foliage per unit bed area. The developed relation of the friction factor for partially submerged vegetation is

$$\lambda = 4C_{d\chi} i_{la} \left(\frac{U}{U_\chi}\right)^{m_\chi} \frac{h}{h_v} \qquad (10.31)$$

where $C_{d\chi}$ is a species-specific drag coefficient, U_χ is a reference velocity, and m_χ is an exponent. The coefficients in Eq. (10.31) for several species are listed in Table 10.2.

Table 10.2 Coefficients in Eq. (10.31) for various vegetation species (Jarvela, 2004)

Species	$C_{d\chi}$	m_χ	U_χ (m·s^{-1})	i_{la}	Data source
Cedar	0.56	−0.55	0.1	1.42	Fathi-Maghadam (1996)
Spruce	0.57	−0.39	0.1	1.31	Fathi-Maghadam (1996)
White pine	0.69	−0.50	0.1	1.14	Fathi-Maghadam (1996)
Austrian pine	0.45	−0.38	0.1	1.61	Fathi-Maghadam (1996)
Willow	0.43	−0.57	0.1	3.2	Jarvela (2002)

Eqs. (10.29) and (10.31) are valid for situations where the trees just cover the channel bed in plan view. In nature, trees may cover only a part of the bed, or the leaves of adjoining trees may overlap. As pointed out by Raupach et al. (1980) and confirmed by Fathi-Maghadam (1996), the pattern or distribution of the trees does not have a significant effect on the friction factor, but the vegetation density is always a dominant parameter. Kouwen and Fathi-Maghadam (2000) suggested the following correction method to consider the effect of vegetation density:

$$\lambda_m = \lambda \frac{A_{tv}}{A_t} \qquad (10.32)$$

where λ_m is the corrected friction factor, λ is the friction factor estimated using Eq. (10.29) or (10.31), A_{tv} is the total top-view area of the channel covered by trees, and A_t is the total top-view area of the channel.

Freeman et al. (2000) conducted experiments on the resistance due to shrubs and woody vegetation in a large 2.44 m-wide flume and a small 0.46 m-wide flume. A total of 20 different species of broadleaf deciduous vegetation commonly found in flood-plains and riparian zones were evaluated. It was observed that the plant leaf mass trailed downstream forming a streamlined, almost teardrop-shaped profile. The leaf shape changed with velocity and became more streamlined with increasing velocity, yielding a significant decrease in the drag coefficient and resistance coefficient with velocity. On the other hand, the resistance increased with depth for partially submerged plants as the blockage area increased with depth until the plants were submerged. The transition between submerged and partially submerged flows occurred at a depth of about 80 percent of the undeflected plant height. Freeman *et al.* (2000) obtained the regression equation for the Manning n in the case of submerged vegetation $(h > 0.8h_v)$:

$$n = 0.183 \left(\frac{EA_s}{\rho A_v U_*^2} \right)^{0.183} \left(\frac{h_v}{h} \right)^{0.243} (N_a A_v)^{0.273} \left(\frac{U_* R}{\nu} \right)^{-0.115} \frac{R^{1/6}}{\sqrt{g}}$$

(10.33)

where A_s is the total cross-sectional area of the stem(s) of an individual plant, measured at $h_v/4$; E is the modulus of plant stiffness $(N \cdot m^{-2})$; and A_v is the frontal blockage area of an individual plant, which is approximated by an equivalent rectangular area of blockage by leaves. It is important to note that the plant characteristics h_v, A_s, and A_v are the initial characteristics of the plants without the effect of flow distortion.

The regression equation for the Manning n in the case of partially submerged vegetation $(h < 0.8h_v)$ is

$$n = 0.00003487 \left(\frac{EA_s}{\rho A_v^* U_*^2} \right)^{0.150} (N_a A_v^*)^{0.166} \left(\frac{U_* R}{\nu} \right)^{0.622} \frac{R^{1/6}}{\sqrt{g}}$$

(10.34)

where A_v^* is the blockage area of the portion of the leaf mass submerged.

The experiment conditions for the data used to develop Eqs. (10.33) and (10.34) were: flow depths from 0.4 to 1.4 m, average flow velocities from 0.15 to 1.1 $m \cdot s^{-1}$, n from 0.04 to 0.14, plant heights from 0.20 to 1.52 m, plant widths from 0.076 to 0.91 m, plant densities from 0.53 to 13 $plants \cdot m^{-2}$, plant moduli of stiffness from 5.3×10^7 to $4.8 \times 10^9 \, N \cdot m^{-2}$, and Reynolds numbers from 1.4×10^5 to 1.6×10^6.

In addition, for vegetation submerged in intermediate flow, Ree and Palmer (1949) presented a set of curves for the Manning n as a function of UR. For both submerged and emergent vegetation, Wu *et al.* (1999) related the drag coefficient and the Manning n to the Reynolds number and the channel (or friction) slope. The obtained relations of $n \sim UR$ or $n \sim (Re, S)$ vary with vegetation species. These relations can be used to determine the roughness coefficient in vegetated channels.

10.1.3 Sediment transport capacity in vegetated channels

How vegetation affects sediment transport is an important issue of concern. Jordanova and James (2003) experimentally investigated the bed-load transport in a flume covered with uniformly distributed, emergent, rigid cylindrical metal rods. The rods were arranged in a staggered pattern, and the median grain size of sediment was 0.45 mm. Jordanova and James used the method of Li and Shen (1973) to determine the effective bed shear stress and proposed the following formula for the bed-load transport rate $(kg \cdot s^{-1} m^{-1})$ in vegetated channels:

$$q_b = 0.017 \, (\tau_b - \tau_c)^{1.05}$$

(10.35)

where τ_b is the effective bed shear stress $(N \cdot m^{-2})$.

Wu *et al.* (2005) applied the formula of Wu *et al.* (2000b) to compute the bed load in vegetated channels, with the effective bed shear stress τ_b determined by

$$\tau_b = \gamma R_s S \qquad (10.36)$$

where S is the channel slope; and R_s is the hydraulic radius, defined in Eq. (10.17), as suggested by Barfield *et al.* (1979). The computed values of bed-load transport rate were compared with the measured data (Series A, $q = 0.0065$ m^{2}s^{-1}) of Jordanova and James (2003), as shown in Fig.10.8. An excellent agreement was observed.

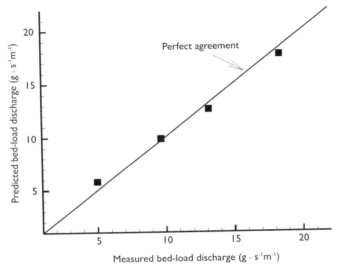

Figure 10.8 Measured and predicted bed-load rates in a vegetated flume.

Okabe *et al.* (1997) investigated the bed-load transport in a flume covered by submerged vegetation. They modeled the vegetation using curved, cylindrical silicone tubes and branched, inclined brass lines. They used a vertical 1-D k-ε turbulent flow model to determine the effective bed shear stress and applied the formula of Ashida and Michiue (1972) for the bed-load transport rate in vegetated channels. The agreement between the measured and predicted values was generally good.

It can be seen from Okabe *et al.* (1997) and Wu *et al.* (2005) that bed-load transport is mainly related to bed shear rather than the drag force exerted on vegetation elements. If the effective bed shear is used, some existing bed-load formulas developed for non-vegetated channels can be extended to the case of vegetated channels.

However, the suspended-load transport in vegetated channels has been little investigated. Because vegetation may reduce the mean flow velocity significantly but intensify the turbulence in a vegetated zone, the effect of vegetation on suspended-load transport is more complex. More experimental and theoretical studies are needed to quantify this effect. As an approximation, one may apply some existing suspended load formulas established under non-vegetated conditions to vegetated channels. Certainly, this application should be done with caution.

Moreover, the flow and sediment transport in channels with large-scale leafy vegetation are strongly three-dimensional. Figs. 10.9(a)–(d) depict the flow and sediment transport in different flow stages, observed qualitatively by Freeman *et al.* (2000). The typical velocity profiles in low and moderate flow conditions are shown in Figs. 10.9(a)–(b), with mild sediment transporting on the bed. When the flow becomes stronger, the leaf mass or foliage canopy diverts the flow beneath the canopy, as shown in Fig. 10.9(c). The bottom flow results in significant velocities along the channel bed, increasing sediment transport and causing scour in open areas. For ground cover vegetation with branches and leaves extending to the bed shown in Fig. 10.9(d), local scour may occur due to three-dimensional vortices that are similar to those typically associated with scour around bridge piers. These complex phenomena need to be investigated quantitatively.

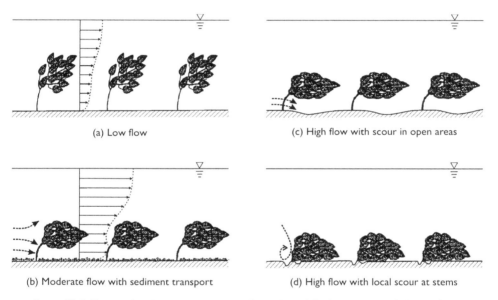

(a) Low flow

(c) High flow with scour in open areas

(b) Moderate flow with sediment transport

(d) High flow with local scour at stems

Figure 10.9 Flow and sediment transport in channels with leafy vegetation (adapted from Freeman *et al.*, 2000).

10.2   SIMULATION OF FLOW IN VEGETATED CHANNELS

10.2.1   Governing equations

3-D equations of flow in vegetated channels

Flow around vegetation usually is three-dimensional, unsteady, and turbulent. However, the mostly considered flow properties in engineering applications are the time- and space-averaged behaviors rather than the detailed features around individual vegetation elements. Time- and space-averaging the Navier-Stokes equations yields the 3-D governing equations for flow in vegetated channels:

$$\frac{\partial[\rho(1-c_{v0})]}{\partial t} + \frac{\partial[\rho(1-c_{v0})u_i]}{\partial x_i} = 0 \tag{10.37}$$

$$\frac{\partial[\rho(1-c_{v0})u_i]}{\partial t} + \frac{\partial[\rho(1-c_{v0})u_iu_j]}{\partial x_j} = \rho(1-c_{v0})F_i - (1-c_{v0})\frac{\partial p}{\partial x_i}$$
$$+ \frac{\partial[(1-c_{v0})\tau_{ij}]}{\partial x_j} - N_a f_{di} \tag{10.38}$$

where $f_{di}(i=1,2,3)$ are the components of drag force per unit vegetation height, defined in Eq. (10.8); and c_{v0} is the local volumetric concentration of vegetation, defined in Eq. (10.5).

Compared with the equations derived by Shimizu and Tsujimoto (1994) and Lopez and Garcia (2001), Eqs. (10.37) and (10.38) include the vegetation concentration c_{v0}, which changes in time and space due to the heterogeneity and seasonal growth and death of vegetation as well as the change of flow conditions. This should be particularly important in the case of high vegetation density. In the case of low vegetation density, $1 - c_{v0} \approx 1$ and, thus, Eqs. (10.37) and (10.38) can be simplified as

$$\frac{\partial u_i}{\partial x_i} = 0 \tag{10.39}$$

$$\frac{\partial u_i}{\partial t} + \frac{\partial(u_iu_j)}{\partial x_j} = F_i - \frac{1}{\rho}\frac{\partial p}{\partial x_i} + \frac{1}{\rho}\frac{\partial \tau_{ij}}{\partial x_j} - \frac{1}{\rho}N_a f_{di} \tag{10.40}$$

The stresses τ_{ij} include the effects of molecular viscosity, turbulence, and non-uniformity of flow velocity around vegetation elements. The last effect causes dispersion, which is often combined with the turbulent effect. Thus, the stresses are calculated using the Boussinesq assumption (7.3), with the eddy viscosity v_t determined by Eq. (2.54) and the turbulent energy k and its dissipation rate ε determined by (Shimizu and Tsujimoto, 1994)

$$\frac{\partial k}{\partial t} + u_j\frac{\partial k}{\partial x_j} = \frac{\partial}{\partial x_j}\left(\frac{v_t}{\sigma_k}\frac{\partial k}{\partial x_j}\right) + P_k + P_{kv} - \varepsilon \tag{10.41}$$

$$\frac{\partial \varepsilon}{\partial t} + u_j\frac{\partial \varepsilon}{\partial x_j} = \frac{\partial}{\partial x_j}\left(\frac{v_t}{\sigma_\varepsilon}\frac{\partial \varepsilon}{\partial x_j}\right) + c_{\varepsilon1}\frac{\varepsilon}{k}(P_k + c_{f\varepsilon}P_{kv}) - c_{\varepsilon2}\frac{\varepsilon^2}{k} \tag{10.42}$$

where P_k is the production of turbulence by shear, defined in Eq. (2.52); and P_{kv} is the generation of turbulence due to vegetation, determined by $P_{kv} = c_{fk}N_a$ $f_{di}u_i/[\rho(1-c_{v0})]$. Shimizu and Tsujimoto (1994) selected coefficients $c_{fk} = 0.07$ and $c_{f\varepsilon} = 0.16$ based on calibration, whereas Lopez and Garcia (2001) determined coefficients $c_{fk} = 1.0$ and $c_{f\varepsilon} = 1.33$ based on a theoretical argument that c_{fk} should be unity if the total turbulence kinetic energy is modeled. More discussion on these two coefficients is given in Section 10.2.3. The values of other coefficients can be found in Section 2.3.

Naot *et al.* (1996) and Neary (2003) applied the algebraic stress model and the k-ω turbulence model, respectively, in the simulation of flow in vegetated channels. The details can be found in their papers.

Depth-averaged 2-D equations of flow in vegetated channels

Integrating Eqs. (10.37) and (10.38) over the flow depth yields the depth-integrated 2-D continuity and momentum equations of flow in vegetated channels (Wu and Wang, 2004b):

$$\frac{\partial[\rho(1-c_v)h]}{\partial t} + \frac{\partial[\rho(1-c_v)hU_x]}{\partial x} + \frac{\partial[\rho(1-c_v)hU_y]}{\partial y} = 0 \qquad (10.43)$$

$$\frac{\partial[\rho(1-c_v)hU_x]}{\partial t} + \frac{\partial[\rho(1-c_v)hU_x^2]}{\partial x} + \frac{\partial[\rho(1-c_v)hU_yU_x]}{\partial y}$$
$$= -\rho g(1-c_v)h\frac{\partial z_s}{\partial x} + \frac{\partial[(1-c_v)h(T_{xx}+D_{xx})]}{\partial x} + \frac{\partial[(1-c_v)h(T_{xy}+D_{xy})]}{\partial y}$$
$$- (1-c_v)\tau_{bx} - N_aF_{dx} \qquad (10.44)$$

$$\frac{\partial[\rho(1-c_v)hU_y]}{\partial t} + \frac{\partial[\rho(1-c_v)hU_xU_y]}{\partial x} + \frac{\partial[\rho(1-c_v)hU_y^2]}{\partial y}$$
$$= -\rho g(1-c_v)h\frac{\partial z_s}{\partial y} + \frac{\partial[(1-c_v)h(T_{yx}+D_{yx})]}{\partial x} + \frac{\partial[(1-c_v)h(T_{yy}+D_{yy})]}{\partial y}$$
$$- (1-c_v)\tau_{by} - N_aF_{dy} \qquad (10.45)$$

where F_{dx} and F_{dy} are the x- and y-components of the drag force on vegetation exerted by the flow, defined in Eq. (10.7); and c_v is the depth-averaged volumetric concentration of vegetation, defined in Eq. (10.4).

In the case of low vegetation density, Eqs. (10.43)–(10.45) are simplified as

$$\frac{\partial h}{\partial t} + \frac{\partial(hU_x)}{\partial x} + \frac{\partial(hU_y)}{\partial y} = 0 \qquad (10.46)$$

$$\frac{\partial(hU_x)}{\partial t} + \frac{\partial(hU_x^2)}{\partial x} + \frac{\partial(hU_yU_x)}{\partial y} = -gh\frac{\partial z_s}{\partial x} + \frac{1}{\rho}\frac{\partial[h(T_{xx}+D_{xx})]}{\partial x}$$
$$+ \frac{1}{\rho}\frac{\partial[h(T_{xy}+D_{xy})]}{\partial y}$$
$$- \frac{1}{\rho}\tau_{bx} - \frac{1}{\rho}N_aF_{dx} \qquad (10.47)$$

$$\frac{\partial(hU_y)}{\partial t} + \frac{\partial(hU_xU_y)}{\partial x} + \frac{\partial(hU_y^2)}{\partial y} = -gh\frac{\partial z_s}{\partial y} + \frac{1}{\rho}\frac{\partial[h(T_{yx}+D_{yx})]}{\partial x}$$
$$+ \frac{1}{\rho}\frac{\partial[h(T_{yy}+D_{yy})]}{\partial y}$$
$$- \frac{1}{\rho}\tau_{by} - \frac{1}{\rho}N_aF_{dy} \tag{10.48}$$

The bed shear stresses τ_{bx} and τ_{by} are determined by Eq. (6.4), with $c_f = gn_b^2/R_s^{1/3}$, in which n_b is the Manning roughness coefficient of the bed and R_s is the hydraulic radius defined in Eqs. (10.17) and (10.22) or simply set as the flow depth h.

The depth-averaged stresses T_{ij} are calculated by Eq. (6.7), with the eddy viscosity v_t determined using Eq. (2.54) and the turbulent energy k and its dissipation rate ε determined by

$$\frac{\partial k}{\partial t} + U_x\frac{\partial k}{\partial x} + U_y\frac{\partial k}{\partial y} = \frac{\partial}{\partial x}\left(\frac{v_t}{\sigma_k}\frac{\partial k}{\partial x}\right) + \frac{\partial}{\partial y}\left(\frac{v_t}{\sigma_k}\frac{\partial k}{\partial y}\right)$$
$$+ P_k + P_{kv} + P_{kb} - \varepsilon \tag{10.49}$$

$$\frac{\partial \varepsilon}{\partial t} + U_x\frac{\partial \varepsilon}{\partial x} + U_y\frac{\partial \varepsilon}{\partial y} = \frac{\partial}{\partial x}\left(\frac{v_t}{\sigma_\varepsilon}\frac{\partial \varepsilon}{\partial x}\right) + \frac{\partial}{\partial y}\left(\frac{v_t}{\sigma_\varepsilon}\frac{\partial \varepsilon}{\partial y}\right)$$
$$+ c_{\varepsilon 1}\frac{\varepsilon}{k}(P_k + c_{f\varepsilon}P_{kv}) + P_{\varepsilon b} - c_{\varepsilon 2}\frac{\varepsilon^2}{k} \tag{10.50}$$

where P_k, P_{kb}, and $P_{\varepsilon b}$ are defined in Eqs. (6.10) and (6.11); and P_{kv} is the generation of turbulence due to vegetation, determined by $P_{kv} = c_{fk}N_a(F_{dx}U_x + F_{dy}U_y)/[\rho(1-c_v)]$.

The dispersion momentum transports D_{ij} due to the non-uniformity of flow velocity along the flow depth are determined using the models introduced in Section 6.3 for curved channels, or combined with the turbulent stresses otherwise.

1-D equations of flow in vegetated channels

Integrating Eqs. (10.43) and (10.44) over the channel width yields the 1-D continuity and momentum equations in vegetated channels:

$$\frac{\partial[\rho(1-c_v)A]}{\partial t} + \frac{\partial[\rho(1-c_v)Q]}{\partial x} = \rho q_l \tag{10.51}$$

$$\frac{\partial[\rho(1-c_v)Q]}{\partial t} + \frac{\partial}{\partial x}\left[\frac{\rho\beta(1-c_v)Q^2}{A}\right] + \rho g(1-c_v)A\frac{\partial z_s}{\partial x} + \rho g(1-c_v)AS_f = \rho q_l v_x \tag{10.52}$$

where c_v is the volumetric concentration of vegetation averaged the cross-section; and S_f is the friction slope, including the effects of bed friction and vegetation drag:

$$S_f = S_{fb} + \frac{B}{\rho g(1-c_v)A}N_aF_d = \frac{Q|Q|}{K^2} \tag{10.53}$$

where S_{fb} is the bed friction slope, F_d is the drag force defined in Eq. (10.7), and K is the conveyance.

When the vegetation concentration c_v is small, Eqs. (10.51) and (10.52) can be simplified as Eqs. (5.1) and (5.2) by eliminating $1 - c_v$.

If the entire cross-section is covered by nearly uniformly distributed vegetation, the conveyance K can be determined by

$$K = \frac{A^{5/3}}{n\chi^{2/3}} \tag{10.54}$$

where the Manning n accounts for the effects of both channel bed friction and vegetation drag, and is determined by one of the relations described in Section 10.1.2.

If the cross-section is partially covered by vegetation or the vegetation density varies along the cross-section, the flow velocity significantly varies in the vegetated and non-vegetated zones or even in different vegetated zones. Thus, the cross-section needs to be divided into a suitable number of subsections, either vegetated or non-vegetated. The conveyance in each subsection is determined by

$$K_j = \frac{A_j^{5/3}}{n_j \chi_j^{2/3}} \tag{10.55}$$

where K_j, A_j, χ_j, and n_j are the conveyance, flow area, wetted perimeter, and Manning roughness coefficient of subsection j, respectively. The total conveyance K can be obtained by summing the conveyances of all subsections as

$$K = \sum_j K_j \tag{10.56}$$

The flow velocity in each subsection is determined using the Manning equation:

$$U_j = \frac{K_j S_f^{1/2}}{A_j} \tag{10.57}$$

Eqs. (10.51) and (10.52) are iteratively solved together with Eqs. (10.53) and (10.55)–(10.57) and a relation between the Manning roughness coefficient and flow conditions introduced in Section 10.1.2. This approach is similar to but more complicated than that used for compound channels in Section 5.1.1.4.

10.2.2 Numerical solutions

The 1-D, 2-D, and 3-D governing equations can be solved using the numerical methods described in Chapters 5–7. For example, the 1-D equations (10.51) and (10.52) can be solved using the Preissmann scheme and the Thomas algorithm described in

Section 5.2.2, with slight modification to consider the variation of Manning n with flow conditions.

The depth-averaged 2-D flow equations (10.43)–(10.45) can be solved using the 2-D SIMPLE(C) algorithm in Section 6.1.3.1 by defining the pressure as

$$p = \rho(1 - c_v)gz_s \qquad (10.58)$$

and the fluxes at cell faces as

$$F_w = [\rho(1 - c_v)h]_w^{n+1}(J\alpha_i^1 \Delta\eta)_w U_{i,w}^{n+1} \qquad (10.59)$$

$$F_s = [\rho(1 - c_v)h]_s^{n+1}(J\alpha_i^2 \Delta\xi)_s U_{i,s}^{n+1} \qquad (10.60)$$

The details on the 2-D SIMPLE(C) algorithm for flow in vegetated channels can be found in Wu and Wang (2004b). Similarly, Eqs. (10.37) and (10.38) can be solved using the 3-D SIMPLE algorithm in Section 7.1.3.2 by replacing ρ with $\rho(1 - c_{v0})$.

For low vegetation density, the 3-D equations (10.39) and (10.40) and 2-D equations (10.46)–(10.48) can be solved directly using the numerical algorithms developed for flow in non-vegetated channels by arranging the drag force terms as source terms.

10.2.3 Examples

Numerous verifications and applications of the vegetation effect models can be found in the literature. Two examples are cited here. One is the simulation of flow in open channels with rigid, submerged vegetation performed by Shimizu and Tsujimoto (1994) using a vertical 2-D model with the k-ε turbulence closure. The experiments were conducted in flumes under uniform flow conditions. The rigid cylinders of equal height and diameter were placed at equal spacings in a square pattern on smooth flume beds. Fig. 10.10 shows the measured and simulated mean flow velocities, Reynolds shear stresses, and streamwise turbulence intensities along the flow depth for the run with a flow depth of 7.47 cm, a depth-averaged flow velocity of 13.87 cm·s⁻¹, an energy slope of 0.00213, a vegetation height of 4.1 cm, a vegetation diameter of 0.1 cm, and a vegetation spacing of 1.0 cm. The flow was retarded by vegetation in the lower layer, and the maximum shear stress and streamwise turbulence intensity occurred at the top of the vegetation elements. The simulated results are in generally good agreement with the measured data.

Note that Shimizu and Tsujimoto (1994) calibrated coefficients $c_{fk} = 0.07$ and $c_{f\varepsilon} = 0.16$ through the above simulation. Lopez and Garcia (2001) also simulated the flow over rigid, submerged vegetation under similar conditions and validated the theoretically-based values $c_{fk} = 1.0$ and $c_{f\varepsilon} = 1.33$. To clarify this difference, Neary (2003) re-simulated the case shown in Fig. 10.10 using the k-ω turbulence model. He found that both sets of c_{fk} and $c_{f\varepsilon}$ values give very close predictions for the mean flow velocity and the Reynolds shear stress, while $c_{fk} = 0.07$ and $c_{f\varepsilon} = 0.16$ give a better prediction for the streamwise turbulence intensity than $c_{fk} = 1.0$ and $c_{f\varepsilon} = 1.33$.

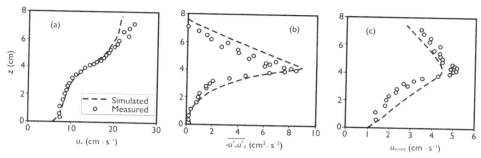

Figure 10.10 (a) Mean velocities, (b) Reynolds shear stresses, and (c) streamwise turbulence intensities of flow over submerged vegetation (Shimizu and Tsujimoto, 1994).

However, because the k-ε and k-ω turbulence models usually may not compute the streamwise turbulence intensity or normal stress accurately, it is necessary to calibrate the two coefficients using the measurement data of other quantities, such as the turbulent energy k, that these turbulence models can handle well. Before this is done, it is the modeler's preference to use which set of c_{fk} and $c_{f\varepsilon}$ values. To this author, it is more comfortable to use $c_{fk} = 1.0$ and $c_{f\varepsilon} = 1.33$ since they are derived theoretically.

The second example cited here is the simulation of flow around alternate vegetation zones done by Wu and Wang (2004b) using the depth-averaged 2-D model described above. The experiments were conducted by Bennett *et al.* (2002) in a 16.5 m long tilting recirculating flume. Six semi-circular vegetation zones with an equal spacing of 2.4 m were distributed alternately to achieve a meandering pattern, as shown in Fig. 10.11(a). The diameter of the vegetation zones was 0.6 m. The model vegetation was emergent wooden dowel with a diameter of 3.2 mm, laid out in a staggered pattern in each vegetation zone. Five vegetation concentrations of 0.04%, 0.2%, 0.6%, 2.5%, and 10% were used. The flow discharge was 0.0043 m³s⁻¹, and the pre-vegetation flow depth was 0.027 m. The slope of the flume was 0.0004. The surface flow velocity was measured using the Particle Image Velocimetry (PIV) technique. The computational

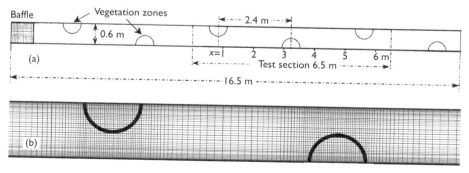

Figure 10.11 (a) Plan view and (b) mesh for experiments of Bennett *et al.* (2002).

mesh consisted of 461 nodes streamwise and 41 nodes laterally, part of which is shown in Fig. 10.11(b). The depth- averaged 2-D k-ε turbulence model with $c_{fk} = 1.0$ and $c_{f\varepsilon} = 1.33$ was used. The drag coefficient C_d was set as 0.8, 1.0, 1.2, 1.8, and 3.0 for the runs with vegetation concentrations of 0.04%, 0.2%, 0.6%, 2.5%, and 10%,

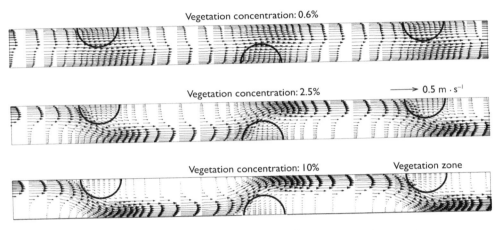

Figure 10.12 Calculated flow fields around alternate vegetation zones.

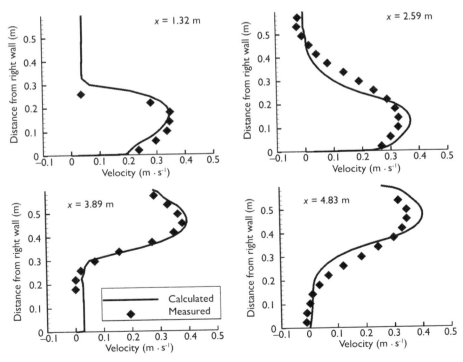

Figure 10.13 Measured and calculated velocities around alternate vegetation zones.

respectively. This is due to that the Reynolds number decreases as the vegetation concentration increases.

Fig. 10.12 shows the simulated flow vectors for vegetation concentrations of 0.6%, 2.5%, and 10%. It can be seen that the vegetation forced the thread of highest velocity to meander. The meandering flow pattern became more obvious, as the vegetation concentration increased. When the vegetation concentration was 10%, a recirculation flow occurred downstream of each vegetation zone. Fig. 10.13 compares the measured and calculated flow velocities along several cross-sections for the case with vegetation concentration of 10%. The simulation results and measurement data match qualitatively well.

10.3   SIMULATION OF SEDIMENT TRANSPORT IN VEGETATED CHANNELS

10.3.1   Sediment transport models in vegetated channels

The total load is separated as bed load and suspended load, as shown in Fig. 2.6. The 3-D transport equation of suspended load in vegetated channels is

$$\frac{\partial[(1 - c_{v0})c_k]}{\partial t} + \frac{\partial[(1 - c_{v0})u_j c_k]}{\partial x_j} - \frac{\partial[(1 - c_{v0})\omega_{sk}\delta_{3j}c_k]}{\partial x_j} = \frac{\partial}{\partial x_j}\left[(1 - c_{v0})\varepsilon_s\frac{\partial c_k}{\partial x_j}\right]$$

$$(10.61)$$

where c_k is the local concentration of the kth size class of suspended load.

To solve Eq. (10.61), the boundary condition at the water surface is given as Eq. (7.44), and the deposition and entrainment rates at the lower boundary of the suspended sediment layer are $D_{bk} = \omega_{sk}c_{bk}$ and $E_{bk} = \omega_{sk}c_{b*k}$.

Integrating Eq. (10.61) over the flow depth yields the depth-averaged 2-D transport equation of suspended load in vegetated channels:

$$\frac{\partial[(1 - c_v)hC_k/\beta_{sk}]}{\partial t} + \frac{\partial[(1 - c_v)hU_x C_k]}{\partial x} + \frac{\partial[(1 - c_v)hU_y C_k]}{\partial y}$$

$$= \frac{\partial}{\partial x}\left[h(1 - c_v)\left(\varepsilon_s\frac{\partial C_k}{\partial x} + D_{sxk}\right)\right] + \frac{\partial}{\partial y}\left[h(1 - c_v)\left(\varepsilon_s\frac{\partial C_k}{\partial y} + D_{syk}\right)\right]$$

$$+ \alpha\omega_{sk}(1 - c_v)(C_{*k} - C_k)$$

$$(10.62)$$

where C_k is the depth-averaged concentration of the kth size class of suspended load.

Integrating Eq. (10.62) over the channel width yields the 1-D transport equation of suspended load in vegetated channels:

$$\frac{\partial[(1 - c_v)AC_k/\beta_{sk}]}{\partial t} + \frac{\partial[(1 - c_v)AUC_k]}{\partial x} = \alpha(1 - c_v)\omega_{sk}B(C_{*k} - C_k) + q_{lsk} \quad (10.63)$$

where C_k is the cross-section-averaged concentration of the kth size class of suspended load.

In the 3-D and depth-averaged 2-D models, the transport equation of bed load in vegetated channels is

$$\frac{\partial[(1-c_{vb})q_{bk}/u_{bk}]}{\partial t} + \frac{\partial[\alpha_{bx}(1-c_{vb})q_{bk}]}{\partial x} + \frac{\partial[\alpha_{by}(1-c_{vb})q_{bk}]}{\partial y}$$

$$= \frac{1}{L}(1-c_{vb})(q_{b*k}-q_{bk}) \tag{10.64}$$

where c_{vb} is the volumetric concentration of vegetation in the bed-load zone.

In the 1-D model, the transport equation of bed load in vegetated channels is

$$\frac{\partial[(1-c_{vb})Q_{bk}/U_{bk}]}{\partial t} + \frac{\partial[(1-c_{vb})Q_{bk}]}{\partial x} = \frac{1}{L}(1-c_{vb})(Q_{b*k}-Q_{bk}) + q_{lbk} \tag{10.65}$$

where Q_{bk} is the transport rate of the kth size class of bed load.

For low vegetation density, Eqs. (10.61)–(10.65) can be simplified by eliminating the vegetation concentration.

The vegetation densities in the water column and channel bed might be different, but this complexity is ignored in the present models. Therefore, the bed change and bed material sorting equations in vegetated channels are the same as those in non-vegetated channels. For example, in the depth-averaged 2-D model, the bed change and the bed-material gradation in the mixing layer in vegetated channels are determined by Eqs. (2.159) and (2.161), respectively.

To close the sediment transport models introduced above, the sediment transport capacities need to be computed using empirical formulas. As described in Section 10.1.3, the work of Okabe et al. (1997) suggests that the Ashida-Michiue (1972) bed-load formula can be used in vegetated channels. According to Wu et al. (2005), the Wu et al. (2000b) formula can be extended to determine the sediment transport capacity in the case of emergent vegetation. In the case of submerged vegetation, the total bottom shear stress determined using Eq. (10.16) with Eq. (10.22) for R_s includes both the shears from the bed and at the top of vegetation elements. Considering only the shear on the bed affects the transport of bed load, the total bottom shear stress should be modified by multiplying a factor h_v/h when the bed-load transport capacity is computed. This factor h_v/h is derived by considering Eq. (10.20) and the first term on the right-hand side of Eq. (10.21). However, this modification is not necessary in the 3-D model that directly calculates the bed shear stress.

The aforementioned sediment transport equations in vegetated channels can be solved by straightforwardly extending the numerical methods presented in Chapters 5–7 for sediment transport models in non-vegetated channels. The factor $1-c_v$ can be eliminated from the left-hand sides of Eqs. (10.61)–(10.65), or treated by replacing ρ with $\rho(1-c_v)$ in the finite volume method, as described in Section 10.2.2. The details can be found in Wu and Wang (2004b) and Wu et al. (2005).

10.3.2 Examples

Case 1. Bed change around a vegetated island

Experiments on the expansion of a vegetation island due to sedimentation were conducted by Tsujimoto (1998). The island was a non-submerged porous body (5 cm wide and 25 cm long), located in the center of a 0.4 m wide straight flume with a sand bed ($d = 1.6$ mm), as shown in Fig. 10.14. The properties of the model vegetation were not reported, but the projected area per unit volume was determined as 0.1 cm^{-1} in the experiments. The slope of the flume was 1/100, and the flow discharge was 0.003 m^{3}s^{-1}. The morphodynamic process around the vegetation island was simulated by Tsujimoto (1998) and Wu and Wang (2004b) using depth-averaged 2-D models. The two models used differ in numerical methods and sediment transport models. Tsujimoto used the Ashida-Michiue (1971) equation for bed load, whereas Wu and Wang used the Wu *et al.* (2000b) formula. In Wu and Wang's simulation, the mesh consisted of 122 × 44 nodes with a refined grid spacing around the vegetation island, the time step was 15 seconds, and C_d was set as 3.0.

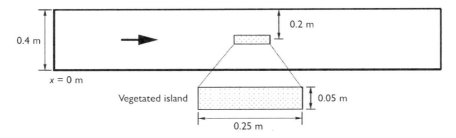

Figure 10.14 Plan view of Tsujimoto's (1998) experiments.

Fig. 10.15 compares the measured and calculated bed elevation changes near the island after 30 minutes for the experimental run in which the inflow sediment discharge was at equilibrium. Deposition happened in front of the island and inside and behind it, whereas erosion occurred on its two sides. Tsujimoto (1998) predicted the pattern of deposition and erosion well but the magnitude less accurately. Wu and Wang (2004b) improved the accuracy, primarily due to the use of different sediment transport capacity formula. Both simulations suggested the growth of the vegetation island, as compared to the measurement.

Case 2. Bed change in a bend in the Little Topashaw Creek

The study reach was a deeply-incised sharp bend in the Little Topashaw Creek, North Central Mississippi, as shown in Fig. 10.16. Five large wood structures made from felled trees were placed along the outside of the study bend in the summer of 2000 in order to stabilize the channel and create aquatic habitats (Shields *et al.*, 2004). The crests of the structures were 1.1 to 3.2 m higher than the bed and were emergent at low flows and submerged at high flows. Logs running transverse to the flow direction were

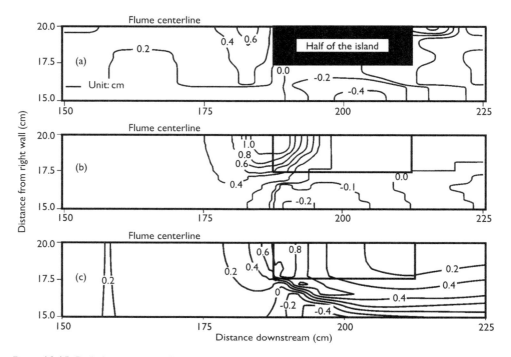

Figure 10.15 Bed changes around vegetated island: (a) measured by Tsujimoto (1998), (b) calculated by Tsujimoto (1998), and (c) calculated by Wu and Wang (2004).

about 6 m long and were anchored into the bank toe. Acoustic Doppler velocimeters were used to measure flow depth and vertically-averaged velocity during high flows.

Conditions during the period from June 2000 to June 2001 were numerically simulated using the depth-averaged 2-D model (Wu et al., 2005). Flow records (15-min interval) from a gage about 1 km upstream from the study reach were used as the inflow condition, and water surface elevations recorded by one of the acoustic Doppler devices were extrapolated to the downstream end of the reach and used as the water level boundary condition. The bed material in the study reach was quite uniform and the median size was about 0.26 mm. The Manning roughness coefficient was estimated as 0.028. The average diameter of the logs was about 0.3 m, and the vegetation concentration was about 20%. The shape factor α_v was set as 0.5 because the large wood structures were irregular and inclined. The product of α_v and C_d was 2.0, which was calibrated using the measured flow velocity. The suspended-load adaptation coefficient α was set as 0.5, and the bed-load adaptation length L_b was 20 m. The dispersion terms in the momentum equations and suspended-load transport equation were evaluated using the algebraic model in Section 6.3, with coefficients β_I and λ_t given 1.0 and 3.0, respectively. The computational time step was 2 minutes.

Fig. 10.17(a) shows the simulated flow field at a discharge of 42.6 m^{3}s^{-1}, which was almost the highest flow in the simulation period. The flow was retarded by the large wood matrices along the left bank and accelerated in the main channel. Fig. 10.18 compares the measured and predicted velocities at locations LTH2A and LTH2B

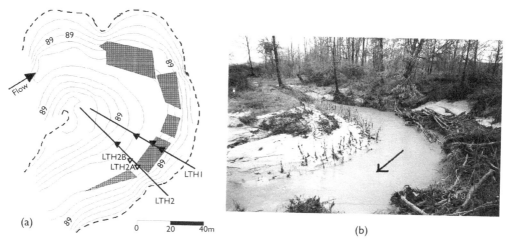

Figure 10.16 (a) Map of study site and (b) photo facing upstream in Little Topashaw Creek (contours represent bed elevation, in m; shaded areas are wood structures).

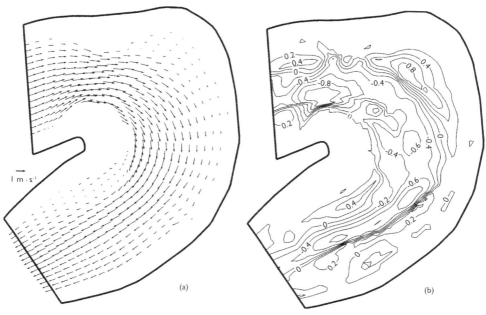

Figure 10.17 (a) Simulated flow at $Q = 42.6$ m^{3}s^{-1} and (b) simulated bed change (m) in 2000–2001.

during a flow event with a peak of 15.5 m^{3}s^{-1}. Instrument LTH2A was secured to the bed immediately downstream from a large wood structure, while instrument LTH2B was secured to the bed along the same cross-section but at the centerline of the base flow channel. Measured velocities were noisy due to turbulent fluctuations, acoustic interference from floating and suspended trash and debris, and factors internal

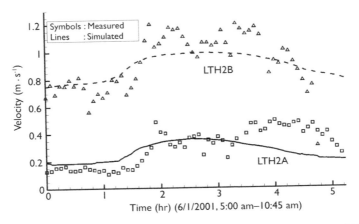

Figure 10.18 Measured and simulated flow velocities with time at LTH2A and LTH2B.

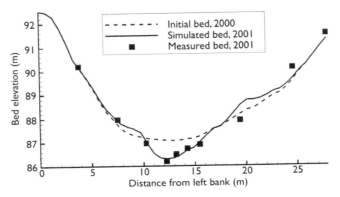

Figure 10.19 Measured and simulated bed profiles at cross-section LTH2.

to the instrument. Due to the large wood structures, the flow velocity at LTH2A was only about one-third that at LTH2B. Fig. 10.17(b) shows the simulated bed elevation change contours in the bend, and Fig. 10.19 compares the simulated and measured bed changes at cross-section LTH2 during the simulation period. Deposition occurred at the structures along the left (outer) bank, with erosion in the main channel. The model reproduced the bed change fairly well.

Chapter 11

Cohesive sediment transport modeling

Fine-grained sediments, such as clay and fine silt, widely exist in rivers, lakes, reservoirs, estuaries, and coastal waters. They generally exhibit cohesive properties and undergo a number of complex mechanical, physicochemical, and biochemical processes. Fundamentals and methodologies for simulation of cohesive sediment transport and the associated morphodynamic processes are presented in this chapter.

11.1 COHESIVE SEDIMENT TRANSPORT PROCESSES

11.1.1 General transport patterns

Fig. 11.1 shows the general transport pattern of cohesive sediments in estuaries and coastal waters. Similar patterns can be found in rivers, reservoirs, and lakes, except for the effects of salinity and tide. Because of the action of electrostatical forces that are comparable to or larger than the gravity forces, fine sediment particles may stick together and form flocs or aggregates when they collide, as shown in Fig. 11.2. This process is called "flocculation." The flocs may be transported by convection (due to river flows, currents, and waves), turbulent diffusion, and gravitational settling. They may move in suspended load or bed load, depending on their sizes; however, suspension is usually presumed to be the main transport mode. Variations in flow conditions may cause sediment erosion and deposition, whereas the settled cohesive deposits may consolidate, due to gravity and the overlying water pressure.

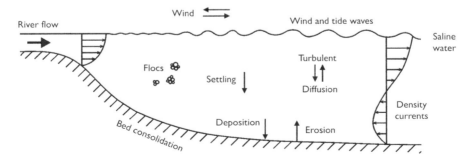

Figure 11.1 Cohesive sediment transport in estuary.

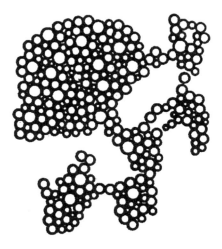

Figure 11.2 Schematic diagram of a typical floc (Tsai *et al.*, 1987).

Flocs experience continuous aggregation and disaggregation during their transport. Finer flocs and single particles may collide and then form larger flocs. On the other hand, larger flocs may be disaggregated into finer flocs and single particles, due to high shear or large eddy ejection and sweeping, in particular near the bottom. Under given flow and sediment conditions, the flocculation and break-up processes may reach an approximately equilibrium state. The typical size distribution of flocs is shown in Fig. 11.3, as compared to that of the corresponding dispersed particles. One can see that the flocs are much coarser than the dispersed particles on average.

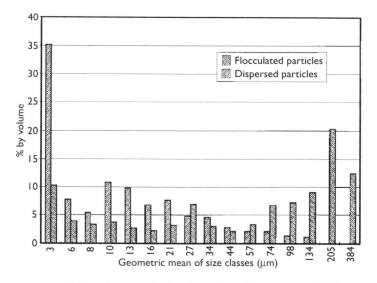

Figure 11.3 Typical size distribution of flocs (Krishnappan, 2000).

11.1.2   Factors affecting flocculation

Flocculation of cohesive sediment particles is affected by sediment size, sediment concentration, salinity, turbulence intensity, temperature, organic matters, etc.

Sediment size

Experiments have shown that flocculation is negligible for sediment particles larger than about 0.03 mm, but it becomes stronger as sediment size reduces. Migniot (1968) defined a flocculation factor as $F_\omega = \omega_{sf}/\omega_{sd}$, in which ω_{sf} and ω_{sd} are the median settling velocities of flocs and the corresponding dispersed sediment particles, respectively. Migniot measured ω_{sf} and ω_{sd} in a settling column using a large number of muddy sediment samples with different compositions at a sediment concentration of $10 \text{ kg} \cdot \text{m}^{-3}$ and a salinity of 30 ppt. As shown in Fig. 11.4, the flocculation factor F_ω varies with the median size d_{50} of the dispersed sediment according to

$$F_\omega = \left(\frac{d_r}{d_{50}}\right)^{n_d} \tag{11.1}$$

where $n_d = 1.8$, and d_r is a reference diameter, about 0.0215 mm.

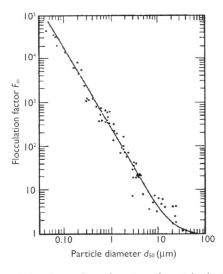

Figure 11.4 Flocculation factor F_ω as function of particle d_{50} (Migniot, 1968).

Qian (1980), Huang (1981), and Dixit et al. (1982) also investigated the flocculation factor F_ω using different sediment samples and obtained Eq. (11.1). As listed in Table 11.1, the exponent n_d in all three studies has almost the same value as that Migniot observed, while the reference diameter d_r exhibits different values perhaps due to difference in flow, salinity, and sediment conditions among these experiments.

Applying the Stokes law to ω_{sd} in Eq. (11.1), one can find that the floc settling velocity is approximately 0.15 to 0.6 $\text{mm} \cdot \text{s}^{-1}$ and does not vary much with the size

Table 11.1 Parameters n_d and d_r of flocculation factor

Authors	Experiment conditions	n_d	d_r (mm)
Migniot (1968)	Muddy sediments, salinity = 30 ppt, sediment concentration = 10 kg·m⁻³	1.8	0.0215
Qian (1980)	River sediments, sediment concentration = 30 kg·m⁻³	2.0	0.011
Huang (1981)	Lianyun Harbor mud, salinity = 30 ppt, sediment concentration = 0.08–1.8 kg·m⁻³	1.9	0.022
Dixit *et al.* (1982)	No salinity, sediment concentration = 1.2–11 kg·m⁻³	1.8	0.012

of the dispersed particles. This range of settling velocity is equivalent to the size range of medium-to-coarse silt particles.

Sediment concentration

Flocculation is highly related to sediment concentration, as shown in Fig. 11.5 (Thorn, 1981; Mehta, 1986). At low sediment concentrations, as the sediment concentration increases, the floc settling velocity increases, due to the intensification of flocculation by increasing the collision probability. As the sediment concentration increases further, the floc settling velocity decreases, due to the effect of hindered settling and the formulation of large floc structures. At very large concentrations, a large number of particles form large-scale floc matrices and, thus, the floc settling velocity becomes very small and even reduces to zero temporarily. The water and sediment mixture tends to be fluidized and becomes a non-Newtonian fluid.

The floc settling velocity in Fig. 11.5 is based on measurements in a settling column using the mud in saltwater from the Severn Estuary, England (Thorn, 1981). The curve

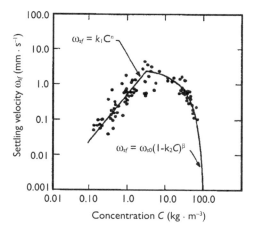

Figure 11.5 Floc settling velocity as function of sediment concentration (Thorn, 1981; Mehta, 1986).

can be expressed as

$$\omega_{sf} = \begin{cases} k_1 C^n & 0 < C \le C_p \\ \omega_{s0}(1 - k_2 C)^\beta & C > C_p \end{cases} \tag{11.2}$$

where C is the sediment concentration in $kg \cdot m^{-3}$; ω_{s0} is a reference settling velocity; n, β, k_1, and k_2 are coefficients; and C_p is the sediment concentration at which the floc settling velocity turns from an increasing to a decreasing trend in the curve of $\omega_{sf} \sim C$. $k_1 = 0.513$, $k_2 = 0.008$, $n = 1.29$, $\beta = 4.65$, $\omega_{s0} = 0.0026 \ m \cdot s^{-1}$, and $C_p = 3.5 \ kg \cdot m^{-3}$ for the curve in Fig. 11.5. However, many experiments have shown that these parameters depend on sediment properties. n ranges from 1 to 2 with a mean value of 1.3, while β ranges between 3 and 5. Huang (1981) evaluated C_p as about $1.5 \ kg \cdot m^{-3}$ for the Lianyun Harbor mud, and Li *et al.* (1994) gave C_p a value of $3.0 \ kg \cdot m^{-3}$ for the Gironde Estuary mud.

Salinity

As observed by Krone (1962), Owen (1970), Huang (1981), and Yue (1983), salinity influences flocculation significantly. Most clay particles have a negative charge. In fresh water, the electrokinetic potential associated with the particles generally is sufficiently large, and as a result, the particles will repel each other. However, in saline water, this potential is reduced below a critical value, the electrical layer associated with the particles collapses; thus, the particles stick together to form flocs, due to the presence of dominating molecular attractive forces (London-Van der Waals forces), electrostatical surface forces (double layer), and chemical forces (hydrogen bonds, cation bonds, and cementation).

Chien and Wan (1983) presented a graphical relation of floc settling velocity and salinity, which is shown in Fig. 11.6. It can be seen that when salinity is low, the floc

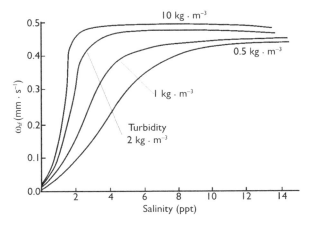

Figure 11.6 Settling velocity as function of salinity (Chien and Wan, 1983).

settling velocity increases rapidly as salinity increases. However, when salinity exceeds a certain value, its influence on floc settling becomes very slight.

Turbulence intensity

The turbulence of flow affects flocculation in two different ways (Owen, 1970; McConnachie, 1991; Haralampides *et al.*, 2003). For low shears, turbulence increases the chance of collision among sediment particles and thus strengthens flocculation; for high shears, strong turbulence may break apart the flocs and attentuate flocculation. Fig. 11.7 shows the relation between the residual turbidity and stirrer speed observed by McConnachie (1991) in a Perspex reactor 0.1 m by 0.1 m in plan view, filled to a depth of 0.1 m. Three types of stirrers were used. It was found that as the stirrer speed increased, the residual turbidity decreased first and then increased. Haralampides *et al.* (2003) performed experiments on the settling of flocs in a rotating circular flume using sediment samples from the St. Clair River near Sarnia, Ontario; they observed that the median size d_{50} of flocs had a maximum value at the bed shear stress of $0.17 \; N \cdot m^{-2}$ and decreased for shear stresses above and below $0.17 \; N \cdot m^{-2}$. These experiments have proven that as the turbulence intensity increases, flocculation will first be strengthened and then attenuated.

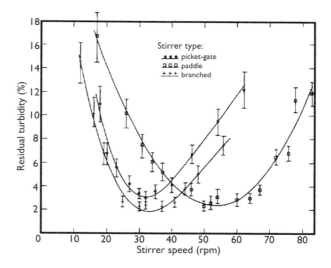

Figure 11.7 Residual turbidity in a Perspex reactor versus stirrer speed (McConnachie, 1991).

Temperature

Temperature affects the thermal motion of ions and, in turn, flocculation. According to Huang's (1981) experiments on the settling of flocs using the Lianyun Harbor mud at temperatures of 6.1, 21.5, and 32°C, the settling velocity of the flocculated sediment is reciprocally proportional to the viscosity of water, i.e.,

$$\frac{\omega_{sfT1}}{\omega_{sfT2}} = \frac{\mu_{T2}}{\mu_{T1}} \tag{11.3}$$

where μ_{T1} and μ_{T2} are the dynamic viscosities of water at temperatures T_1 and T_2, respectively.

In addition, organic matters also have significant influence on flocculation. Organic matters usually have a positive charge, which enlarges the binding forces among sediment particles and thus intensifies flocculation. However, quantification of this effect needs to be investigated further.

11.1.3   Formulas of floc settling velocity

The Migniot formula (11.1) and the Thorn formula (11.2) can be used to determine the settling velocity of flocs. However, each of them only considers the effect of a single factor. Yue (1983) proposed a formula of floc settling velocity that considers the effects of sediment concentration, size, and non-uniformity, as well as salinity:

$$\omega_{sf} = 0.18 \zeta \beta^{-1/6} C_{sa}^{3/4} C^{1/3} d_{50}^{1/4} \tag{11.4}$$

where C_{sa} is the salinity (ppt), C is the sediment concentration by volume, $\beta = 1 + 0.14 d_{50}^{-1/2}$, and $\zeta = (d_{80}/d_{20})^{1/2}$.

Eq. (11.4) introduces a monotonous function between the floc settling velocity and sediment concentration and thus cannot represent the general trend shown in Fig. 11.5.

Lick and Lick (1988) and Gailani *et al.* (1991) proposed a formula to determine the floc diameter:

$$d_f = \left(\frac{\alpha_0}{CG} \right)^{1/2} \tag{11.5}$$

where d_f is the median diameter of flocs (cm), C is the sediment concentration $(g \cdot cm^{-3})$, G is the fluid shear stress $(dynes \cdot cm^{-2})$, and α_0 is an experimentally determined coefficient. For fine-grained, cohesive sediments in freshwater, $\alpha_0 = 10^{-8}$.

According to Burban *et al.* (1990), for the same floc diameter, a larger settling velocity is observed for the floc produced at a higher fluid shear because the effective density and shape of the floc are affected by the conditions in which it is produced. Burban *et al.* proposed a formula for the floc settling velocity ω_{sf} $(cm \cdot s^{-1})$ based on experiments on flocculated, cohesive sediments in freshwater:

$$\omega_{sf} = a d_f^b \tag{11.6}$$

where $a = B_1(CG)^{-0.85}$, $b = -[0.8 + 0.5 \log(CG - B_2)]$, $B_1 = 9.6 \times 10^{-4}$, and $B_2 = 7.5 \times 10^{-6}$ (Gailani *et al.*, 1991).

After the floc diameter is calculated using Eq. (11.5), the floc settling velocity can be determined using Eq. (11.6). Eqs. (11.5) and (11.6) consider the effects of sediment concentration and flow shear on flocculation. It can be seen that the settling velocity increases as either C or G increases. It seems that these two equations are applicable only for low sediment concentrations and low shears.

The formula proposed by Peng (1989) based on the Yangtze Estuary mud considers the influences of sediment size, sediment concentration, salinity, and turbulence intensity on flocculation. Zhang (1999) modified the Peng formula for a study in the Yellow River mouth as

$$\frac{\omega_{sf}}{\omega_{sd}} = 0.274 a f(C) \frac{C_{sa}^{0.03} I^{0.22}}{d_{50}^{0.58}} \tag{11.7}$$

where I is the turbulence intensity, defined as $I = \sqrt{UhS_e/\nu}$ with S_e being the energy slope; a is a coefficient between 1 and 1.5 to be calibrated using observed data; and $f(C)$ is a function of sediment concentration C:

$$f(C) = \begin{cases} C^{0.48} & C \leq 15 \, \text{kg} \cdot \text{m}^{-3} \\ [15/(C-14)]^{0.48} & C > 15 \, \text{kg} \cdot \text{m}^{-3} \end{cases} \tag{11.8}$$

The exponent of salinity in Eq. (11.7) is 0.03, which is very small, as compared with that in the Yue formula (11.4). Eq. (11.7) introduces a monotonous relation between floc settling velocity and turbulence intensity. This does not agree with the observations by Owen (1970), McConnachie (1991), and Haralampides et al. (2003). The exponent of I should be a variable rather than a constant, and there should be a threshold value of I at which the exponent of I turns from positive to negative.

Eqs. (11.4), (11.6), and (11.7) were obtained under certain conditions, and thus, their applicability should be restricted somehow. For more general applications, the following formula was suggested by Wu and Wang (2004c):

$$\frac{\omega_{sf}}{\omega_{sd}} = K_d K_s K_{sa} K_t \tag{11.9}$$

where K_d, K_s, K_{sa}, and K_t are the correction factors accounting for the influences of sediment size, sediment concentration, salinity, and turbulence intensity, respectively. Note that the effect of temperature is considered through ω_{sd}.

Following Migniot (1968), Qian (1980), Huang (1981), and Dixit et al. (1982), one can evaluate the correction factor K_d as

$$K_d = \left(\frac{d_r}{d_{50}}\right)^{n_d}, \quad d_{50} \leq d_r \tag{11.10}$$

Eq. (11.10) is only applied to the range of $d_{50} \leq d_r$. For $d_{50} > d_r$, K_d is set as 1.0. This means that no flocculation occurs for coarse sediments.

Following Thorn (1981) and Mehta (1986), one can have the following formulation for the correction factor K_s:

$$K_s = \begin{cases} k_1 C^n & 0 < C \leq C_p \\ k(1 - k_2 C)^\beta & C > C_p \end{cases} \tag{11.11}$$

where $k = k_1 C_p^n / (1 - k_2 C_p)^\beta$.

Note that Eq. (11.11) adopts the original formulation of Eq. (11.2), but the coefficient k_1 should be adjusted accordingly. The reason is that the effects of sediment size, salinity, and turbulence intensity on flocculation are accounted for by three correction factors in Eq. (11.9), whereas these effects are lumped into k_1 in Eq. (11.2).

Fig. 11.8 shows the relation between K_{sa} and salinity based on Huang's data for the Lianyun Harbor mud. The trend can be approximated by

$$K_{sa} = \begin{cases} (C_{sa}/C_{sap})^{n_{sa}} & C_{sa,\min} < C_{sa} \leq C_{sap} \\ 1 & C_{sa} > C_{sap} \end{cases} \tag{11.12}$$

where n_{sa} is an empirical exponent, C_{sap} is the salinity at which the influence of salinity tends to be saturated, and $C_{sa,\min}$ is a small threshold value of salinity above which Eq. (11.12) is valid.

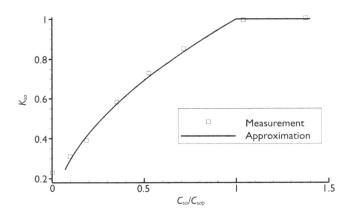

Figure 11.8 Relation of K_{sa} and salinity based on Huang's (1981) data (Wu and Wang, 2004c).

It can be seen from Fig. 11.6 that C_{sap} and n_{sa} are related to sediment concentration. According to the investigations of Owen (1970), Huang (1981), and Yue (1983), C_{sap} has a value of about 30 ppt. The approximation curve in Fig. 11.8 is obtained from Eq. (11.12) with $C_{sap} = 28$ ppt and $n_{sa} = 0.53$. The exponent n_{sa} is 0.75 in the Yue (1983) formula (11.4) and 0.03 in the modified Peng formula (11.7). This difference may be because their experiments are in different ranges of salinity.

The correction factor K_t is determined by

$$K_t = \begin{cases} 1 + k_{t1}(\tau_b/\tau_p)^{n_{t1}} & 0 < \tau_b \leq \tau_p \\ (1 + k_{t1})(\tau_b/\tau_p)^{-n_{t2}} & \tau_b > \tau_p \end{cases} \qquad (11.13)$$

where k_{t1} is an empirical coefficient, n_{t1} and n_{t2} are empirical exponents, τ_b is the bed shear stress in $N \cdot m^{-2}$, and τ_p is the threshold bed shear stress at which K_t attains the maximum value.

The first expression of (11.13) gives K_t a value of 1 in quiescent conditions. According to the experiments of Haralampides *et al.* (2003), τ_p is about 0.17 $N \cdot m^{-2}$ at the maximum floc d_{50}; however, like many other parameters, τ_p may depend on the properties of sediment. In accordance with the modified Peng formula (11.7), n_{t1} has a value of about 0.165. However, n_{t2} has not been investigated well. From McConnachie's (1991) experiments, n_{t2} and n_{t1} might have values close to each other. Further investigation is needed to quantify k_{t1}, n_{t1}, and n_{t2}.

11.1.4 Deposition of cohesive sediments

Cohesive sediments deposit as flocs as long as the flocs are strong enough to settle through the bottom region of high shear (Mehta and Partheniades, 1975). Some flocs may break up as they approach the bed where the shear is stronger and return to the flow.

Krone (1962) and Mehta and Partheniades (1975) investigated the deposition process of fine sediments and proposed formulas to determine the deposition rate. Their formulas can be written as

$$D_b = \alpha \omega_{sf} C \qquad (11.14)$$

where α is the deposition probability coefficient between 0 and 1, which is related to the bed shear stress τ_b and approximated as (Mehta and Partheniades, 1975)

$$\alpha = \begin{cases} 1 & \tau_b < \tau_{bd,\min} \\ 1 - (\tau_b - \tau_{bd,\min})/(\tau_{bd,\max} - \tau_{bd,\min}) & \tau_{bd,\min} \leq \tau_b \leq \tau_{bd,\max} \\ 0 & \tau_b > \tau_{bd,\max} \end{cases} \qquad (11.15)$$

where $\tau_{bd,\min}$ is the critical bed shear stress below which all sediment particles have a full probability to deposit on the bed, and $\tau_{bd,\max}$ is the critical bed shear stress above which all sediment particles remain in suspension yielding a zero deposition rate.

The parameters $\tau_{bd,\min}$ and $\tau_{bd,\max}$ are related to sediment properties. According to Krone (1962), $\tau_{bd,\min} = 0$, whereas Mehta and Partheniades (1975) found that $\tau_{bd,\min}$ might be larger than zero (0.2 $N \cdot m^{-2}$ in their experiments). For the sediment having uniform properties sampled in the San Francisco Bay, Krone found $\tau_{bd,\max} = 0.078$ $N \cdot m^{-2}$ when the initial sediment concentration ranged from 0.3 to 10 $kg \cdot m^{-3}$. For the sediment with a broad size distribution, Mehta and Partheniades found that $\tau_{bd,\max}$ might vary from 0.18 to 1.1 $N \cdot m^{-2}$.

11.1.5 Erosion of cohesive sediments

Erosion modes

Erosion of cohesive sediments is affected by flow conditions, sediment properties, and bed configurations. The first erosion mode is surface or floc erosion, in which sediment is eroded from the bed in particles, due to the breaking of inter-particle electrochemical bonds under the action of the flow that exceeds a critical shear stress. The second mode is mass erosion, in which sediment is eroded in layers, due to bed failures along planes below the bed surface when the applied shear stress exceeds the bed bulk strength. The third mode is sediment entrainment due to bed fluidization followed by the destabilization of the water-sediment interface (Mehta, 1986).

Critical conditions for incipient motion and erosion

The critical velocity or shear stress for erosion (or incipient motion) of cohesive sediments is related to the properties of bed materials, such as plasticity index, void ratio, water content, and yield stress, but general accepted relationships are not available, especially for consolidated muds. Determination of the critical flow conditions must be based on laboratory or in-situ field tests using the natural muds of study.

Dou (1960) and Zhang (1961) studied the incipient motion of newly deposited cohesive sediments and proposed several empirical formulas for the critical depth-average velocity. The Zhang formula is

$$U_c = \left(\frac{h}{d}\right)^{0.14}\left(1.8\frac{\gamma_s - \gamma}{\gamma}gd + 0.000000605\frac{10 + h}{d^{0.72}}\right)^{1/2} \tag{11.16}$$

where U_c is the critical depth-averaged velocity for the incipient motion of sediment $(\text{m}\cdot\text{s}^{-1})$, and d is the sediment size in meters. When the sediment size is larger, the last term on the right-hand side is negligible and Eq. (11.16) reduces to Eq. (3.28), which is for the incipient motion of non-cohesive sediments.

The newly deposited mud is a kind of Bingham fluid. The critical shear stress for erosion is related to the yield stress by Migniot (1968) as follows:

$$U_{*c} = \begin{cases} 0.95\tau_B^{1/4} & \tau_B < 15 \\ 0.50\tau_B^{1/2} & \tau_B \geq 15 \end{cases} \tag{11.17}$$

where U_{*c} is the critical shear velocity for incipient motion $(\text{cm}\cdot\text{s}^{-1})$, and τ_B is the Bingham yield stress $(\text{dynes}\cdot\text{cm}^{-2})$.

According to many experiments, Otsubo and Muraoko (1988) related the critical shear stresses for surface erosion (τ_{ce1}) and mass erosion (τ_{ce2}) to the yield stress τ_B as

$$\tau_{ce1} = 0.27\tau_B^{0.56}, \quad \tau_{ce2} = 0.79\tau_B^{0.94} \tag{11.18}$$

Erosion rate

According to Partheniades (1965), the surface erosion rate is a linear function of the dimensionless excess shear stress:

$$E_b = M \left(\frac{\tau_b - \tau_{ce}}{\tau_{ce}} \right) \tag{11.19}$$

where M is the erodibility coefficient, and τ_{ce} is the critical shear stress for erosion. Both are related to dry density, mineral composition, organic material, salinity, temperature, pH value, Sodium Absorption Ratio (SAR), and so on.

However, Raudkivi and Hutchison (1974) and Mehta *et al.* (1982) established exponential relations between the surface erosion rate and dimensionless excess shear stress. The formula of Mehta *et al.* reads

$$E_b = E_0 e^{\alpha_0(\tau_b - \tau_{ce})/\tau_{ce}} \tag{11.20}$$

where E_0 is the value of E_b at $\tau_b = \tau_{ce}$, and α_0 is a coefficient.

Normally, the exponential relation is valid for partly consolidated beds, whereas the linear relation is valid for fully consolidated beds in which soil properties do not vary with time and over depth (Ariathurai and Mehta, 1983; Mehta, 1986).

Gailani *et al.* (1991) and Ziegler and Nisbet (1995) found a power function between the erosion rate and dimensionless excess shear stress:

$$E_b = \frac{a_0}{t_d^m} \left(\frac{\tau_b - \tau_{ce}}{\tau_{ce}} \right)^n \tag{11.21}$$

where a_0 is a site-specific coefficient, t_d is the time after deposition, m is an exponent of about 2, and n is between 2 and 3. The parameters a_0, m, and n are dependent upon sediment properties and deposition environments.

11.1.6 Consolidation of cohesive bed materials

Consolidation process

Consolidation is a compaction process of deposited materials under the influence of gravity and water pressure with a simultaneous expulsion of pore water and a gain in strength of bed materials. According to Hamm and Migniot (1994), consolidation can be described as a three-stage process. The first stage is the settlement of flocs to form a particle-supported matrix or fluid mud, which happens perhaps within several hours of deposition. The second stage is the elimination of interstitial water, which occurs in one to two days. The third stage is the gelling of clays, which is a very slow process. According to the degree of consolidation, the deposited mud is classified into fluid mud, plastic mud, or solid mud. The fluid mud is the unconsolidated deposit, the plastic mud is the deposit undergoing consolidation, and the solid mud is the older consolidated deposit.

The three distinct stages of consolidation were also observed in Nedeco's (1965) experiments on the consolidation process of natural mud from Bangkok Bar Channel in saline water. The mud surface sank linearly with time t in the initial stage, with $t^{0.5}$ in the second stage, and with $\log(t)$ in the third stage. Several initial suspension heights and wet sediment densities were tested. It was found that the mud with low initial density gradually tended to attain the same density as the sample with high initial density, and the wet sediment density increased toward the bottom of the cylinder, due to larger pressure in the deeper layer.

The degree of consolidation depends on sediment size, mineralogical composition, deposit layer thickness, etc. According to van Rijn's (1989) experiments on the consolidation process in a layer of pure kaolinite material in saline water, a thin layer of mud consolidates faster than a thick layer with the same initial concentration because in the latter case the pore water has a larger travel distance to the mud surface.

Variation of bed density

As experimentally observed by Owen (1975), Dixit (1982), and Hayter (1983), the dry bed density varies along the depth below the bed surface. The general trend can be approximated by (Hayter, 1983)

$$\frac{\rho_d}{\bar{\rho}_d} = a \left(\frac{H - z}{H} \right)^m \tag{11.22}$$

where ρ_d is the dry density of bed, $\bar{\rho}_d$ is the mean dry density of bed, H is the bed thickness, z is the depth below the bed surface, and a and m are coefficients dependent on soil properties and consolidation time.

Fig. 11.9 shows the mean dry bed density varying with consolidation time for the Avonmouth mud (Owen, 1975), commercial grade kaolinite in salt water (salinity = 35 ppt) (Parchure, 1980), and kaolinite in tap water (no salinity) (Dixit,

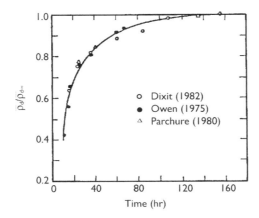

Figure 11.9 Variation of mean bed density with consolidation time (Dixit, 1982).

1982). Noteworthy is the very rapid increase in $\bar{\rho}_d$ in approximately the first 48 hours, after which the increase was much less rapid, and the almost asymptotic approach to the final mean bed density, $\bar{\rho}_{d\infty}$. The temporal variation of the mean dry bed density for all three muds can be approximated by (Hayter, 1983)

$$\frac{\bar{\rho}_d}{\bar{\rho}_{d\infty}} = 1 - a_\rho e^{-pt} \qquad (11.23)$$

where a_ρ and p are empirical coefficients.

Lane and Koelzer (1953) proposed a formula to determine the dry density of bed material in the consolidation process, which seems to coincide with the third stage of consolidation observed in the experiments of Nedeco (1965) and Hamm and Migniot (1994). The Lane-Koelzer formula is

$$\rho_d = \rho_{d0} + \beta \log t \qquad (11.24)$$

where ρ_d is the dry density ($kg \cdot m^{-3}$) at time t, ρ_{d0} is the dry density after 1 year of consolidation, t is the consolidation time (years), and β is a coefficient. ρ_{d0} and β depend on sediment size and reservoir operation conditions, as given in Table 11.2.

Table 11.2 ρ_{d0} and β in Eq. (11.24) for dry density of reservoir deposits (Lane and Koelzer, 1953)

Reservoir operation	Sand		Silt		Clay	
	ρ_{d0}	β	ρ_{d0}	β	ρ_{d0}	β
Sediment always submerged or nearly submerged	1489	0	1041	91	480	256
Normally a moderate reservoir drawdown	1489	0	1185	43	737	171
Normally considerable reservoir drawdown	1489	0	1265	16	961	96
Reservoir normally empty	1489	0	1313	0	1249	0

Influence of consolidation on bed shear strength and erosion rate

Consolidation significantly influences the bed shear strength and, in turn, the erosion rate. Fig. 11.10 shows the relation between dry bed density and shear strength for statically deposited beds of the Avonmouth mud observed by Owen (1975). A regression relation between τ_{ce} (in $N \cdot m^{-2}$) and ρ_d (in $kg \cdot m^{-3}$) can be obtained as

$$\tau_{ce} = \varsigma \rho_d^\beta \qquad (11.25)$$

with $\varsigma = 6.85 \times 10^{-6}$ and $\beta = 2.44$.

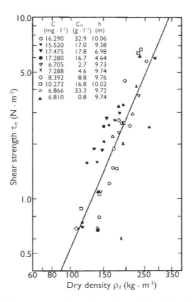

Figure 11.10 Bed shear strength as function of bed density (Owen, 1975).

Thorn and Parsons (1980) likewise found a power relation between τ_{ce} and ρ_d at the bed surface for the Grangemouth, Belawan, and Brisbane muds in saline water. They obtained $\varsigma = 5.42 \times 10^{-6}$ and $\beta = 2.28$.

Nicholson and O'Connor (1986) approximated the relation between τ_{ce} and ρ_d obtained by Thorn (1981) for the mud from Scheldt, Belgium, as follows:

$$\tau_{ce} = \tau_{ce0} + k_\tau(\rho_d - \rho_{d0})^{n_\tau} \tag{11.26}$$

where ρ_d is the dry density of bed material, τ_{ce0} and ρ_{d0} are the critical shear stress and bed density in the initial period of bed formation, k_τ is a coefficient of 0.00037, and n_τ is an exponent of about 1.5.

11.2 MULTIPLE-FLOC-SIZE MODEL OF COHESIVE SEDIMENT TRANSPORT

As described in Section 11.1.1, the size distribution of flocs varies with time, due to the consecutive aggregation and disaggregation processes. Even for the flocs generated from uniform dispersed sediment particles, their size distribution is quite non-uniform and dynamic. For the convenience of analysis, the size distribution of flocs is often represented by a discrete number of size fractions, as shown in Fig. 11.3. The number of flocs in size fraction k per unit volume is denoted as n_k. The governing equation for

n_k considering the aggregation and disaggregation of flocs is (see Tsai *et al.*, 1987)

$$
\frac{\partial n_k}{\partial t} + \frac{\partial(u_i n_k)}{\partial x_i} - \frac{\partial(\omega_{sf,k}\delta_{i3}n_k)}{\partial x_i} = \frac{\partial}{\partial x_i}\left(\varepsilon_s \frac{\partial n_k}{\partial x_i}\right) + \frac{1}{2}\sum_{j+l=k}\alpha_{lj}\beta_{lj}n_l n_j
$$

$$
- \sum_{j=1}^{\infty}\alpha_{jk}\beta_{jk}n_j n_k + \sum_{j>k}\gamma_{kj}n_j
$$

$$
- \sum_{j<k}\gamma_{jk}n_k \tag{11.27}
$$

where β_{ij} is the frequency function for collision between flocs of size fractions i and j, α_{ij} is the probability of cohesion after collision, and γ_{ij} is the frequency function of disaggregation of flocs from size fraction j to i. The second term on the right-hand side of Eq. (11.27) accounts for the formation of size fraction k by collision of smaller flocs. The third term quantifies the loss of size fraction k due to collision with other flocs. The fourth term represents the generation of size fraction k due to disaggregation of larger flocs. The fifth term denotes the loss of size fraction k due to disaggregation to smaller flocs.

Particle collision may be caused by Brownian motion, fluid shear, and differential settling. The original collision theory was due to Smoluchowski (1917), and additional work was done by Camp and Stein (1943). The frequency function β_{ij} for different collisions is presented below (see Tsai *et al.*, 1987).

For Brownian motion,

$$
\beta_{ij} = \frac{2}{3}\frac{kT}{\mu}\frac{(d_i + d_j)^2}{d_i d_j} \tag{11.28}
$$

where k is the Boltzmann constant (1.38×10^{-23} N·m·°K^{-1}), T is the absolute temperature (°K), μ is the dynamic viscosity of the fluid, and d_i and d_j are the diameters of the two colliding particles.

For fluid shear,

$$
\beta_{ij} = \frac{G}{6}(d_i + d_j)^3 \tag{11.29}
$$

where G is the mean velocity gradient in the fluid. For a turbulent fluid, G can be approximated by $(\varepsilon/v)^{1/2}$, in which ε is the energy dissipation rate and v is the kinematic viscosity of the fluid (Saffman and Turner, 1956).

For differential settling,

$$
\beta_{ij} = \frac{\pi g}{72\mu}(\rho_s - \rho_f)(d_i + d_j)^2(d_i^2 - d_j^2) \tag{11.30}
$$

Fig. 11.11 shows the frequency functions for particles of different sizes colliding with a 1-μm particle and a 25-μm particle due to Brownian motion, fluid shear, and

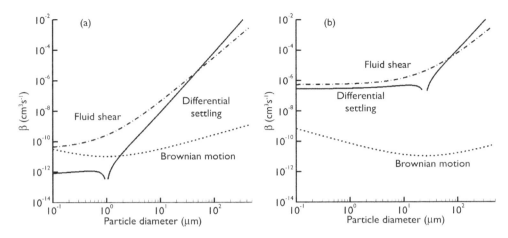

Figure 11.11 Collision function β versus particle size: (a) collision with a 1 μm particle and (b) collision with a 25 μm particle (Tsai *et al.*, 1987).

differential settling. The conditions were $T = 20°C$ (293°K), $G = 200$ s^{-1} (2 dynes·cm^{-2}), $\rho_s = 2650$ kg·m^{-3}, and $\rho_f = 1000$ kg·m^{-3}. As can be seen, the Brownian motion is only important for collision of small particles, the differential settling is the main cause for collision of large particles, and the fluid shear is important for collision of medium particles.

The multiple-floc-size model was tested by Lick and Lick (1988) in simple cases, but it encounters difficulties in quantification of α_{ij}, γ_{ij}, and the size distribution of flocs entrained from a cohesive bed.

11.3 SINGLE-FLOC-SIZE MODEL OF COHESIVE SEDIMENT TRANSPORT

11.3.1 Governing equations

As described in the previous section, it is difficult to simulate the aggregation and diaggregation processes using the multiple-floc-size model in the present time. More often used is the single-floc-size model, which does not resolve the details of aggregation and disaggregation processes but considers flocculation in a lumped form through a representative floc settling velocity ω_{sf} that varies with flow and sediment conditions (Nicholson and O'Connor, 1986; Li *et al.*, 1994; Chen *et al.*, 1999; Le Normant, 2000; Wu and Wang, 2004c).

In the 3-D model, the sediment transport equation is

$$\frac{\partial c}{\partial t} + \frac{\partial (u_x c)}{\partial x} + \frac{\partial (u_y c)}{\partial y} + \frac{\partial (u_z c)}{\partial z} - \frac{\partial (\omega_{sf} c)}{\partial z} = \frac{\partial}{\partial x}\left(\varepsilon_s \frac{\partial c}{\partial x}\right) + \frac{\partial}{\partial y}\left(\varepsilon_s \frac{\partial c}{\partial y}\right)$$
$$+ \frac{\partial}{\partial z}\left(\varepsilon_s \frac{\partial c}{\partial z}\right) \qquad (11.31)$$

with the boundary condition at the bed

$$\left(\omega_{sf}c + \varepsilon_s \frac{\partial c}{\partial z}\right)_{z=z_b} = D_b - E_b \tag{11.32}$$

where c is the local sediment concentration; D_b is the deposition rate determined using Eq. (11.14); E_b is the erosion rate determined using one of Eqs. (11.19)–(11.21); and ω_{sf} is determined using one of Eqs. (11.1), (11.2), (11.4), (11.6), (11.7), and (11.9).

In the depth-averaged 2-D model, the sediment transport equation is

$$\frac{\partial (hC)}{\partial t} + \frac{\partial (hU_xC)}{\partial x} + \frac{\partial (hU_yC)}{\partial y} = \frac{\partial}{\partial x}\left(E_{s,x}h\frac{\partial C}{\partial x}\right) + \frac{\partial}{\partial y}\left(E_{s,y}h\frac{\partial C}{\partial y}\right)$$
$$+ E_b - D_b \tag{11.33}$$

where C is the depth-averaged sediment concentration.

One can see similarity between Eqs. (7.43) and (11.31) and between Eqs. (6.53) and (11.33). The width-averaged 2-D and 1-D cohesive sediment transport equations are similar to Eqs. (6.123) and (5.27), which are not repeated here. The difference between cohesive and non-cohesive sediment transport models lies in the determination of ω_{sf}, D_b, and E_b, as well as bed consolidation.

The rate of change in bed elevation, $\partial z_b/\partial t$, is determined by

$$(1 - p'_m)\frac{\partial z_b}{\partial t} = D_b - E_b \tag{11.34}$$

Gibson *et al.* (1967) proposed a theory to describe the bed consolidation process. The model based on this theory determines the evolution of the void ratio of a soil layer using the following 1-D equation in the vertical direction at each horizontal computational point:

$$\frac{\partial e}{\partial t} + \left(\frac{\rho_s}{\rho_f} - 1\right)\frac{d}{de}\left(\frac{k}{1+e}\right)\frac{\partial e}{\partial z'} + \frac{\partial}{\partial z'}\left(\frac{1}{\rho_f g}\frac{k}{1+e}\frac{d\sigma'}{de}\frac{\partial e}{\partial z'}\right) = 0 \tag{11.35}$$

where e is the void ratio of bed soil, z' is the reduced material coordinate deduced from the vertical coordinate z by $\delta z = \delta z'(1 + e)$, k is the permeability, and σ' is the effective stress.

Constitutive relationships for the permeability k and the effective stress σ' as functions of the void ratio are needed to close Eq. (11.35). These constitutive relationships are rarely available; thus, it is difficult to apply this approach in the solution of real-life problems. Therefore, a simpler approach is often used, in which the evolution of bed density is determined by empirical functions, such as Eqs. (11.23) and (11.24).

The overall mass of sediment should be conserved in the consolidation process, i.e.,

$$\frac{\partial}{\partial t}\int_{z_0}^{z_b}\rho_d dz = 0 \tag{11.36}$$

where z_0 is the elevation of the uncompactable layer of bed material.

By defining the bed thickness as $H = z_b - z_o$ and the average dry density of the bed as $\bar{\rho}_d = \int_{z_0}^{z_b} \rho_d dz/H$, Eq. (11.36) can be written as

$$\frac{\partial}{\partial t}(H\bar{\rho}_d) = 0 \tag{11.37}$$

The bed change due to consolidation is then determined by

$$\frac{\partial H}{\partial t} = -\frac{H}{\bar{\rho}_d}\frac{\partial \bar{\rho}_d}{\partial t} \tag{11.38}$$

11.3.2 Numerical solutions

Numerical discretization

Eqs. (11.31) and (11.33) can be discretized using the same numerical methods as those for Eq. (7.43) and (6.53). Usually, the time-derivative terms are discretized using the forward or backward difference scheme to establish an explicit or implicit time-marching procedure. The convection terms should be discretized using an upwind scheme, while the diffusion terms are discretized using the central difference scheme or a similar scheme. The settling term in Eq. (11.31) and the exchange term in Eq. (11.33) are often treated as source terms. The resulting algebraic equations can be solved using the Gauss-Seidel, ADI, or SIP method introduced in Section 4.5.

In the 3-D model, the near-bed boundary condition (11.32) can be discretized using a scheme similar to Eq. (7.54) or (7.55).

The bed change equation (11.34) is discretized in time as

$$\Delta z_b = \frac{\Delta t}{1 - p'_m}(D_b - E_b) \tag{11.39}$$

Eq. (11.38) can be discretized in time to calculate the bed change due to consolidation. However, to consider the heterogeneous properties of bed materials deposited in different times, the following multiple-layer model is often used. The bed soil from the bed surface to the uncompactable layer is divided into a suitable number of layers in the vertical direction, as shown in Fig. 11.12. Each layer is characterized by its dry density and residence time. The top layer holds the newly deposited mud, and a bed formation time (such as 2 hours) may apply to this layer. The evolution of dry densities at other layers is determined by Eq. (11.23) or (11.24). The mass of sediment at each layer is conserved during the consolidation process, i.e.,

$$\frac{\partial}{\partial t}(\delta_j \rho_{dj}) = 0 \tag{11.40}$$

where δ_j and ρ_{dj} are the thickness and dry density of the jth layer of bed material.

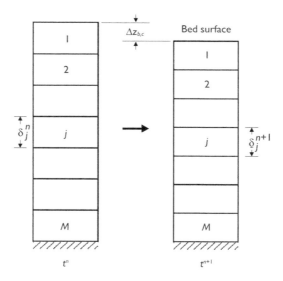

Figure 11.12 Multiple-layer model of bed consolidation.

Discretizing Eq. (11.40) yields

$$\delta_j^{n+1} = \delta_j^n \frac{\rho_{dj}^n}{\rho_{dj}^{n+1}} \tag{11.41}$$

Then, the overall bed change due to consolidation is calculated as

$$\Delta z_{b,c} = \sum_{j=1}^{J} (\delta_j^{n+1} - \delta_j^n) = \sum_{j=1}^{J} \delta_j^n \left(\frac{\rho_{dj}^n}{\rho_{dj}^{n+1}} - 1 \right) \tag{11.42}$$

where J is the total number of the divided bed material layers.

Calculation procedure

As described in Section 5.1, the flow and sediment calculations may be coupled or decoupled, depending on the intensity of sediment transport. If the sediment concentration is low and the bed varies slowly, the effect of sediment transport on the flow is negligible and thus a decoupled model is applicable; otherwise, a coupled model should be used. However, in the sediment module (or loop), sediment transport, bed change, and bed consolidation are correlated, and thus it is usually required to solve them simultaneously in an iteration form. The following calculation procedure is used by Wu and Wang (2004c):

(1) Initialize flow, salinity, and sediment fields;
(2) Calculate the flow field using a flow solver;

(3) Solve the salinity transport equation, if needed;
(4) Guess the bed density, etc., and calculate the sediment settling velocity, deposition rate, and erosion rate;
(5) Solve the sediment transport equation;
(6) Calculate the bed change due to sediment transport;
(7) Calculate the bed consolidation and obtain a new bed density;
(8) Return to step (4) and repeat the iteration until a convergent solution for sediment transport is obtained;
(9) Adjust the bed topography, and return to step (2) for the next time step.

11.3.3 Examples

Many applications of cohesive sediment transport models can be found in Nicholson and O'Connor (1986), Li *et al.* (1994), Chen *et al.* (1999), Le Normant (2000), Wu and Wang (2004c), and so on. As one example, the simulation of cohesive sediment transport in the Gironde Estuary, France, conducted by Wu and Wang (2004c) using the depth-averaged 2-D model introduced above is cited here. This estuary is partially mixed and macrotidal, with a tidal period of 12 hours and 25 minutes and a tidal amplitude of 1.5 to 5 m at the mouth. The simulation domain was about 80 km

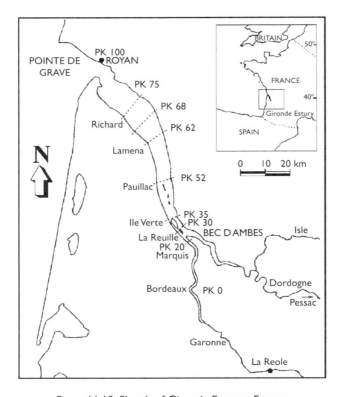

Figure 11.13 Sketch of Gironde Estuary, France.

long, starting from the mouth to the Garonne and Dordogne Rivers (see Fig. 11.13). The computational mesh was uniform, with a size of $500 \, \text{m} \times 250 \, \text{m}$ for each cell. The tidal flow and cohesive sediment transport during May 19–22, 1974 were simulated. To consider the effect of salinity on sediment flocculation, salinity transport was also simulated using the depth-averaged 2-D model introduced in Section 12.1. The computational time step was 30 minutes.

The sediment parameters were selected carefully, according to suggestions of Li *et al.* (1994). For the sand bottom, extending from the estuary mouth to about 15 km upstream from Royan, τ_{ce} was set to be $2.0 \, \text{N} \cdot \text{m}^{-2}$, and for the reach covered by cohesive sediments, τ_{ce} was given values between 1.3 and $1.5 \, \text{N} \cdot \text{m}^{-2}$. M was given $0.002 \, \text{kg} \cdot \text{m}^{-2} \text{s}^{-1}$. $\tau_{bd,\text{min}}$ was set as zero, and $\tau_{bd,\text{max}}$ was prescribed as $0.3 \, \text{N} \cdot \text{m}^{-2}$. Because only a few days of sediment transport were simulated, the consolidation process was not considered. The settling velocity was calculated using Eq. (11.9). For the correction factor of sediment diameter in Eq. (11.10), $n_d = 1.8$ and $d_r = 0.022$ mm, which were calibrated using Migniot's measurement data. The parameters in the correction factor of sediment concentration in Eq. (11.11) were $k_1 = 2.5$, $k_2 = 0.008$, $C_p = 3.0 \, \text{kg} \cdot \text{m}^{-3}$, $n = 1.3$, and $r = 4.65$, which were used by Li *et al.* (1994) except that the value of k_1 was adjusted accordingly. For the correction factor of salinity in Eq. (11.12), $n_{sa} = 0.5$ and C_{sap} was given $30 \, \text{kg} \cdot \text{m}^{-3}$. For the correction factor of turbulence shear in Eq. (11.13), τ_p was given $0.17 \, \text{N} \cdot \text{m}^{-2}$, n_{t1} and n_{t2} were 0.165, and k_{t1} was 1.5.

Fig. 11.14 shows the simulated flow fields in flood and ebb tides, and Fig. 11.15 shows the simulated and measured water levels and velocities. The amplitudes and phases of water level and velocity were predicted well by the numerical model. No obvious phase difference existed between measurement and simulation. Fig. 11.16 shows the sediment transport rate per unit cross-sectional area, i.e., UC, the product of flow velocity and sediment concentration. The simulated sediment transport rates match the measured data generally well.

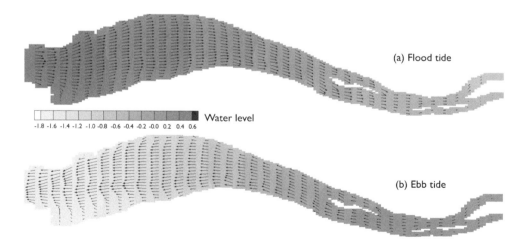

Water level
-1.8 -1.6 -1.4 -1.2 -1.0 -0.8 -0.6 -0.4 -0.2 -0.0 0.2 0.4 0.6

(a) Flood tide

(b) Ebb tide

Figure 11.14 Calculated flow patterns in Gironde Estuary (Wu and Wang, 2004c).

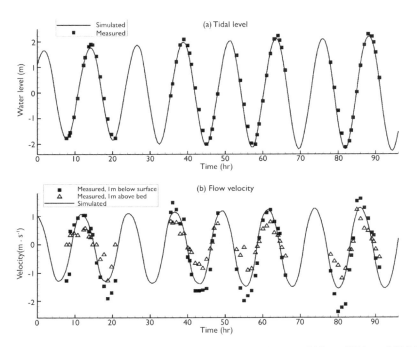

Figure 11.15 Measured and calculated water levels and velocities (Wu and Wang, 2004c).

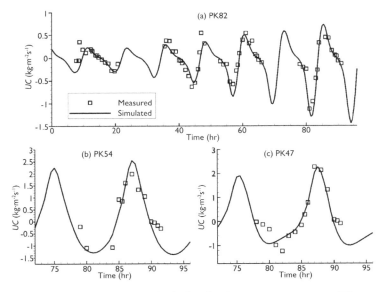

Figure 11.16 Measured and calculated sediment transport rates UC.

11.4 SIMULATION OF TRANSPORT OF COHESIVE AND NON-COHESIVE SEDIMENT MIXTURES

When the fraction of clay and fine silt is larger than about 10%, a sediment mixture composed of cohesive and non-cohesive particles may exhibit cohesive properties. The flocculation and consolidation of the cohesive particles may affect the erosion, deposition, and transport of the non-cohesive particles significantly. In particular, when the fraction of cohesive particles is appreciable and the sediment concentration is high, the floc structures formed by the cohesive particles will involve the non-cohesive particles.

Interactions between cohesive and non-cohesive particles should be taken into account in the simulation of the mixed cohesive and non-cohesive sediment transport, but this is difficult because the interaction mechanisms are little understood. In the case of low sediment concentration, such interactions may be ignored and, thus, the following modeling framework is often used (Ziegler and Nisbet, 1995; Ziegler *et al.*, 2000; Wu and Vieira, 2002).

The entire sediment mixture is divided into a suitable number of size classes. The depth-averaged 2-D transport equation of the kth size class of sediment is

$$\frac{\partial(hC_k)}{\partial t} + \frac{\partial(hU_xC_k)}{\partial x} + \frac{\partial(hU_yC_k)}{\partial y} = \frac{\partial}{\partial x}\left(E_{s,x}h\frac{\partial C_k}{\partial x}\right) + \frac{\partial}{\partial y}\left(E_{s,y}h\frac{\partial C_k}{\partial y}\right)$$
$$+ E_{bk} - D_{bk} \qquad (11.43)$$

where C_k is the depth-averaged concentration of the kth size class of sediment. For brevity, the 1-D, width-averaged 2-D, and 3-D transport equations are omitted here.

The deposition rate D_{bk} is determined by

$$D_{bk} = \alpha_k \omega_{sf,k} C_k \qquad (11.44)$$

where the coefficient α_k is determined using Eq. (11.15) for the cohesive size classes and the methods introduced in Section 2.5 for the non-cohesive size classes; and $\omega_{sf,k}$ is the settling velocity of size class k, determined using one of Eqs. (11.1), (11.2), (11.4), (11.6), (11.7), and (11.9) for the cohesive size classes and one of the formulas introduced in Section 3.1 for the non-cohesive size classes.

The erosion rate E_{bk} is determined by

$$E_{bk} = p_{bk}E_b^{(k)} \qquad (11.45)$$

where p_{bk} is the fraction of size class k in the surface layer of bed material, and $E_b^{(k)}$ is the potential erosion rate of size class k.

If the cohesive portion is dominant in the bed (surface layer), all cohesive and non-cohesive particles are usually eroded simultaneously in flocs (even blocks) and have the same potential erosion rate; thus, $E_b^{(k)}$ is actually the total erosion rate and can be determined using an erosion model of cohesive sediments, such as Eqs. (11.19)–(11.21). If the cohesive portion is not dominant in the bed, the non-cohesive particles are eroded in dispersed form and the cohesive particles may be eroded in flocs. Because there may be hiding, exposure, and armoring among different size classes, $E_b^{(k)}$ may vary with size classes; it should be determined using the entrainment models introduced in Section 2.5 for the non-cohesive size classes and Eqs. (11.19)–(11.21) for the cohesive size classes.

The fractional bed change is determined by

$$(1 - p'_m)\left(\frac{\partial z_b}{\partial t}\right)_k = D_{bk} - E_{bk} \qquad (11.46)$$

The temporal variation of bed-material gradation can be simulated using the multiple layer model introduced in Section 2.7.2.

The consolidation of non-uniform bed material composed of cohesive and non-cohesive particles can be determined using Eq. (11.37) or (11.40). However, the temporal evolution of the dry densities of cohesive size classes should be determined using Eq. (11.23), (11.24), or a more complex model, while the overall bed dry density is determined using the Colby formula (2.18) or a similar formula.

The modeling framework introduced above has been applied in many case studies, including the fine-grained sediment transport in the Watts Bar Reservoir by Ziegler and Nisbet (1995), the sediment transport dynamics in Thompson Island Pool, Upper Hudson River by Ziegler *et al.* (2000), and the sedimentation process in the Danjiangkou Reservoir by Wu and Vieira (2002) shown in Section 5.6. The Watts Bar Reservoir case is introduced briefly below as an example.

In the Watts Bar Reservoir (Fig. 11.17), suspended sediment particles ranged from 0.001 to 0.25 mm in size, and clay and silt are dominant (about 90%). Two size

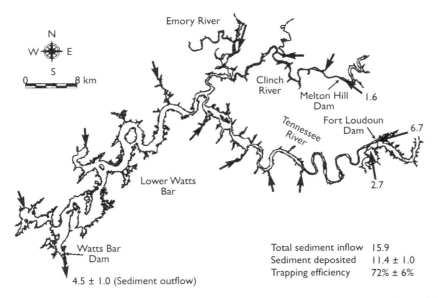

Figure 11.17 Sketch of Watts Bar Reservoir (numbers are sediment loads during 1961–1991 in millions of tons) (Ziegler and Nisbet, 1995).

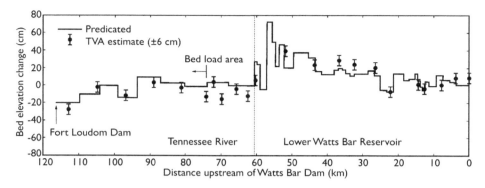

Figure 11.18 Bed changes in Watts Bar Reservoir from 1961 to 1991 (Ziegler and Nisbet, 1995).

classes separated by size 0.062 mm were used in the simulation. The settling velocity was determined using the Burban *et al.* formula (11.6) for the fine class and estimated at 5 mm·s^{-1} for the coarse class. The deposition rate was computed using the Krone formula and a non-cohesive model for the two size classes. The total erosion rate was estimated using the Gailani *et al.* formula (11.21), and the fractional one was then obtained by Eq. (11.45). The bed consolidation was simulated by a 3-D model. The sediment loads in major tributaries in a 30-year period from 1961 to 1991 are shown in Fig. 11.17, and the simulated bed changes in this period are presented in Fig. 11.18. The simulation and measurement are in generally good agreement.

Chapter 12

Contaminant transport modeling

This chapter presents models for the fate and transport of water-quality constituents, such as temperature, salinity, dissolved oxygen (DO), nitrogen, phosphorus, carbons, and toxicants, in aquatic systems. In particular, heat exchange across the water surface, interactions among DO, nitrogen, phosphorus, carbons and phytoplankton in the eutrophication system, and sorption of contaminants on sediment particles are discussed.

12.1 HEAT AND SALINITY TRANSPORT MODEL

Water temperature and salinity are among the key environmental conditions that significantly affect the physical, chemical, and biological processes in aquatic systems. Thus, the simulation of heat and salinity transport is essential to water quality modeling.

12.1.1 Governing equations

The presence of heat and salinity may induce stratification and density currents. To adequately model these effects, the most preferable approach is the 3-D hydrodynamic model, coupled with heat and salinity transport calculations. As described in Section 2.4.4, the 3-D hydrodynamic equations are (2.112)–(2.115) with the water density ρ varying with temperature and salinity. Under the hydrostatic pressure assumption, the momentum equations (2.113)–(2.115) are simplified to Eqs. (2.117) and (2.118). A further simplification may be made by assuming the temporal and spatial variations of water density in the continuity equation (2.112) to be negligible, yielding (Sheng, 1983; Blumberg and Mellor, 1987)

$$\frac{\partial u_x}{\partial x} + \frac{\partial u_y}{\partial y} + \frac{\partial u_z}{\partial z} = 0 \qquad (12.1)$$

$$\frac{\partial u_x}{\partial t} + \frac{\partial (u_x^2)}{\partial x} + \frac{\partial (u_y u_x)}{\partial y} + \frac{\partial (u_z u_x)}{\partial z} = -\frac{1}{\rho}\left(\rho_0 g \frac{\partial z_s}{\partial x} + g \int_z^{z_s} \frac{\partial \rho}{\partial x} dz\right) + \frac{1}{\rho}\frac{\partial \tau_{xx}}{\partial x}$$
$$+ \frac{1}{\rho}\frac{\partial \tau_{xy}}{\partial y} + \frac{1}{\rho}\frac{\partial \tau_{xz}}{\partial z} + f_c u_y \qquad (12.2)$$

$$\frac{\partial u_y}{\partial t} + \frac{\partial (u_x u_y)}{\partial x} + \frac{\partial (u_y^2)}{\partial y} + \frac{\partial (u_z u_y)}{\partial z} = -\frac{1}{\rho}\left(\rho_0 g \frac{\partial z_s}{\partial y} + g \int_z^{z_s} \frac{\partial \rho}{\partial y} dz\right) + \frac{1}{\rho}\frac{\partial \tau_{yx}}{\partial x}$$

$$+ \frac{1}{\rho}\frac{\partial \tau_{yy}}{\partial y} + \frac{1}{\rho}\frac{\partial \tau_{yz}}{\partial z} - f_c u_x \qquad (12.3)$$

where ρ_0 is the water density at the water surface, and f_c is the Coriolis coefficient.

As a simplified approximation, the width-averaged 2-D model can also be used to study the stratified flows due to heat and salinity transport in rivers and reservoirs. The width-averaged 2-D hydrodynamic equations are (2.122)–(2.124) in general. However, for the vertically well-mixed water bodies, the effect of stratification is negligible and, thus, the 1-D and depth-averaged 2-D models are applicable. In such cases, the depth-averaged 2-D shallow water equations are (2.119)–(2.121) and the 1-D equations are (2.126) and (2.127) without the bed change terms. In analogy to Eq. (12.1), the 1-D and 2-D continuity equations (2.119), (2.122), and (2.126) can be simplified by ignoring the temporal and spatial variations. The resulting 1-D and 2-D hydrodynamic equations can also be derived by integrating Eqs. (12.1)–(12.3) over the cross-section, depth, and width of flow, respectively. The details are left to the interested reader.

The 3-D heat transport equation is

$$\frac{\partial T}{\partial t} + \frac{\partial (u_x T)}{\partial x} + \frac{\partial (u_y T)}{\partial y} + \frac{\partial (u_z T)}{\partial z} = \frac{\partial}{\partial x}\left(\varepsilon_{T,x}\frac{\partial T}{\partial x}\right) + \frac{\partial}{\partial y}\left(\varepsilon_{T,y}\frac{\partial T}{\partial y}\right)$$

$$+ \frac{\partial}{\partial z}\left(\varepsilon_{T,z}\frac{\partial T}{\partial z}\right) + \frac{q_T}{\rho c_p} \qquad (12.4)$$

where T is the local temperature (usually in degree Celsius, °C), $\varepsilon_{T,i}(i = x, y, z)$ are the turbulent diffusivities of heat, c_p is the specific heat, and q_T is the heat source rate per unit volume.

The 3-D salinity transport equation is

$$\frac{\partial C_{sa}}{\partial t} + \frac{\partial (u_x C_{sa})}{\partial x} + \frac{\partial (u_y C_{sa})}{\partial y} + \frac{\partial (u_z C_{sa})}{\partial z} = \frac{\partial}{\partial x}\left(\varepsilon_{sa,x}\frac{\partial C_{sa}}{\partial x}\right) + \frac{\partial}{\partial y}\left(\varepsilon_{sa,y}\frac{\partial C_{sa}}{\partial y}\right)$$

$$+ \frac{\partial}{\partial z}\left(\varepsilon_{sa,z}\frac{\partial C_{sa}}{\partial z}\right) \qquad (12.5)$$

where C_{sa} is the local salinity (usually in parts per thousand, ppt), and $\varepsilon_{sa,i}(i = x, y, z)$ are the turbulent diffusivities of salinity.

The width-integrated 2-D heat and salinity transport equations are

$$\frac{\partial (bT)}{\partial t} + \frac{\partial (bU_x T)}{\partial x} + \frac{\partial (bU_z T)}{\partial z} = \frac{\partial}{\partial x}\left(E'_{T,x} b \frac{\partial T}{\partial x}\right) + \frac{\partial}{\partial z}\left(E'_{T,z} b \frac{\partial T}{\partial z}\right)$$

$$+ \frac{1}{\rho c_p}\left(bq_T - \sum_{i=1}^{2} m_i q_{ni}\right) \qquad (12.6)$$

$$\frac{\partial(bC_{sa})}{\partial t} + \frac{\partial(bU_xC_{sa})}{\partial x} + \frac{\partial(bU_zC_{sa})}{\partial z} = \frac{\partial}{\partial x}\left(E'_{sa,x}b\frac{\partial C_{sa}}{\partial x}\right) + \frac{\partial}{\partial z}\left(E'_{sa,z}b\frac{\partial C_{sa}}{\partial z}\right)$$

$$(12.7)$$

where T and C_{sa} are the width-averaged temperature and salinity, respectively; $E'_{T,i}$ and $E'_{sa,i}$ are the effective diffusivities (mixing coefficients) of heat and salinity in the longitudinal section, respectively; and $q_{ni}(i = 1, 2)$ are the heat fluxes per unit of bank surface area due to conduction, seepage flow, etc., across the two banks.

The depth-integrated 2-D heat and salinity transport equations are

$$\frac{\partial(bT)}{\partial t} + \frac{\partial(bU_xT)}{\partial x} + \frac{\partial(bU_yT)}{\partial y} = \frac{\partial}{\partial x}\left(E_{T,x}b\frac{\partial T}{\partial x}\right) + \frac{\partial}{\partial y}\left(E_{T,y}b\frac{\partial T}{\partial y}\right) + \frac{J_T}{\rho c_p}$$

$$(12.8)$$

$$\frac{\partial(bC_{sa})}{\partial t} + \frac{\partial(bU_xC_{sa})}{\partial x} + \frac{\partial(bU_yC_{sa})}{\partial y} = \frac{\partial}{\partial x}\left(E_{sa,x}b\frac{\partial C_{sa}}{\partial x}\right) + \frac{\partial}{\partial y}\left(E_{sa,y}b\frac{\partial C_{sa}}{\partial y}\right)$$

$$(12.9)$$

where T and C_{sa} are the depth-averaged temperature and salinity, respectively; $E_{T,i}$ and $E_{sa,i}$ are the horizontal effective diffusivities of heat and salinity, respectively; and J_T is the net flux across the water and bed surfaces.

The 1-D heat and salinity transport equations are

$$\frac{\partial(AT)}{\partial t} + \frac{\partial(QT)}{\partial x} = \frac{\partial}{\partial x}\left(E_{T,L}A\frac{\partial T}{\partial x}\right) + \frac{1}{\rho c_p}\left(BJ_T - \sum_{i=1}^{2}\bar{m}_ib\bar{q}_{ni}\right) \quad (12.10)$$

$$\frac{\partial(AC_{sa})}{\partial t} + \frac{\partial(QC_{sa})}{\partial x} = \frac{\partial}{\partial x}\left(E_{sa,L}A\frac{\partial C_{sa}}{\partial x}\right) \quad (12.11)$$

where T and C_{sa} are the temperature and salinity averaged in the cross-section, respectively; $E_{T,L}$ and $E_{sa,L}$ are the longitudinal effective diffusivities of heat and salinity, respectively; $\bar{q}_{ni}(i = 1, 2)$ are the average heat fluxes per unit bank surface area due to conduction, seepage flow, etc., across the two banks; and $\bar{m}_i(i = 1, 2)$ are the ratios of the wetted bank slope lengths to the flow depth.

Water density varies with temperature and salinity. This can be described by the equation of state (Crowley, 1968):

$$\rho = 1000 + (28.14 - 0.0735T - 0.00469T^2) + (0.802 - 0.002T)(C_{sa} - 35)$$

$$(12.12)$$

where ρ is in $kg \cdot m^{-3}$, C_{sa} is in ppt, and T is in $°C$.

The boundary conditions for heat and salinity transport are similar to those for sediment transport. In an inflow boundary, the values of water temperature and salinity should be specified. In a wall boundary, the gradient of salinity in the direction normal to the wall is specified as zero. If the heat exchange across the wall is not considered, the gradient of water temperature in the direction normal to the wall is zero; however, a general heat exchange flux model can be applied at the wall boundary. In an outflow boundary, the gradients of temperature and salinity in the flow direction can be set as zero.

12.1.2   Effects of buoyancy on vertical turbulent transport

The vertical turbulent transport of mass, momentum, and heat is strongly influenced by buoyancy effects; in particular, the eddy viscosity and diffusivity are reduced by stable stratification (Rodi, 1993). To account for the buoyancy effects, damping functions are usually applied to the eddy viscosity and diffusivity:

$$\nu_t = \nu_{t0}(1 + \alpha_1 \mathrm{Ri})^{\alpha_2} \tag{12.13}$$

$$\varepsilon_t = \varepsilon_{t0}(1 + \beta_1 \mathrm{Ri})^{\beta_2} \tag{12.14}$$

where Ri is the gradient Richardson number:

$$\mathrm{Ri} = -\frac{g}{\rho}\frac{\partial\rho/\partial z}{(\partial U/\partial z)^2} \tag{12.15}$$

which is the ratio of gravity to inertial forces and characterizes the importance of buoyancy effects. ν_{t0} and ε_{t0} are the eddy viscosity and diffusivity, respectively, for the neutrally stratified flow (Ri = 0). According to Munk and Anderson (1948), the values of coefficients $\alpha_1 = 10$, $\alpha_2 = -0.5$, $\beta_1 = 3.33$, and $\beta_2 = -1.5$.

From Eqs. (12.13) and (12.14), the effect of buoyancy on the turbulent Prandtl/Schmidt number $\sigma_t = \nu_t/\varepsilon_t$ can be determined.

The mixing length is also altered by buoyancy effects. For the stably stratified flow (Ri > 0), the following Monin-Obukhov relation is mostly used:

$$l_m = l_{m0}(1 - \gamma_1 \mathrm{Ri}) \tag{12.16}$$

where l_{m0} is the mixing length for the neutrally stratified flow, and γ_1 is a coefficient ranging from 5 to 10 and having a mean value of about 7 (Busch, 1972). For the unstably stratified flow (Ri < 0), the following relation is usually employed:

$$l_m = l_{m0}(1 - \gamma_2 \mathrm{Ri})^{-1/4} \tag{12.17}$$

with $\gamma_2 \approx 14$ (Busch, 1972).

The 3-D k- and ε-equations considering the buoyancy effects are written as (Rodi, 1993)

$$\frac{\partial k}{\partial t} + u_i \frac{\partial k}{\partial x_i} = \frac{\partial}{\partial x_i}\left(\frac{v_t}{\sigma_k}\frac{\partial k}{\partial x_i}\right) + P_k + G_k - \varepsilon \tag{12.18}$$

$$\frac{\partial \varepsilon}{\partial t} + u_i \frac{\partial \varepsilon}{\partial x_i} = \frac{\partial}{\partial x_i}\left(\frac{v_t}{\sigma_\varepsilon}\frac{\partial \varepsilon}{\partial x_i}\right) + c_{\varepsilon 1}\frac{\varepsilon}{k}(P_k + c_{\varepsilon 3}G_k) - c_{\varepsilon 2}\frac{\varepsilon^2}{k} \tag{12.19}$$

where $c_{\varepsilon 3}$ is a coefficient; and G_k is the buoyancy product of turbulence, determined by

$$G_k = \beta g_i \varepsilon_\phi \frac{\partial \phi}{\partial x_i} \tag{12.20}$$

where β is the volumetric expansion coefficient; g_i is the gravitational body force per unit mass in the x_i-direction; and ϕ denotes the scalar quantity, such as temperature and salinity.

When $G_k > 0$, $c_{\varepsilon 3}$ is given a value of 1.0; and when $G_k < 0$, $c_{\varepsilon 3}$ is in the range of 0–0.2 (ASCE Task Committee, 1988).

12.1.3   Effective diffusivities

The 1-D and 2-D heat and salinity transport equations should have dispersion terms when they are derived by integrating the corresponding 3-D transport equations. These dispersion terms are herein combined with the turbulent diffusion terms, so the effective diffusivity (mixing coefficient) includes the molecular diffusivity ε_m, turbulent diffusivity ε_t, and dispersion coefficient ε_d. Usually, the molecular diffusivity is approximately equal to the kinematic viscosity of water. The turbulent diffusivity is related to flow conditions, as described in Section 6.1. Normally, we have $\varepsilon_t \propto v_t$. The dispersion effect results from the non-uniformity of flow velocity and constituent concentration along the flow depth and/or the channel width. Elder (1959) estimated the longitudinal dispersivity as

$$\varepsilon_d = 5.86 h U_* \tag{12.21}$$

It can be seen that $\varepsilon_m \ll \varepsilon_t \ll \varepsilon_d$. The magnitude of the longitudinal effective diffusivity E is

$$E = \varepsilon_m + \varepsilon_t + \varepsilon_d \approx 6.0 h U_* \tag{12.22}$$

Iwasa and Aya (1991) related the longitudinal effective diffusivity to the channel width/depth ratio, B/h, as

$$\frac{E}{h U_*} = 2.0 \left(\frac{B}{h}\right)^{1.5} \tag{12.23}$$

which applies in the range of $2 < B/h < 20$ for laboratory channels and $10 < B/h < 100$ in waterways.

Fischer *et al.* (1979) proposed the following relation for natural waterways:

$$E = 0.011 \frac{B^2 U^2}{h U_*} \tag{12.24}$$

Eqs. (12.22)–(12.24) give the magnitude of the longitudinal effective diffusivity, which may be used in the 1-D model. However, the effective diffusivity usually is anisotropic. In particular, the dispersion effect is strongly related to the flow direction. Sladkevich *et al.* (2000) proposed a method to convert the longitudinal effective diffusivity to the depth-averaged 2-D components in the Cartesian coordinate axes. The effective diffusivity is treated as a tensor, E_{ij}, which is determined by

$$E_{ij} = \varepsilon_t \delta_{ij} + \varepsilon_d \frac{q_i q_j}{q^2} \tag{12.25}$$

where q is the flow discharge per unit width, q_i is the component of q in the i-direction, and δ_{ij} is the Kronecker delta.

An alternative is the relation suggested by Holly and Usseglio-Polatera (1984):

$$
\begin{aligned}
E_{xx} &= E_s \cos^2 \theta + E_n \sin^2 \theta \\
E_{xy} &= (E_s - E_n) \sin \theta \cos \theta \\
E_{yy} &= E_s \sin^2 \theta + E_n \cos^2 \theta
\end{aligned}
\tag{12.26}
$$

where E_s and E_n are the longitudinal and transverse effective diffusivities, respectively; and θ is the angle (positive counter-clockwise) of the longitudinal direction from the x-axis. Note that E_{xy} is applied in the cross-derivative diffusion terms, which need to be added in Eqs. (12.8) and (12.9) for more general applications.

As described in Section 6.3, the helical flow motion in curved channels induces a kind of dispersion and significantly affects the main flow, sediment transport, and channel morphological change. It might also affect heat and salinity transport. Therefore, the application of formulas (12.22)–(12.24) in strongly curved channels should be done with caution. Estimation using on-site measurement data is recommended.

12.1.4   Heat transfer across water and bed surfaces

Surface heat exchange plays an important role in the thermodynamic processes in aquatic systems. As shown in Fig. 12.1, heat generally transfers across the water surface by short-wave radiation, long-wave radiation, evaporation, condensation, convection, and conduction. The short-wave radiation is a penetrative effect that distributes its heat through a significant range of the water column, while the others occur only at the water surface.

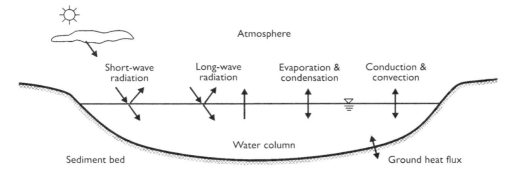

Figure 12.1 Heat budget in water column.

Short-wave radiation

Radiation emitted by the sun is termed short-wave or solar radiation. The net short-wave radiation $(W \cdot m^{-2})$ penetrating the water can be determined by (TVA — Tennessee Valley Authority, 1972; Jacquet, 1983)

$$J_{Tsw} = J_{Tsw,clear}(1 - 0.65 C_{cloud}^2)(1 - R_{tsw})(1 - f_{shade}) \qquad (12.27)$$

where $J_{Tsw,clear}$ is the short-wave radiation $(W \cdot m^{-2})$ that would reach the water surface in a clear day after atmospheric attenuation, C_{cloud} is the fractional cloud cover between 0 and 1, R_{tsw} is the dimensionless surface reflexivity, and f_{shade} is the shading factor of riparian vegetation.

The short-wave radiation reaching the water surface on a clear day, $J_{Tsw,clear}$, is related to the distance between the sun and earth, solar declination, and latitude of the local meridian, and is affected by atmosphere scattering and absorption (TVA, 1972; Brown and Barwell, 1987; Deas and Lowney, 2000). Solar radiation may be measured relatively inexpensively, and is reported by some weather stations. If the measured data are available, the cloud cover term in Eq. (12.27) may be eliminated (but not the reflectivity and shading factor).

The reflection coefficient R_{tsw} depends on cloud cover and altitude of the sun, and the shading factor f_{shade} depends on vegetation height, bearing of the sun, etc. The details may be found in Brown and Barwell (1987) and Deas and Lowney (2000).

Long-wave radiation

Radiation emitted by terrestrial objects and atmosphere is termed long-wave radiation. The net long-wave radiation at the water surface is the result of two processes: the downward radiation from the atmosphere and the upward radiation emitted by the water surface. The long-wave radiation largely depends on air temperature, humidity, and cloud cover. Hodges (1998) rewrote the formulas of long-wave heat flux proposed by Fischer *et al.* (1979), Imberger and Patterson (1981), Jacquet (1983), and TVA (1972) as

$$J_{Tlw} = \varepsilon_{air}\sigma T_{air}^4(1 + 0.17C_{cloud}^2)(1 - R_{tlw}) - \varepsilon_{water}\sigma T_{water}^4 \qquad (12.28)$$

where σ is the Stefan-Boltzman constant (5.669×10^{-8} W$\cdot$m$^{-2}\cdot$ $^\circ K^{-4}$); T_{air} is the air temperature in Kelvins ($^\circ K = {}^\circ C + 273.15$), measured two meters above the water surface; T_{water} is the water surface temperature in $^\circ K$; R_{tlw} is the reflectivity of the water surface for long-wave radiation, which is generally small and ≈ 0.03 (TVA, 1972; Brown and Barwell, 1987; Chapra, 1997); ε_{water} is the emissivity of water, which is between 0.95 and 0.963 corresponding to the temperature range of 0° and 100°C (Reynolds and Perkins, 1977), but is given 0.97 by TVA; and ε_{air} is the emissivity of air, determined by (Swinbank, 1963)

$$\varepsilon_{air} = 0.938 \times 10^{-5}T_{air}^2 \qquad (12.29)$$

Latent heat flux

The latent heat flux per unit surface area (W$\cdot$m^{-2}) due to evaporation and condensation can be modeled as

$$J_{Te} = LE \qquad (12.30)$$

where L is the latent heat of evaporation (J$\cdot$kg^{-1}), which is related to temperature (TVA, 1972; Jacquet, 1983; Blanc, 1985), but given a constant value of 2.5×10^6 J$\cdot$kg^{-1} by Gill (1982); and E is the water vapor flux (kg$\cdot$s^{-1}m^{-2}), determined by (Imberger and Patterson, 1981)

$$E = C_W U_{wind}\rho_{air}(q_{air} - q_{surface}) \qquad (12.31)$$

where U_{wind} is the wind speed; C_W is the dimensionless bulk transfer coefficient for evaporation (primarily due to wind), given as 1.4×10^{-3}; ρ_{air} is the density of air at the surface; q_{air} is the specific humidity in the air (unitless); and $q_{surface}$ is the specific humidity at the water surface (unitless). Note that 1 W $= 1$ J$\cdot s^{-1}$.

Edinger *et al.* (1974) determined the latent heat flux as a function of wind speed and water vapor:

$$J_{Te} = f(U_{wind})(e_{air} - e_s) \qquad (12.32)$$

where e_s is the saturation vapor pressure (mb) at the water surface temperature, e_{air} is the air vapor pressure (mb), and $f(U_{wind})$ is a function of wind speed. Various formulations were examined by Edinger *et al.* (1974), and one choice was $f(U_{wind}) = 6.9 + 0.345U_{wind,7}^2$ (W$\cdot$m^{-2}mb^{-1}) with $U_{wind,7}$ being the wind speed (m$\cdot$s^{-1}) measured 7 m above the water surface.

The saturation vapor pressure e_s is the highest pressure of water vapor that can exist in equilibrium with a plane, free water surface at a given temperature. It can be approximated by the Tetens formula:

$$e_s = a_* \exp\left(\frac{b_* T}{T + c_*}\right) \tag{12.33}$$

where T is the water temperature in °C. For temperatures above freezing, the coefficients are $a_* = 6.108$ mb, $b_* = 17.27$, and $c_* = 237.3$°C. The air vapor pressure e_{air} can be calculated using Eq. (12.33) by substituting T with the dew point temperature.

Sensible heat flux

Sensible heat flux is due to conduction and convection. It can be in either direction, depending on the temperature difference between air and water. Edinger *et al.* (1974) determined the sensible heat flux as

$$J_{Ts} = C_b f(U_{wind})(T_{air} - T_{water}) \tag{12.34}$$

where C_b is the Bowen coefficient ($0.62 \text{ mb} \cdot °\text{K}^{-1}$), and $f(U_{wind})$ is the wind speed function defined in Eq. (12.32).

An alternative formula for the sensible heat flux is (Imberger and Patterson, 1981)

$$J_{Ts} = C_b c_{p,air} \rho_{air} U_{wind}(T_{air} - T_{water}) \tag{12.35}$$

where C_b is the bulk coefficient of sensible heat flux, about 1.4×10^{-3}; and $c_{p,air}$ is the specific heat capacity at constant pressure, approximately $1003 \text{ J} \cdot \text{kg}^{-1} \cdot °\text{C}^{-1}$ for typical air temperatures in the near surface region.

Net heat flux in water column

It is generally presumed that the long-wave radiation (J_{Tlw}), latent heat flux (J_{Te}), and sensible heat flux (J_{Ts}) are non-penetrative; thus, they would appropriately be modeled by the surface boundary condition:

$$\varepsilon_T \frac{\partial T}{\partial z} = \frac{1}{\rho c_P}(J_{Tlw} + J_{Te} + J_{Ts}) \tag{12.36}$$

The short-wave radiation is penetrative and has an exponential decay distribution along the flow depth:

$$J_{Tsw}(z) = J_{Tsw}(z_s)e^{-\lambda(z_s - z)} \tag{12.37}$$

where $J_{Tsw}(z)$ is the short-wave radiation absorbed at height z, $J_{Tsw}(z_s)$ is the net short-wave radiation penetrating the water surface, and λ is the bulk extinction coefficient determined by Eqs. (12.69) and (12.70) in Section 12.2.2.

Normally, there is very little heat transfer across the bed surface for deep-water bodies. However, for shallow, transparent lakes and reservoirs, part of the heat flux may penetrate into the sediment bed and should be excluded (Tsay et al., 1992). A complete heat budget model in the sediment bed is preferable, which would consider the absorption and reflection of the short-wave radiation by the sediment bed as well as the heat exchange flux due to conduction, seepage flow, etc., at the wetted perimeter. However, the following simple approach may be used (Hodges, 1998):

$$J_{Bsw}(z) = \beta_r J_{Tsw}(z_b)e^{-\lambda(z-z_b)} \tag{12.38}$$

where $J_{Bsw}(z)$ is the heat flux returned to the water column at height z, $J_{Tsw}(z_b)$ is the short-wave radiation that reaches the bottom boundary, β_r represents the fraction of the short-wave radiation returned to the water column, and z_b is the bed surface elevation.

For the heat source from short-wave radiation, q_T is determined by

$$q_T = \frac{\partial}{\partial z}[J_{Tsw}(z) - J_{Bsw}(z)] \tag{12.39}$$

where $J_{Tsw}(z)$ and $J_{Bsw}(z)$ are determined using Eqs. (12.37) and (12.38), respectively. Note that the heat fluxes $J_{Tsw}(z)$ and $J_{Bsw}(z)$ in Eq. (12.39) transfer in opposite directions. The determined q_T is used in the source terms of Eqs. (12.4) and (12.6) in the 3-D and width-averaged 2-D models.

The net heat flux absorbed in the water column is

$$J_T = J_{Tlw} + J_{Te} + J_{Ts} + J_{Tsw}(z_s) - (1 - \beta_r)J_{Tsw}(z_b) \tag{12.40}$$

which is used as the heat source rate in Eqs. (12.8) and (12.10) in the depth-averaged 2-D and 1-D models. The last term on the right-hand side of Eq. (12.40) represents the flux penetrating into the bed, which is excluded in the heat budget in the water column.

12.1.5   Numerical solutions

The numerical methods introduced in Chapters 4–7 can be extended to solve the afore-mentioned flow, heat and salinity transport equations. For example, the SIMPLE(C) algorithms described in Sections 6.1.3.1, 7.1.3.2 and 7.2.4 can be straightforwardly applied to solve the 2-D and 3-D hydrodynamic equations here, since the flow density has been considered in the formulations.

The heat and salinity transport equations are similar to the suspended sediment transport equation. They are typical convection-diffusion equations and can be solved easily. If the finite volume method is used, Eqs. (12.4) and (12.5) are discretized as

$$\frac{\Delta V_P}{\Delta t}(T_P^{n+1} - T_P^n) = a_W T_W^{n+1} + a_E T_E^{n+1} + a_S T_S^{n+1} + a_N T_N^{n+1} + a_B T_B^{n+1}$$

$$+ a_T T_T^{n+1} - a_P T_P^{n+1} + \frac{1}{\rho c_p}[\Delta A_t(J_{Tsw,t} - J_{Bsw,t})$$

$$- \Delta A_b(J_{Tsw,b} - J_{Bsw,b})] \tag{12.41}$$

$$\frac{\Delta V_P}{\Delta t}(C_{sa,P}^{n+1} - C_{sa,P}^n) = a_W C_{sa,W}^{n+1} + a_E C_{sa,E}^{n+1} + a_S C_{sa,S}^{n+1} + a_N C_{sa,N}^{n+1} + a_B C_{sa,B}^{n+1}$$

$$+ a_T C_{sa,T}^{n+1} - a_P C_{sa,P}^{n+1} + S_{sa} \qquad (12.42)$$

where ΔA_t and ΔA_b are the areas of cell faces t and b projected on the horizontal plane, respectively; and $J_{Tsw,t}$ and $J_{Bsw,t}$ are the short-wave radiations penetrating to the water surface and reflected from the bottom surface, respectively.

It should be noted that the finite volume method and finite difference method handle the surface heat fluxes differently. In the finite volume method, when integrating Eq. (12.4) over the control volume near the water surface shown in Fig. 4.22, the long-wave radiation and latent and sensible heat fluxes are specified directly at t-face and arranged into the source term, and then the coefficient a_T is set to be zero. In the finite difference method, Eq. (12.36) is often used to determine the temperature at the water surface. However, Eq. (12.36) has been reported to be inefficient. More recently, many finite difference models also arrange the surface heat fluxes into the source term, following the approach used in the finite volume method.

Because flow density is influenced by temperature and salinity, the above heat and salinity transport equations should be solved with the flow model in a coupled form. For example, the SIMPLE algorithm for the full 3-D hydrodynamic model is described below:

(1) Guess the salinity, temperature, and pressure p^*;
(2) Calculate the flow density ρ^* using the state equation (12.12);
(3) Solve the momentum equations to obtain u_i^*;
(4) Solve the p' equation (7.14);
(5) Calculate p^{n+1} by adding p' to p^*;
(6) Calculate u_i^{n+1} using the velocity-correction relation (7.9) and the intercell fluxes using Eqs. (7.10)–(7.12);
(7) Solve the transport equations (12.41) and (12.42);
(8) Treat the corrected pressure p as a new guessed p^*, and repeat the procedure from step 2 to 6 until a converged solution is obtained;
(9) Calculate other water quality constituents, if needed; and
(10) Conduct the calculation of next time step if the unsteady flow is concerned.

Nevertheless, in the well-mixed cases, the effects of temperature and salinity on the flow are often neglected so that the hydrodynamic model may be decoupled from the computations of heat and salinity transport.

12.2 WATER QUALITY MODEL

Pollutants from municipal and industrial wastes (point sources) and from agricultural fields, urban and suburban runoff, groundwater and atmosphere (nonpoint sources) significantly affect the water quality in aquatic systems. They may be conservative or non-conservative, transport through convection and diffusion, and transform through a large number of physical, chemical and biological processes that interact with one another. Modeling the fate and transport of these pollutants and the resulting water quality in aquatic systems is an important task in environmental engineering. Some important aspects of water quality modeling are briefly described here. More details may be found in Thomann and Mueller (1987), Huber (1993), and Chapra (1997).

12.2.1 Kinetics and rate coefficients

The production or loss of a constituent, with or without interaction with other constituents, is a kinetic process (Huber, 1993). Examples of kinetic processes include decay of bacteria, oxidation of carbonaceous materials, and oxidation of nitrogen compounds. Such processes are usually quantified through the following equation:

$$\frac{DC_i}{Dt} = S_c = f(C_i, C_j, T) \quad j = 1, 2, \ldots \tag{12.43}$$

where C_i is the concentration of constituent i, T is the water temperature, DC_i/Dt denotes the rate of change in concentration of constituent i. In the 1-D model, DC/Dt is defined as

$$\frac{DC}{Dt} = \frac{1}{A}\left[\frac{\partial(AC)}{\partial t} + \frac{\partial(QC)}{\partial x} - \frac{\partial}{\partial x}\left(E_L A \frac{\partial C}{\partial x}\right)\right] \tag{12.44}$$

where C is the constituent concentration averaged over the cross-section, and E_L is the longitudinal effective diffusivity (mixing coefficient).

In the depth-averaged 2-D model, DC/Dt is

$$\frac{DC}{Dt} = \frac{1}{h}\left[\frac{\partial(hC)}{\partial t} + \frac{\partial(hU_xC)}{\partial x} + \frac{\partial(hU_yC)}{\partial y} - \frac{\partial}{\partial x}\left(E_x h \frac{\partial C}{\partial x}\right) - \frac{\partial}{\partial y}\left(E_y h \frac{\partial C}{\partial y}\right)\right] \tag{12.45}$$

where C is the depth-averaged constituent concentration, and $E_i(i = x, y)$ are the horizontal effective diffusivities.

In the width-averaged 2-D model, DC/Dt is

$$\frac{DC}{Dt} = \frac{1}{b}\left[\frac{\partial(bC)}{\partial t} + \frac{\partial(bU_xC)}{\partial x} + \frac{\partial(bU_zC)}{\partial z} - \frac{\partial}{\partial x}\left(E'_x b \frac{\partial C}{\partial x}\right) - \frac{\partial}{\partial z}\left(E'_z b \frac{\partial C}{\partial z}\right)\right] \tag{12.46}$$

where C is the width-averaged constituent concentration, and E'_i $(i = x, z)$ are the effective diffusivities in the longitudinal section.

In the 3-D model, DC/Dt is

$$\frac{DC}{Dt} = \frac{\partial C}{\partial t} + \frac{\partial(u_x C)}{\partial x} + \frac{\partial(u_y C)}{\partial y} + \frac{\partial(u_z C)}{\partial z} - \frac{\partial}{\partial x}\left(\varepsilon_x \frac{\partial C}{\partial x}\right)$$
$$- \frac{\partial}{\partial y}\left(\varepsilon_y \frac{\partial C}{\partial y}\right) - \frac{\partial}{\partial z}\left(\varepsilon_z \frac{\partial C}{\partial z}\right) \qquad (12.47)$$

where C is the local constituent concentration, and ε_i $(i = x, y, z)$ are the turbulent diffusivities.

A common kinetic model that is adequate for many processes is the first-order (linear) kinetics:

$$\frac{DC}{Dt} = -KC \qquad (12.48)$$

where K is the first-order rate coefficient (day^{-1}).

A more complex formulation is the Michaelis-Menten or Monod kinetics:

$$\frac{DC}{Dt} = -\frac{k_s C}{k_{1/2} + C} \qquad (12.49)$$

where k_s is the limiting reaction rate when $C \gg k_{1/2}$; and $k_{1/2}$ is called the half-saturation constant, because DC/Dt is half the limiting value when $C = k_{1/2}$.

The Michaelis-Menten kinetics may be written as Eq. (12.48), with the rate coefficient:

$$K = \frac{k_s}{k_{1/2} + C} = K_0 \frac{k_{1/2}}{k_{1/2} + C} \qquad (12.50)$$

where K_0 is the first-order rate coefficient, defined as $K_0 = k_s/k_{1/2}$. Both linear and Michaelis-Menten kinetics are depicted in Fig. 12.2. One can see that when $C \ll k_{1/2}$, the Michaelis-Menten kinetics becomes the first-order kinetics.

Eq. (12.49) is the general formulation of the Michaelis-Menten kinetics. Its variants for different species can be found in the next subsection.

The rate coefficient K usually depends on temperature. This is often described with reference to the rate at 20°C:

$$K(T) = K(20)\theta^{T-20} \qquad (12.51)$$

where T is in degree Celsius, and θ is a coefficient that is typically in the range of 1.01 to 1.10.

Eq. (12.51) implies that the reaction rate increases with temperature, as shown in the dashed line (Theta) in Fig. 12.3. However, for many species, such as phytoplankton, the temperature dependence is zero at a minimum temperature, increases to a peak

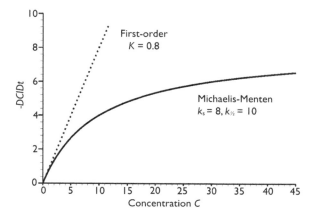

Figure 12.2 Linear and Michaelis-Menten kinetics.

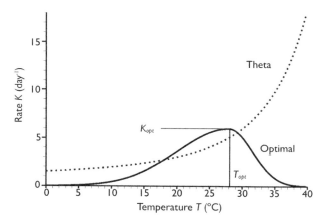

Figure 12.3 Rate coefficient as function of temperature.

growth rate at an optimal temperature, and then decreases at higher temperatures. Several models have been used in the literature to represent this trend (Chapra, 1997, p.605). For example, Cerco and Cole (1994) suggested the following formulation based on the normal distribution:

$$K(T) = \begin{cases} K_{opt}e^{-\kappa_1(T-T_{opt})^2} & T \leq T_{opt} \\ K_{opt}e^{-\kappa_2(T_{opt}-T)^2} & T > T_{opt} \end{cases} \qquad (12.52)$$

where κ_1 and κ_2 are the shape factors for the relationships of growth to temperatures blow and above the optimal temperature T_{opt}, respectively; and K_{opt} is the optimal rate coefficient at T_{opt}. The relation of Eq. (12.52) is depicted in the solid line in Fig. 12.3.

12.2.2 Constituent reactions and interrelationships

Maintaining an adequate dissolved oxygen (DO) concentration is essential to aquatic ecosystems. As shown in Fig. 12.4, the DO concentration is affected by many factors, such as atmospheric reaeration, photosynthesis, plant and animal respiration, biochemical oxygen demand, nitrification, and benthal demand. These factors also interact with one another. According to their interrelationships, the kinetic processes in the water column are usually divided into DO cycle, carbon cycle, nitrogen cycle, and phosphorus cycle. All cycles are more or less related to phytoplankton growth and respiration in an eutrophication system. In addition, similar cycles exist in the benthic sediment, and flux exchanges occur between the water column and the benthic sediment.

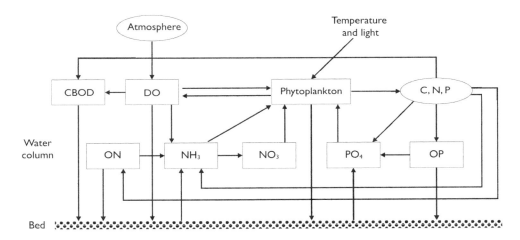

Figure 12.4 Major kinetic processes in water column.

Dissolved oxygen balance

The dissolved oxygen in aquatic systems is reaerated through exchange with the atmosphere, produced by photosynthesis of aquatic plants, and consumed by respiration of plants, animals and bacteria, oxidation of carbonaceous matters, oxidation of nitrogen compounds, and sediment oxygen demand. In the 1-D and depth-averaged 2-D models, the differential equation describing the effects of these processes on DO concentration is

$$\frac{DC_{DO}}{Dt} = \frac{K_L}{h}(C_{sDO} - C_{DO}) + P - R - K_d C_{CBOD} - K_N C_{NBOD} - \frac{S_{SOD}}{h}$$

$$(12.53)$$

where C_{DO} is the DO concentration ($gO_2 \cdot m^{-3}$, i.e., gram of oxygen per cubic meter); C_{sDO} is the saturation DO concentration ($gO_2 \cdot m^{-3}$); K_L is the liquid-film coefficient ($m \cdot day^{-1}$); P is the rate of DO production due to photosynthesis ($gO_2 \cdot m^{-3} day^{-1}$);

R is the rate of DO loss due to respiration $(gO_2 \cdot m^{-3}day^{-1})$; C_{CBOD} is the concentration of carbonaceous biochemical oxygen demand (CBOD) $(gO_2 \cdot m^{-3})$; K_d is the deoxygenation coefficient (day^{-1}); C_{NBOD} is the concentration of nitrogenous biochemical oxygen demand (NBOD) $(gO_2 \cdot m^{-3})$; K_N is the rate coefficient for NBOD (day^{-1}); S_{SOD} is the sediment oxygen demand (SOD), defined as the rate of DO loss per unit (horizontal) bed area $(gO_2 \cdot m^{-2}day^{-1})$; and h is the flow depth (m).

The terms on the right-hand side of Eq. (12.53) represent atmospheric reaeration, phytoplankton photosynthesis (P), respiration (R), CBOD oxidation, NBOD oxidation, and sediment oxygen demand, respectively. Note that the exchanges fluxes $K_L(C_{sDO} - C_{DO})$ and S_{SOD} at the water and bed surfaces are considered through boundary conditions in the 3-D and width-averaged 2-D models; thus, the 3-D or width-integrated DO balance equation is

$$\frac{DC_{DO}}{Dt} = P - R - K_d C_{CBOD} - K_N C_{NBOD} \qquad (12.54)$$

The determination of all terms in Eqs. (12.53) and (12.54) is discussed in the following subsections.

Atmospheric reaeration

The exchange of oxygen across the water surface is affected by temperature, flow conditions, and atmospheric conditions. The effect of temperature can be accounted for using Eq. (12.51), with a value of about 1.024 for the coefficient θ.

In the absence of wind, O'Connor and Dobbins (1958) suggested the reaeration coefficient of oxygen (day^{-1}) as

$$K_a = \frac{K_L}{h} = 3.9 \frac{U^{1/2}}{h^{3/2}} \qquad (12.55)$$

where U is the average flow velocity $(m \cdot s^{-1})$, and h is the average flow depth (m).

O'Connor and Dobbins's relation may underestimate the reaeration coefficient for small streams. Two equations that performed well in a comparison between predicted and measured reaeration rates (Rathbun, 1977) are those by Padden and Gloyna (1971):

$$K_a = 4.55 \frac{U^{0.703}}{h^{1.054}} \qquad (12.56)$$

and by Tsivoglou and Wallace (1972):

$$K_a = 13648 US \qquad (12.57)$$

where K_a is in day^{-1} at $20°C$, U is in $m \cdot s^{-1}$, h is in m, and S is the stream slope (unitless).

In small-roughness, uniform channels, the reaeration rate can be described by the following formula derived from the small-eddy model for oxygen interfacial transfer (Moog, 1995; Moog and Jirka, 1999):

$$K_L = 0.161 \text{Sc}^{-1/2} (\varepsilon v)^{1/4} \tag{12.58}$$

where Sc is the Schmidt number; and ε is the turbulent dissipation rate per mass near the water surface, which may be estimated as $\varepsilon = u_*^3/h$.

In channels with large-scale form roughness, the reaeration rate is enhanced by the large bed variations that produce depth-scale form drag. Moog and Jirka (1999) proposed a modification for Eq. (12.58) to take this enhancement into account.

For standing water bodies, such as lakes, lagoons, and bays, reaeration is affected by wind. Banks and Herrera (1977) suggested the following relationship:

$$K_L = 0.728 U_w^{1/2} - 0.317 U_w + 0.0372 U_w^2 \tag{12.59}$$

where K_L has units of $m \cdot day^{-1}$, and U_w is the wind speed ($m \cdot s^{-1}$) at 10 m above the water surface.

The saturation DO concentration, C_{sDO}, is a function of temperature and salinity. In fresh and saline waters, C_{sDO} is approximated by (Benson and Krause, 1984; see Huber, 1993)

$$C_{sDO} = \exp\left[c_0 + \frac{c_1}{T} + \frac{c_2}{T^2} + \frac{c_3}{T^3} + \frac{c_4}{T^4} + C_{sa}\left(c_5 + \frac{c_6}{T} + \frac{c_7}{T^2}\right)\right] \tag{12.60}$$

where C_{sDO} is in $g \cdot m^{-3}$; T is in $°K$; C_{sa} is the salinity in ppt, which is related to chlorinity or chloride concentration C_{chl} by $C_{sa} = 1.80655 C_{chl}$, with both in ppt; and the coefficients: $c_0 = -139.34411$, $c_1 = 1.575701 \times 10^5$, $c_2 = -6.642308 \times 10^7$, $c_3 = 1.243800 \times 10^{10}$, $c_4 = -8.621949 \times 10^{11}$, $c_5 = -0.017674$, $c_6 = 10.754$, and $c_7 = 2140.7$.

Photosynthesis and respiration of phytoplankton

The presence of aquatic plants, such as phytoplankton, weeds, and algae, can significantly affect the DO concentration in a water body through photosynthesis. These plants containing chlorophyll can utilize the radiant energy from the sun, convert water and carbon dioxide into glucose, and release oxygen. The photosynthesis reaction can be written as

$$6CO_2 + 6H_2O \xrightarrow{\text{photosynthesis}} C_6H_{12}O_6 + 6O_2 \tag{12.61}$$

Because the photosynthetic process is dependent on solar radiant energy, the production of oxygen proceeds only during daylight hours. Concurrently with this production, however, the algae require oxygen for respiration, which can be considered to proceed continuously. These two processes result in a diurnal variation in DO concentration. Minimum values of DO usually occur in the early morning, predawn hours, and maximum values occur in the early afternoon. The likely magnitude can be measured via light- and dark-bottle BOD tests, and simulated using simple cosine source/sink functions (Dresnack and Dobbins, 1968) or a more complicated model coupled with phytoplankton (or algae) growth and mortality. The latter simulation approach is used in recent water quality models, such as QUAL2E (Brown and Barnwell, 1987) and WASP (Wool et al., 1995). This is demonstrated below.

In the 1-D and depth-averaged 2-D models, the phytoplankton population is governed by

$$\frac{DC_{Phy}}{Dt} = K_{Phy}C_{Phy} - K_M C_{Phy} - \frac{\omega_{Phy}}{h}C_{Phy} \tag{12.62}$$

where C_{Phy} is the biomass concentration of phytoplankton, defined in carbon ($gC \cdot m^{-3}$, i.e., gram of carbon per cubic meter) or chlorophyll; K_{Phy} is the growth rate coefficient of phytoplankton (day^{-1}); K_M is the mortality rate coefficient of phytoplankton (day^{-1}), which is affected by temperature, as described in Eq. (12.51); and ω_{Phy} is the settling velocity of phytoplankton ($m \cdot day^{-1}$).

In the 3-D and width-averaged 2-D models, the settling term has a different formulation, and thus the phytoplankton population equation is

$$\frac{DC_{Phy}}{Dt} = K_{Phy}C_{Phy} - K_M C_{Phy} - \frac{\partial(\omega_{Phy}C_{Phy})}{\partial z} \tag{12.63}$$

The growth of phytoplankton is affected by temperature, solar radiation (light), and nutrient availability. Thus, the growth rate coefficient is assumed as

$$K_{Phy} = K_{Phy,m}\theta_{Phy}^{T-20}f_N f_L \tag{12.64}$$

where $K_{Phy,m}$ is the optimal growth rate coefficient of phytoplankton at 20°C, θ_{Phy} is the temperature coefficient for phytoplankton growth, f_N is the nutrient limitation factor, and f_L is the light limitation factor.

An initial estimate of the optimal growth rate $K_{Phy,m}$ can be obtained from studies of phytoplankton dynamics and refined through calibration. Low concentration of either inorganic nitrogen or phosphorus would affect the growth of phytoplankton, so the nutrient limitation factor is determined by

$$f_N = \min\left[\left(\frac{C_{NH3} + C_{NO3}}{k_{N,1/2} + C_{NH3} + C_{NO3}}\right), \left(\frac{C_{PO4}}{k_{P,1/2} + C_{PO4}}\right)\right] \tag{12.65}$$

or

$$f_N = \left(\frac{C_{NH3} + C_{NO3}}{k_{N,1/2} + C_{NH3} + C_{NO3}}\right) \cdot \left(\frac{C_{PO4}}{k_{P,1/2} + C_{PO4}}\right) \tag{12.66}$$

where C_{NH3} is the ammonia nitrogen concentration ($gN \cdot m^{-3}$, i.e., gram of nitrogen per cubic meter); C_{NO3} is the nitrate nitrogen concentration ($gN \cdot m^{-3}$); C_{PO4} is the inorganic (dissolved) phosphorus concentration ($gP \cdot m^{-3}$, i.e., gram of phosphorus per cubic meter); and $k_{N,1/2}$ and $k_{P,1/2}$ are the Michaelis-Menten half-saturation nitrogen and phosphorus concentrations for phytoplankton growth, respectively. C_{NH3}, C_{NO3}, and C_{PO4} are determined using Eqs. (12.78), (12.79), and (12.86), respectively.

Phytoplankton growth is a function of light intensity, until an optimal value is reached. The light limitation factor can be determined by Smith's (1936), Steele's (1962), or the half-saturation approach. The half-saturation approach gives

$$f_L = \frac{I_z}{k_{L,1/2} + I_z} \tag{12.67}$$

where $k_{L,1/2}$ is the Michaelis-Menten half-saturation light intensity for phytoplankton growth; and I_z is the light intensity at a given height z and varies with z according to Beer's law:

$$I_z = I_0 \exp[-\lambda(z_s - z)] \tag{12.68}$$

where I_0 is the light intensity at the water surface, and λ is the light extinction coefficient. Note that the energy source for photosynthesis is the light in the range of 400- to 700-nanometer wavelengths. It is called the photosynthetically active radiation (PAR). PAR is different from the insolation J_{Tsw} in Eq. (12.27), which is in the entire spectrum of wavelengths (see Rounds *et al.*, 1999).

The light extinction coefficient λ is affected by phytoplankton, suspended sediments, etc., in the water column. The following linear relation between λ and C_{Phy} is often used:

$$\lambda = \lambda_0 + k_r a_{CChl} C_{Phy} \tag{12.69}$$

where λ_0 is the light extinction coefficient without phytoplankton, k_r is a coefficient for light attenuation by phytoplankton, and a_{CChl} is the conversion factor of carbon to chlorophyll of phytoplankton.

To consider the effects of both phytoplankton and suspended sediments on light attenuation, Stefan *et al.* (1983) suggested the following relation:

$$\lambda = \lambda_0 + 0.025 a_{CChl} C_{Phy} + 0.043 C_s \tag{12.70}$$

where λ and λ_0 are in $l \cdot m^{-1}$, C_{Phy} is in $mgC \cdot m^{-3}$, and C_s is the suspended sediment concentration in $g \cdot m^{-3}$.

The depth-averaged light limitation factor is obtained by integrating Eq. (12.67) over the flow depth as

$$f_{L,av} = \frac{1}{\lambda h} \ln \left(\frac{k_{L,1/2} + I_0}{k_{L,1/2} + I_0 e^{-\lambda h}} \right) \tag{12.71}$$

A byproduct of phytoplankton growth is dissolved oxygen. An additional source of oxygen from phytoplankton growth occurs when the available ammonia nutrient resource is exhausted and the phytoplankton begins to use the available nitrate. For nitrate uptake, the initial step is a reduction to ammonia that produces oxygen. Thus, the total rate of oxygen production rate due to phytoplankton growth is (Wool *et al.*, 1995)

$$P = K_{Phy}\left[\frac{32}{12} + \frac{48}{14}a_{NC}(1 - p_{NH3})\right]C_{Phy} \qquad (12.72)$$

where p_{NH3} is the preference factor for ammonia uptake, determined by Eq. (12.81); and a_{NC} is the phytoplankton nitrogen-carbon ratio. The stoichiometric constant 32/12 arises because 32/12 g of oxygen corresponds to 1 g of phytoplankton produced by the growth, and the constant 48/14 arises because 48/14 g of oxygen is produced for 1 g of phytoplankton nitrate reduced. Note that the P determined by Eq. (12.72) is substituted into Eq. (12.53).

Oxygen is diminished in the water column as a result of phytoplankton respiration, which is basically the reverse process of photosynthesis. Thus, the rate of oxygen loss due to phytoplankton respiration is

$$R = \frac{32}{12}K_M C_{Phy} \qquad (12.73)$$

which is subsitiuted into Eq. (12.53).

In addition, phytoplankton may be predated by zooplankton. This can be modeled by considering the predator-prey relation and the nutrients/food chain interaction. The details can be found in Chapra (1997).

Carbonaceous BOD

CBOD represents the oxygen demand by bacteria in the oxidation of organic (carbonaceous) matters present in a waste. CBOD is exerted by the presence of heterotrophic organisms that are capable of deriving the energy for oxidation from an organic carbon substrate. A large number of these heterotrophic organisms are contained in municipal sewage as well as most rivers, estuaries, and lakes.

CBOD may be particulate or dissolved in the water. The change in concentration of CBOD usually results from both settling of the particulate CBOD and oxidation of the dissolved CBOD. The oxidation of the dissolved CBOD can be represented by a first-order kinetic process. In addition, CBOD is generated as a result of phytoplankton death and loss due to denitrification reaction under low DO conditions. Thus, the kinetic rate of CBOD can be determined by

$$\frac{DC_{CBOD}}{Dt} = \frac{32}{12}K_M C_{Phy} - \frac{\omega_{CBOD}}{h}(1 - f_{CBOD,d})C_{CBOD}$$

$$- K'_d f_{CBOD,d}C_{CBOD} - \frac{5}{4}\frac{32}{14}K_{NO3}C_{NO3} \qquad (12.74)$$

where $f_{CBOD,d}$ and $1 - f_{CBOD,d}$ are the fractions of the dissolved and particulate CBOD in the total CBOD, respectively; ω_{CBOD} is the settling velocity of particulate CBOD;

K'_d is the CBOD decay rate coefficient; and K_{NO3} is the denitrification rate coefficient of nitrate (day^{-1}), determined by Eq. (12.84).

The terms on the right-hand side of Eq. (12.74) represent the production of CBOD due to phytoplankton death, settling of the particulate CBOD, oxidation of the dissolved CBOD, and sink of CBOD due to denitrification, respectively. The stoichiometric constant 32/12 arises from the conversion between oxygen and phytoplankton (as carbon) concentration. The constant $(5/4) \cdot (32/14)$ appears because for each g of nitrate nitrogen reduced, $(5/4) \cdot (12/14)$ g of carbon are consumed, which reduces CBOD by $(5/4) \cdot (12/14) \cdot (32/12)$ g.

Note that the settling term in Eq. (12.74) is valid only for the 1-D and depth-averaged 2-D models. In the 3-D and width-averaged 2-D models, the settling process should be represented by the settling term similar to that in Eq. (12.63).

The effective deoxygenation coefficient K_d is usually used to replace $K'_d f_{CBOD,d}$. Typical values of K_d are in the range from 0.1 to 4.0 day^{-1}, with larger values for untreated wastewater and smaller values for treated wastewater and natural waters. Wright and McDonnell (1979) suggested the following relationship for K_d (day^{-1}) at 20°C:

$$K_d(20) = 1.80Q^{-0.49} \tag{12.75}$$

where Q is the flow discharge (m^{3}s^{-1}).

The deoxygenation coefficient K_d is also affected by water temperature and DO concentration. These effects are considered by

$$K_d = K_d(20)\theta_d^{T-20}\left(\frac{C_{DO}}{k_{BOD,1/2} + C_{DO}}\right) \tag{12.76}$$

where θ_d is the temperature coefficient for deoxygenation, with a value of about 1.047; and $k_{BOD,1/2}$ is the Michaelis-Menten half-saturation DO concentration for deoxygenation, with a value of about 0.5 gO$_2 \cdot$ m^{-3}.

Nitrogen cycle

In natural aerobic waters, the nitrogen cycle consists of several steps of transformation. Organic nitrogen (ON, contained in organic wastes and algae) is converted first to ammonia nitrogen (NH$_3$_N), which is then oxidized to nitrite (NO$_2$_N) and nitrate (NO$_3$_N). Oxygen is required for the oxidation of ammonia to nitrite and then to nitrate. In addition, phytoplankton utilize ammonia and nitrate nitrogen, and recycling occurs to the organic forms as they die.

The sum of organic nitrogen and ammonia nitrogen is called the total Kjeldahl nitrogen (TKN) in a laboratory analysis procedure. Most of the NBOD is due to TKN, because the nitrite concentration in most wastewater streams and ambient waters is very low, less than 0.1 mg$\cdot$l^{-1}. Early models simulate TKN collectively (Thomann and Mueller, 1987; Huber, 1993); however, many recent models, such as QUAL2E and WASP, compute the nitrogen components individually. Introduced below is a modeling framework for the nitrogen cycle, which is essentially similar to the WASP model. Because the nitrite nitrogen usually changes to nitrate very quickly, a direct process from ammonia nitrogen to nitrate nitrogen is assumed and only three states (organic nitrogen, ammonia nitrogen, and nitrate nitrogen) are modeled. Their kinetic processes are described by

$$\frac{DC_{ON}}{Dt} = K_M a_{NC} f_{Phy,ON} C_{Phy} - K_{ON} C_{ON} - \frac{\omega_{ON}}{h}(1 - f_{ON,d}) C_{ON} - \frac{S_{ON}}{h}$$

$$\tag{12.77}$$

$$\frac{DC_{NH3}}{Dt} = K_M a_{NC}(1 - f_{Phy,ON}) C_{Phy} + K_{ON} C_{ON} - K_{Phy} a_{NC} p_{NH3} C_{Phy}$$
$$- K_{NH3} C_{NH3} - \frac{S_{NH3}}{h} \tag{12.78}$$

$$\frac{DC_{NO3}}{Dt} = K_{NH3} C_{NH3} - K_{Phy} a_{NC}(1 - p_{NH3}) C_{Phy} - K_{NO3} C_{NO3} - \frac{S_{NO3}}{h}$$

$$\tag{12.79}$$

where C_{ON} is the concentration of organic nitrogen $(\text{gN} \cdot \text{m}^{-3})$; K_{ON} is the mineralization rate coefficient of organic nitrogen (day^{-1}); K_{NH3} is the nitrification rate coefficient of ammonia (day^{-1}); $f_{Phy,ON}$ and $1 - f_{Phy,ON}$ are the fractions of respired phytoplankton recycled to the organic and ammonia nitrogen pools, respectively; $f_{ON,d}$ is the fraction of the dissolved organic nitrogen; and S_{ON}, S_{NH3}, and S_{NO3} are the organic nitrogen, ammonia, and nitrate fluxes from the sediment bed, respectively.

The first terms on the right-hand sides of Eqs. (12.77) and (12.78) represent the production of nitrogen due to phytoplankton death and respiration. $f_{Phy,ON}$ of the produced nitrogen is organic, while $1 - f_{Phy,ON}$ is in the inorganic form of ammonia. DiToro and Matystik (1980) assigned 0.5 for $f_{Phy,ON}$ in the Great Lakes model.

The second terms on the right-hand sides of Eqs. (12.77) and (12.78) represent the change of organic nitrogen to ammonia nitrogen due to mineralization. Nonliving organic nitrogen must undergo mineralization or bacterial decomposition into ammonia nitrogen before utilization by phytoplankton. This process is affected by temperature and phytoplankton population, and thus the following relation for K_{ON} is often used

$$K_{ON} = K_{ON}(20)\theta_{ON}^{T-20}\left(\frac{C_{Phy}}{k_{mPc} + C_{Phy}}\right) \tag{12.80}$$

where θ_{ON} is the temperature coefficient for mineralization (about 1.08), and k_{mPc} is the half-saturation phytoplankton concentration for mineralization.

The third term on the right-hand side of Eq. (12.77) represents the settling of the particulate organic nitrogen in the 1-D and depth-averaged 2-D models. Note that it has a different formulation in the 3-D and width-averaged 2-D models, as shown in Eq. (12.63).

The third term on the right-hand side of Eq. (12.78) and the second term on the right-hand side of Eq. (12.79) represent the nitrogen uptake for phytoplankton growth.

As phytoplankton grows, the dissolved inorganic nitrogen is taken up and incorporated into biomass. Both ammonia and nitrate are available for uptake but, for physiological reasons, the preferred form is ammonia nitrogen. The ammonia preference factor p_{NH3} is given as

$$p_{NH3} = \frac{C_{NH3}C_{NO3}}{(k_{mN} + C_{NH3})(k_{mN} + C_{NO3})} + \frac{C_{NH3}k_{mN}}{(C_{NH3} + C_{NO3})(k_{mN} + C_{NO3})}$$

$$(12.81)$$

where k_{mN} is the Michaelis-Menten limitation.

Eq. (12.81) shows that when the nitrate concentration is zero, the preference for ammonia is 1.0; when the ammonia concentration is zero, the preference for ammonia is zero. When both ammonia and nitrate are abundant, preference is given to ammonia and the factor approaches 1.0.

The fourth term on the right-hand side of Eq. (12.78) and the first term on the right-hand side of Eq. (12.79) represent the change of ammonia nitrogen to nitrate nitrogen due to nitrification. Nitrification is a two-step reaction carried out by aerobic autotrophs. Nitrosomonas bacteria catalyze the first reaction converting ammonia to nitrite, and in the second reaction Nitrobacter bacteria convert nitrite to nitrate. The process of nitrification in natural waters is complex, and depends on temperature, DO, pH, and flow conditions. The following relation for K_{NH3} is often used:

$$K_{NH3} = K_{NH3}(20)\theta_{NH3}^{T-20} \left(\frac{C_{DO}}{k_{NIT} + C_{DO}} \right) \qquad (12.82)$$

where θ_{NH3} is the temperature coefficient for nitrification (about 1.08), and k_{NIT} is the half-saturation DO concentration for nitrification.

In the nitrification process, the dissolved oxygen is required. The NBOD concentration is related to the ammonia nitrogen concentration by

$$C_{NBOD} = \frac{64}{14}C_{NH3} \qquad (12.83)$$

where the stoichiometric constant 64/14 arises because 64/14 g of oxygen is required to convert 1 g of ammonia to nitrate nitrogen. Eq. (12.83) is inserted into Eq. (12.53), and K_N is set as K_{NO3}.

The third term on the right-hand side of Eq. (12.79) represents the loss of nitrate due to denitrification. Denitrification refers to the reduction of nitrate (or nitrite) to N_2 and other gaseous products. This process is carried out by a large number of heterotrophic, facultative anaerobes. Denitrification is not a significant loss in the water column, but can be important in anaerobic benthic conditions. The denitrification process depends on temperature, DO, etc.; thus, the denitrification rate coefficient K_{NO3} is determined by

$$K_{NO3} = K_{NO3}(20)\theta_{NO3}^{T-20} \left(\frac{k_{DNI}}{k_{DNI} + C_{DO}} \right) \tag{12.84}$$

where θ_{NO3} is the temperature coefficient for denitrification (about 1.045), and k_{DNI} is the half-saturation DO concentration for denitrification (about 0.1 $gO_2 \cdot m^{-3}$).

The last terms on the right-hand side of Eqs. (12.77)–(12.79) account for the organic, ammonia, and nitrate nitrogen fluxes across the bed surface. Note that to be consistent with the water/bed exchange fluxes in Eq. (12.53) a "–" sign is used in these terms. This implies a positive value directs from the water column to the benthic sediment.

Phosphorus cycle

Modeling the phosphorus cycle is essentially analogous to the approach used for the nitrogen cycle. Phosphorus is divided into organic and inorganic forms. In some models, the organic phosphorus is further partitioned into particulate and dissolved forms that are separately simulated, and the inorganic phosphorus is partitioned into dissolved and sorbed forms when sediment sorption is considered. Here, two single kinetics equations are introduced for organic phosphorus (OP) and inorganic phosphorus (orthophosphate, PO_4) without further partition. Their kinetics are described by

$$\frac{DC_{OP}}{Dt} = K_M a_{PC} f_{Phy,OP} C_{Phy} - K_{OP} C_{OP} - \frac{\omega_{OP}}{h}(1 - f_{OP,d}) C_{OP} - \frac{S_{OP}}{h} \tag{12.85}$$

$$\frac{DC_{PO4}}{Dt} = K_M a_{PC}(1 - f_{Phy,OP}) C_{Phy} + K_{OP} C_{OP} - K_{Phy} a_{PC} C_{Phy} - \frac{S_{PO4}}{h} \tag{12.86}$$

where C_{OP} and C_{PO4} are the concentrations of organic and inorganic phosphorus ($gP \cdot m^{-3}$), respectively; K_{OP} is the mineralization rate coefficient of organic phosphorus (day^{-1}); $f_{Phy,OP}$ and $1 - f_{Phy,OP}$ are the fractions of dead and respired phytoplankton recycled to the organic and inorganic phosphorus pools, respectively; $f_{OP,d}$ is the fraction of the dissolved organic phosphorus; a_{PC} is the phytoplankton phosphorus-carbon ratio; and S_{OP} and S_{PO4} are the organic and inorganic phosphorus fluxes from the sediment bed ($gP \cdot m^{-2} day^{-1}$), respectively.

The terms on the right-hand side of Eq. (12.85) represent the production of organic phosphorus due to phytoplankton death and respiration, loss due to mineralization, settling of the particulate organic nitrogen, and exchange at the bed surface, respectively. The terms on the right-hand side of Eq. (12.86) represent the production of inorganic phosphorus due to phytoplankton death and respiration, gaining due to mineralization, uptake for phytoplankton growth, and exchange at the bed surface, respectively.

Mineralization from organic to inorganic phosphorus is affected by temperature and phytoplankton population, and thus the following relation for K_{OP} is used:

$$K_{OP} = K_{OP}(20)\theta_{OP}^{T-20} \left(\frac{C_{Phy}}{k_{mPc} + C_{Phy}} \right) \qquad (12.87)$$

where θ_{OP} is the temperature coefficient for organic phosphorus mineralization, about 1.08.

Again, the settling term in Eq. (12.85) is valid only in the 1-D and depth-averaged 2-D models, and it has a different form in the 3-D and width-averaged 2-D models, as shown in Eq. (12.63).

Flux exchanges between water column and benthic sediment

Settled waste materials (sludge), phytoplankton, dead aquatic plant roots and leaves, etc., that accumulate in the bed usually undergo decomposition, releasing nutrients to the sediment interstitial water and removing oxygen from the overlying water. As a result, the benthic sediment can be a substantial nutrient source and/or oxygen sink to the overlying water column.

SOD and nutrient fluxes from the bed depend on the extent of organic materials and the nature of the benthic community. Table 12.1 summarizes the range of SOD values at 20°C. Temperature effect in the 10–30°C range can be accounted for by

$$S_{SOD}(T) = S_{SOD}(20)\theta_{sod}^{T-20} \qquad (12.88)$$

where θ has a reported range of 1.040 to 1.130 (Zison *et al.*, 1978) and a typical value of 1.065. However, Eq. (12.88) probably overestimates below 10°C, and SOD approaches zero in 0–5°C.

Many early models give lumped values for the oxygen and nutrient fluxes, without resolving the details of the benthic processes. More recently, the following diffusion

Table 12.1 Sediment oxygen demand ranges (Huber, 1993)

Location	SOD range $gO_2 \cdot m^{-2}day^{-1}$	Source
Municipal sewage sludge, outfall vicinity	2–10	Thomann and M. (1987)
Municipal sewage sludge, aged, d/s of outfall	1–2	Thomann and M. (1987)
Estuarine mud	1–2	Thomann and M. (1987)
Sandy bottom	0.2–1	Thomann and M. (1987)
Mineral soils	0.05–0.1	Thomann and M. (1987)
Measured in rivers and streams	0.02–44	Bowie (1985)
Measured in estuaries and ocean	0.1–11	Bowie (1985)
Measured in lakes and reservoirs	0.004–9	Bowie (1985)

model has been adopted to quantify the exchange flux at the bed surface:

$$S_{BED} = K_{BED}(C_W - C_{BED}) \qquad (12.89)$$

where K_{BED} is the diffusional exchange coefficient $(m \cdot day^{-1})$, C_W denotes the concentration of a constituent (DO, NH_3, PO_4, etc.) in the water column, and C_{BED} is the concentration of the corresponding constituent in the sediment bed. This flux exchange model is often used when the fate and transport of DO, nitrogen, phosphorus, etc., in the benthic sediment are computed. The details can be found in Wool et al. (1995) and DiToro (2001).

Many experiments have been conducted to measure the coefficient K_{BED} in lake systems (DiToro, 2001). Steinberger and Hondzo (1999) investigated the factors affecting K_{BED} of DO and established an empirical relation:

$$\frac{K_{BED}h}{D} = 0.012 \left(\frac{Uh}{\nu}\right)^{0.89} \left(\frac{\nu}{D}\right)^{0.33} \qquad (12.90)$$

where D is the molecular diffusivity.

In addition, the development of bed forms, such as sand ripples and dunes, will affect the mass transfer at the bed surface. Under unsteady flow conditions, the pore water in the sediment bed will move in and out, and thus induce additional mass transfer.

12.2.3   Other biochemical processes

Many other biochemical processes of non-conservative constituents can be modeled by considering the first-order decay, gravitational settling, and flux exchange at the bed surface as

$$\frac{DC}{Dt} = -KC - \frac{\omega_c}{h}C - \frac{S_{BED}}{h} \qquad (12.91)$$

where ω_c is the settling velocity of the constituent, and S_{BED} is the flux of the constituent entering from the water column to the sediment bed.

For example, first-order decay has been a very good assumption in many modeling studies of coliform bacteria. The decay coefficient ranges from 0.0004 to 1.1 hr^{-1} (Bowie, 1985), but most values are in the range from 0.02 to 0.1 hr^{-1} and a median rate of 0.04 hr^{-1} for total coliforms (Huber, 1993). The decay of bacteria is affected by temperature, salinity, light, etc. The effect of temperature can be accounted for through Eq. (12.51), with θ being about 1.07. Mancini (1978) shows that coliform mortality increases with the percent seawater:

$$\frac{K_{sea}}{K_{fresh}} = 0.8 + 0.006 p_{sea} \qquad (12.92)$$

where K_{sea} and K_{fresh} are the first-order decay coefficients in saline and fresh waters, respectively; and p_{sea} is the percent seawater (e.g., $p_{sea} = 100$ percent in the ocean).

12.3   SIMULATION OF SEDIMENT-BORNE
CONTAMINANT TRANSPORT

Sediment is a major source of pollutants in aquatic systems. Not only does sediment itself significantly affect aquatic systems by erosion and deposition, but also it sorbs contaminants that degrade the quality of receiving water bodies. Traditionally, modeling of contaminant transport and water quality has mainly focused on the role of water flow and paid less attention to the effect of sediment. It is necessary to establish numerical models to investigate the transport of both water-borne and sediment-borne contaminants and their impacts on water quality. This has been studied by Lang and Chapra (1982) for lake systems. A generalized modeling framework for various aquatic systems is presented in this section.

12.3.1   Sorption and desorption of contaminants
on sediment particles

Sorption is a process whereby a dissolved substance is transferred to and becomes associated with solid materials (Chapra, 1997). It includes both adsorption and absorption. Adsorption is a surface phenomenon in which the dissolved substance is accumulated on the surface of solids, whereas absorption is a bulk phenomenon in which the dissolved substance interpenetrates or intermingles with solids. Desorption is the reverse process of sorption, in which a sorbed substance is released from solid particles. Sorption and desorption are encountered in diverse situations of contaminant transport. Many contaminant species, such as phosphorus, heavy metals, nuclides, bacteria, and viruses, can be transferred from the dissolved phase to the sorbed phase associated with sediment particles and then transported with sediment by the flow.

Consider a control volume consisting of a water and sediment mixture in the water column or sediment bed, as shown in Fig. 2.4. A contaminant constituent is either dissolved in water or sorbed on sediment particles. The concentrations of the dissolved and sorbed parts are defined as

$$C_d = \frac{M_d}{V_t}, \quad C_s = \frac{M_s}{V_t} \tag{12.93}$$

and the total contaminant concentration is

$$C_t = C_d + C_s = \frac{M_d + M_s}{V_t} \tag{12.94}$$

where V_t is the total volume of the water and sediment mixture (m^3); M_d and M_s are the masses (mg) of the dissolved and sorbed contaminants in the control volume, respectively; C_d and C_s are the concentrations (mg $\cdot$ m^{-3}) of the dissolved and sorbed contaminants, respectively; and C_t is the total contaminant concentration (mg $\cdot$ m^{-3}).

Note that some constituents, such as phosphorus, also exist in a separate particulate form (not necessarily sorbed to sediment particles). This may be considered by adding a particulate component in Eq. (12.94). However, for simplicity, only the dissolved and sorbed components are included here. The details of handling the particulate contaminant can be found in Lang and Chapra (1982).

Sorption and desorption of contaminants on sediment particles are often described by the linear isotherm:

$$R_{ad} = k_{ad}\rho_s s \frac{C_d}{1 - s} = k_{ad}r_{sw}C_d \tag{12.95}$$

$$R_{de} = k_{de}C_s \tag{12.96}$$

where R_{ad} and R_{de} are the sorption and desorption rates ($mg \cdot m^{-3}s^{-1}$), respectively; k_{ad} is the sorption rate coefficient ($m^3 \cdot kg^{-1}s^{-1}$); k_{de} is the desorption rate coefficient (s^{-1}); ρ_s is the sediment density ($kg \cdot m^{-3}$); s is the volumetric concentration of sediment (unitless); and r_{sw} is the sediment-to-water phase ratio ($kg \cdot m^{-3}$), defined as $r_{sw} = \rho_s s/(1 - s)$. In the water column, usually $s \ll 1$, so $r_{sw} = \rho_s s$. In the sediment bed, s is equal to $1 - p'_m$, and thus $r_{sw} = \rho_s(1 - p'_m)/p'_m$. Here, p'_m is the porosity of bed material.

In the equilibrium state, the sorption and desorption rates in Eqs. (12.95) and (12.96) should be equal, thus yielding

$$\frac{C_s}{C_d} = \frac{k_{ad}r_{sw}}{k_{de}} = k_D r_{sw} \tag{12.97}$$

where k_D is the equilibrium partition coefficient, defined as $k_D = k_{ad}/k_{de}$ (m^3kg^{-1}).

Using Eqs. (12.94) and (12.97) yields

$$f_d = \frac{C_d}{C_t} = \frac{1}{1 + k_D r_{sw}}, \quad f_s = \frac{C_s}{C_t} = \frac{k_D r_{sw}}{1 + k_D r_{sw}} \tag{12.98}$$

where f_d and f_s are the fractions of the dissolved and sorbed contaminants in the equilibrium state, respectively.

In fact, other models, such as the Langmuir isotherm and Freundlich isotherm, are also commonly used. The sorption rate in Eq. (12.95) can be replaced by

$$R_{ad} = k_{ad}r_{sw}C_d \left(1 - \frac{C_s}{C_{s,max}}\right) \tag{12.99}$$

where $C_{s,max}$ is the maximum concentration of contaminant sorbed to sediment.

Eq. (12.99) implies that the sorption process is limited by the maximum concentration of the sorbed contaminant. In the equilibrium state, equating the sorption and desorption rates in Eqs. (12.99) and (12.96) leads to

$$C_s = C_{s,max} \frac{C_d}{C_d + C_{s,max}/(k_D r_{sw})} = C_{s,max} \frac{aC_d}{1 + aC_d} \tag{12.100}$$

where a is the Langmuir sorption coefficient, defined as $a = k_D r_{sw}/C_{s,max}$.

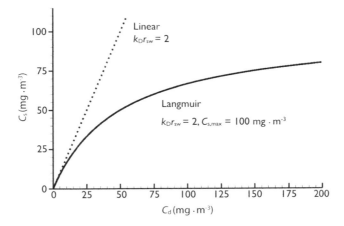

Figure 12.5 Linear and Langmuir isotherms of sorption.

Eq. (12.100) is the general form of the Langmuir isotherm. Both linear and Langmuir isotherms are depicted in Fig. 12.5. One can see that when $C_s \ll C_{s,\max}$, Eq. (12.99) reduces to Eq. (12.95) and the Langmuir isotherm becomes the linear isotherm.

Using Eqs. (12.94) and (12.100) yields the fractions of the dissolved and sorbed contaminants in the equilibrium state:

$$f_d = \frac{1}{2}\left[1 - \frac{C_{s,\max}}{C_t}\left(1 + \frac{1}{k_D r_{sw}}\right)\right]$$

$$\pm \sqrt{\frac{C_{s,\max}}{C_t}\frac{1}{k_D r_{sw}} - \frac{1}{4}\left[1 - \frac{C_{s,\max}}{C_t}\left(1 + \frac{1}{k_D r_{sw}}\right)\right]^2}$$

$$f_s = 1 - f_d \tag{12.101}$$

The parameters k_{ad}, k_{de}, k_D, and $C_{s,\max}$ vary with contaminant species, sediment properties, and water conditions. They are usually measured through sorption and desorption experiments. The estimation of these parameters can be found in Thomann and Mueller (1987), Chapra (1997), Furumai and Ohgaki (1989), Chao *et al.* (2006), etc.

12.3.2 Contaminant transport in water column

12.3.2.1 Non-equilibrium partition model

Fig. 12.6 shows the general transport and transformation patterns of contaminant constituents in both water column and sediment bed. Changes in concentrations of the dissolved and sorbed contaminants in the water column are caused by advection, diffusion, external loading, sorption, desorption, and decay. Additionally, the settling of

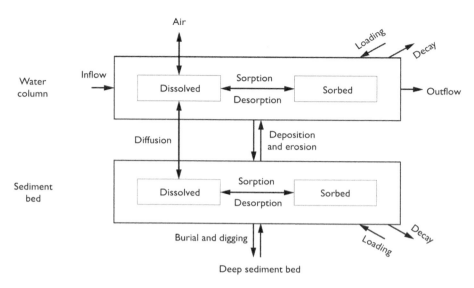

Figure 12.6 Contaminant transport with water and sediment.

sediment particles also contributes to the variation of the sorbed contaminant concentration. Adopting the linear sorption/desorption isotherm and the first-order kinetics yields the following transport equations for the dissolved and sorbed contaminants in the water column in the 1-D and depth-averaged 2-D models:

$$\frac{DC_{dw}}{Dt} = \frac{J_{d,aw}}{h} + q_{dw} - k_{ad,w}r_{sw,w}C_{dw} + k_{de,w}C_{sw} - k_{dw}C_{dw}$$

$$+ \frac{k_{dbw}}{h}(C_{db} - C_{dw}) + \frac{q_{d,ex}}{h} \qquad (12.102)$$

$$\frac{DC_{sw}}{Dt} = q_{sw} + k_{ad,w}r_{sw,w}C_{dw} - k_{de,w}C_{sw} - k_{sw}C_{sw} + \frac{q_{s,ex}}{h} \qquad (12.103)$$

where C_{dw} and C_{sw} are the cross-section-averaged or depth-averaged concentrations of the dissolved and sorbed contaminants in the water column, respectively; $k_{ad,w}$ and $k_{de,w}$ are the sorption and desorption rate coefficients, respectively; $r_{sw,w}$ is the sediment-to-water ratio; k_{dw} and k_{sw} are the decay coefficients of the dissolved and sorbed contaminants, respectively; q_{dw} and q_{sw} are the loading rates of the dissolved and sorbed contaminants per unit volume, respectively; $J_{d,aw}$ is the flux across the water surface per unit surface area; C_{db} is the concentration of contaminant dissolved in the bed surface layer; k_{dbw} is the diffusional transfer coefficient of the dissolved contaminant across the bed surface; and $q_{d,ex}$ and $q_{s,ex}$ are the exchange rates of the dissolved and sorbed contaminants due to sedimentation, respectively.

In the 3-D and width-averaged 2-D models, the mass transfers at the water and bed surfaces are handled through boundary conditions, so the transport equations of the dissolved and sorbed contaminants in the water column are

$$\frac{DC_{dw}}{Dt} = q_{dw} - k_{ad,w}r_{sw}C_{dw} + k_{de,w}C_{sw} - k_{dw}C_{dw} \quad (12.104)$$

$$\frac{DC_{sw}}{Dt} = q_{sw} + k_{ad,w}r_{sw}C_{dw} - k_{de,w}C_{sw} - k_{sw}C_{sw} - \frac{\partial(\omega_s C_{sw})}{\partial z} \quad (12.105)$$

where ω_s is the settling velocity of sediment particles.

In each of the above models, two differential equations are used to determine the concentrations of the dissolved and sorbed contaminants. This approach is generally applied to simulate the non-equilibrium sorption and desorption processes, and thus is called the non-equilibrium partition model.

The loading rates q_{dw} and q_{sw} in Eqs. (12.102)–(12.105) account for external loading or loss due to side inflow, rainfall, infiltration, seepage flow, biochemical interaction with other constituents, etc.

Many toxic substances, such as PCBs, transfer across the water surface by volatilization. The interfacial transfer rate of a dissolved gas is modeled by (see Chapra, 1997)

$$J_{d,aw} = v_v \left(\frac{p_g}{H_e} - C_{dw} \right) \quad (12.106)$$

where v_v is the net transfer velocity across the water surface $(m \cdot s^{-1})$, p_g is the partial pressure of gas in the air over the water (atm), and H_e is Henry's constant $(atm \cdot m^3 mg^{-1})$. As described in Section 12.2.2, for dissolved oxygen, Eq. (12.106) can be rewritten as $J_{d,aw} = K_L(C_{sDO} - C_{DO})$. For substances that are not abundant in the atmosphere, p_g can be assumed to be zero.

Mass may exchange at the bed surface through a variety of mechanisms. The most common one is the diffusion flux across the bed surface. In particular, in the case of lakes and reservoirs, the flow is slow, so diffusion is the main mode of bed release. This diffusion flux across the bed surface is modeled in Eq. (12.102) by the term $k_{dbw}(C_{db} - C_{dw})$, which is similar to Eq. (12.89).

The sorbed contaminant exchanges at the bed as sediment erosion or deposition occurs. This exchange rate is determined by

$$q_{s,ex} = \max(E_b - D_b, 0)\frac{C_{sb}}{1 - p'_m} + \min(E_b - D_b, 0)\frac{C_{sw}}{s} \quad (12.107)$$

where D_b and E_b are the sediment deposition and erosion rates $(m \cdot s^{-1})$, defined in Eqs. (11.14) and (11.19), respectively; s is the volumetric concentration of sediment in the water column; and p'_m is the porosity of sediment in the bed surface layer. The two terms on the right-hand side of Eq. (12.107) represent the cases of net erosion and net deposition, respectively. The factors $1/(1 - p'_m)$ and $1/s$ are introduced in Eq. (12.107) to convert the net sediment flux $E_b - D_b$ to the volumes of the corresponding water and sediment mixtures in the sediment bed and water column, respectively, in which C_{sb} and C_{sw} are defined.

As sediment exchanges at the net rate $E_b - D_b$ at the bed, the pore water in the sediment bed also exchanges with the water column at a rate $(E_b - D_b)p'_m/(1 - p'_m)$. Consequently, the dissolved contaminant exchanges at a rate:

$$q_{d,ex} = \max(E_b - D_b, 0)\frac{C_{db}}{1 - p'_m} + \min(E_b - D_b, 0)\frac{p'_m}{1 - p'_m}\frac{C_{dw}}{1 - s}$$

$$(12.108)$$

The factors $1/p'_m$ and $1/(1 - s)$ are used in Eq. (12.108) to convert the net water flux $(E_b - D_b)p'_m/(1 - p'_m)$ to the volumes of the corresponding water and sediment mixtures in the sediment bed and water column, respectively, in which C_{db} and C_{dw} are defined. The parameter p'_m is cancelled in the first term on the right-hand of Eq. (12.108).

Note that the porosity p'_m of the bed surface layer in Eq. (12.108) is assumed to have the same value in cases of erosion and deposition. However, different values might be used, because the newly deposited material is unconsolidated and the eroded material may be partially- or fully-consolidated in the case of cohesive sediment.

In addition, the expulsion of pore water due to consolidation of cohesive bed material, infiltration, and seepage flow also results in the dissolved contaminant flux between the water column and the sediment bed. These effects may be included in the loading terms q_{dw} and q_{sw}, as discussed above.

12.3.2.2 Equilibrium partition model

In cases where the time scale of sorption and desorption processes is much faster than that of flow and sediment transport, the sorption and desorption may be assumed to reach the equilibrium state instantaneously. The fractions of the dissolved and sorbed contaminants can then be determined using Eq. (12.98) or (12.101). This is called the equilibrium partition model. Because Eq. (12.98) or (12.101) provides an algebraic relation between C_{dw} and C_{sw}, it is only necessary to compute the total concentration of contaminant, which is determined in the 1-D and depth-averaged 2-D models by

$$\frac{DC_{tw}}{Dt} = \frac{J_{d,aw}}{h} + q_{tw} - k_{tw}C_{tw} + \frac{k_{dbw}}{h}(f_{db}C_{tb} - f_{dw}C_{tw}) + \frac{q_{t,ex}}{h}$$

$$(12.109)$$

where C_{tw} is the total concentration of contaminant in the water column, k_{tw} is the total contaminant decay coefficient, q_{tw} is the total loading rate of contaminant per unit volume, f_{dw} is the fraction of the dissolved contaminant over the total contaminant in the water column, f_{db} is the fraction of the dissolved contaminant over the total contaminant in the bed surface layer, and $q_{t,ex}$ is the total exchange rate of contaminant due to sediment erosion and deposition.

Eq. (12.109) can be derived by summing Eqs. (12.102) and (12.103). Similarly, the 3-D or width-averaged 2-D equation for the total concentration of contaminant can be derived by summing Eqs. (12.104) and (12.105) as

$$\frac{DC_{tw}}{Dt} = q_{tw} - k_{tw}C_{tw} - \frac{\partial(\omega_s f_{sw} C_{tw})}{\partial z} \qquad (12.110)$$

where f_{sw} is the fraction of the sorbed contaminant over the total contaminant in the water column. Thus, the following relations exist:

$$q_{tw} = q_{dw} + q_{sw} \qquad (12.111)$$

$$k_{tw} = f_{dw}k_{dw} + f_{sw}k_{sw} \qquad (12.112)$$

and

$$\begin{aligned} q_{t,ex} &= q_{d,ex} + q_{s,ex} \\ &= \max\left(E_b - D_b, 0\right)\frac{C_{tb}}{1 - p'_m} \\ &\quad + \min\left(E_b - D_b, 0\right)\left(\frac{p'_m}{1 - p'_m}\frac{f_{dw}}{1 - s} + \frac{f_{sw}}{s}\right)C_{tw} \qquad (12.113) \end{aligned}$$

The equilibrium partition model has been used in many studies (Thomann *et al.*, 1991; Schrestha and Orlob, 1996; Ji *et al.*, 2002), because it is simpler than the non-equilibrium one.

12.3.3 Contaminant transport in sediment bed

In the sediment bed, contaminants are transported by the pore water flow. In general, the pore water flow and contaminant transport can be simulated using a subsurface flow model. It is of great interest to couple the surface and subsurface flow models to investigate the flux exchanges between the water column and the sediment bed. However, because the pore water velocity in the sediment bed is much slower than the flow velocity in the water column, the transport of contaminants in the sediment bed is often simulated using the multiple-layer flux model described below, which is generalized from those of Lang and Chapra (1982) and DiToro (2001).

The sediment bed is divided into several layers. Definition of the layers in sediment and contaminant transport models may be different. However, for convenience, the definition in the sediment model, which is shown in Fig. 2.9, is used directly in the contaminant model. In the first layer, which is the mixing layer in the sediment model and the aerobic layer in the contaminant model, the variation in concentration of the dissolved contaminant is attributed to external loading, adsorption, desorption, decay, diffusional transfer at the bed surface, exchange due to sediment erosion and deposition, diffusional transfer at the interface between bed layers 1 and 2, and exchange due to burial and digging (rising and lowering) at the interface between bed layers 1 and 2. The variation in concentration of the sorbed contaminant is attributed to these effects, except for the diffusional transfers at the bed surface and between bed layers 1 and 2. The dissolved and sorbed contaminant balances in bed layer 1 are described by

$$\frac{\partial(\delta_1 C_{db,1})}{\partial t} = Q_{db,1} - k_{ad,b1} r_{sw,b1} \delta_1 C_{db,1} + k_{de,b1} \delta_1 C_{sb,1} - k_{db,1} \delta_1 C_{db,1}$$
$$- k_{dbw}(C_{db,1} - C_{dw}) - q_{d,ex} + k_{db12}(C_{db,2} - C_{db,1}) + q_{db12}$$

$$(12.114)$$

$$\frac{\partial(\delta_1 C_{sb,1})}{\partial t} = Q_{sb,1} + k_{ad,b1} r_{sw,b1} \delta_1 C_{db,1} - k_{de,b1} \delta_1 C_{sb,1}$$
$$- k_{sb,1} \delta_1 C_{sb,1} - q_{s,ex} + q_{sb12}$$

$$(12.115)$$

where $C_{db,i}$ and $C_{sb,i}$ are the concentrations of the dissolved and sorbed contaminants in bed layer i ($i = 1, 2, \ldots$), respectively; $Q_{db,i}$ and $Q_{sb,i}$ are the loading rates of the dissolved and sorbed contaminants in bed layer i, respectively; $k_{ad,bi}$ and $k_{de,bi}$ are the adsorption and desorption rate coefficients in bed layer i, respectively; $r_{sw,bi}$ is the sediment-to-water ratio in bed layer i; $k_{db,i}$ and $k_{sb,i}$ are the decay coefficients of the dissolved and sorbed contaminants in bed layer i, respectively; k_{db12} is the diffusional transfer coefficient of the dissolved contaminant between layers 1 and 2; q_{db12} and q_{sb12} are the exchange rates of the dissolved and sorbed contaminants between bed layers 1 and 2 due to interface lowering or rising; and δ_i is the thickness of bed layer i.

The exchange rates q_{db12} and q_{sb12} between bed layers 1 and 2 due to interface lowering and rising are determined by

$$q_{db12} = -\left(\frac{\partial z_b}{\partial t} - \frac{\partial \delta_1}{\partial t}\right) C_{db,x}, \quad q_{sb12} = -\left(\frac{\partial z_b}{\partial t} - \frac{\partial \delta_1}{\partial t}\right) C_{sb,x} \quad (12.116)$$

where $C_{db,x} = C_{db,1}$ and $C_{sb,x} = C_{sb,1}$, when $\partial z_b/\partial t - \partial \delta_1/\partial t \geq 0$, i.e., the interface between layers 1 and 2 rises; $C_{db,x} = C_{db,2}$ and $C_{sb,x} = C_{sb,2}$, when $\partial z_b/\partial t - \partial \delta_1/\partial t < 0$, i.e., the interface between layers 1 and 2 lowers.

The dissolved and sorbed contaminant balances in bed layer 2 are described by

$$\frac{\partial(\delta_2 C_{db,2})}{\partial t} = Q_{db,2} - k_{ad,b2} r_{sw,b2} \delta_2 C_{db,2} + k_{de,b2} \delta_2 C_{sb,2} - k_{db,2} \delta_2 C_{db,2}$$
$$- k_{db12}(C_{db,2} - C_{db,1}) - q_{db12} + k_{db23}(C_{db,3} - C_{db,2}) + q_{db23}$$

$$(12.117)$$

$$\frac{\partial(\delta_2 C_{sb,2})}{\partial t} = Q_{sb,2} + k_{ad,b2} r_{sw,b2} \delta_2 C_{db,2} - k_{de,b2} \delta_2 C_{sb,2}$$
$$- k_{sb,2} \delta_2 C_{sb,2} - q_{sb12} + q_{sb23}$$

$$(12.118)$$

where k_{db23} is the diffusional transfer coefficient of the dissolved contaminant between bed layers 2 and 3; and q_{db23} and q_{sb23} are the exchange rates of the dissolved and sorbed contaminants between bed layers 2 and 3 due to interface lowering or rising.

The contaminant mass balance equations for the underlying layers can be derived similarly, which are not presented here.

Note that Eqs. (12.114), (12.115), (12.117), and (12.118) are the governing equations of the non-equilibrium partition model in the sediment bed. In analogy to the equilibrium partition model in the water column described in Section 12.3.2.2, the equilibrium partition model in the sediment bed can be established for the cases where the sorption and desorption processes proceed much faster than other processes. Thus, the fractions of the dissolved and sorbed contaminants in the sediment bed are determined using Eq. (12.98) or (12.101), and the equations for the total concentration of contaminant in bed layers 1 and 2 are obtained by summing Eqs. (12.114) and (12.115) and summing Eqs. (12.117) and (12.118):

$$\frac{\partial (\delta_1 C_{tb,1})}{\partial t} = Q_{tb,1} - k_{tb,1}\delta_1 C_{tb,1} - k_{dbw}(f_{db,1}C_{tb,1} - f_{dw}C_{tw}) - q_{t,ex}$$
$$+ k_{db12}(f_{db,2}C_{tb,2} - f_{db,1}C_{tb,1}) + q_{tb12} \tag{12.119}$$

$$\frac{\partial (\delta_2 C_{tb,2})}{\partial t} = Q_{tb,2} - k_{tb,2}\delta_2 C_{tb,2} - k_{db12}(f_{db,2}C_{tb,2} - f_{db,1}C_{tb,1}) - q_{tb12}$$
$$+ k_{db23}(f_{db,3}C_{tb,3} - f_{db,2}C_{tb,2}) + q_{tb23} \tag{12.120}$$

where $C_{tb,i}$ is the total concentration of contaminant in bed layer i; $Q_{tb,i}$ is the total contaminant loading rate in bed layer i; $k_{tb,i}$ is the total contaminant decay coefficient in bed layer i; q_{tb12} and q_{tb23} are the total exchange rates of contaminant between bed layers 1 and 2 and between layers 2 and 3, respectively, due to interface lowering or rising; and $f_{db,i}$ is the fraction of the dissolved contaminant in bed layer i. The following relations exist:

$$q_{tb12} = q_{db12} + q_{sb12} \tag{12.121}$$

$$Q_{tb,i} = Q_{db,i} + Q_{sb,i} \tag{12.122}$$

$$k_{tb,i} = f_{db,i}k_{db,i} + f_{sb,i}k_{sb,i} \tag{12.123}$$

where $f_{sb,i}$ is the fraction of the sorbed contaminant in bed layer i.

Note that the above multiple-layer flux model is simple, but its application should be limited to the situations where the effects of groundwater flow are negligible or can be lumped into the diffusion process between the bed layers.

References

Abbott, M.B. (1966) *An Introduction to the Method of Characteristics*, Thames and Hudson, London, and American Elsevier, New York.

Abe, K., Kondoh, T. and Nagano, Y. (1994) 'A new turbulence model for predicting fluid flow and heat transfer in separating and reattaching flows – I. Flow field calculations', *Int. J. Heat Mass Transfer*, 37(1), 139–151.

Ackers, P. and White, W.R. (1973) 'Sediment transport: A new approach and analysis', *J. Hydr. Div.*, ASCE, 99(HY11), 2041–2060.

Ahrens, J.P. (2000) 'A fall-velocity equation', *J. Waterway, Port, Coastal, and Ocean Eng.*, ASCE, 126(2), 99–102.

Alam, M.S. and Kennedy, J.F. (1969) 'Friction factors for flow in sand-bed channels', *J. Hydr. Div.*, ASCE, 95(6), 1973–1992.

Alcrudo, F. and Garcia-Navarro, P. (1993) 'A high-resolution Godunov-type scheme in finite volumes for the 2D shallow-water equations', *Int. J. Numer. Methods Fluids*, 16, 489–505.

Alonso, C.V. (1980) 'Selecting a formula to estimate sediment transport capacity in non-vegetated channels', *CREAMS* (A Field Scale Model for Chemicals, Runoff, and Erosion from Agricultural Management System), W.G. Knisel, ed., USDA, Conservation Research Report, No. 26, Chapter 5, pp. 426–439.

Ariathurai, R. and Mehta, A.J. (1983) 'Fine sediments in waterway and harbor shoaling problems', *Proc. Int. Conf. on Coastal and Port Eng. in Developing Countries*, Colombo, Sri Lanka, 1094–1108.

Armanini, A. and di Silvio, G. (1986) Discussion on the paper 'A depth-integrated model for suspended sediment transport' by G. Galappatti and C.B. Vreugdenhil, *J. Hydr. Res.*, 24(5), 437–441.

Armanini, A. and di Silvio, G. (1988) 'A one-dimensional model for the transport of a sediment mixture in non-equilibrium conditions', *J. Hydr. Res.*, IAHR, 26(3), 275–292.

Arnold, J.G., Allen, P.M. and Bernhardt, G. (1993) 'A comprehensive surface-groundwater flow model', *J. Hydrology*, 142, 47–69.

Arulanandan, K., Gillogley, E. and Tully, R. (1980) 'Development of a quantitative method to predict critical shear stress and rate of erosion of naturally undisturbed cohesive soils', *Rep. GL-80-5*, U.S. Army Engineers Waterway Experiment Station, Vicksburg, MS, USA.

ASCE Task Committee (1988) 'Turbulence modeling of surface water and transport', *J. Hydraulic Eng.*, ASCE, 114(9), 970–1073.

Ashida, K. and Michiue, M. (1971) 'An investigation of river bed degradation downstream of a dam', *Proc. 14th Congress of IAHR*, Paris, France, 3, 247–256.

Ashida, K. and Michiue, M. (1972) 'Study on hydraulic resistance and bed-load transport rate in alluvial stream', *Proc. JSCE*, No. 206, 59–69 (in Japanese).

Bagnold, R.A. (1966) 'An approach to the sediment transport problem from general physics', *Professional Paper 422-J*, USGS, Washington D.C., USA.

Bagnold, R.A. (1973) 'The nature of saltation and of bed load transport in water', *Proc. Royal Society*, Ser. A, 332, 473–504.

Banks, R.B. and Herrera, E.F. (1977) 'Effect of wind and rain on surface reaeration', *J. Env. Eng. Div.*, ASCE, 103(EE3), 489–504.

Barfield, B.J., Tollner, E.W. and Hayes, J.C. (1979) 'Filtration of sediment by simulated vegetation, I. Steady-state flow with homogeneous sediment', *Trans.*, ASAE, 540–556.

Barfield, B.J., Fogle, A., Froelich, D. and Rohlf, R. (1991) 'Deterministic models of channel headwall erosion: Initiation and propagation', University of Kentucky, Lexington, USA.

Barnes, H.H.Jr. (1967) 'Roughness characteristics of natural channels', *USGS Water Supply Paper 1849*, Washington D.C., USA.

Bayazit, N. (1975) 'Simulation of armor coat formation and destruction', *Proc. XVIth IAHR Congress*, Sao Paulo, Brazil, 73–80.

Bell, R.G. (1980) 'Non-equilibrium bedload transport by steady and non-steady flows', *Ph.D. Dissertation*, University of Canterbury, Christchurch, New Zealand.

Bell, R.G. and Sutherland, A.J. (1983) 'Non-equilibrium bed load transport by steady flows', *J. Hydraulic Eng.*, ASCE, 109(3), 353–367.

Bennett, J.P. and Nording, C.F. (1977) 'Simulation of sediment transport and armouring', *Hydrolog. Sci. Bull.*, XXII(4), 555–569.

Bennett, S.J., Alonso, C.V., Prasad, S.N. and Romkens, M.J.M. (1997) 'Dynamics of headcuts in upland concentration flows', *Proc. Conf. on Management of Landscapes Disturbed by Channel Incision*, Mississippi, USA, pp. 510–515.

Bennett, S.J., Pirim, T. and Barkdoll, B.D. (2002) 'Using simulated emergent vegetation to alter stream flow direction within a straight experimental channel', *Geomorphology*, 44, 115–126.

Benson, B.B. and Krause, D.Jr. (1984) 'The concentration and isotopic fractionation of oxygen dissolved in freshwater and seawater in equilibrium with the atmosphere', *Limnology and Oceanography*, 29(3), 620–632.

Bingner, R.L., Alonso, C.V., Arnold, J.G. and Garbrecht, J. (1997) 'Simulation of fine sediment yield within a DEC watershed', *Proc. Conf. on Management of Landscapes Disturbed by Channel Incision*, Mississippi, USA, pp. 1106–1110.

Bingner, R.L. and Theurer, F.D. (2001) 'AGNPS 98: A suite of water quality models for watershed use', *Proc. 7th Federal Interagency Sedimentation Conference*, Reno, NV, USA.

Blanc, T.V. (1985) 'Variation of bulk-derived surface flux, stability, and roughness results due to the use of different transfer coefficient schemes', *J. Phys. Oceanogr.*, 15, 650–669.

Blumberg, A.F. (1977) 'Numerical model of estuarine circulation', *J. Hydr. Div.*, ASCE, 103(3), 295–310.

Blumberg, A.F. and Mellor, G.L. (1987) 'A description of a three-dimensional coastal ocean circulation model', *Three-Dimensional Coastal Ocean Models*, N.S. Heaps (ed.), American Geophysical Union, Washington D.C., USA.

Borah, D.K, Alonso, C.V. and Prasad, S.N. (1982) 'Routing graded sediments in streams: Formulations', *J. Hydr. Div.*, ASCE, 108(12), 1486–1503.

Bowie, G.L. (1985) 'Rates, constants and kinetics formulations in surface water quality modeling', Second Edition, *EPA/600/3-85/040*, Environmental Protection Agency, Athens, GA, USA.

Bridge, J.S. and Dominic, D.F. (1984) 'Bed load grain velocities and sediment transport rates', *Water Resources Res.*, 20(4), 476–490.

Bridge, J.S. and Bennett, S.J. (1992) 'A model for the entrainment and transport of sediment grains of mixed sizes, shapes and densities', *Water Resources Res.*, 28(2), 337–363.

Brooks, N.H. (1963), Discussion of 'boundary shear stresses in curved trapezoidal channels' by A.T. Ippen and P.A. Drinker, *J. Hydr. Div.*, ASCE, 89(HY3), 327–333.

Brown, L.C. and Barnwell, T.O. (1987) 'The enhanced stream water quality models QUAL2E and QUAL2E-UNCAS: documentation and user manual', *EPA/600/3-87/007*, Environmental Protection Agency, Athens, GA, USA.

Brownlie, W.R. (1981) 'Prediction of flow depth and sediment discharge in open channels' and 'Compilation of fluvial channel data: laboratory and field', *Rep. No. KH-R-43A & B*, W.M. Keck Lab. of Hydr. and Water Resources, California Institute of Technology, Pasadena, California, USA.

Brownlie, W.R. (1983) 'Flow depth in sand-bed channels', *J. Hydraulic Eng.*, ASCE, 109(7), 959–990.

Brush, L.M., Ho, H.W. and Singamsetti, S.R. (1962) 'A study of sediment in suspension', *Pub. No. 59*, Inter. Assoc. Sci. Hydraul., Commiss. Land Erosion (Bari).

Burban, P.Y., Xu, Y.J., McNeil, J. and Lick, W. (1990) 'Settling speeds of flocs in fresh water and seawater', *J. Geophys. Res.*, AGU, 95(C10), 18.213–18.220.

Busch, N.E. (1972) 'On the mechanics of atmospheric turbulence', *Workshop on Micrometeorology*, Am. Met. Soc., pp. 1–65.

Camp, T.R. and Stein, P.C. (1943) 'Velocity gradients and internal work in fluid motion', *J. Boston Society of Civil Engineers*, 30(4), 219–237.

Cao, Z. (1999) 'Equilibrium near-bed concentration of suspended sediment', *J. Hydraulic Eng.*, ASCE, 125(12), 1270–1278.

Cao, Z., Day, R. and Egashira, S. (2002) 'Coupled and decoupled numerical modeling of flow and morphological evolution in alluvial rivers', *J. Hydraulic Eng.*, ASCE, 128(3), 306–321.

Cao, Z., Pender, G., Wallis, S. and Carling, P. (2004) 'Computational dam-break hydraulics over erodible sediment bed', *J. Hydraulic Eng.*, ASCE, 130(7), 689–703.

Capart, H. and Young, D.L. (1998) 'Formation of jump by the dam-break wave over a granular bed', *J. Fluid Mech.*, 372, 165–187.

Casulli, V. (1990) 'Numerical simulation of shallow water flow', *Computational Methods in Surface Hydrology*, C.A. Brebbia, W.G. Gray and G.W. Pinder (eds.), Springer-Verlag, Berlin, 13–22.

Casulli, V. and Cheng, R.T. (1992) 'Semi-implicit finite difference methods for three-dimensional shallow water flow', *Int. J. Numer. Methods in Fluids*, 15, 629–648.

Cebeci, T. and Bradshaw, P. (1977) *Momentum Transfer in Boundary Layers*, Hemisphere, Washington D.C.

Celik, I. and Rodi, W. (1988) 'Modelling suspended sediment transport in non-equilibrium situations', *J. Hydraulic Eng.*, ASCE, 114, 1157–1191.

Cerco, C.F. and Cole, T. (1994) 'Three-dimensional eutrophication model of Chesapeake Bay', *Tech. Report EL-94-4*, Vol. 1, Main Report, U.S. Army Corps of Engineers, Waterways Experiment Station, Vicksburg, MS, USA.

Chang, H.H. (1988) *Fluvial Processes in River Engineering*, Krieger Publishing Company.

Chao, X., Jia, Y., Cooper, C.M., Shields, F.D.Jr. and Wang, S.S.Y. (2006) 'Development and application of a phosphorus model for a shallow oxbow lake', *J. Envirn. Eng.*, ASCE, 132(11), 1498–1507.

Chapra, S.C. (1997) *Surface Water Quality Modeling*, WCB/McGraw Hill, New York.

Chen, C.J. and Li, P. (1980) 'Finite analytic numerical methods for steady and unsteady heat transfer problems', *ASME Paper No. 80-HT-86, 19th ASME-AICHE National Heat Transfer Conf.*, Orlando, USA, pp. 1–10.

Chen, Y., Wai, O.W.H., Li, Y.S. and Lu, Q. (1999) 'Three-dimensional numerical modeling of cohesive sediment transport by tidal current in Pearl River estuary', *Int. J. Sediment Res.*, 14(2), 107–123.

Chen, Y.-S. and Kim, S.-M. (1987) 'Computation of turbulent flows using an extended k-ε turbulence closure model', *CR-179204*, NASA, USA, p. 21.

Cheng, N.S. (1997) 'Simplified settling velocity formula for sediment particle', *J. Hydraulic Eng.*, ASCE, 123(2), 149–152.

Cheng, N.S. (2004) 'Analysis of velocity lag in sediment-laden open channels flows', *J. Hydraulic Eng.*, ASCE, 130(7), 657–666.

Chien, K.Y. (1982) 'Prediction of channel and boundary layer flows with a low-Reynolds-number turbulence model', *AIAA J.*, 20(1), 33–38.

Chien, N. (1980) 'Comparison of bed-load formulas', *Chinese J. Hydraulic Eng.*, Chinese Assoc. of Hydr. Eng., No. 4, 1–11.

Chien, N. and Wan, Z.H. (1983) *Mechanics of Sediment Movement*, Science Press, Beijing, China (in Chinese).

Chinnarasri, C., Tingsanchali, T., Weesakul, S. and Wongwises, S. (2003) 'Flow patterns and damage of dike overtopping', *Int. J. Sediment Res.*, 18(4), 301–309.

Chorin, A.J. (1968) 'Numerical solution of the Navier-Stokes equations', *Math. Comp.*, 22, 745–762.

Chow, V.T. (1959) *Open-Channel Hydraulics*, McGraw-Hill, New York.

Chung, T.J. (1978) *Finite Element Analysis in Fluid Dynamics*, McGraw-Hill, New York.

Colby, B.R. and Hembree, C.H. (1955) 'Computation of total sediment discharge, Niobrara River near Cody, Nebraska', USGS Water Supply, Paper 1357.

Colby, B.R. (1963) Discussion of 'Sediment transportation mechanics: Introduction and properties of sediment', Progress Report by the Task Committee on Preparation of Sediment Manual of the Committee on Sedimentation of the Hydraulics Division, V.A. Vanoni, Chmn., *J. Hydr. Div.*, ASCE, 89(1), 266–268.

Corey, A.T. (1949) 'Influence of shape on the fall velocity of sand grains', *Master's Thesis*, Colorado A & M College, Colorado, USA.

Correia, L., Krishnappan, B.G. and Graf, W.H. (1992) 'Fully coupled unsteady mobile boundary flow model', *J. Hydraulic Eng.*, ASCE, 118(3), 476–494.

Courant, R., Issacson, E. and Rees, M. (1952) 'On the solution of non-linear hyperbolic differential equations by finite differences', *Comm. Pure Appl. Math.*, 5, p. 243.

Crowley, W.P. (1968) 'A global numerical ocean model: Part I', *J. Comp. Phys.*, 3, 111–147.

Cunge, J.A. and Perdreau, N. (1973) 'Mobile bed fluvial mathematical models', *La Houille Blance*, Grenoble, France, 7, 562–580.

Cunge, J.A., Holly, F.M.Jr. and Verwey, A. (1980) *Practical Aspects of Computational River Hydraulics*, Pitman Publishing Inc., Boston, MA.

Damgaard, J.S., Whitehouse, R.J.S. and Soulsby, R.L. (1997) 'Bed-load sediment transport on steep longitudinal slopes', *J. Hydraulic Eng.*, ASCE, 123(12), 1130–1138.

Darby, S.E. (1999) 'Effect of riparian vegetation on flow resistance and flood potential', *J. Hydraulic Eng.*, ASCE, 125(5), 443–454.

Daubert, A. and Lebreton, J.C. (1967) 'Etude experimentale et sur modele mathematique de quelques aspects des processus d'erosion des lits alluvionnaires, en regime permanent et non permanent', *Proc. 12th IAHR Congress*, 3, Fort Collins, USA.

Day, T.J. (1980) 'A study of the transport of graded sediments', *Report N. IT 190*, HR Wallingford, U.K.

Deas, M.L. and Lowney, C.L. (2000) 'Water temperature modeling review', Central Valley, California, USA.

Dee, D.P., Toro, F.M. and Wang, S.S.Y. (1992) 'Numerical model verification by prescribed solution forcing – A test case', *Hydraulic Engineering, Proc. Hydraulic Engineering Sessions at Water Forum '92*, edited by M. Jennings and N.G. Bhowmik, ASCE, pp. 416–421.

Delis, A.I., Skeels, C.P. and Ryrie, S.C. (2000) 'Implicit high-resolution methods for modeling one-dimensional open channel flow', *J. Hydraulic Eng.*, ASCE, 38(5), 369–382.

De Ploey, J. (1989) 'A model for headcut retreat in rills and gullies', *CATENA*, Supplement 14, pp. 81–86, Cremlingen, Germany.

De Saint Venant, B. (1871) 'Théorie du mouvement non-permanent des eaux avec application aux crues des rivières et à l'introduction des marées dans leur lit', *Acad. Sci. Comptes rendus*, 73, 148–154, 237–240.

De Sutter, R., Huygens, M. and Verhoeven, R. (2001) 'Sediment transport experiments in unsteady flows', *Int. J. Sediment Res.*, 16(1), 19–35.

De Vriend, H.J. (1977) 'A mathematical model of steady flow in curved shallow channels', *J. Hydr. Res.*, 15(1), 37–54.

De Vriend, H.J. (1981a) 'Steady flow in shallow channel bends', *Communications on Hydraulics*, 81-3, Delft University of Technology, Delft, The Netherlands.

De Vriend, H.J. (1981b) 'Flow measurements in a curved rectangular channel, Part II: rough bottom', Laboratory of Fluid Mechanics, Delft University of Technology, The Netherlands.

De Vries, M. (1981) 'Morphological computations', *Lecture Notes, F10a.I*, Technische Hogeschool, Afdelingder Civele Techniek, Delft, The Netherlands.

Dietrich, W.E. (1982) 'Settling velocity of natural particles', *Water Resources Res.*, 18(6), 1615–1626.

Ding, J.S. and Qiu, F.L. (1981) 'Calculation of water and sediment diversion ratios in braided channels', *J. Sediment Res.* (in Chinese).

Dingman, S. (1984) *Fluvial Hydrology*, W. Freeman Co., New York.

DiToro, D.M. and Matystik, W.F. (1980) 'Mathematical models of water quality in large lakes, Part I: Lake Huron and Saginaw Bay', *Report EPA-600/3-80-056*, U.S. EPA Environmental Research Lab, Duluth, MN, USA.

DiToro, D.M. (2001) *Sediment Flux Modeling*, Wiley-Interscience.

Dixit, J.G. (1982) 'Resuspension potential of deposited kaolinite beds', *M.S. Thesis*, University of Florida, Gainesville, Florida, USA.

Dixit, J.G., Mehta, A.J. and Partheniades, E. (1982) 'Redepositional properties of cohesive sediments deposited in a long flume', *UFL/COEL-82/002*, Coastal and Oceanographic Engineering Department, University of Florida, Gainesville, Florida, USA.

Dou, G.R. (1960) 'On the incipient motion of sediment', *Chinese J. Hydraulic Eng.*, Chinese Assoc. of Hydr. Eng., No. 4, 44–60.

Dou, G.R. (1964) 'Bed-load transport', Nanjing Hydraulic Research Institute, China (in Chinese).

Dou, X. (1997) 'Numerical simulation of three-dimensional flow field and local scour at bridge crossings', *Ph.D. Dissertation*, University of Mississippi, USA.

Dresnack, R. and Dobbins, W.E. (1968) 'Numerical analysis of BOD and DO profiles', *J. San. Eng. Div.*, ASCE, 94(SA3), 789–807.

Duan, J.G., Wang, S.S.Y. and Jia, Y. (2001) 'The application of the enhanced CCHE2D model to study the alluvial channel migration processes', *J. Hydr. Res.*, IAHR, 39(5), 469–480.

Duboys, M.P. (1879) 'Études du Regime et l' Action Excercée par les Eaux sur un Lit à Fond de Graviers Indefiniment Affouilable', *Annales de Ponts et Chaussés*, Ser. 5, 18, 141–195.

Edinger, J.E., Brady, D.K. and Greyer, J.C. (1974) 'Heat exchange and transport in the environment', *Rep. No. 14, Cooling Water Res. Project (RP-49)*, Electric Power Research Institute, Palo Alto, California, USA.

Edinger, J.E., Buchak, E.M. and Zhang, Z. (1994) 'Alternative formulation of laterally averaged hydrodynamics', *J. Hydraulic Eng.*, ASCE, 120(7), 886–890.

Egiazaroff, I.V. (1965) 'Calculation of nonuniform sediment concentration', *J. Hydr. Div.*, ASCE, 91(HY4), 225–247.

Einstein, H.A. (1942) 'Formulas for the transportation of bed load', *Trans.*, ASCE, 107, 561–573.

Einstein, H.A. (1950) 'The bed-load function for sediment transportation in open channel flows', *Technical Bulletin No. 1026*, U.S. Department of Agriculture, Soil Conservation Service, Washington D.C., USA.

Einstein, H.A. and Barbarossa, N.L. (1952) 'River channel roughness', *Trans.*, ASCE, 117, 1121–1132.

Einstein, H.A. and Chien, N. (1954) 'Second approximation to the solution of the suspended-load theory', *MRD Series Report No. 3*, Univ. of California and Missouri River Division, U.S. Corps of Engineers, Omaha, Nebr., USA.

Einstein, H.A. and Chien, N. (1955) 'Effects of heavy sediment concentration near the bed on velocity and sediment distribution', *MRD Series Report No. 8*, Univ. of California and Missouri River Division, U.S. Corps of Engineers, Omaha, Nebr., USA.

Elder, J.W. (1959) 'The dispersion of marked fluid in turbulent shear flow', *J. Fluid Mechanics*, 5(4), 544–560.

Engelund, F. (1966 & 67) 'Hydraulic resistance of alluvial streams', *Proc.* ASCE, 92(HY2) & 93(HY4).

Engelund, F. and Hansen, E. (1967) *A Monograph on Sediment Transport in Alluvial Streams*, Teknisk Vorlag, Copenhagen, Denmark.

Engelund, F. (1974) 'Flow bed topography in channel bend', *J. Hydr. Div.*, ASCE, 100(11), 1631–1648.

Engelund, F. and Fredsøe, J. (1976) 'A sediment transport model for straight alluvial channels', *Nordic Hydrology*, 7, 293–306.

Ettema, R. (1980) 'Scour at bridge piers', University of Auckland, New Zealand.

Fang, H.W. and Rodi, W. (2000) 'Three dimensional calculations of flow and suspended sediment transport in the neighborhood of the dam for the Three Gorges Project (TGP) reservoir in the Yangtze River', *Report 762*, Institute for Hydromechanics, Karlsrhe University, Germany; *J. Hydr. Res.*, IAHR, 2003, 41(4), 379–394.

Fang, H.W. and Wang, G.Q. (2000) 'Three-dimensional mathematical model of suspended-sediment transport', *J. Hydraulic Eng.*, ASCE, 126(8), 578–592.

Fasken, G.B. (1963) 'Guide for selecting roughness coefficient n values for channels', Soil Conservation Service, Department of Agriculture, USA.

Fathi-Maghadam, M. (1996) 'Momentum absorption in non-rigid, non-submerged, tall vegetation along rivers', *Ph.D. Dissertation*, University of Waterloo, Canada.

Fennema, R.J. and Chaudhry, M.H. (1990) 'Explicit methods for 2-D transient free-surface flows', *J. Hydraulic Eng.*, ASCE, 116(8), 1013–1034.

Ferguson, R.I. (1983) 'Kinematic model of meander migration', *River Meandering*, New Orleans, Louisiana, USA.

Ferreira, R. and Leal, J. (1998) '1D mathematical modeling of the instantaneous dam-break flood wave over mobile bed: application of TVD and flux-splitting schemes', *Proc. European Concerted Action on Dam-Break Modeling*, Munich, Germany, 175–222.

Ferziger, J.H. and Peric, M. (1995) *Computational Methods for Fluid Dynamics*, Springer-Verlag.

FHWA (1995) 'Evaluating scour at bridges', *HEC-18*, Federal Highway Administration, Third Edition, U.S. Department of Transportation.

Fischer, H.B., List, E.J., Koh, R.C.Y., Imberger, J. and Brooks, N.H. (1979) *Mixing in Inland and Coastal Waters*, Academic Press, New York.

Fletcher, C.A.J. (1991) *Computational Techniques for Fluid Dynamics*, Vols. 1 and 2, Springer-Verlag.

Flokstra, C. (1977) 'Generation of two-dimensional horizontal secondary current', *Report S163, Part II*, Delft Hydraulics Laboratory, The Netherlands.

Fraccarollo, L. and Armanini, A. (1998) 'A semi-analytical solution for the dam-break problem over a movable bed', *Proc. European Concerted Action on Dam-Break Modeling*, Munich, Germany, 145–152.

Fraccarollo, L. and Capart, H. (2002) 'Riemann wave description of erosional dam-break flows', *J. Fluid Mech.*, 461, 183–228.

Francis, J.R.D. (1973) 'Experiments on the motion of solitary grains along the bed of water stream', *Proc. R. Soc. London, Ser. A*, 332(1591), 443–471.

Freeman, G.E., Rahmeyer, W.J. and Copeland, R.R. (2000) 'Determination of resistance due to shrubs and woody vegetation', *ERDC/CHL TR-00-25*, Coastal and Hydraulics Laboratory, Engineer Research and Development Center, US Army Corps of Engineers.

French, R.H. (1985) *Open-Channel Hydraulics*, McGraw-Hill Book Company, New York.

Furumai, H. and Ohgaki, S. (1989) 'Adsorption-desorption of phosphorus by lake sediments under anaerobic conditions', *Water Research*, 23(6), 677–683.

Gailani, J., Ziegler, C.K. and Lick, W. (1991) 'Transport of suspended solids in the Lower Fox River', *J. Great Lakes Res.*, 17(4), 479–494.

Galappatti, G. and Vreugdenhil, C.B. (1985) 'A depth-integrated model for suspended sediment transport', *J. Hydr. Res.*, IAHR, 23(4), 359–377.

Garbrecht, J., Kuhnle, R.A. and Alonso, C.V. (1995) 'A sediment transport formulation for large channel networks', *J. Soil and Water Conservation*, 50(5), 527–529.

Garbrecht, J. and Martz, L.W. (1995) 'An automated digital landscape analysis tool for topographic evaluation, drainage identification, watershed segmentation and subcatchment parameterization', *Report No. NAWQL 95-1*, National Agricultural Water Quality Laboratory, USDA, Agricultural Research Service, Durant, Oklahoma, USA.

Garcia, M. and Parker, G. (1991) 'Entrainment of bed sediment into suspension', *J. Hydraulic Eng.*, ASCE, 117(4), 414–433.

Garcia-Navarro, P., Alcrudo, F. and Saviron, J.M. (1992) '1-D open-channel flow simulation using TVD-McCormack scheme', *J. Hydraulic Eng.*, ASCE, 118(10), 1359–1372.

Gessler, J. (1971) 'Aggradation and degradation', *River Mechanics*, H.W. Shen ed., Vol. 1, Fort Collins, Colorado, USA.

Ghanem, A.H. (1995) 'Two-dimensional finite element modeling of flow in aquatic habitats', *Ph.D. Thesis*, University of Alberta, Alberta, Canada.

Gibson, R.E., England, G.L. and Hussey, M.J. (1967) 'The theory of one-dimensional consolidation of saturated clays', *Geotechnique*, 17, 216–263.

Gill, A.E. (1982), *Atmosphere-Ocean Dynamics*, Academic Press.

Glaister, P. (1988) 'Approximate Riemann solutions of the shallow water equations', *J. Hydr. Res.*, IAHR, 26(3), 293–306.

Godunov, S.K. (1959) 'Finite difference methods for the computation of discontinuous solutions of the equations of fluid dynamics', *Mathematicheskii Sbornik*, 47(3), 271–306.

Goldstein, S. (1929) 'The steady flow of viscous fluid past a fixed spherical obstacle at small Reynolds numbers', *Proc. Roy. Soc. London*, 123A, 225–235.

Golub, G. and Ortega, J.M. (1993) *Scientific Computing, An Introduction with Parallel Computing*, Academic Press.

Graf, W.H. (1971) *Hydraulics of Sediment Transport*, McGraw-Hill, New York.

Graf, W.H. and Altinakar, M.S. (1998) *Fluvial Hydraulics*, John Wiley & Sons Ltd.

Greimann, B.P. and Holly, F.M.Jr. (2001) 'Two-phase flow analysis of concentration profiles', *J. Hydraulic Eng.*, ASCE, 127(9), 753–762.

Guo, J. (2002) 'Logarithmic matching and its application in computational hydraulics and sediment transport', *J. Hydr. Res.*, IAHR, 40(5), 555–565.

Guy, H.P., Simons, D.B. and Richardson, E.V. (1966) 'Summary of alluvial channel data from flume experiments, 1956–61', *USGS Professional Paper 426-1*, U. S. Department of the Interior, Washington D.C., USA.

Hageman, L.A. and Young, D.M. (1981) *Applied Iterative Methods*, Wiley, New York.

Hagerty, D.J. (1991) 'Piping/sapping erosion I: Basic consideration', *J. Hydraulic Eng.*, ASCE, 117, 991–1008.

Hallermeier, R.J. (1981) 'Terminal settling velocity of commonly occurring sand grains', *Sedimentology*, 28(6), 859–865.

Hamm, L. and Migniot, C. (1994) 'Elements of cohesive sediment deposition, consolidation, and erosion', *Coastal, Estuarial, and Harbor Engineer's Reference Book*, M.B. Abbott and W.A. Price (eds.), Chapman and Hall, London, 93–106.

Han, Q. (1980) 'A study on the non-equilibrium transportation of suspended load', *Proc. First Int. Symp. on River Sedimentation*, Beijing, China.

Han, Q.W., Wang, Y.C. and Xiang, X.L. (1981) 'Initial dry density of sediment deposit', *J. Sediment Res.*, No. 1 (in Chinese).

Han, Q.W. and He, M.M. (1984) *Stochastic Theory of Sediment Movement*, Science Press, Beijing, China (in Chinese).

Han, Q.W., He, M.M. and Sun, W.D. (1986) 'The calculation of suspended-load deposition in Three Gorges Reservoir (150 m scheme)', *Compilation of Research Reports on the Sedimentation Problems of Three Gorges Project (150m Scheme)*, Beijing, China (in Chinese).

Hanson, G.J., Robinson, K.M. and Cook, K.R. (1997) 'Headcut migration analysis of a compacted soil', *Trans.*, ASAE, 40(2), 335–361.

Haralampides, K., McCorquodale, J.A. and Krishnappan, B.G. (2003) 'Deposition properties of fine sediment', *J. Hydraulic Eng.*, ASCE, 129(3), 230–234.

Harlow, F.H. and Welch, J.E. (1965) 'Numerical calculation of time dependent viscous incompressible flow with free surface', *Phys. Fluids*, 8, 2182–2189.

Harten, A. (1983) 'A high resolution scheme for the computation of weak solutions of hyperbolic conservation laws', *J. Comput. Phys.*, 49, 357–393.

Harten, A., Lax, P.D. and van Leer, B. (1983) 'On upstream differencing and Godunov-type schemes for hyperbolic conservation laws', *SIAM Review*, 25(1), 35–61.

Hayashi, T.S., Ozaki and Ichibashi, T. (1980) 'Study on bed load transport of sediment mixture', *Proc. 24th Japanese Conf. on Hydraulics*, Japan.

Hayter, E.J. (1983) 'Prediction of cohesive sediment movement in estuarial waters', *Ph.D. Dissertation*, University of Florida, Gainesville, Florida, USA.

HEC (1997) 'HEC-RAS, river analysis system – hydraulic reference manual', Hydrologic Engineering Center, Army Corps of Engineers, USA.

Hey, R.D. (1979) 'Flow resistance in gravel-bed rivers', *J. Hydr. Div.*, ASCE, 105(4), 365–379.

Hicks, D.M. and Mason, P.D. (1991) 'Roughness characteristics of New Zealand rivers', *Water Resources Survey*, DSIR Marine and Freshwater, New Zealand.

Hinatsu, M. (1992) 'Numerical simulation of unsteady viscous nonlinear waves using moving grid system fitted on a free surface', *J. Kansai Soc. N. A.*, Japan, No. 217, 1–11.

Hirano, M. (1971) 'River bed degradation with armoring', *Trans.*, JSCE, 3(2), 55–65.

Hirsch, C. (1988) *Numerical Computation of Internal and External Flows*, Vols. 1 & 2, Wiley.

Hirt, C.W. and Nichols, B.D. (1981) 'Volume of fluid (VOF) method for dynamics of free boundaries', *J. Computational Physics*, 39, 201–221.

Hodges, B. (1998) 'Heat budget and thermodynamics at a free surface', Center for Water Research, The University of Western Australia, Australia, p. 14.

Holly, F.M.Jr. and Usseglio-Polatera, J.-M. (1984) 'Dispersion simulation in two-dimensional tidal flow', *J. Hydraulic Eng.*, ASCE, 110(7), 905–926.

Holly, F.M.Jr. and Rahuel, J.L. (1990) 'New numerical/physical framework for mobile-bed modeling, Part I: Numerical and physical principles', *J. Hydr. Res.*, IAHR, 28(4), 401–416.

Holly, F.M.Jr., Yang, J.C., Schwarz, J., Schaefer, J., Hsu, S.H. and Einhellig, R. (1990) 'CHARIMA: numerical simulation on unsteady water and sediment movement in multiply connected networks of movable-bed channels', *IIHR Rep. No. 343*, Iowa Institute of Hydraulic Research, University of Iowa, USA.

Howard, A.D. and Knutson, T.R. (1984) 'Sufficient conditions for river meandering: A simulation approach', *Water Resources Res.*, 20(11), 1659–1667.

Hsu, E.M. and Holly, F.M. (1992) 'Conceptual bed-load transport model and verification for sediment mixtures', *J. Hydraulic Eng.*, ASCE, 118(8), 1135–1152.

Hu, H.M. and Wang, K.H. (1999) 'Entrainment function of suspended sediment in open channels', *Int. J. Sediment Res.*, 14(3), 1–8.

Huang, J.W. (1981) 'Experimental study of settling properties of cohesive sediment in still water', *J. Sediment Res.*, No. 2, 30–41 (in Chinese).

Huber, W.C. (1993) 'Contaminant transport in surface water', *Handbook of Hydrology*, D.R. Maidment (eds.), McGraw-Hill, pp. 14.1–14.50.

Imberger, J. and Patterson, J.C. (1981) 'A dynamic reservoir simulation model – DYRESM: 5', *Transport Models for Inland and Coastal Waters*, H.B. Fischer, ed., Academic Press, pp. 310–361.

Issa, R.I. (1982) 'Solution of the implicitly discretised fluid flow equations by operator-splitting', *Fluids Section Report FS/82/15*, Mech. Eng. Dept., Imperial College, London.

Iwasa, Y. and Aya, S. (1991) 'Predicting longitudinal dispersion coefficients in open-channel flow', *Environmental Hydraulics*, J. Lee *et al.*, eds., Balkema, Rotterdam, The Netherlands.

Jacquet, J. (1983) 'Simulation of thermal regime of rivers', *Mathematical Modeling of Water Quality: Streams, Lakes and Reservoirs*, G.T. Orlob, ed., Wiley-Interscience, pp. 150–176.

Jankowski, J.A., Malcherek, A. and Zielke, W. (1994) 'Numerical modeling of sediment transport processes caused by deep sea mining discharge', *Proc. OCEANS 94 Conference*, Brest, IEEE/SEE, 3, 269–276.

Jarvela, J. (2002) 'Determination of flow resistance of vegetated channel banks and floodplains', *River Flow 2002*, D. Bousmar and Y. Zech (eds.), Swets & Zeitlinger, Lisse, 311–318.

Jarvela, J. (2004) 'Determination of flow resistance caused by non-submerged woody vegetation', *Int. J. River Basin Management*, 2(1), 61–70.

Jha, A.K., Akiyama, J. and Ura, M. (2000) 'Flux-difference splitting scheme for 2D flood flows', *J. Hydraulic Eng.*, ASCE, 126(1), 33–42.

Ji, Z.-G., Hamrick, J.H. and Pagenkopf, J. (2002) 'Sediment and metals modeling in shallow river', *J. Environ. Eng.*, ASCE, 128(2), 105–119.

Jia, Y. and Wang, S.S.Y. (1999) 'Numerical model for channel flow and morphological change studies', *J. Hydraulic Eng.*, ASCE, 125(9), 924–933.

Jia, Y., Kitamura, T. and Wang, S.S.Y. (2001) 'Simulation of scour process in plunging pool of loose bed-material', *J. Hydraulic Eng.*, ASCE, 127(3), 219–229.

Jia, Y., Wang, S.S.Y. and Xu, Y. (2002) 'Validation and application of a 2D model to channels with complex geometry', *Int. J. Comput. Eng. Sci.*, 3(1), 57–71.

Jimenez, J.A. and Madsen, O.S. (2003) 'A simple formula to estimate settling velocity of natural sediments', *J. Waterway, Port, Coastal, and Ocean Eng.*, ASCE, 129(2), 70–78.

Jin, Y.C. and Steffler, P.M. (1993) 'Predicting flow in curved open channels by depth-averaged method', *J. Hydraulic Eng.*, ASCE, 119(1), 109–124.

Johannesson, H. and Parker, G. (1989) 'Linear theory of river meander', Pages 181–213 in *River Meandering*, edited by S. Ikeda and G. Parker, American Geophysical Union, Washington D.C., USA.

Jones, W.P. and Launder, B.E. (1972) 'The prediction of laminarization with a two-equation model of turbulence', *Int. J. Heat and Mass Transfer*, 15, 301–314.

Jordanova, A.A. and James, C.S. (2003) 'Experimental study of bed load transport through emergent vegetation', *J. Hydraulic Eng.*, ASCE, 129(6), 474–478.

Julien, P.Y. (1995) *Erosion and Deposition*, Cambridge University Press, Cambridge, UK.

Karim, F. and Kennedy, J.F. (1982) 'IALLUVIAL: A computer-based flow- and sediment-routing model for alluvial streams and its application to the Missouri river', *Technical Report No. 250*, IIHR, University of Iowa, USA.

Karim, F. (1995) 'Bed configuration and hydraulic resistance in alluvial-channel flows', *J. Hydraulic Eng.*, ASCE, 121(1), 15–25.

Karim, F. (1998) 'Bed material discharge prediction for nonuniform bed sediments', *J. Hydraulic Eng.*, ASCE, 124(6), 597–604.

Karpik, S.R. and Raithby, G.D. (1990) 'Laterally averaged hydrodynamics model for reservoir predictions', *J. Hydraulic Eng.*, ASCE, 116(6), 783–798.

Kassem, A. and Chaudhry, M.H. (1998) 'Comparison of coupled and semicoupled numerical models for alluvial channels', *J. Hydraulic Eng.*, ASCE, 124(8), 794–802.

Khan, A.A. (2000) 'Modeling flow over an initially dry bed', *J. Hydr. Res.*, IAHR, 38(5), 383–388.

Kikkawa, H., Ikeda, S. and Kitagawa, A. (1976) 'Flow and bed topography in curved open channels', *J. Hydr. Div.*, ASCE, 102(9), 1327–1342.

Kim, C.W., Yoon, T.H., Cho, Y.S. and Kim, S.T. (2003) 'A two-dimensional conservative finite difference model in nonorthogonal coordinate system', *J. Hydr. Res.*, IAHR, 41(4), 395–403.

Kitamura, T., Jia, Y., Wang, S.S.Y. and Tsujimoto, T. (1999) 'A model for bed-scour induced head-cut', *J. Hydroscience & Hydraulic Eng.*, JSCE, 43(2), 611–616.

Klaassen, G.J. and Zwaard, J.J. (1974) 'Roughness coefficients of vegetated floodplains', *J. Hydr. Res.*, 12(1), 43–63.

Komura, S. (1963) Discussion of 'Sediment transportation mechanics: – Introduction and properties of sediment', Progress Report by the Task Committee on Preparation of Sedimentation Manual of the Committee on Sedimentation of the Hydraulics Division, V.A. Vanoni, Chmn., *J. Hydr. Div.*, ASCE, 89(HY1), 263–266.

Kordulla, W. and Vinokur, M. (1983) 'Efficient computation of volume in flow predictions', *AIAA J.*, 21, 917–918.

Kouwen, N. and Li, R.M. (1980) 'Biomechanics of vegetative channel linings', *J. Hydr. Div.*, ASCE, 106(HY6), 1085–1103.

Kouwen, N. (1988) 'Field estimation of the biomechanical properties of grass', *J. Hydr. Res.*, IAHR, 26(5), 559–568.

Kouwen, N. and Fathi-Moghadam, M. (2000) 'Friction factors for coniferous trees along rivers', *J. Hydraulic Eng.*, ASCE, 126(10), 732–740.

Kramer, H. (1935) 'Sand mixtures and sand movement in fluid models', *Trans.*, ASCE, Vol. 100, Paper No. 1909, 798–878.

Krishnappan, B.G. (2000) 'In situ size distribution of suspended particles in the Fraser River', *J. Hydraulic Eng.*, ASCE, 126(8), 561–569.

Krone, R.B. (1962) 'Flume studies on the transport of sediment in estuarine shoaling processes', Hydraulic Engineering Laboratory, University of Berkeley, California, USA.

Krumbein, W.C. (1942) 'Settling velocities and flume behavior of non-spherical particles', *Transactions*, American Geophysical Union, pp. 621–633.

Kuhnle, R.A. (1993) 'Fluvial transport of sand and gravel mixtures with bimodal size distributions', *Sedimentary Geology*, 85, 17–24.

Kuhnle, R.A., Garbrecht, J. and Alonso, C.V. (1996) 'A transport algorithm for variable sediment sizes: Application to wide sediment size distributions', *Proc. Sixth Federal Interagency Sedimentation Conf.*, Las Vegas, Nevada, pp. VI-1–VI-7.

Kuipers, J. and Vreugdenhil, C.B. (1973) 'Calculations of two-dimensional horizontal flow', *Report S 163*, Part I, Delft Hydraulics Laboratory, The Netherlands.

Kutija, V. and Hewett, C.J. (2002) 'Modelling of supercritical flow conditions revisited: NewC scheme', *J. Hydr. Res.*, IAHR, 40(2), 145–152.

Lane, E.W. and Kalinske, A.A. (1941) 'Engineering calculations of suspended sediment', *Trans. Amer. Geophy. Union*, 20(3), 603–607.

Lane, E.W. and Koelzer, V.A. (1953) 'Density of sediments deposited in reservoirs', *Report No. 9, A Study of Methods Used in Measurement and Analysis of Sediment Loads in Streams*, Engineering District, St. Paul, MN, USA.

Lang, G.A. and Chapra, S.C. (1982) 'Documentation of SED – A sediment/water column contaminant model', *NOAA Technical memorandum ERL GLERL-41*, Great Lakes Environmental Research Laboratory, Ann Arbor, Michigan, USA.

Langendoen, E.J. (1996) 'Discretization diffusive wave model', *Technical Report No. CCHE-TR-96-1*, Center for Computational Hydroscience and Engineering, The University of Mississippi, USA.

Launder, B.E. and Spalding, D.B. (1974) 'The numerical computation of turbulent flows', *Comput. Mecth. Appl. Mech. Eng.*, 3, 269–289.

Laursen, E.M. (1958) 'The total sediment load of streams', *J. Hydr. Div.*, ASCE, 84(1), 1–36.

Le Normant, C. (2000) 'Three-dimensional modeling of cohesive sediment transport in the Loire estuary', *Hydrological Processes*, 14, 2231–2243.

Lee, H.Y. and Hsu, I.S. (1994) 'Investigation of saltating particle motions', *J. Hydraulic Eng.*, ASCE, 120(7), 831–845.

Leonard, B.P. (1979) 'A stable and accurate convective modelling procedure based on quadratic interpolation', *Comput. Meths. Appl. Mech. Eng.*, 19, 59–98.

Leister, H.-J. and Peric, M. (1994) 'Vectorized strongly implicit solving procedure based on quatratic upstream interpolation', *Comput. Meth. Appl. Mech. Eng.*, 19, 59–98.

Li, C.H. and Liu, J.M. (1963) 'Resistance of alluvial rivers', Nanjing Hydraulic Research Institute, China (in Chinese).

Li, R.M. and Shen, H.W. (1973) 'Effect of tall vegetation on flow and sediment', *J. Hydr. Div.*, ASCE, 99(5), 793–814.

Li, W. and Yang, Y.S. (1990) 'The hybrid finite analytic methods of steady Navier-Stokes equations and its application', *J. Wuhan Univ. of Hydraulic and Electric Eng.*, 23, 29–35 (in Chinese).

Li, Z.H., Nguyen, K.D., Brun-Cottan, J.C. and Martin, J.M. (1994) 'Numerical simulation of the turbidity maximum transport in the Gironde Estuary (France)', *Oceanologica Acta*, 17(5), 479–500.

Lick, W. and Lick, J. (1988) 'Aggregation and disaggregation of fine-grained lake sediments', *J. Great Lakes Res.*, 14(4), 514–523.

Lin, B.L. and Falconer, R.A. (1996) 'Numerical modelling of three-dimensional suspended sediment for estuarine and coastal waters', *J. Hydr. Res.*, IAHR, 34(4), 435–456.

Lin, B.N. (1984) 'Current study of unsteady transport of sediment in China', *Proc. Japan-China Bilateral Seminar on River Hydraulics and Engineering Experiences*, July, Tokyo-Kyoto-Saporo, Japan, 337–342.

Lindner, K. (1982) 'Der Stroemungswiderstand von Pflanzenbestaenden', Mitteilungen 75, Leichtweiss-Institut fuer Wasserbau, TU Braunschweig.

Liu, D.Y. (1993) *Fluid Dynamics of Two-Phase Systems*, High Education Publish (in Chinese).

Liu, X.L. (1986) 'Nonuniform bed load transport rate and coarsening stabilization', *Master's Thesis*, Chengdu Univ. of Technology, China (in Chinese).

Lopez, F. and Garcia, M. (2001) 'Mean flow and turbulence structure of open-channel flow through non-emergent vegetation', *J. Hydraulic Eng.*, 127(5), 392–402.

Lu, J.A. and Si, G. (1990) 'A kind of finite analytic methods for convection-diffusion equation', *Chinese J. Computational Physics*, 7(2), 179–188.

Lu, Y. and Zhang, H. (1993) 'Numerical simulation of 2-D river-bed deformation', *J. Hydrodynamics*, Ser. A, 8(3), 273–284 (in Chinese).

Lumley, J.L. (1970) 'Toward a turbulent constitutive equation', *J. Fluid Mech.*, 41, 413–434.

Luque, R.F. and van Beek, R. (1976) 'Erosion and transport of bed-load sediment', *J. Hydr. Res.*, IAHR, 14(2), 127–144.

Lyn, D.A. and Goodwin, P. (1987) 'Stability of a general Preissmann scheme', *J. Hydraulic Eng.*, ASCE, 113(1), 16–28.

Mahmood, K. and Yevjevich, V. (1975) *Unsteady Flow in Open Channels*, Vols. I–II, Water Resources Publications, USA.

Majumdar, H. and Carstens, M.R. (1967) 'Diffusion of particles by turbulence: Effect on particle size', *WRC-0967*, Water Res. Center, Georgia Inst. Techn., Atlanta, USA.

Majumdar, S. (1988) 'Role of underrelaxation in employing momentum interpolation practice for calculation of flow with non-staggered grids', *Num. Heat Transfer*, 13, 125–132.

Majumdar, S., Rodi, W. and Zhu, J. (1992) 'Three-dimensional finite-volume method for incompressible flows with complex boundaries', *J. Fluids Eng.*, 114, 496–503.

Mancini, J.L. (1978) 'Numerical estimates of coliform mortality rates under various conditions', *J. Water Pollut. Control Fed.*, 50(11), 2477–2487.

Martinelli, L. and Jameson, A. (1988) 'Validation of a multigrid method for the Reynolds averaged equations', *AIAA paper 88-0414*.

Matyukhin, V.J. and Prokofyev, O.N. (1966) 'Experimental determination of the coefficient of vertical turbulent diffusion in water for settling particles', *Soviet Hydrol.* (Amer. Geophy. Union), No. 3.

McAnally, W.H., Letter, J.V. and Thomas, W.A. (1986) 'Two- and three-dimensional modeling systems for sedimentation', *Proc. Third Int. Symp. on River Sedimentation*, The University of Mississippi, USA, pp. 400–411.

McConnachie, G.L. (1991) 'Turbulence intensity of mixing in relation to flocculation', *J. Environ. Eng.*, ASCE, 117(6), 731–750.

McNown, J.S., Malaika, J. and Pramanik, R. (1951) 'Particle shape and settling velocity', *Transactions, The 4th Meeting of IAHR*, Bombay, India, 511–522.

McNown, J.S. and Lin, P.N. (1952) 'Sediment concentration and fall velocity', *Proc. 2nd Midwestern Conf. on Fluid Mechanics*, Ohio State University, Columbus, Ohio, pp. 402–411.

Mehta, A.J. and Partheniades, E. (1975) 'An investigation of the depositional properties of flocculated fine sediment', *J. Hydr. Res.*, IAHR, 13(4), 361–381.

Mehta, A.J., Parchure, T.M., Dixit, J.G. and Ariathurai, R. (1982) 'Resuspension potential of deposited cohesive sediment beds', *Estuarine Comparisons*, V. S. Kennedy (ed.), Academic Press, New York, pp. 591–609.

Mehta, A.J. (1986) 'Characterization of cohesive sediment properties and transport processes in estuaries', *Estuarine Cohesive Sediment Dynamics*, A.J. Mehta (ed.), Springer-Verlag, pp. 290–325.

Meselhe, E.A. and Holly, F.M.Jr. (1993) 'Simulation of unsteady flow in irrigation canals with dry bed', *J. Hydraulic Eng.*, ASCE, 119(9), 1021–1039.

Meyer-Peter, E. and Mueller, R. (1948) 'Formulas for bed-load transport', *Report on Second Meeting of IAHR*, Stockholm, Sweden, 39–64.

Migniot, C. (1968) 'A study of the physical properties of different very fine sediments and their behavior under hydrodynamic action', *La Houille Blanche*, No. 7, 591–620 (in French).

Minh Duc, B. (1998) 'Berechnung der stroemung und des sedimenttransports in fluessen mit einem tieffengemittelten numerischen verfahren', *Ph.D. Dissertation*, Karlsruhe University, Germany.

Minh Duc, B., Wenka, T. and Rodi, W. (2004) 'Numerical modeling of bed deformation in laboratory channels', *J. Hydraulic Eng.*, ASCE, 130(9), 894–904.

Misri, R.L., Ranga Raju, K.G. and Garde, R.J. (1984) 'Bed load transport of coarse nonuniform sediments', *J. Hydraulic Eng.*, ASCE, 110(3), 312–328.

Molls, T. and Chaudhry, M.H. (1995) 'Depth-averaged open-channel flow model', *J. Hydraulic Eng.*, ASCE, 121(6), 453–465.

Moog, D.B. (1995) 'Stream reaeration and the effects of large-scale roughness and bedforms', *Ph.D. Thesis*, Cornell Univ., Ithaca, New York.

Moog, D.B. and Jirka, G.H. (1999) 'Stream reaeration in nonuniform flow: Macroroughness enhancement', *J. Hydraulic Eng.*, ASCE, 125(1), 11–16.

Munk, W.H. and Anderson, E.R. (1948) 'Notes on the theory of the thermocline', *J. Mar. Res.*, 7, 276–295.

Muste, M. and Patel, V.C. (1997) 'Velocity profiles for particles and liquid in open-channel flow with suspended sediment', *J. Hydraulic Eng.*, ASCE, 123(9), 742–751.

Nagata, N., Hosoda, T. and Muramoto, Y. (2000) 'Numerical analysis of river channel processes with bank erosion', *J. Hydraulic Eng.*, ASCE, 126(4), 243–252.

Nagata, N., Hosoda, T., Nakata, T. and Muramoto, Y. (2005) 'Three-dimensional numerical model for flow and bed deformation around river hydraulic structures', *J. Hydraulic Eng.*, ASCE, 131(12), 1074–1087.

Nakagawa, H. and Tsujimoto, T. (1980) 'Sand bed instability due to bed load motion', *J. Hydr. Div.*, ASCE, 106(12), 2029–2051.

Nakagawa, H., Tsujimoto, T. and Murakami, S. (1986) 'Non-equilibrium bed load transport along side slope of an alluvial stream', *Proc. Third Int. Symp. on River Sedimentation*, University of Mississippi, USA, pp. 885–893.

Nakato, T. (1990) 'Tests of selected sediment-transport formulas', *J. Hydraulic Eng.*, ASCE, 116(3), 362–379.

Naot, D., Nezu, I. and Nakagawa, H. (1996) 'Hydrodynamic behavior of partly vegetated open channels', *J. Hydraulic Eng.*, ASCE, 122(11), 625–633.

Neary, V.S. (2003) 'Numerical solution of fully developed flow with vegetative resistance', *J. Eng. Mechanics*, ASCE, 129(5), 558–563.

Nedeco (1965) 'Siltation Bangkok port channel', The Hague, The Netherlands.

Nezu, I. and Nakagawa, H. (1993) *Turbulence in Open-Channel Flows*, IAHR Monograph, Rotterdam, The Netherlands.

Ni, J., Wang, G. and Zhang, H. (1991) *Basic Theories of Liquid-Solid Two-Phase Flow and Their Latest Applications*, Science Press, Beijing, China (in Chinese).

Nicholson, J. and O'Connor, B.A. (1986) 'Cohesive sediment transport model', *J. Hydraulic. Eng.*, ASCE, 112(7), 621–639.

Niklas, K.J. and Moon, F.C. (1988) 'Flexural stiffness and modulus of elasticity of flower stalks from *Allium sativum* as measured by multiple resonance frequency spectra.' *Am. J. Botany*, 75(10), 1517–1525.

O'Connor, D.J. and Dobbins, W.E. (1958) 'Mechanism of reaeration in natural streams', *Trans.*, ASCE, 123, 641–684.

Odgaard, A.J. (1986) 'Meander flow model, I: Development', *J. Hydraulic Eng.*, ASCE, 112(12), 1117–1136.

Odgaard, A.J. and Bergs, M.A. (1988) 'Flow processes in a curved alluvial channel', *Water Resources Res.*, AGU, 24(1), 45–56.

Odgaard, A.J. (1989) 'River-meander model, I: Development; II: Application', *J. Hydraulic Eng.*, ASCE, 115(11), 1433–1464.

Okabe, T., Yuuki, T. and Kojima, M. (1997) 'Bed-load rate on movable beds covered by vegetation', *Proc. 27th Congress of IAHR*, San Francisco, USA, Vol. 2, 809–814.

Oliver, D.R. (1961) 'The sedimentation and suspension of closely-sized spherical particles', *Chem. Eng. Science*, 15, 230–242.

Olsen, N.R.B. and Kjellesvig, H.M. (1998) 'Three-dimensional numerical flow modeling for estimation of maximum local scour depth', *J. Hydr. Res.*, IAHR, 36(4), 579–590.

Olsen, N.R.B. (2003) 'Three-dimensional CFD modeling of self-forming meandering channel', *J. Hydraulic Eng.*, ASCE, 129(5), 366–372.

Oseen, C. (1927) *Hydrodynamik*, Akademische Verlagsgesellschaft, Leipzig.

Osher, S. and Solomon, F. (1982) 'Upwind difference schemes for hyperbolic conservation laws', *Math. Comp.*, 38, 339–374.

Osher, S. and Sethian, J.A. (1988) 'Fronts propagating with curvature-dependent speed: Algorithms based on Hamilton-Jacobi formulations', *J. Computation Physics*, 79, 12–49.

Osman, A.M. (1985) 'Channel width response to changes in flow hydraulics and sediment load', *Ph.D. Dissertation*, Colorado State University, Fort Collins, USA.

Osman, A.M. and Thorne, C.R. (1988) 'Riverbank stability analysis, I: Theory', *J. Hydraulic Eng.*, ASCE, 114(2), 134–150.

Otsubo, K. and Muraoka, K. (1988) 'Critical shear stress of cohesive bottom sediments', *J. Hydraulic Eng.*, ASCE, 114(10), 1214–1256.

Owen, M.W. (1970) 'A detailed study of the settling velocities of an estuary mud', *Report No. INT 78*, Hydraulics Research Station, Wallingford, UK.

Owen, M.W. (1975) 'Erosion of Avonmouth mud', *Report No. INT 150*, Hydraulics Research Station, Wallingford, UK.

Padden, T.J. and Gloyna, E.F. (1971) 'Simulation of stream processes in a model river', *Report No. EHE-7-23*, CRWR-72, University of Texas, Austin, USA.

Paintal, A.S. (1971) 'Concept of critical shear stress in loose boundary open channel', *J. Hydr. Res.*, IAHR, 9(1), 90–113.

Parchure, T.M. (1980) 'Effect of bed shear stress on the erosional characteristics of kaolinite', *M. S. Thesis*, University of Florida, Gainesville, Florida, USA.

Parker, G., Kilingeman, P.C. and McLean, D.G. (1982) 'Bed load and size distribution in paved gravel-bed streams', *J. Hydr. Div.*, ASCE, 108(4), 544–571.

Parker, G. (1984) Discussion of 'Lateral bed load transport on side slopes' by S. Ikeda, *J. Hydraulic Eng.*, ASCE, 110(2), 197–203.

Parker, G. (1990) 'Surface-based bedload transport relation for gravel rivers', *J. Hydr. Res.*, IAHR, 28(4), 417–436.

Partheniades, E. (1965) 'Erosion and deposition of cohesive soils', *J. Hydr. Div.*, ASCE, 91(HY1), 105–139.

Pasche, E. and Rouve, G. (1985) 'Overbank flow with vegetatively roughened flood plains', *J. Hydraulic Eng.*, ASCE, 111(9), 1262–1278.

Patankar, S.V. and Spalding, D.B. (1972) 'A calculation procedure for heat, mass and momentum transfer in three-dimensional parabolic flows', *Int. J. Heat Mass Transfer*, 15, 1787–1806.

Patankar, S.V. (1980) *Numerical Heat Transfer and Fluid Flow*, Taylor & Francis.

Patel, P.L. and Ranga Raju, K.G. (1996) 'Fractionwise calculation of bed load transport', *J. Hydr. Res.*, IAHR, 34(3), 363–379.

Peaceman, D.W. and Rachford, H.H. (1955) 'The numerical solution of parabolic and elliptic differential equations', *J. Soc. Ind. Appl. Math.*, 3, 28–41.

Peng, R.Z. (1989) 'Experimental study on flocculation fall of sediment particles in Yangtze estuary', *Technical Report*, IWHR, Beijing, China.

Peric, M. (1985) 'A finite volume method for the prediction of three-dimensional fluid flow in complex ducts', *Ph.D. Thesis*, University of London, UK.

Petryk, S. and Bosmajian, G.III. (1975) 'Analysis of flow through vegetation', *J. Hydr. Div.*, ASCE, 99(5), 793–814.

Peyret, R. and Taylor, T.D. (1983) *Computational Methods for Fluid Flow*, Springer, Berlin, Heidelberg.

Pezzinga, G. (1994) 'Velocity distribution in compound channel flows by numerical modeling', *J. Hydraulic Eng.*, ASCE, 120(10), 1176–1198.

Phillips, B.C. and Sutherland, A.J. (1989) 'Spatial lag effects in bed load sediment transport', *J. Hydr. Res.*, IAHR, 27(1), 115–133.

Phillips, B.C. and Sutherland, A.J. (1990) 'Temporal lag effect in bed load sediment transport', *J. Hydr. Res.*, IAHR, 28(1), 5–23.

Prager, W. (1961) *Introduction to Mechanics of Continua*, Ginn, Boston.

Preissmann, A. (1961) 'Propagation des intumescences dans les canaux et Les Rivieres', *1 Congres de l'Association Francaise de Calcule*, Grenoble, France.

Proffitt, G.T. and Sutherland., A.J. (1983) 'Transport of nonuniform sediment', *J. Hydr. Res.*, IAHR, 21(1), 33–43.

Qian, Y.Y. (1980) 'Basic properties of hyper-concentrated flow', *Proc. First Int. Symp. on River Sedimentation*, Beijing, China.

Qin, Y.Y. (1980) 'Incipient motion of nonuniform sediment', *J. Sediment Res.*, No. 1 (in Chinese).

Rahuel, J.L., Holly, F.M., Chollet, J.P., Belleudy, P.J. and Yang, G. (1989) 'Modeling of riverbed evolution for bedload sediment mixtures', *J. Hydraulic Eng.*, ASCE, 115(11), 1521–1542.

Rastogi, A.K. and Rodi, W. (1978) 'Predictions of heat and mass transfer in open channels', *J. Hydr. Div.*, ASCE, 104(HY3), 397–420.

Rathbun, R.E. (1977) 'Reaeration coefficients of streams – State-of-the-Art', *J. Hydraulics Div.*, ASCE, 103(HY4), 409–424.

Raudkivi, A.J. and Hutchison, D.L. (1974) 'Erosion of kaolinite clay by flowing water', *Proc. Royal Society*, London, England, Series A, 337, 537–544.

Raudkivi, A.J. and Ettema, R. (1983) 'Clear water scour at cylindrical piers', *J. Hydr. Div.*, ASCE, 109(10), 1209–1213.

Raudkivi, A.J. (1990) *Loose Boundary Hydraulics*, 3rd ed., Pergamon Press, Inc., Tarrytown, N.Y.

Raupach, M.R., Thom, A.S. and Edwards, I. (1980) 'A wind-tunnel study of turbulent flow close to regularly arrayed rough surfaces', *Bound. Layer Meteor.*, 18, 373–397.

Ree, W.O. and Palmer, V.J. (1949) 'Flow of water in channels protected by vegetative linings', *Tech. Bull. No. 967*, Soil Conservation Service, U.S. Department of Agriculture, Washington D.C., USA.

Reynolds, W.C. and Perkins, H.C. (1977) *Engineering Thermodynamics*, Second Edition, McGraw-Hill.

Rhie, T.M. and Chow, A. (1983) 'Numerical study of the turbulent flow past an isolated airfoil with trailing-edge separation', *AIAA J.*, 21, 1525–1532.

Ribberink, J.S., Blom, A. and van der Sheer, P. (2002) 'Multi-fraction techniques for sediment transport and morphological modeling in sand-gravel rivers', *River Flow 2002*, Bousmar and Zech (eds.), Swets & Zeitinger, Lisse, 731–739.

Richardson, E.V. and Simons, D.B. (1967) 'Resistance to flow in sand channels', *Proc. 12th IAHR Congress*, Vol. 1, Fort Collins, Colorado, USA.

Richardson, J.F. and Zaki, W.N. (1954) 'Sedimentation and fluidisation, part I', *Trans.*, Inst. Chem. Engrs., 32(1), 35–53.

Richtmyer, R.D. and Morton, K.W. (1967) *Difference Methods for Initial Value Problems*, Wiley, New York.

Robinson, K.M. and Hanson, G.J. (1994) 'A deterministic head-cut advance model', *Trans.*, ASAE, 37(5), 1437–1443.

Robinson, K.M. (1996) 'Gully erosion and headcut advance', *Ph.D. Thesis*, Oklahoma State University, USA.

Rodi, W. (1971) 'On the equation governing the rate of turbulent energy dissipation', *Mech. Eng. Rep. TWF/TN/A/14*, Imperial College, London, UK.

Rodi, W. (1976) 'A new algebraic relation of calculating the Reynolds stresses', *ZAMM*, 56, 1219–1221.

Rodi, W. (1993) *Turbulence Models and Their Applications in Hydraulics*, 3rd ed., IAHR Monograph, Rotterdam, The Netherlands.

Roe, P.L. (1981) 'Approximate Riemann solvers, parameter vectors and difference schemes', *J. Comput. Phys.*, 43, 357–372.

Romanovskii, B.B. (1972) 'Experiments on settling velocity of sediment', Chinese Translation by S.L. Zhang and Y.Y. Qian, Yellow River Commission, Zhengzhou, China.

Rounds, S.A., Wood, T.M. and Lynch, S.D. (1999) 'Modeling discharge, temperature, and water quality in the Tualatin River, Oregon', *USGS Water Supply Paper 2465-B*, Reston, Virginia.

Rozovskii, I.L. (1957) *Flow of water in bends of open channel*, Academy of Sciences of the Ukrainian SSR, Kiev.

Rouse, H. (1937) 'Modern conceptions of the mechanics of turbulence', *Trans.*, ASCE, 102, 463–543.

Rouse, H. (1938) *Fluid Mechanics for Hydraulic Engineers*, Dover, New York.

Rubey, W. (1933) 'Settling velocities of gravel, sand and silt particles', *Amer. J. Sci.*, 225, 325–338.

Saffman, P.G. and Turner, J.T. (1956) 'On the collision of drops in turbulent clouds', *J. Fluid Mechanics*, 1, 16–30.

Saffman, P.G. (1965) 'The lift force on a sphere in a slow shear flow', *J. Fluid Mech.*, 22, 385–400.

Saffman, P.G. (1976) 'Development of a computer model for the calculation of turbulent shear flow', *Proc. 1976 Symp. on Turbulence and Dynamical Systems*, Durham, NC, USA.

Saiedi, S. (1997) 'Coupled modeling of alluvial flows', *J. Hydraulic Eng.*, ASCE, 123(5), 440–446.

Samaga, B.R., Ranga Raju, K.G. and Garde R.J. (1986a) 'Bed load transport rate of sediment mixture', *J. Hydraulic Eng.*, ASCE, 112(11), 1003–1018.

Samaga, B.R., Ranga Raju, K.G. and Garde R.J. (1986b) 'Suspended load transport rate of sediment mixture', *J. Hydraulic Eng.*, ASCE, 112(11), 1019–1038.

Schneider, G.E. and Zedan, M. (1981) 'A modified strongly implicit procedure for the numerical solution of field problems', *Num. Heat Transfer*, 4, 1–19.

Schoklitsch, A. (1930) *Handbuch des Wasserbaues*, Springer Vienna (2nd ed., 1950), English Translation by S. Shulits (1937).

Schulz, E.F., Wilde, R.H. and Albertson, M.L. (1954) 'Influence of shape on the fall velocity of sedimentary particles', *Missouri River Division Sedimentation Series Report No. 5*, Corps of Engineers, U.S. Army, Omaha, Nebraska, USA.

Seal, R., Parker, G., Paola, C. and Mullenbach, B. (1995) 'Laboratory experiments on downstream fining of gravel, narrow channel runs 1 through 3: supplemental methods and data', *External Memorandum M-239*, St. Anthony Falls Hydraulic Lab., University of Minnesota, USA.

Sekine, M. and Parker, G. (1992) 'Bed-load transport on transverse slope. I', *J. Hydraulic Eng.*, ASCE, 118(4), 513–535.

Sha, Y.Q. (1965) *Introduction to Sediment Dynamics*, Industry Press, Beijing, China, p. 302 (in Chinese).

Shanhar, N.J., Chan, E.S. and Zhang, Q.Y. (2001) 'Three-dimensional numerical simulation for an open channel flow with a constriction', *J. Hydr. Res.*, IAHR, 39(2), 187–201.

Shen, H.W. and Hung, C.S. (1983) 'Remodified Einstein procedure for sediment load', *J. Hydraulic Eng.*, ASCE, 109(4), 565–578.

Sheng, Y.P. (1983) 'Mathematical modeling of three-dimensional coastal currents and sediment dispersion: Model development and application', *Technical Report CERC-83-2*, Aeronautical Research Associates of Princeton, Inc., N.J., USA.

Sheppard, D.M., Odeh, M. and Glasser, T. (2004) 'Large scale clear-water local pier scour experiments', *J. Hydraulic Eng.*, ASCE, 130(10), 957–963.

Shields, A. (1936) 'Anwendung der Aechlichkeitsmechanik und der Turbulenz Forschung auf die Geschiebebewegung', *Mitteilungen der Pruessischen Versuchsanstalt fuer Wasserbau and Schiffbau*, Berlin, Germany.

Shields, F.D.Jr., Morin, N. and Cooper, C.M. (2004) 'Large woody debris structures for sand bed channels', *J. Hydraulic Eng.*, ASCE, 130(3), 208–217.

Shimizu, Y. and Tsujimoto, T. (1994) 'Numerical analysis of turbulent open-channel flow over a vegetation layer using a k-ε turbulence model', *J. Hydroscience and Hydraulic Eng.*, JSCE, 11(2), 57–67.

Shrestha, P.A. and Orlob, G.T. (1996) 'Multiphase distribution of cohesive sediments and heavy metals in esturine systems', *J. Environ. Eng.*, 122(8), 730–740.

Shyy, W., Udaykumar, H.S., Rao, M.M. and Smith, R.W. (1996) *Computational Fluid Dynamics with Moving Boundaries*, Taylor and Francis.

Shyy, W., Thakur, S.S., Quyang, H., Liu, J. and Blosch, E. (1997) *Computational Techniques for Complex Transport Phenomenon*, Cambridge University Press.

Simon, A., Curini, A., Darby, S.E. and Langendoen, E.J. (2000) 'Bank and near-bank processes in an incised channel', *Geomorphology*, 35, 193–217.

Simons, D.B. and Senturk, F. (1992) *Sediment Transport Technology – Water and Sediment Dynamics*, Water Resources Publications, Colorado, U.S.

Sladkevich, M., Militeev, A.N., Rubin, H. and Kit, E. (2000) 'Simulation of transport phenomena in shallow aquatic environment', *J. Hydraulic Eng.*, ASCE, 126(2), 123–136.

Smith, E.L. (1936) 'Photosynthesis in relation to light and carbon dioxide', *Proc., National Academy of Sciences*, 22, 504–510.

Smith, J.D. and McLean, S.R. (1977) 'Spatially averaged flow over a wavy surface', *J. Geoph. Res.*, 82(12), 1735–1746.

Smoluchowski, M. (1917) 'Versuch einer Mathematischen Theorie der Koagulations-Kinetik Kolloid Losungen', *Zeitschrift fur Physikalische Chemie*, 13, 421–427.

Song, T. and Graf, W.H. (1996) 'Experimental study of bedload transport in unsteady open-channel flows.' *Int. J. Sediment Res.*, 12(3), 63–71.

Soo, S.L. (1967) *Fluid Dynamics of Multiphase Systems*, Blaisdell, Waltham, MA.

Soulsby, R.L. (1996) 'Dynamics of marine sands', *Rep. No. SR 466*, HR Wallingford, U.K.

Spalding, D.B. (1972) 'A novel finite-difference formulation for differential expressions involving both first and second derivatives', *Int. J. Num. Meth. Engrg.*, 4, 551–559.

Spasojevic, M. and Holly, F.M.Jr. (1990) '2-D bed evolution in natural watercourses – New simulation approach', *J. Waterway, Port, Coastal and Ocean Eng.*, ASCE, 116(4), 425–443.

Spasojevic, M. and Holly, F.M.Jr. (1993) 'Three-dimensional numerical simulation of mobile-bed hydrodynamics', *Technical Report No. 367*, Iowa Institute of Hydraulic Research, The University of Iowa, USA.

Speziale, C.G. (1987) 'On nonlinear k-l and k-ε models of turbulence', *J. Fluid Mech.*, 178, 459–475.

Srikanth, V.C. and Majumdar, S. (1992) 'Relative performance of different low Reynolds number trubulence models for prediction of fully developed channel flow', *Project Doc. CF 9224*, National Aeronautical Laboratory, Bangalore, India.

Steele, J.H. (1962) 'Environmental control of photosynthesis in the sea', *Limnology and Oceanography*, 7, 137–150.

Stefan, H.G., Cardoni, J.J., Schiebe, F.R. and Cooper, C.M. (1983) 'Model of light penetration in a turbid lake', *Water Resources Res.*, AGU, 19(1), 109–120.

Steffler, P.M. (1984) 'Turbulent flow in a curved rectangular channel', *Ph.D. Thesis*, University of Alberta, Alberta, Canada.

Steinberger, N. and Hondzo, M. (1999) 'Diffusional mass transfer at sediment-water interface', *J. Environ. Eng.*, ASCE, 125(2), 192–200.

Stone, B.M. and Shen, H.T. (2002) 'Hydraulic resistance of flow in channels with cylindrical roughness', *J. Hydraulic Eng.*, ASCE, 128(5), 500–506.

Stone, H.L. (1968) 'Iterative solution of implicit approximation of multidimensional partial differential equations', *SIAM J. on Numerical Analysis*, 5, 530–558.

Strickler, A. (1923) 'Beitraezo zur Frage der Gerschwindigheits Formel und der Rauhigkeitszahlen fuer Strome kanale und Geschlossene Leitungen', Mitteilungen des Eidgenossischer Amtes fuer Wasserwirtschaft, Bern.

Struiksma, N., Olsen, K.W., Flokstra, C. and de Vriend, H.J. (1985) 'Bed deformation in curved alluvial channels', *J. Hydr. Res.*, IAHR, 23(1), 57–79.

Sweby, P.K. (1984) 'High resolution schemes using flux limiters for hyperbolic conservation laws', *SIAM J. Numer. Anal.*, 21, 995–1011.

Swinbank, W.C. (1963) 'Longwave radiation from clear skies', *Q. J. R. Meteorol. Soc.*, 89, 339–448.

Tao, J. (1984) 'Numerical modeling of wave run up and break on seashore', *Acta Oceanologica Sinica*, 6(5) (in Chinese).

Temple, D.M. (1987) 'Closure of 'Velocity distribution coefficients for grass-lined channels', *J. Hydraulic Eng.*, ASCE, 113(9), 1224–1226.

Temple, D.M. (1992) 'Estimating flood damage to vegetated deep soil spillways', *Applied Engineering in Agriculture*, 8(2), 237–242.

Thomann, R.V. and Mueller, J.A. (1987) *Principles of Surface Water Quality Modeling and Control*, Harper & Row Publishers, New York.

Thomann, R.V., Mueller, J.A., Wingield, R.P. and Huang, C.-R. (1991) 'Model of fate and accumulation of PCB homologue in Hudson Estuary', *J. Environ. Eng.*, ASCE, 117(2), 161–177.

Thomas, W.A. (1982) 'Chapter 18: Mathematical modeling of sediment movement', *Gravel-Bed Rivers*, R.D. Hey *et al.*, eds., John Wiley & Sons, Ltd., New York.

Thompson, T.J., Warsi, Z.U.A. and Mastin, C.W. (1985), *Numerical Grid Generation*, North-Holland.

Thorn, M.F.C. and Parsons, J.G. (1980) 'Erosion of cohesive sediments in estuaries: An engineering guide', *Proc. Third Intl. Symp. on Dredging Technology*, Paper F1, Bordeaux, France, 349–358.

Thorn, M.F.C. (1981) 'Physical processes of siltation in tidal channels', *Proc. Hydraulic Modelling Applied to Maritime Eng. Problems*, ICE, London, UK, 47–55.

Thorne, C.R. and Tovey, N.K. (1981) 'Stability of composite river banks', *Earth Surf. Proc. Landforms*, 6, 469–484.

Thuc, T. (1991) 'Two-dimensional morphological computations near hydraulic structures', *Doctoral Dissertation*, Asian Institute of Technology, Bangkok, Thailand.

Tingsanchali, T. and Chinnarasri, C. (1999) 'Mathematical model of dam surface evolution due to flow overtopping', *Proc. 1999 Int. Water Resources Engineering Conf.*, Seattle, USA, (on CD-Rom).

Toffaletti, F.B. (1968) 'A procedure for computation of the total river sand discharge and detailed distribution, bed to surface', *Technical Report No. 5*, US Army Corps of Engineers, Vicksburg, Mississippi, USA.

Toro, E.F. (1989) 'A weighted average flux method for hyperbolic conservation laws', *Proc. Roy. Soc. London*, A423, 401–418.

Toro, E.F. (1992) 'Riemann problems and the WAF method for solving two-dimensional shallow water equations', *Phil. Trans. Roy. Soc. London*, A338, 43–68.

Toro, E.F., Spruce, M. and Speares, W. (1994) 'Restoration of the contact surface in the HLL-Riemann solver', *Shock Waves*, 4, 25–34.

Toro, E.F. (2001) *Shock-Capturing Methods for Free-Surface Shallow Flows*, Wiley.

Tsai, C.H., Iacobellis, S. and Lick, W. (1987) 'Flocculation of fine-grained lake sediments due to a uniform shear stress', *J. Great Lakes Res.*, 13(2), 135–146.

Tsay, T.K., Ruggaber, G.J., Effler, S.W. and Driscoll, C.T. (1992) 'Thermal stratification modeling of lakes with sediment heat flux', *J. Hydraulic Eng.*, ASCE, 118(3), 407–419.

Tsinghua University (1996) 'The report on the sediment scale model experiment of the neighborhood of the dam for Three Gorges Project (TGP)', Sediment Research Laboratory, Tsinghua University, Beijing, China.

Tsivoglou, E.C. and Wallace, J.R. (1972) 'Characterization of stream reaeration capacity', *EPA-R3-72-012 (NTIS PB-214649)*, Environmental Protection Agency, Washington D.C., USA.

Tsujimoto, T., Graf, W.H. and Suszka, L. (1988) 'Bed-load transport in unsteady flow', *Proc. 6th Congress of Asian and Pacific Regional Division of IAHR*, Kyoto, Japan.

Tsujimoto, T. (1998) 'Development of sand island with vegetation in fluvial fan river under degradation', *Proc. Water Resource Engineering '98*, S.R. Abt, J. Young-Pezeshk and C.C. Watson (eds.), ASCE, Vol. 1, pp. 574–579.

Tsujimoto, T. and Kitamura, T. (1998) 'A model for flow over flexible vegetation-covered bed', *Water Resources Engineering '98*, S.R. Abt, J. Young-Pezeshk and C.C. Watson (eds.), ASCE, Vol. 1, 556–561.

TVA (Tennessee Valley Authority, 1972) 'Heat and mass transfer between a water surface and the atmosphere', *Water Resources Lab Report No. 14*, Norris, TN, USA.

U.S. Interagency Committee (1957) 'Some fundamentals of particle size analysis, A study of methods used in measurement and analysis of sediment loads in streams', *Report No. 12*, Subcommittee on Sedimentation, Interagency Committee on Water Resources, St. Anthony Falls Hydraulic Laboratory, Minneapolis, Minnesota, USA.

Van Doormal, J.P. and Raithby, G.D. (1984) 'Enhancements of the SIMPLE method for predicting incompressible fluid flows', *Num. Heat Transfer*, 7, 147–163.

Van Leer, B. (1982) 'Flux vector splitting for the Euler equations', *Proc. 8th Int. Conf. on Numerical Methods in Fluid Dynamics*, E. Krause (ed.), Springer, Berlin, Germany, 507–512.

Van Niekerk, A., Vogel, K.R., Slingerland, R.L. and Bridge, J.S. (1992) 'Routing of heterogeneous sediments over movable bed: Model development', *J. Hydraulic Eng.*, ASCE, 118(2), 246–262.

Van Rijn, L.C. (1984a) 'Sediment transport, part I: bed load transport', *J. Hydraulic Eng.*, ASCE, 110(10), 1431–1456.

Van Rijn, L.C. (1984b) 'Sediment transport, part II: suspended load transport', *J. Hydraulic Eng.*, ASCE, 110(11), 1613–1641.

Van Rijn, L.C. (1984c) 'Sediment transport, part III: bed forms and alluvial roughness', *J. Hydraulic Eng.*, ASCE, 110(12), 1733–1754.

Van Rijn, L.C. (1987) 'Mathematical modelling of morphological processes in the case of suspended sediment transport', *Delft Hydraulics Communication No. 382*, The Netherlands.

Van Rijn, L.C. (1989) 'Handbook: sediment transport by current and waves', *Report H 461*, Delft Hydraulics, The Netherlands.

Vanoni, V.A. and Brooks, N.H. (1957) 'Laboratory studies of the roughness and suspended load of alluvial streams', *Report E-68*, Sedimentation Laboratory, California Institute of Technology, Pasadena, California, USA.

Vanoni, V.A. (ed.) (1975) *Sedimentation Engineering*, ASCE Manuals and Reports on Engineering Practice, No. 54, N.Y., USA.

Vasiliev, O.F. (2002) 'Vertical two-dimensional hydrodynamic models for water bodies: The state of the art and current issues', *Proc. 5th Int. Conf. on Hydroscience and Eng.*, Warsaw, Poland (on CD Rom).

Venutelli, M. (2002) 'Stability and accuracy of weighted four-point implicit finite difference schemes for open channel flow', *J. Hydraulic Eng.*, ASCE, 128(3), 281–288.

Vieira, D.A. (1995) 'Coupling of one- and two-dimensional free surface flow models.' *Master's Thesis*, University of Mississippi, USA.

Vieira, D.A. and Wu, W. (2002) 'One-dimensional channel network model CCHE1D version 3.0 – user's manual', *Technical Report No. NCCHE-TR-2002-2*, National Center for Computational Hydroscience and Engineering, University of Mississippi, USA.

Wang, J.S., Ni, H.G. and He, Y.S. (2000) 'Finite-difference TVD scheme for computation of dam-break problems', *J. Hydraulic Eng.*, ASCE, 126(4), 253–262.

Wang, S.S.Y. and Adeff, S.E. (1986) 'Three-dimensional modeling of river sedimentation processes', *Proc. 3rd Int. Symp. on River Sedimentation*, University of Mississippi, USA.

Wang, S.S.Y. and Hu, K.K. (1993) 'Improved methodology for formulating finite-element hydrodynamic models', *Finite Element in Fluids*, edited by T.J. Chung, Vol. 8, Hemisphere Publication Cooperation, pp. 457–478.

Wang, S.S.Y. and Wu, W. (2005) 'Computational simulation of river sedimentation and morphology – A review of the state of the art', *Int. J. Sediment Res.*, 20(1), 7–29.

Wellington, N.W. (1978) 'A sediment-routing model for alluvial streams', *M. Eng. Sc. Thesis*, University of Melbourne, Australia.

Wenka, T. (1992) 'Numerische berechnung von stroemungsvorgaengen in naturnahen flusslaeufen mit einen tiefengemitelten model', *Ph.D. Dissertation*, Karlsruhe University, Germany.

White, F.M. (1991) *Viscous Fluid Flow*, McGraw-Hill.

Wilcock, P.R. and Southard, J.B. (1988) 'Experimental study of incipient motion in mixed-size sediment', *Water Resources Res.*, 24(7), 1137–1151.

Wilcock, P.R. (1993) 'Critical shear stress of natural sediments', *J. Hydraulic Eng.*, ASCE, 119(4), 491–505.

Wilcock, P.R. and McArdell, B.W. (1993) 'Surface-based fractional transport rate: mobilization thresholds and partial transport of a sand-gravel sediment', *Water Resources Res.*, 29(24), 1297–1312.

Wilcox, D.C. and Rubesin, M.W. (1980) 'Progress in turbulence modeling for complex flow fields including effects of incompressibility', *NASA TP-1519*, NASA.

Wilcox, D.C. (1993) 'Turbulence modeling for CFD', DCW Industries Inc., La Canada, CA, USA.

Wilde, R.H. (1952) 'Effect of shape on the fall-velocity of gravel-sized particles', *Master's Thesis*, Colorado A & M College, Colorado, USA.

Williams, G.P. (1970) 'Flume width and water depth effects in sediment transport experiments', *Prof. Paper No. 562-H*, USGS, Washington D.C., USA.

Williams, G.P. and Rosgen, D.L. (1989) 'Measured total sediment load (suspended loads and bed loads) for 93 United States streams', *Open-File Report 89-67*, USGS.

Woo, H. and Yoo, K. (1991) Discussion on 'Test of selected sediment-transport formulas', *J. Hydraulic Eng.*, ASCE, 117, 1233–1234.

Wool, T.A., Ambrose, R.B., Martin, J.L. and Comer, E.A. (1995) 'Water quality analysis simulation program (WASP)', Version 6.0, User's Manual, EPA, USA.

Wright, R.M. and McDonnell, A.J. (1979) 'In-stream deoxygenation rate prediction', *J. Environ. Eng. Div.*, ASCE, 105(EE2), 323–335.

Wu, F.C., Shen, H.W. and Chou Y.J. (1999) 'Variation of roughness coefficient for unsubmerged and submerged vegetation', *J. Hydraulic Eng.*, ASCE, 125(9), 934–942.

Wu, W. (1991) 'The study and application of 1-D, horizontal 2-D and their nesting mathematical models for sediment transport', *Ph.D. Dissertation*, Wuhan University of Hydraulic and Electric Eng., Wuhan, China (in Chinese).

Wu, W. (1992) 'A laterally-averaged vertical two-dimensional mathematical model of water flow', *J. Wuhan Univ. of Hydraulic and Electric Eng.*, Supplement, 14–19 (in Chinese).

Wu, W. and Li, Y. (1992) 'One- and two-dimensional nesting mathematical model for river flow and sedimentation', *Proc. Fifth Int. Symp. on River Sedimentation*, Karlsruhe, Germany, Vol. 1, 547–554.

Wu, W. (1993) 'High-order difference schemes of one-dimensional convection, diffusion and convection-diffusion equations', *Proc. 1993 Chinese Conference on Hydrodynamics*, Qinghuangdao, China, pp. 53–59.

Wu, W., Hu, C. and Yang, G. (1995) 'Horizontal two-dimensional numerical model for water flow and sediment transport', *Chinese J. Hydraulic Eng.*, Chinese Assoc. of Hydr. Eng., No. 10, 40–46.

Wu, W. (1996a) 'Unsteady coordinate transformations and adaptive grid techniques for movable boundary flows', *Proc. Third Asian-Pacific Conference on Computational Mechanics*, Seoul, Korea.

Wu, W. (1996b) 'Numerical methods on irregular quadrilateral grid', *Proc. Second Int. Conf. on Hydrodynamics*, Hong Kong.

Wu, W. and Wenka, T. (1998) '3-D calculation of bed morphology in the case of bed load transport', *Proc. Third Int. Conf. on Hydroscience and Eng.*, Cottbus/Berlin, Germany (on CD-ROM).

Wu, W. and Wang, S.S.Y. (1999) 'Movable bed roughness in alluvial rivers', *J. Hydraulic Eng.*, ASCE, 125(12), 1309–1312.

Wu, W. and Wang, S.S.Y. (2000) 'Mathematical models for liquid-solid two-phase flow', *Int. J. Sediment Res.*, 15(3), 288–298.

Wu, W., Rodi, W. and Wenka, T. (2000a) '3D numerical modeling of flow and sediment transport in open channels', *J. Hydraulic Eng.*, ASCE, 126(1), 4–15.

Wu, W., Wang, S.S.Y. and Jia, Y. (2000b) 'Nonuniform sediment transport in alluvial rivers', *J. Hydr. Res.*, IAHR, 38(6), 427–434.

Wu, W. and Vieira, D.A. (2002) 'One-dimensional channel network model CCHE1D 3.0 – Technical manual', *Technical Report No. NCCHE-TR-2002-1*, National Center for Computational Hydroscience and Engineering, University of Mississippi, USA.

Wu, W. (2004) 'Depth-averaged 2-D numerical modeling of unsteady flow and nonuniform sediment transport in open channels', *J. Hydraulic Eng.*, ASCE, 135(10), 1013–1024.

Wu, W., Vieira, D.A. and Wang. S.S.Y. (2004a) 'One-dimensional numerical model for nonuniform sediment transport under unsteady flows in channel networks', *J. Hydraulic Eng.*, ASCE, 130(9), 914–923.

Wu, W., Wang, P. and Chiba, N. (2004b) 'Comparison of five depth-averaged 2-D turbulence models for river flows', *Archives of Hydro-Eng. and Environ. Mechanics*, Polish Academy of Science, 51(2), 183–200.

Wu, W. and Wang, S.S.Y. (2004a) 'Depth-averaged 2-D calculation of flow and sediment transport in curved channels', *Int. J. Sediment Res.*, 19(4), 241–257.

Wu, W. and Wang, S.S.Y. (2004b) 'Depth-averaged numerical modeling of flow and sediment transport in open channels with vegetation', *Riparian Vegetation and Fluvial Geomorphology*, S.J. Bennett and A. Simon (eds.), AGU, 253–265.

Wu, W. and Wang, S.S.Y. (2004c) 'Depth-averaged 2-D calculation of tidal flow, salinity and cohesive sediment transport in estuaries', *Int. J. Sediment Res.*, 19(3), 172–190.

Wu, W., Shields, F.D. Jr., Bennett, S.J. and Wang, S.S.Y. (2005) 'A depth-averaged 2-D model for flow, sediment transport and bed topography in curved channels with riparian vegetation', *Water Resources Res.*, AGU, 41, W03015.

Wu, W. and Wang, S.S.Y. (2005) 'Empirical-numerical analysis of headcut migration', *Int. J. Sediment Res.*, 20(3), 233–243.

Wu, W. and Wang, S.S.Y. (2006) 'Formulas for sediment porosity and settling velocity', *J. Hydraulic Eng.*, ASCE, 132(8), 858–862.

Wu. W., Altinakar, M. and Wang, S.S.Y. (2006) 'Depth-average analysis of hysteresis between flow and sediment transport under unsteady conditions', *Int. J. Sediment Res.*, 21(2), 101–112.

Wu, W. and Wang, S.S.Y. (2007) 'One-dimensional modeling of dam-break flow over movable beds', *J. Hydraulic Eng.*, ASCE, 133(1), 48–58.

Yakhot, V., Orszag, S.A., Thangam, S., Gatski, T.B. and Speziale, C.G. (1992) 'Development of turbulence models for shear flows by a double expansion technique', *Phys. Fluids A*, 4(7).

Yalin, M.S. (1972) *Mechanics of Sediment Transport*, Pergamon Press.

Yalin, M.S. and Finlayson, G.D. (1972) 'On the velocity distribution of the flow-carrying sediment in suspension', *Symp. to Honor Prof. H.A. Einstein*, H.W. Shen, ed., pp. 8–11.

Yalin, M.S. and Karahan, E. (1979) 'Inception of sediment transport', *J. Hydr. Div.*, ASCE, 105(11), 1433–1443.

Yanenko, N.N. (1971) *The Method of Fractional Steps*, Translated by M. Holt, Springer-Verlag, New York, pp. 17–41.

Yang, C.T. (1973) 'Incipient motion and sediment transport', *J. Hydr. Div.*, ASCE, 99(HY10), 1679–1704.

Yang, C.T. (1984) 'Unit stream power equation for gravel', *J. Hydraulic Eng.*, ASCE, 110(12).

Yang, C.T. (1995) *Sediment Transport: Theory and Practice*, McGraw-Hill.

Yang, C.T. and Greimann, B.P. (1999) 'Dambreak unsteady flow and sediment transport', *Proc., European Concerted Action on Dam-Break Modeling*, Zaragoza, 327–365.

Yang, C.T., Trevino, M.A. and Simoes, F.J.M. (1998) 'User's manual for GSTARS 2.0 (generalized stream tube model for alluvial river simulation version 2.0)', Sedimentation and River Hydraulics Group, Technical Service Center, Bureau of Reclamation, U.S. Department of the Interior, Denver, Colorado, USA.

Yang, G.L. and Cunge, J.A. (1989) 'High-order numerical schemes of convection equation', *Chinese J. Hydraulic Eng.*, Chinese Assoc. of Hydr. Eng., No. 10.

Yanmaz, A.M. and Altinbilek, H.D. (1991) 'Study of time-dependent local scour around bridge piers', *J. Hydraulic Eng.*, ASCE, 117(10), 1247–1268.

Ye, J. and McCorquodale, J.A. (1997) 'Depth-averaged hydrodynamic model in curvilinear collocated grid', *J. Hydraulic Eng.*, ASCE, 123(5), 380–388.

Yee, H.C. (1987) 'Construction of explicit and implicit symmetric TVD schemes and their applications', *J. Comput. Phys.*, 68, 151–179.

Yeh, K.-C., Li, S.-J. and Chen, W.-L. (1995) 'Modeling non-uniform-sediment fluvial process by characteristics method', *J. Hydraulic Eng.*, ASCE, 121(2), 159–170.

Yen, C.L. and Lee, K.T. (1995) 'Bed topography and sediment sorting in channel bend with unsteady flow', *J. Hydraulic Eng.*, ASCE, 121(8), 591–599.

Ying, X., Khan, A.A. and Wang, S.S.Y. (2004) 'Upwind conservative scheme for the Saint Venant equations', *J. Hydraulic Eng.*, ASCE, 130(10), 977–987.

Ying, X. and Wang, S.S.Y. (2004) 'Two-dimensional numerical simulation of Malpasset dam-break wave propagation', *Proc. 6th Int. Conf. on Hydroscience and Eng.*, Brisbane, Australia (on CD-Rom).

Yue, P.J. (1983) 'Preliminary study of flocculation formed by cohesive sediment and its influence on rheologic properties of slurry', *J. Sediment Res.*, No. 1, 25–35 (in Chinese).

Zanke, U. (1977) 'Berechnung der Sinkgeschwindigkeiten von Sedimenten', Mitt. des Franzius-Instituts fuer Wasserbau, Heft 46, Seite 243, Technical University, Hannover, Germany.

Zeng, J., Constantinescu, G. and Weber, L. (2005) 'Validation of a computational model to predict suspended and bed load sediment transport and equilibrium bed morphology in open channels', *Proc. 31st IAHR Congress*, Seoul, Korea, (on CD-Rom).

Zhang, R.J. (1961) *River Dynamics*, Industry Press, Beijing, China (in Chinese).

Zhang, R.J., Xie, J.H., Wang, M.F. and Huang, J.T. (1989) *Dynamics of River Sedimentation*, Water and Power Press, Beijing, China (in Chinese).

Zhang, R.J. and Xie, J.H. (1993) *Sedimentation Research in China, Systematic Selections*, Water and Power Press, Beijing, China.

Zhang, S.Q. (1999) 'One-D and two-D combined model for estuary sedimentation', *Int. J. Sediment Res.*, 14(1), 37–45.

Zhou, J. and Lin, B. (1998) 'One-dimensional mathematical model for suspended sediment by lateral integration', *J. Hydraulic Eng.*, ASCE, 124(7), 712–717.

Zhu, J. (1991) 'A low diffusive and oscillation-free convection scheme', *Communication in Applied Numerical Methods*, 7, 225–232.

Zhu, J. and Rodi, W. (1991) 'A low dispersion and bounded convection scheme', *Comput. Meths. Appl. Mech. Eng.*, 92, 87–96.

Zhu, J. (1992a) 'FAST2D: A computer program for numerical simulation of two-dimensional incompressible flows with complex boundaries', Institute of Hydromechanics, Karlsruhe University, Germany.

Zhu, J. (1992b) 'An introduction and guide to the computer program FAST3D', Institute for Hydromechanics, Karlsruhe University, Germany.

Ziegler, C.K. and Nisbet, B.S. (1995) 'Long-term simulation of fine-grained sediment transport in large reservoir', *J. Hydraulic Eng.*, ASCE, 121(11), 773–781.

Ziegler, C.K., Israelsson, P.H. and Connolly, J.P. (2000) 'Modeling sediment transport dynamics in Thompson Island Pool, upper Hudson River', *Water Quality and Ecosystem Modeling*, Kluwer Academic Publishers, 1, 193–222.

Zienkiewicz, O.C. and Taylor, R.L. (2000) *Finite Element Method*, 5th Edition, Vol. 3 – *Fluid Dynamics*, Elsevier.

Zimmermann, C. and Kennedy, J.F. (1978) 'Transverse bed slope in curved alluvial streams', *J. Hydr. Div.*, ASCE, 104(1), 33–48.

Zison, S.W., Mills, W.B., Diemer, D. and Chen, C.W. (1978) 'Rates, constants and kinetic formulations in surface water quality modeling', *Report EPA-600-3-78-105*, Environmental Protection Agency, Athens, GA, USA.

Zyserman, J.A. and Fredsøe, J. (1994) 'Data analysis of bed concentration of suspended sediment', *J. Hydraulic Eng.*, ASCE, 120(9), 1021–1042.

Index

For Product Safety Concerns and Information please contact our EU representative GPSR@taylorandfrancis.com Taylor & Francis Verlag GmbH, Kaufingerstraße 24, 80331 München, Germany

T - #0132 - 160425 - C0 - 246/174/27 - PB - 9780415449601 - Gloss Lamination